Meso-scale Atmospheric Circulations

Meso-scale Atmospheric Circulations

B. W. ATKINSON

Queen Mary College, University of London

1981

ACADEMIC PRESS
A Subsidiary of Harcourt Brace Jovanovich, Publishers
London New York Toronto Sydney San Francisco

ACADEMIC PRESS INC. (LONDON) LTD
24/28 Oval Road
London NW1 7DX

United States Edition published by
ACADEMIC PRESS INC.
111 Fifth Avenue
New York, New York 10003

Copyright © 1981 by
ACADEMIC PRESS INC. (LONDON) LTD

All Rights Reserved
No part of this book may be reproduced in any form by photostat, microfilm or any other means, without written permission from the publishers

British Library Cataloguing in Publication Data
Atkinson, B. W.
   Meso-scale atmospheric circulations.
   1. Atmospheric—circulation
   I. Title
   551.5′17   QC880.4.A8

ISBN 0-12-065960-3

LCCCN 81 6638 6.

Printed in Great Britain by
Thomson Litho Ltd, East Kilbride, Scotland

To
My Mother, Robert,
Pamela and Heather

Preface

In common with virtually all sciences the study of atmospheres has flourished over the last three decades. Certainly meteorological study of the Earth's atmosphere has proceeded apace, unravelling a whole suite of atmospheric circulations ranging in size from those with characteristic horizontal dimensions of a few millimetres to the global atmospheric circulation. This tremendous width of the spectrum of motions has fostered a shift of interest from the apparently well-known mid-latitude, synoptic scale systems to both larger and smaller systems—the general atmospheric circulation and meso-scale circulations respectively. Over the last three decades these two topics have taken an increasing role in meteorological research and are likely to continue to do so for at least the next decade.

Meso-meteorology was first so called in the early 1950s. Since then the scope of the sub-discipline, its research priorities and the nature of the circulations that comprise its major subject matter have been fashioned by an extensive literature. Yet so relentless has been the research output that we have no overall view of our understanding of meso-scale atmospheric circulations, in sharp contrast to our well-established knowledge of synoptic scale or micro-scale circulations. In the light of the growing appreciation of the significance of meso-scale circulations it appears sensible to try to record past achievements before a new decade's enthusiasm generates a literature which will no doubt become increasingly difficult to digest. Such a review of our present knowledge of meso-scale atmospheric circulations is the aim of this book. Its scope is purposely restricted. Only those circulations that are intrinsically of the meso-scale are included: others typically of a different scale but which, with appropriate forcing, may be of the meso-scale are not covered in this monograph. Similarly many studies with the prefix "meso-" in their title were omitted from this review because they merely added more detail to our knowledge of a larger scale system rather than providing information on real meso-scale systems. Their results were more a function of meso-scale observational resolution than of meso-scale atmospheric structure. Application of these closely related criteria means that some circulations, such as low level jets, fronts, squall lines, urban circulations, clear air turbulence and Kelvin–Helmholtz waves, are not covered here. The restriction to circulations arises from the belief that atmospheric motion is fundamental to all weather and climate, a belief in accord with the central philosophy of dynamical meterology. As a result of this restriction, meso-scale studies of the scalar meteorological elements, such as comprise most of urban climatology, have not been treated in this book except in so far as they contribute to an understanding of air flows. Finally, some circulations, such as tropical cloud clusters, are not included because at the time of writing they had been only recently discovered and consequently are not, as yet, well understood.

The book comprises three Parts: Part I contains an introductory chapter on the meso-scale; Part II deals with topographically induced circulations; and Part III is concerned with free atmosphere circulations. Ultimately of course all atmospheric circulations are influenced by topography, so Part II treats those that result from topographical variations. For example sea breezes result from thermal differences between land and sea; lee waves result from air passing over orography. Other systems, for example those resulting from meso-scale cellular convection, owe their origin to heat transfer from an apparently uniform surface (such as the ocean) but their distinct morphology and dynamics are due to atmospheric characteristics rather than topographical variations. For this reason the chapter on shallow cellular convection lies in Part III of the book. We should note that topographically induced systems are not necessarily boundary-layer systems.

Both Parts II and III cover, in turn, mechanically and thermally induced circulations and in all chapter headings unnecessary words are omitted. Thus, for example, it is understood that the downslope winds in Chapter 3 are both meso-scale and mechanically induced by topography. Similarly, the circulations reviewed in Chapter 9 are meso-scale, convective and occur in the free atmosphere. We should note, however, that although the circulations treated in this book are intrinsically of the meso-scale, there are, for example, some circulations in cyclones (Chapter 10) and in wakes (Chapter 4) which may not be meso-scale. Such circulations are not treated in this monograph.

Wherever possible the chapters have a similar fourfold structure, which covers, in order, a qualitative account of the mechanism of the particular circulation, followed by reviews of observational, theoretical and hardware studies. Within both the observational and theoretical sections a consistent mode of presentation is attempted. In the theoretical treatments the analytical results have been quite fully presented but physical and mathematical details of numerical studies have of necessity been somewhat abbreviated. A narrative style is used in the presentation of the theoretical results, most of which ultimately derive from the equations of thermodynamics and motion, so that the evolution and improvement in theory may be more readily followed.

The need for the book remained in the forefront of my mind throughout its preparation. Rarely in my forays into the literature did I find that most useful of publications, the comprehensive review article. Consequently, the scope and organization of the chapters owe little to the efforts of other reviewers. Of the wide literature consulted and ultimately specifically mentioned in the book most was in English, but German, French, Norwegian, Spanish, Russian and Japanese writings have also played a part. By far the greater part of the literature cited was published after 1950, but where appropriate I have reached back as far as the nineteenth century. In attempting to present a coherent picture I have standardized all units to conform with recognized metric practice in meterology.

I am most grateful to Dr A. A. White of the Meteorological Office, Bracknell, for both his careful reading of several chapters and his helpful comments. Drs A. J. Gadd and P. A. White of the Meteorological Office and Drs J. S. A. Green and M. W. Moncrieff of Imperial College, London also helped, even if they were not aware of it, by being ready to discuss some of the many questions that arose in my mind. The diagrams were produced in the Cartographic Unit at Queen Mary College, mainly by L. Cooper-Grundy, R. Hines, G. McCarthy, S. Pratt and D. Shewan. Photographic work was undertaken by P. A. Newman and R. Crundwell. The manuscript was typed by Mrs L. Agombar and Mrs M. Putt. For all this technical assistance I am most thankful.

Queen Mary College, B. W. Atkinson
University of London
January 1981

Contents

Preface   vii
Symbols   xv

Part I
Introduction

1. The Meso-scale   3
   I. Introduction   3
   II. Observation   3
   III. Theory   13
   References   20

Part II
Topographically Induced Circulations
A. Mechanically Induced Circulations

2. Lee Waves   25
   I. Introduction   25
      A. General mechanism of lee waves   28
   II. Observation   30
      A. Wave characteristics   31
      B. Atmospheric conditions suitable for waves   40
   III. Theory   42
      A. Small amplitude theory   42
      B. Large amplitude theory   67
   IV. Hardware models   69
   V. Momentum transfer by lee waves   73
   References   76

3. Downslope winds   80
   I. Introduction   80
   II. Observation   83
      A. Surface characteristics   84
      B. Climatology   87
      C. Synoptic background of downslope winds   90
      D. Concluding remarks   92
   III. Theory   93
   IV. Hardware models   105
   References   106

4. Circulations in Wakes — 109
  I. Introduction — 109
  II. Observation — 110
     A. Meso-scale lee lows — 110
     B. Meso-scale lee vortices — 112
  III. Theory — 115
     A. Meso-scale lee lows — 115
     B. Meso-scale lee vortices — 117
  IV. Hardware models — 120
  References — 121

B. Thermally Induced Circulations

5. Sea/Land Breeze Circulation — 125
  I. Introduction — 125
     A. The mechanism of the sea and land breeze — 125
  II. Observation — 127
     A. Climatology — 127
     B. Mean surface characteristics of the tropical sea/land breeze — 134
     C. The sea/land breeze circulation — 138
     D. Factors affecting the sea/land breeze circulation — 151
  III. Theory — 158
     A. Analytical results — 158
     B. Numerical results — 178
  IV. Hardware models — 206
  V. The role of sea/land breezes — 207
  References — 209

6. Slope and Valley Wind Circulation — 215
  I. Introduction — 215
     A. Mechanism of the slope and valley wind — 217
  II. Observation — 219
     A. Climatology of slope and valley winds — 219
     B. Surface characteristics of the slope and valley winds — 229
     C. Slope wind circulation — 233
     D. Valley wind circulation — 238
     E. Mountain–plain circulation — 249
     F. Effects of stability and gradient wind — 250
  III. Theory — 251
     A. Downslope winds — 255
     B. Upslope winds — 259
     C. Valley and mountain winds — 267
  IV. Hardware models — 274
  V. The role of slope and valley winds — 276
  References — 277

Part III
Free Atmosphere Circulations
A. Non-convective Circulations

7. Moving Gravity Waves — 283
   I. Introduction — 283
   II. Observation — 285
   III. Theory — 298
   References — 307

B. Convective Circulations

8. Severe Local Storms — 313
   I. Introduction — 313
   II. Observation — 316
      A. Climatology — 316
      B. Extra-tropical severe local storms — 318
      C. Tropical severe local storms — 362
   III. Theory — 365
      A. Extra-tropical severe local storms — 367
      B. Tropical severe local storms — 387
   References — 389

9. Shallow Cellular Circulations — 399
   I. Introduction — 399
   II. Observation — 401
      A. Two-dimensional cells — 401
      B. Three-dimensional cells — 403
   III. Theory — 411
      A. Medium at rest — 416
      B. Flowing medium — 418
   References — 419

10. Circulations in Cyclones — 421
   I. Introduction — 421
   II. Observation — 423
      A. Extra-tropical circulations — 423
      B. Tropical circulations — 450
   III. Theory — 464
      A. Extra-tropical circulations — 464
      B. Tropical circulations — 470
   IV. Hardware models — 473
   References — 474

Author index — 479
Subject index — 487

Symbols

The following is a list of the symbols used frequently in the text. Other symbols are defined in the text after they are first used.

c	speed of sound in air, unless otherwise specified
C	circulation
c_p, c_v	specific heat capacities at constant pressure and constant volume
f	coriolis parameter
$F_{x,y,z}$	frictional force per unit mass in directions x, y and z
Fr	Froude number
g	acceleration due to gravity
i	$(-1)^{1/2}$
k	wave number, unless otherwise specified
k_c	thermal conductivity
k_f	constant expressing intensity of frictional force in Goldman–Mohn representation (ku)
k_v	von Kármán's constant
$K^H_{x,y,z,n}$	eddy diffusivity of heat in direction x, y, z or n
$K^M_{x,y,z,n}$	eddy viscosity or exchange coefficient in turbulent transfer of momentum in direction x, y, z or n
$K^W_{x,y,z,n}$	eddy diffusivity of water in direction x, y, z or n
L	a characteristic length; length of a boundary
L_{vw}	latent heat of evaporation
N	Brunt–Väisäla frequency
p	air pressure
Pr	Prandtl number
P_x, P_y	components of large scale pressure gradient force
q	specific humidity of air
q_v	mixing ratio for water vapour

q_w	liquid water content of air
Q	heat
R	specific gas constant
Ra	Rayleigh number
Re	Reynolds number
Ri	Richardson number
Real	real part of
St	Strouhal number
t	time
T	absolute temperature, period
u	component of air velocity along x-axis
u_g	component of geostrophic wind velocity along x-axis
u_*	friction velocity
U	a representative wind speed
v	component of air velocity along y-axis
v_g	component of geostrophic wind velocity along y-axis
w	component of air velocity along z-axis; any other direction is specified
x	Cartesian coordinate distance; on global scale refers to eastward coordinate in a tangent plane system; any other direction is specified
y	Cartesian coordinate distance; on global scale refers to northward coordinate; any other direction is specified
z	Cartesian coordinate distance; vertical distance in the tangent plane system; elevation above a specified datum base
$\sim$	order of magnitude of
$\simeq$	approximately equal to
α	coefficient of thermal expansion; angle
β	coefficient of stability defined as $(1/\theta)(\partial\theta/\partial z)$; Lin's parameter
γ	lapse rate of temperature; ratio of specific heat capacities at constant pressure and volume
Γ	adiabatic vertical temperature lapse rate (subscript d for dry; subscript s for saturated)
ζ	displacement of a streamline from its undisturbed position; relative vorticity about the vertical
η	lateral component of relative vorticity around y-axis
θ	potential temperature; angle
λ	wavelength
ν	kinematic viscosity
ξ	lateral component of relative vorticity around x-axis
ρ	air density

σ	Stefan–Boltzmann constant
ϕ	latitude; stream function in the (y, z) plane
ψ	stream function in the (x, z) plane
Ω	angular velocity of rotation of the Earth

Part I

Introduction

1
The Meso-scale

I. Introduction

Most meteorologists agree that a major problem in atmospheric science is the full explanation of atmospheric motion. This problem has confronted physical scientists for over three centuries and even today is not fully solved. Over these centuries the general atmospheric circulation has been shown to be exceedingly complicated and indeed the complexity seems to increase with every new massive observational onslaught on the atmosphere. In the face of such complexity it is perhaps natural that meteorologists have attempted to "break down" the global atmosphere into parts that may be more easily examined and understood. These "parts" are frequently known as "motion systems", configurations of motion which have different sizes and life-times—or alternatively different scales. This appreciation of scale soon became recognized as vital to any real understanding of atmospheric circulation. It is fundamental to the contents of this book and the remainder of this chapter reviews the scale concept in the light of both observational and theoretical evidence, with particular reference to the meaning of the traditional threefold classification of macro-, meso- and micro-sales. Space scales and time scales are described in terms of wavelengths and periods respectively. Thus the "wavelength" is the average distance from one updraught to the next updraught, or from one high pressure centre to the next. Similarly, the "period" is the average time from one gust to the next gust, or from one temperature maximum to the next.

II. Observation

The most immediately obvious air movements are those that make themselves felt to the individual, whether it be in the observation of curling cigarette smoke or in the risk of being blown over by a strong wind. Such movements are generally recognized by meteorologists to be of "micro-scale", with characteristic dimensions ranging from millimetres to hundreds of metres. More familiar meteorological features such as cumuliform clouds and valley fogs have come to be known as "local" phenomena, with characteristic dimensions of a few kilometres. The introduction of instruments, and more particularly, networks of instruments, allowed the identification of larger motion systems, notably cyclones and anticyclones. Such systems have characteristic horizontal dimensions of over at least 1000 km and are known as "synoptic scale" systems. Many meteorologists recognize a yet larger type of system, typified by the Rossby wave, with characteristic horizontal dimensions of 3000–6000 km.

Appreciation of the above hierarchy of motion systems was consolidated in the period 1920–50. In the first of those three decades the Bergen school of meteorologists presented its classical model of the synoptic scale, frontal cyclone (Bjerknes and Solberg, 1921, 1922). At the same time micro-scale motions came under close scrutiny with the opening of the British Chemical Defence Establishment (CDE) near Porton Down, Wiltshire, UK. One of the important projects of CDE was to increase our understanding of diffusion mechanisms in the atmosphere and largely under this stimulus the foundations of micro-meteorology were laid. The emphasis was upon the diffusion of particles or gases over distances of the order of tens of metres and consequently much of the research was directed towards the nature of small scale turbulence. Since the initiation of CDE effort, many institutions in many countries have become interested in turbulence in its own right as a vital mechanism in transferring heat, water and momentum within the atmosphere and between the earth and atmosphere. The theory and practice of micro-meteorology up to 1950 was well summarized by Sutton (1953) and Pasquill (1974) has covered developments since that date.

At the other extreme, the planetary waves, so brilliantly analysed by Rossby (1939, 1940), were at last clearly revealed in the 1940s. Rossby's work, together with that of Sutcliffe (1938, 1947), Bjerknes and Holmboe (1944) and Eady (1949) elucidated some of the links between the planetary waves and the smaller cyclones and anticyclones. By 1950 meteorology had the scientific and technological knowledge to sketch an initial description and dynamical understanding of the large scale flows in the mid-latitude atmosphere.

In contrast to the major developments outlined above, "local" studies changed little in both their number and character. They remained essentially observational boundary-layer studies, usually over areas extending horizontally for no more than a few kilometres: typical topics of interest were frost hollows and urban areas (the latter particularly in central Europe). The main reason for this interest in the boundary layer at the expense of small scale, free atmospheric circulations was probably the ease of observing the former when compared with the latter. So, although meteorologists prior to 1950 were aware that many local or slightly larger scale features, other than those in the boundary layer, awaited their full attention, they were largely prevented from analysing them for want of the appropriate observational tools. After 1950 such tools became available.

Within the last 30 years four new methods of observation have facilitated the identification and analysis of motion systems of a size between the large, or macro-scale, and the small, or local and micro-scale, i.e. the middle, or meso-scale. The four new tools are radar, instrumented aircraft, satellites and very dense, surface instrumental networks. Of these, satellites are the most recent and probably the least useful. In contrast, weather radars became available shortly after World War II and, despite their early crudity for meteorological purposes, immediately made accessible to analysis precisely those systems that were beyond the reach of pre-War meteorologists. In a review of the achievements of and prospects for radar in meteorology Ligda (1951, p. 1281) wrote:

> It is anticipated that radar will provide useful information concerning the structure and behaviour of that portion of the atmosphere which is not covered by either micro- or synoptic meteorological studies. We have already observed with radar that precipitation formations which are undoubtedly of significance occur on a scale too gross to be observed from a single station, yet too small to appear even on sectional synoptic charts. Phenomena of this size might well be designated meso-meteorological.

By 1953 Swingle and Rosenberg (1953) used the adjective in the title of a paper given at the first weather radar conference and by the mid-1950s the prefix "meso-" was well established, replacing "micro-" as used by Fujita (1951) in what was essentially a meso-scale study.

The accumulation of evidence from radars and special surface observational networks, particularly from the Severe Local Storms Unit of the US Weather Bureau (the forerunner of today's National Oceanic and Atmospheric Administration), clearly substantiated Ligda's claim, leading to the production of a handbook on meso-analysis by Fujita, Newstein and Tepper (1956). This was followed up by Tepper (1959) who spelled out to a wider public the importance of "Meso meteorology—the link between macroscale atmospheric motions and local weather", to quote the title of his paper. He appreciated that his threefold division (see Table 1(a)) into macro- meso- and micro-scale was somewhat crude, and indeed to some extent arbitrary, but found it convenient to make his point that (Tepper, 1959, p. 57) "the emphasis on the larger scale motions and the deliberate disregard of the smaller scale motions has become well ingrained among meteorologists". With direct reference to the work of Sutcliffe and Charney, Tepper claimed that motions smaller than macro-scale are not "meteorologically insignificant" nor "meteorological noise". Rather, they are vital to the local forecaster and probably to the flow of energy within the whole circulation of the atmosphere. Tepper's words are rather harsh if applied to today's forecasting techniques. The smaller scale motions are not "deliberately disregarded" in current numerical forecast models: even if they are not dealt with explicitly, their effects are represented in various ways. Nevertheless two decades ago Tepper's point had some validity and five years later, in a Presidential Address to the Royal Meteorological Society, Sawyer (1964) re-iterated the case, particularly in the context of severe convective storms, in a masterly appraisal of meteorological analytical techniques on all scales. In many subsequent articles the essence of Tepper's threefold subdivision reappeared, exemplified by Fiedler and Panofsky's (1970) results shown in Table 1(b).

Table 1 Three scales of atmospheric motion

Scale	Macro-	Meso-	Local
(a) Characteristic dimension (km)†‡	>483	16–160	<8

Scale	Synoptic	Meso-	Micro-
(b) Period (h)§	>48	1–48	<1
Wavelength (km)§	>500	20–500	<20

† *Note:* Dimensions have been converted from non-metric units.
‡ *Source:* Tepper (1959).
§ *Source:* Fiedler and Panofsky (1970).

The somewhat arbitrary "feel" for a threefold division of scale of atmospheric motion has been further investigated by the methods of spectral analysis (see Panofsky, 1955). The spectrum measures the distribution of variance of a variable over wavelengths or periods. If the variable is a velocity component, the spectrum also describes the distribution of kinetic energy over wavelengths or periods. Two types of spectra are useful: space spectra and time spectra. In the first case observations are made at many places simultaneously, or nearly so; the spectrum (or power density) is then plotted as a function of wavelength, or its reciprocal, the wave number. In the second case, observations are made at a point, at frequent intervals, and the spectrum is plotted as a function of period, or its reciprocal, the frequency. If large ranges of period or wavelength are to be considered, it is convenient to plot period or wavelength as abscissa on a logarithmic scale, from right to left, and the ratio of the computed spectrum to period or wavelength as ordinate (Fig. 1). In this case, the area under the curve between two periods or wavelengths

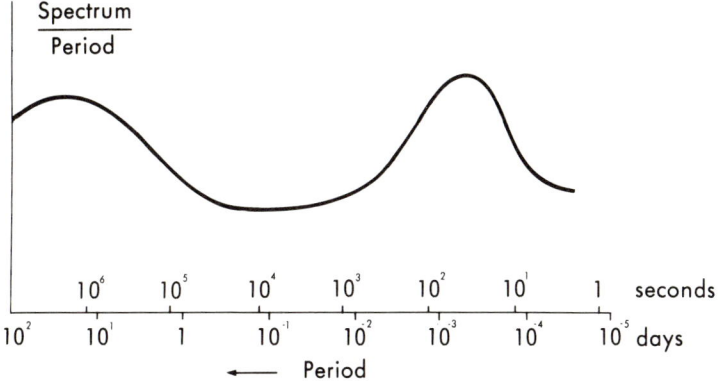

FIG. 1. Schematic frequency spectrum. (After Fiedler and Panofsky, 1970.)

represents the variance contributed by the interval between these periods or wavelengths to the total variance. In the schematic Fig. 1, most of the variance is due to fluctuations with periods between 10 s and 1000 s or between 12 h and 100 days; but very little of the variance comes from between 1000 s and 12 h.

Time spectra and space spectra are not independent of each other. Frequently, particularly for the smaller scales, it is possible to interpret variations at a point as a function of time by assuming that a pattern varying in space is moving past the observer with a known velocity, say c. Then the periods in the time spectrum are given by the wavelengths in the space spectrum divided by c. For the smaller scales, the speed c is essentially the wind speed.

With the aid of spectral analysis, van der Hoven (1957) appeared to provide substantial observational support for the threefold scale division. He produced a variance spectrum of horizontal wind speed in the frequency range from 0.0007 to 900 cycles per hour (periods of 2 months to 4 s) which showed two main peaks, one at periods of 3–4 days and one at periods of a few seconds separated by a minimum of variance at periods from 10 min to 5 h. This result appeared to confirm those of the earlier paper by Panofsky and van der Hoven (1955) which had paid particular attention to meso-scale periods and associated wavelengths. Van der Hoven produced six more spectra from different locations, all revealing a minimum of variance at the same order of magnitude of period. As the variance of wind speed is proportional to the kinetic energy of the speed fluctuations, van der Hoven (1957, p. 162) concluded that "the lack of physical process which could support eddy energy in the atmosphere is... the reason for the spectral gap in this range". This conclusion led to the construction of many spectra in later years in an attempt to establish the existence or otherwise of the so-called

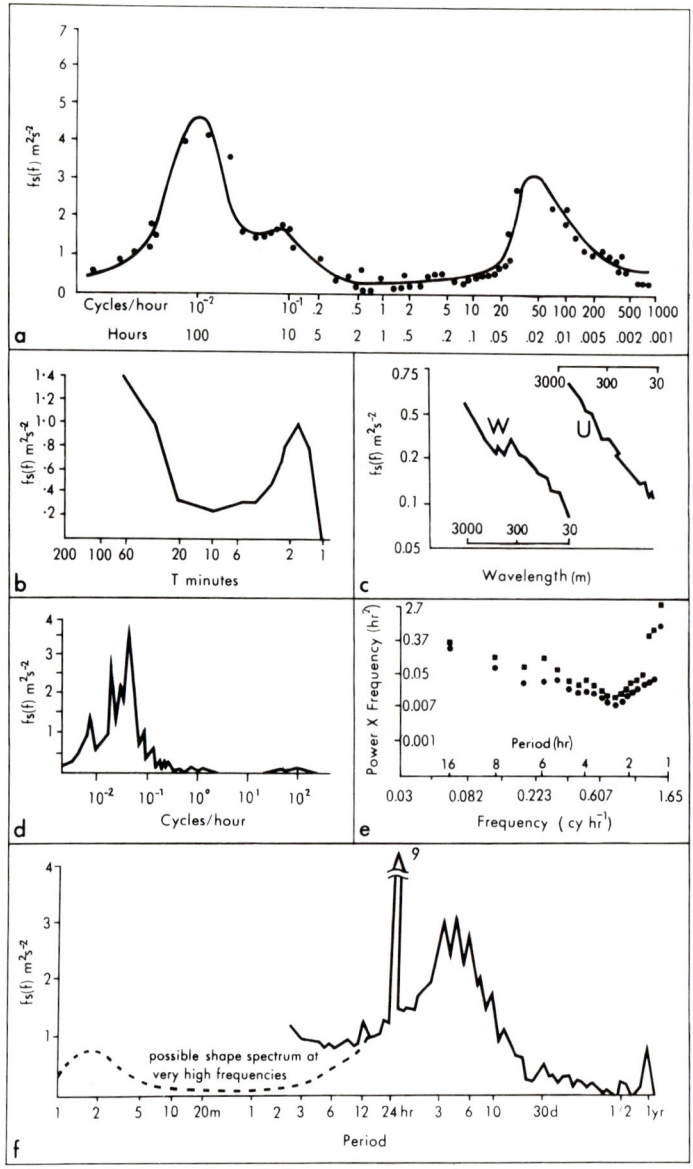

FIG. 2. Typical spectra of horizontal wind speed that show a "gap". (a) Brookhaven National Laboratory at a height of about 100 m. (After van der Hoven, 1957.) (b) Minnesota at a height of about 9 km. (After Mantis, 1963.) (c) Australia at heights between 9 and 11 km. (After Reiter and Burns, 1965.) (d) Oregon at surface. (After Frye et al., 1972.) (e) Nevada at height of about 450 m—data from both a tower and radar-tracked balloons. (After Cornett and Brundidge, 1970.) (f) Maine at surface. (After Oort and Taylor, 1969.)

"spectral gap". Support for van der Hoven's results appears in papers by Mantis (1963), Reiter and Burns (1965), Oort and Taylor (1969), Cornett and Brundidge (1970), Hess and Clarke (1973), Smedman-Hogström and Hogström (1975) and Frye *et al.* (1972) (Fig. 2). Although these spectra were taken in different parts of the atmosphere in different conditions, each one contains a suggestion of a minimum of energy density on time scales from 10 min to 1 day and space scales from 5 to 150 km or so, depending upon whether frequency or wave number was used for the construction of the spectrum.

In direct contrast, Goldman (1968) refuted the results of van der Hoven, claiming that the maximum of energy at high frequencies was primarily due to van der Hoven's use of data collected in hurricane conditions. Goldman's carefully collected data from a tower in Oklahoma suggested that the

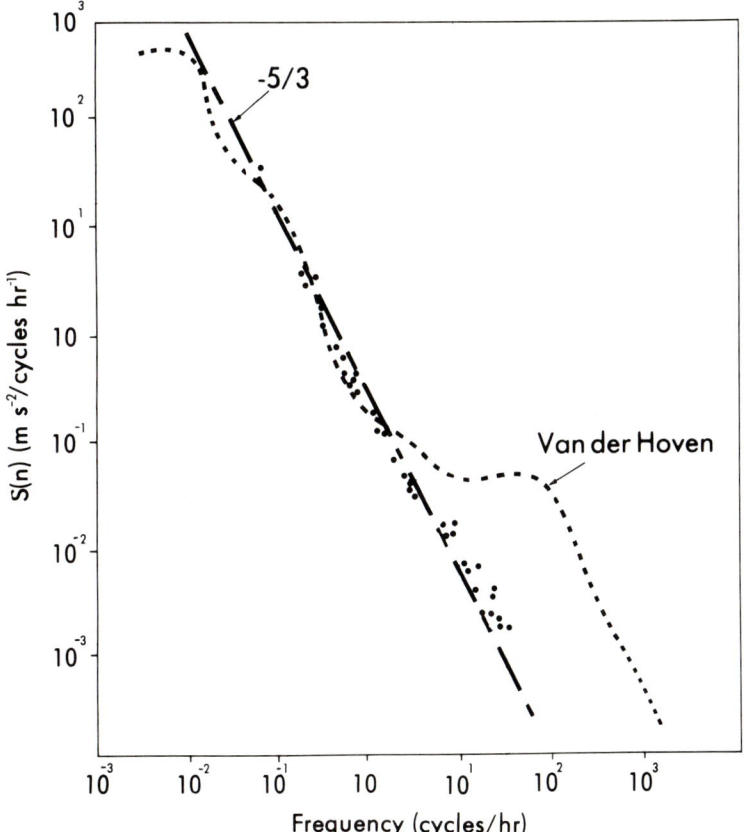

FIG. 3. Spectrum of horizontal wind speed—Oklahoma at height of 177 m. (After Goldman, 1968.)

variance curve follows the "$-5/3$ power law" predicted by Kolmogorov (1941) (Fig. 3). Further observational evidence to support Goldman's results was provided by Kao and Woods (1964), Pinus *et al.* (1967), Hseuh (1968), Hwang (1970), Vinnichenko and Dutton (1969) and Vinnichenko (1970).

Clearly the shape of spectra of both horizontal and vertical winds is not yet firmly established. Sufficient spectra have now been computed to show that they vary with location, season of the year, altitude, synoptic situation and values of indicative parameters such as the Richardson number. Vinnichenko (1970) combined several spectra in a further attempt to summarize the energy distribution of the atmosphere (Fig. 4) and he recognized the difference of opinion about the gap. Figure 4 illustrates the spectrum on both a log–log plot (Fig. 4(a)) and semi-log plot (Fig. 4(b)), the latter showing only that spectrum involving curve a in Fig. 4(a). The energy density of the mean wind in the free atmosphere is about two orders of magnitude larger than near the ground. In contrast to that in the free atmosphere, the energy density of micro-scale turbulence is small in all parts of the atmosphere. The meso-scale variations in the free atmosphere contain much more energy than variations near the ground so that, even if a micro-scale maximum exists, it is not separated from synoptic motions by so deep and wide a meso-scale gap as occurs near the ground.

It is tempting to believe wholeheartedly in the spectral gap as it so conveniently divides the atmosphere into three types of motion, depending on different types of dynamics dominant at the three scales: in particular, synoptic scales (quasi-geostrophic and hydrostatic), local and micro-scale (non-geostrophic, non-hydrostatic and turbulent) and meso-scale (non-geostrophic and hydrostatic). From another viewpoint, "the synoptic scales represent the scales of horizontal heterogeneities in radiation and the micro-scales represent the scales of the vertical heterogeneities of the temperature and wind fields" (Panofsky, 1969, p. 1101). But two closely associated changes of thinking have occurred since van der Hoven concluded that the meso-scale gap was empty for lack of suitably sized atmospheric circulations to fill it. First, in an analysis of the spectral distribution of energy and atmospheric predictability, Robinson (1967, p. 417) noted: "...I find unconvincing the argument that disturbances on scales between the cyclone and the thunderstorm do not exist because we do not regularly see them on synoptic charts". Robinson saw meso-scale circulations as a vital link in the atmospheric energy cascade from very large to very small scales. Second, Robinson made the important point that most spectra, rather than displaying an anomalous minimum in the meso-scale range, display an anomalous maximum in the micro-scale range. This apparently simple, yet fundamentally different mode of thought was soon appreciated in the literature. As mentioned earlier, Goldman (1968) could find no "hump" and therefore no gap, but Bretherton (1969) recognized that gaps do appear, as shown above, in certain types of

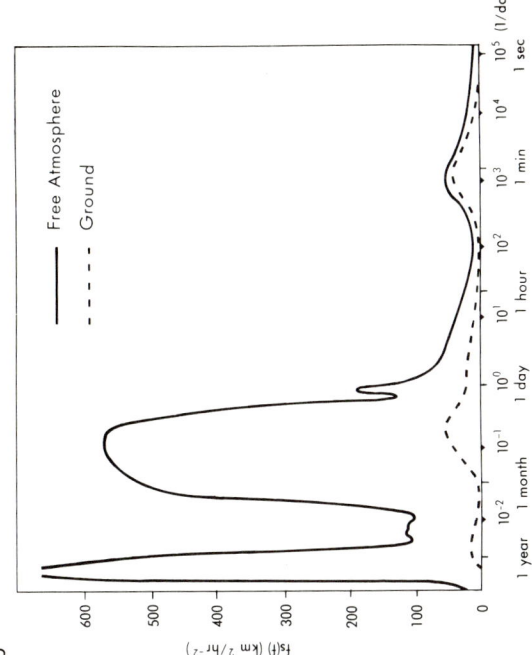
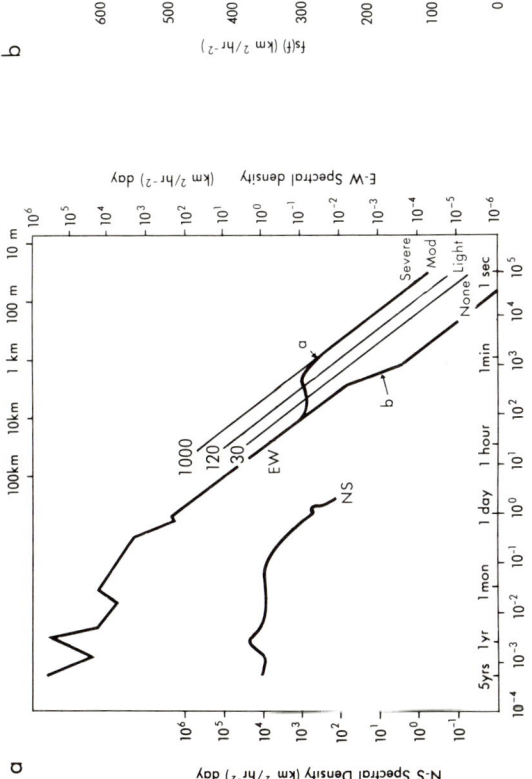

FIG. 4. (a) Log–log plot of spectra of zonal (E–W) and meridional (N–S) atmospheric motion in upper troposphere. Note that the scales for zonal and meridional flow are offset. The lines with a slope $-5/3$ shown in the high frequency part of the spectrum correspond to values of energy dissipation rate equal to 30, 120 and 1000 cm^2 s^{-3}, which give approximate separations between regions of "no", "light", "moderate" and "severe turbulence" as generally reported on basis of reactions of aircraft encountering turbulence. (After Vinnichenko, 1970.) (b) Semi-log plot of curve a of zonal flow part (a). Solid line, free atmosphere from 3 to 20 km; dashed line, near the ground. (After Vinnichenko, 1970.)

spectra. In the same year Vinnichenko and Dutton (1969) showed that the spectrum in the meso-scale region spanning periods from 1 day to approximately 10 min followed a 5/3 slope fairly well. The two branches of the high frequency part of the spectrum (Fig. 4(a)) fit well with Robinson's ideas. Vinnichenko and Dutton claimed that the observational evidence shows that the quasi-two-dimensional flow at the meso-scale is usually stable (e.g. see Pohle et al., 1965) in the cloudless free atmosphere and therefore the turbulence at higher frequencies had negligibly small energy (curve b in Fig. 4(a)). On the other hand, when wind shear and temperature stratification are favourable for the breakdown of meso-scale motion, a relatively large local maximum of energy occurs at scales from 1 to 5 km or periods of 1–3 min, leading to moderate and severe turbulence (curve a in Fig. 4(a)). In general, Vinnichenko and Dutton's evidence pointed towards a smooth spectrum declining in energy into the higher frequencies, except when marked activity (mainly in the vertical) at these frequencies produces a "hump" there.

In a review sympathetic to the "gap" "in its own right", Fiedler and Panofsky (1970) saw it as an important influence on atmospheric predictability. They were of the opinion that a gap would necessitate the parameterization of small scale motion in terms of the characteristics of the large scale motions on the other side of the gap, but conceded that free atmosphere spectra exhibit a gap only at those times when micro-scale motion is strong. In contrast to all previous literature, they recognize the gap *between* micro- and meso-scales, rather than it *comprising* the meso-scale range. Thus (Fiedler and Panofsky, 1970, p. 1118): "Whereas gaps occur in the free atmosphere between micro-scale and meso-scale, there is no corresponding gap between meso-scale and macro-scale. Yet meso-scale motions in the free atmosphere are ubiquitous, in contrast to the situation close to the ground". To a certain extent this "mental shifting" of the gap inverts the use of the spectrum. Initially it appeared to be useful in broadly identifying the meso-scale and now, with more refined spectra (or more experience of meso-meteorology), the gap is slipped in between the meso- and micro-scales. It will be interesting to see if it is moved again as more spectra become available.

It appears unlikely that the spectral gap, even if it is a real feature, is a universal one. Nevertheless, in those cases where it has been observed it serves as a useful guide to the size of meso-scale circulations or, at worst, provides a lower limit to this size, as noted by Fiedler and Panofsky. Although the gap suggests that meso-scale circulations have small energy densities, this does not mean that they are unimportant atmospheric features when compared to the macro- and micro-scale features. They may have a critical role in transferring energy between the macro- and micro-scale circulations and, to use an analogy of Dr A. A. White (personal communication), "at any one time there is not much water in the outflow pipe from a bath, but it is inconvenient if it gets blocked".

III. Theory

In common with other sciences meterology has developed through the parallel progression of observation and theory. Indeed the real birth-place of the subject lies in the development of physical science as a whole, the distinctive meteorological flavour becoming evident only in the nineteenth century. As meteorological observations accumulated in the latter half of the nineteenth century it became increasingly evident that the atmosphere was a hydro- and thermodynamical system that could be analysed by equations developed over the previous 200 years. This (by now very familiar) concept was a truly immense breakthrough in the study of meteorology, leading eventually to realistic treatment of the subject as "an exact science" (Sutton, 1954). It meant that the familiar weather elements, such as temperature, cloud and precipitation, could be profitably analysed in terms of the air motions which caused them: and air motion itself was understandable in terms of familiar Newtonian mechanics. In addition to this tremendous diagnostic breakthrough was the bonus of potential prognostic ability. The relevant equations of motion contain time derivatives and so future states of the system can be predicted from initial states.

The equations of motion for the three components u, v and w of a velocity vector along the three orthogonal axes x, y, z, the last-mentioned being in the vertical, are

$$\frac{du}{dt} = \frac{\partial u}{\partial t} + u\frac{\partial u}{\partial x} + v\frac{\partial u}{\partial y} + w\frac{\partial u}{\partial z}$$
$$= 2\Omega v \sin\phi - 2w\Omega \cos\phi - \frac{1}{\rho}\frac{\partial p}{\partial x} + K_z^M \frac{\partial^2 u}{\partial z^2}, \quad (1)$$

$$\frac{dv}{dt} = \frac{\partial v}{\partial t} + u\frac{\partial v}{\partial x} + v\frac{\partial v}{\partial y} + w\frac{\partial v}{\partial z} = -2\Omega u \sin\phi - \frac{1}{\rho}\frac{\partial p}{\partial y} + K_z^M \frac{\partial^2 v}{\partial z^2}, \quad (2)$$

$$\frac{dw}{dt} = \frac{\partial w}{\partial t} + u\frac{\partial w}{\partial x} + v\frac{\partial w}{\partial y} + w\frac{\partial w}{\partial z} = 2\Omega u \cos\phi - \frac{1}{\rho}\frac{\partial p}{\partial z} - g + K_z^M \frac{\partial^2 w}{\partial z^2}. \quad (3)$$

These fundamental equations are derived in most standard books on dynamical meteorology (e.g. Hess, 1959). Analytical experience indicates that each term in these equations has a characteristic order of magnitude at the synoptic scale as shown below. In equations (1) and (2), all four parts of the acceleration terms are of the order 10^{-4}. In equation (3) the equivalent terms are of the order of 10^{-7}. On the right-hand side of equations (1)–(3) the coriolis terms involving u and v are of order 10^{-3} whereas that involving w is of order 10^{-5}. The pressure gradient term in equations (1) and (2) is of order 10^{-3} and in equation (3) of order 10^3. In this last case the large vertical pressure gradient is largely balanced by the force of gravity of order 10^3. The final term in all three equations accounts for frictional effects and is usually of order 10^{-6}–10^{-8} and as such is frequently ignored in many analyses of these equations. Similarly the component of the coriolis force due to the vertical velocity in equation (1) is often omitted. Bearing in mind the relative sizes of the terms in equation (3), it is frequently justified in large scale analysis to ignore all but the pressure gradient and gravity forces, reducing the equation to

$$\frac{\partial p}{\partial z} = -\rho g. \qquad (4)$$

This is known as the hydrostatic equation.

In addition to the equations of motion, the continuity, thermodynamic and moisture equations, together with the equation of state for ideal gases, are fundamentally important to the analysis of atmospheric motion. In turn they are

$$\frac{\partial \rho}{\partial t} + \frac{\partial (\rho u)}{\partial x} + \frac{\partial (\rho v)}{\partial y} + \frac{\partial (\rho w)}{\partial z} = 0, \qquad (5)$$

$$\frac{\partial T}{\partial t} + u\frac{\partial T}{\partial x} + v\frac{\partial T}{\partial y} + w\frac{\partial T}{\partial z} = \frac{RT}{pc_p}\left(\frac{\partial p}{\partial t} + u\frac{\partial p}{\partial x} + v\frac{\partial p}{\partial y} + w\frac{\partial p}{\partial z}\right), \qquad (6)$$

$$\frac{\partial q}{\partial t} + u\frac{\partial q}{\partial x} + v\frac{\partial q}{\partial y} + w\frac{\partial q}{\partial z} = 0, \qquad (7)$$

$$p = R\rho T. \qquad (8)$$

In equation (6) radiation and conduction have been neglected and in equation (7) the effects of eddy diffusion, evaporation and condensation have been omitted.

In a classical analysis of the equations of motion, Jeffreys (1922) produced a threefold classification of winds based upon their dynamics. In addition Jeffreys also showed that certain types of wind were associated with circulations of certain sizes. Both these important contributions are reviewed below.

Jeffreys employed only one equation of motion, that for the north–south component of the wind (equation (2)). He argued that the pressure gradient force is an important term, being the force responsible for most air motion relative to the Earth's surface. If the coriolis term far exceeds in value both the accelerational and frictional terms, then the pressure gradient term is nearly balanced by the coriolis terms, and Jeffreys used Sir Napier Shaw's adjective "geostrophic" to describe such winds. If the coriolis and frictional terms are small in comparison with the acceleration term, then the latter balances the pressure gradient term to give what Jeffreys called "Eulerian" winds. In such winds, each particle of air moves with an accelerated velocity corresponding to the horizontal pressure gradient. If the frictional terms exceed the coriolis and accelerational terms, then friction must balance the pressure gradient. The wind blows along the pressure gradient and the friction, assumed to act opposite to the flow, is sufficient to prevent the velocity of a particular mass of air from increasing steadily throughout its journey. Such a wind was called "antitriptic" by Jeffreys.

A further alternative, not considered by Jeffreys, is the case of "inertial" motions where the acceleration terms and coriolis terms are dominant in the presence of negligible pressure gradient forces. In this situation each parcel of air moves like a particle under the coriolis force (circular trajectories) nearly independent of neighbouring parcels. Such trajectories were described by the GHOST balloons in the southern hemisphere.

Jeffreys recognized the comparative simplicity of his scheme as follows:

> Besides these main divisions there will be many cases where the number of terms in the equations of motion comparable with the pressure term is two or three. In such cases the motion is composite in character and intermediate between the three main types. The elucidation of the simpler cases must, however, afford some assistance towards the understanding of more complex ones. (Jeffreys, 1922, p. 34).

Nevertheless, he went on to show the value of his scheme to an appreciation of the many types of air movement experienced in reality; and to give a glimpse of the importance of the effect of a circulation size on its own behaviour.

The magnitude of the acceleration term dv/dt is comparable with the ratio of the velocity to the time it takes this velocity to vary by a large fraction of its maximum value; and this time is the time it takes a particle to travel a quarter

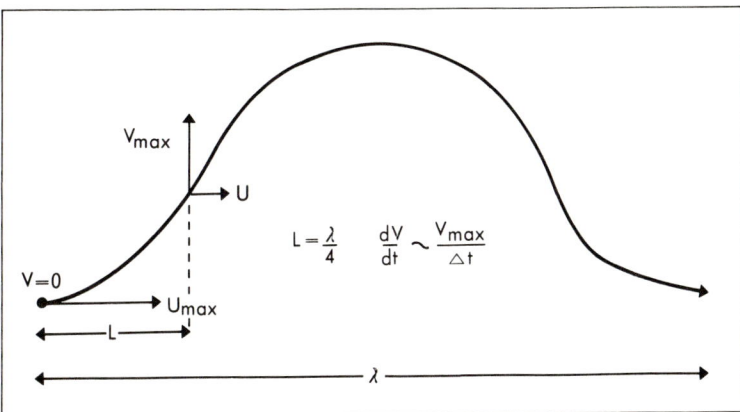

FIG. 5. Schematic disturbance to illustrate characteristic velocities and lengths. λ is the wavelength and $L = \lambda/4$.

of the way round a disturbed region or circulation system (see Fig. 5). If the distance over which the maximum change in v occurs is L, the term dv/dt will be of the order u^2/L, where $u \simeq v$, and this can be compared with the magnitude of the other terms, particularly the coriolis ones. Thus, if the accelerational terms are small compared with the coriolis terms, $v^2/L \ll 2\Omega u \sin\phi$, i.e. $L \gg v/\Omega \sin\phi$, assuming $v \simeq u$. If $v \approx 10\,\text{m s}^{-1}$, L must be greater than 70 km (at the poles) and 400 km (10° from the equator) for the coriolis term to have greater effects on the motion than the accelerational terms.

By 1920 there were sufficient wind observations to show that the free atmosphere appeared to be organized into large cyclonic and anticyclonic vortices, troughs and ridges, monsoonal reversals and other large circulations around which the air, particularly in the extra-tropical and free atmosphere, flowed in accord with geostrophic behaviour rather than so-called Eulerian or antitriptic behaviour. Jeffreys (1926) further analysed the dynamics of geostrophic winds, showing how they explained the gross features of the general atmospheric circulation and pointing out the necessity for cyclones and anticyclones as transfer mechanisms within the general circulation. In this context cyclones and anticyclones were defined in the general sense of deviations of the flow from zonal symmetry; and the transfers were both horizontal and vertical, although Jeffreys considered only the horizontal, meridional transfer of angular momentum.

A quarter of a century later Jeffreys' results were extended by Charney (1948) in his classical paper entitled "On the scale of atmospheric motions". Recognizing that the equations of motion permit wave solutions of many types, Charney concentrated his attention on "meteorologically significant motions", which he interpreted as synoptic or larger in scale. In using dimensional analysis similar to that applied by Jeffreys, Charney defined his scale properties as follows:

S – the mean horizontal distance between points at which the velocity components take extreme values. In effect this means that S is a quarter of a wavelength within a streamline pattern (equivalent to L in Fig. 5)

H – the corresponding mean vertical distance, usually the height of the tropopause

V – the mean magnitude of the horizontal velocity component

W – the mean magnitude of the vertical velocity component

C – the mean speed of the horizontal streamline pattern

g – acceleration due to gravity.

The orders of magnitude of these properties were chosen in accord with synoptic analytical experience (excluding g of course) as follows: $S \sim 10^6$ m, $H \sim 10^4$ m, $C \sim 10\,\text{m s}^{-1}$, $V \sim 10\,\text{m s}^{-1}$, $g \sim 10\,\text{m s}^{-2}$. Using these properties and appropriate values, Charney was able to evaluate the orders of magnitude of all quantities appearing in the equations of motion, which he wrote as follows:

$$\frac{du}{dt} - fv + jw = -\frac{1}{\rho}\frac{\partial p}{\partial x}, \quad (9)$$

$$\frac{dv}{dt} + fu = -\frac{1}{\rho}\frac{\partial p}{\partial y}, \quad (10)$$

$$\frac{dw}{dt} - ju + g = -\frac{1}{\rho}\frac{\partial p}{\partial z}, \quad (11)$$

where j is the y component of the coriolis parameter. Initially he replaced differentials by finite increments and expressed the incremental ratios in terms of S, H, C and V. Thus, to determine the order $\partial u/\partial s$, s being a horizontal distance coordinate, he replaced $\partial u/\partial s$ by $\Delta u/\Delta s$ and chose Δs equal to S. Then, by definition of S, Δu has the same order of magnitude as u itself (approximately equal to $U_{max} - U_{min}$) and we have

$$\frac{\partial u}{\partial s} \simeq \frac{\Delta u}{\Delta s} \sim \frac{V}{S}.$$

Charney also applied this argument to the v and w components of velocity and then consciously retained the order of magnitude ($10 \, \text{m s}^{-1}$) set for C. This was done

> because the specification of this order of C is precisely what distinguishes the meteorologically significant motions from the several varieties of theoretically possible motions having the same values of S, H and V that may exist. Thus C has the order $10^2 \, \text{m s}^{-1}$ in external gravity waves and order $10^3 \, \text{m s}^{-1}$ in tidal waves and by assigning the order $10 \, \text{m s}^{-1}$ we exclude such motions (Charney, 1948, p. 7).

Although some internal gravity waves have $C \sim 10 \, \text{m s}^{-1}$, Charney excluded them by the way he applied the geostrophic approximation.

As a result of this analysis the hydrodynamical equations take on the form

$$\frac{C}{S}V + fV + jW \sim \frac{1}{\rho}\frac{\partial p}{\partial x}, \quad (12)$$

$$\frac{C}{S}V + fV \sim \frac{1}{\rho}\frac{\partial p}{\partial y}, \quad (13)$$

$$\frac{C}{S}W + jV + g \sim \frac{1}{\rho}\frac{\partial p}{\partial z}, \quad (14)$$

and, from the first two,

$$\frac{\text{Horizontal acceleration}}{\text{Horizontal coriolis force}} \sim \frac{C/S}{f} = \frac{10/10^6}{10^{-4}} = \frac{1}{10},$$

which shows that the horizontal acceleration is one order of magnitude less than that of the horizontal coriolis force. To quote Charney (1948, pp. 7–8), "We may therefore regard the geostrophic approximation to be substantiated for the primary large scale of perturbations of the atmosphere—as manifested, for example, in the isobaric patterns on the upper level pressure maps". Also from the above equations,

$$\frac{\text{Vertical coriolis force}}{\text{Acceleration of gravity}} \sim \frac{jV}{g} = \frac{10^{-4} \times 10}{10} = 10^{-4},$$

$$\frac{\text{Vertical acceleration}}{\text{Acceleration of gravity}} \sim \frac{CW}{gS} \leqslant \frac{C^2 H}{gS^2} \sim \frac{10^2 \times 10^4}{10 \times 10^{12}} = 10^{-7},$$

which together serve to justify the hydrostatic approximation which is quite basic to simplifying the analysis of large scale meteorological systems.

To some extent Charney was restating Jeffreys' results of 20 years earlier, but Charney took the analysis further. He showed that the geostrophic approximation could prove a useful substitution for the horizontal velocity in the equations of motion in that it filtered out smaller scale motions, large scale inertial waves and very long "sound" waves to give realistic solutions from both a simulation and forecasting point of view. Clearly, by 1950 there were exciting prospects ahead—particularly with the advent of computers—for the solution of the equations of motion with a view to forecasting large scale atmospheric motions and simulating the general atmospheric circulation.

Table 2   Dimensional parameters

Parameter	Form	Characteristic ratio
Froude number	$U^2/gL(\Delta\rho/\rho)$	(Inertial force)/(Buoyancy force)
Rayleigh number	$gL^3(\Delta\rho/\rho)/\alpha\nu$	(Buoyancy force)/(Dissipative "force")
Reynolds number	UL/ν	(Inertial force)/(Viscous force)
Richardson number	$g(\Delta\rho/\rho)/L(\partial U/\partial z)^2$	(Hydrostatic stability)/(Shear)^2
Rossby number	$U/\Omega L$	(Inertial force relative to coordinate rotation)/(Coriolis force)
Strouhal number	U/fL	(Convective acceleration)/(Local acceleration relative to a fixed observer)

Source: Hidy (1967).

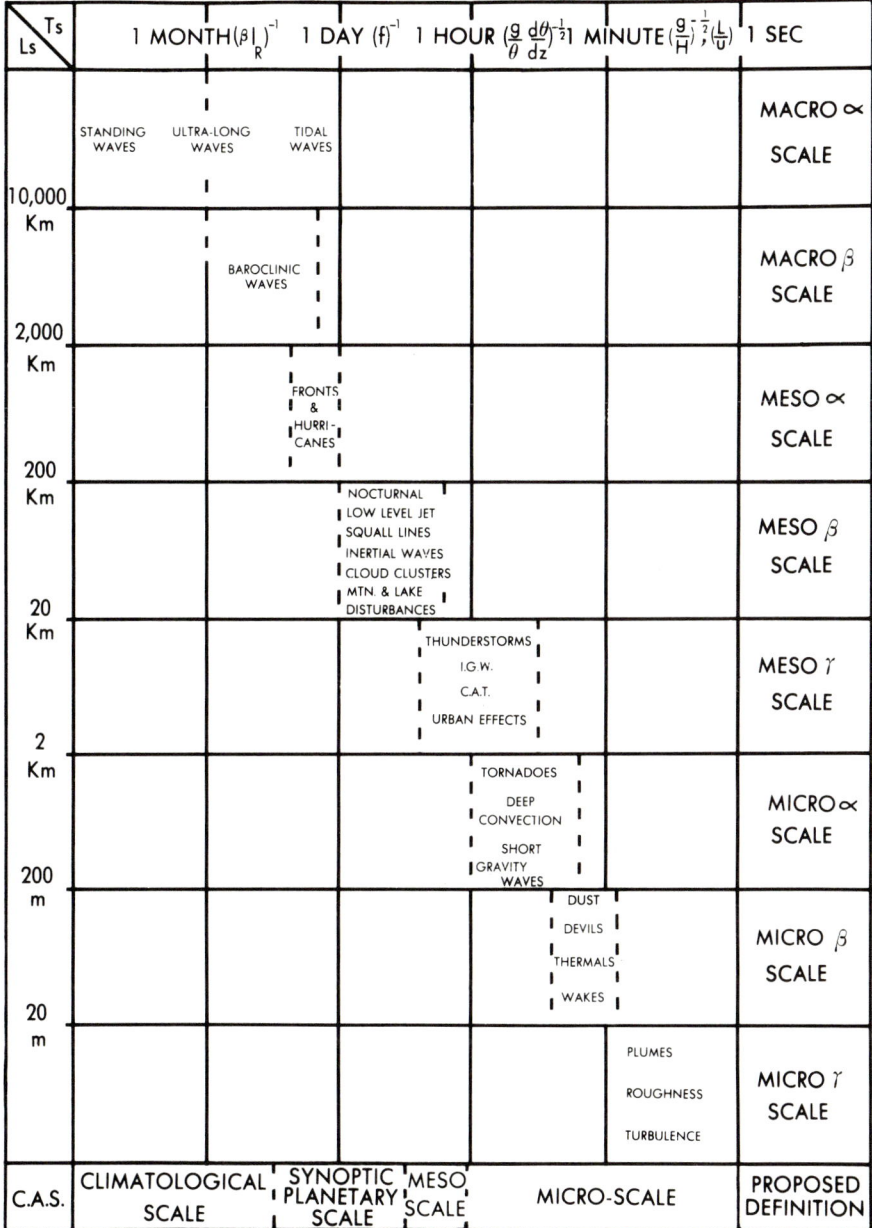

FIG. 6. Scale definitions and different processes with characteristic time and horizontal scales. (After Orlanski, 1975.)

The works of Charney and Jeffreys showed that motion systems of certain sizes were primarily due to certain types of force within the equations of motion. This result is well summarized by the formulation of dimensionless numbers which relate forces of different types (Table 2). In atmospheric motion the Reynolds number is always very large. The Froude and Richardson numbers are respectively important in dealing with gravity waves in the atmosphere and in studying the relative stability of stratified media and its effects upon the onset of turbulence. The Rossby number serves as a criterion for the importance of the effects of the Earth's rotation on large scale airflow. The Strouhal number gives an indication of the importance of the two parts of force acceleration of a fluid element. The Rayleigh number is useful in the analysis of convection. Most of these numbers appear at some stage throughout this book.

The above discussion of both observational and theoretical work on scale is well summarized by Orlanski's (1975) "A rational subdivision of scales for atmospheric processes". The left part of Fig. 6 shows the definitions of a few groups, based upon the horizontal scale of motion. A "core" meso-scale size of 2–200 km emerges. In the right part of the figure scales are defined by both length and time. As Orlanski stresses, processes with similar times and horizontal sizes need not have any dynamical similarity. The terms in parentheses along the time scale row are physical parameters known to be controlling each particular range of time scales. Scales between one month and one day are governed by a time scale which is the inverse of the product of the Rossby radius of deformation, $L_R = (H/f)[(g/\theta)(d\theta/dz)]^{1/2}$, and the variation of the Earth's rotation with latitude, β. H is the depth of a homogeneous atmosphere. The local effective period (f^{-1}) of the Earth's rotation is a controlling factor for motions with scales of one day to a few hours. The static stability of the atmosphere, measured by the Brunt–Väisäla period (Brunt, 1927), is the time scale for motions of a few hours' duration. The Brunt–Väisäla period is given by $N^{-1} = [(g/\theta)(d\theta/dz)]^{-1/2}$. For shorter time scales Orlanski suggests two characteristic parameters depending on the scale of the motion: $(g/H)^{-1/2}$ for external gravity waves, or L/u, which is the advective time for turbulent motion.

This book is largely concerned with the meso β and meso γ scales of Fig. 6, more particularly the former. It has two main parts: the first deals with circulations initiated by topographic irregularities, either mechanical or thermal; the second deals with free atmosphere meso-scale circulations, most of which are ultimately convective in origin.

References

Bjerknes, J. and Holmboe, J. (1944). On the theory of cyclones. *J. Met.*, **1**, 1–22.
Bjerknes, J. and Solberg, H. (1921). Meteorological conditions for the formation of rain. *Geofys. Publr*, **2** (3).

Bjerknes, J. and Solberg, H. (1922). Life cycle of cyclones and the polar front theory of atmospheric circulation. *Geofys. Publr*, **3** (1).
Bretherton, F. P. (1969). The spectral gap. *Radio Sci.*, **4**, 1361–1363.
Brunt, D. (1927). The period of simple vertical oscillations in the atmosphere. *Q. Jl R. met. Soc.*, **53**, 30–32.
Charney, J. (1948). On the scale of atmospheric motions. *Geofys. Publr*, **17** (2).
Cornett, J. S. and Brundidge, K. C. (1970). A comparison of the wind spectra calculated from data obtained by independent sampling methods. *Mon. Weath. Rev.*, **98**, 233–237.
Eady, E. T. (1949). Long waves and cyclone waves. *Tellus*, **1**, 33–52.
Fiedler, F. and Panofsky, H. A. (1970). Atmospheric scales and spectral gaps. *Bull. Am. met. Soc.*, **51**, 1114–1119.
Frye, D. E., Pond, S. and Elliott, W. P. (1972). Note on the kinetic energy spectrum of coastal winds. *Mon. Weath. Rev.*, **100**, 671–673.
Fujita, T. (1951). Microanalytical study of a thundernose. *Geophys. Mag.*, **22**, 71–88.
Fujita, T., Newstein, H. and Tepper, M. (1956). Mesoanalysis—an important scale in the analysis of weather data. US Weather Bureau, Research Paper no. 39.
Goldman, J. L. (1968). *The Power Spectrum in the Atmosphere below Macroscale*, Institute of Storm Research, University of St Thomas, Houston, Texas.
Hess, G. D. and Clarke, R. H. (1973). Time-spectra and cross-spectra of kinetic energy in the planetary boundary layer. *Q. Jl R. met. Soc.*, **99**, 130–153.
Hess, S. L. (1959). *Introduction to Theoretical Meteorology*, Henry Holt & Co., Inc., New York.
Hidy, G. M. (1967). Adventures in atmospheric simulation. *Bull. Am. met. Soc.*, **48**, 143–161.
Hsueh, Y. (1968). Mesoscale turbulence spectra over the Indian Ocean. *J. atmos. Sci.*, **25**, 1025–1057.
Hwang, H. J. (1970). Power density spectrum of surface wind speed on Palmyra Island. *Mon. Weath. Rev.*, **98**, 70–74.
Jeffreys, H. (1922). On the dynamics of wind. *Q. Jl R. met. Soc.*, **48**, 29–46.
Jeffreys, H. (1926). On the dynamics of geostrophic winds. *Q. Jl R. met. Soc.*, **52**, 85–103.
Kao, S.-K. and Woods, H. D. (1964). Energy spectra of meso-scale turbulence along and across the jet stream. *J. atmos. Sci.*, **21**, 513–519.
Kolmogorov, A. (1941). The local structure of turbulence in incompressible viscous fluid for very large Reynolds numbers, *C. r. Dokl. Acad. Sci. U.R.S.S.*, **30**, 301–305.
Ligda, M. G. H. (1951). Radar storm detection. In *Compendium of Meteorology* (T. F. Malone, ed.), American Meteorological Society, Boston, pp. 1265–1282.
Mantis, H. T. (1963). The structure of winds of the upper troposphere at mesoscale. *J. atmos. Sci.*, **20**, 94–106.
Oort, A. H. and Taylor, A. (1969). On the kinetic energy spectrum near the ground. *Mon. Weath. Rev.*, **97**, 623–636.
Orlanski, I. (1975). A rational subdivision of scales for atmospheric processes. *Bull. Am. met. Soc.*, **56**, 527–530.
Panofsky, H. A. (1955). Meteorological applications of power-spectrum analysis. *Bull. Am. met. Soc.*, **36**, 163–166.
Panofsky, H. A. (1969). Spectra of atmospheric variables in the boundary layer. *Radio Sci.*, **4**, 1101–1109.
Panofsky, H. A. and van der Hoven, I. (1955). Spectra and cross-spectra of velocity components in the meso-meteorological range. *Q. Jl R. met. Soc.*, **81**, 603–606.
Pasquill, F. (1974). *Atmospheric Diffusion*, 2nd edn, Ellis Horwood, Chichester.
Pinus, N. Z., Reiter, E. R., Shur, G. N. and Vinnichenko, N. K. (1967). Power spectra of turbulence in the free atmosphere. *Tellus*, **19**, 206–213.

Pohle, J. F., Blackadar, A. D. and Panofsky, H. A. (1965). Characteristics of horizontal mesoscale eddies. *J. atmos. Sci.*, **22**, 219–221.

Reiter, E. R. and Burns, A. (1965). Atmospheric structure and clear air turbulence. Atmospheric Science Technical Paper no. 65, Colorado State University, Fort Collins.

Robinson, G. D. (1967). Some current projects for global meteorological observation and experiment. *Q. Jl R. met. Soc.*, **93**, 409–418.

Rossby, C. G. (1939). Relation between variations in the intensity of the zonal circulation of the atmosphere and the displacement of the semi-permanent centres of action. *J. marine Res.*, **2**, 38–55.

Rossby, C. G. (1940). Planetary flow patterns in the atmosphere. *Q. Jl R. met. Soc.*, **66**, Supplement, 68–87.

Sawyer, J. S. (1964). Meteorological analysis—a challenge for the future. *Q. Jl R. met. Soc.*, **90**, 227–247.

Smedman-Högström, A.-S. and Högström, U. (1975). Spectral gap in surface-layer measurements. *J. atmos. Sci.*, **32**, 340–350.

Sutcliffe, R. C. (1938). On development in the field of barometric pressure. *Q. Jl R. met. Soc.*, **64**, 495–509.

Sutcliffe, R. C. (1947). A contribution to the problem of development. *Q. Jl R. met. Soc.*, **73**, 370–384.

Sutton, O. G. (1953). *Micrometeorology*, McGraw-Hill, New York.

Sutton, O. G. (1954). The development of meteorology as an exact science. *Q. Jl R. met. Soc.*, **80**, 328–338.

Swingle, D. M. and Rosenberg, C. (1953). Mesometeorological analysis of a cold front passage using radar weather data. *Proc. Conf. Radar Met., Austin, Texas*, Chap. XI-5.

Tepper, M. (1959). Mesometeorology—the link between macro-scale atmospheric motions and local weather. *Bull. Am. met. Soc.*, **40**, 56–72.

van der Hoven, I. (1957). Power spectrum of horizontal wind speed in the frequency range from 0.0007 to 900 cycles per hour. *J. Met.*, **14**, 160–164.

Vinnichenko, N. K. (1970). The kinetic energy spectrum in the free atmosphere—1 second to 5 years. *Tellus*, **22**, 158–166.

Vinnichenko, N. K. and Dutton, J. A. (1969). Empirical studies of atmospheric structure and spectra in the free atmosphere. *Radio Sci.*, **4**, 115–1126.

Part II

Topographically induced circulations

A. Mechanically induced circulations

2
Lee Waves

I. Introduction

Atmospheric circulations resulting from airflow over hills and mountains exist on many scales, ranging from planetary waves to minor perturbations over very small obstacles. Clearly the size of a given circulation (measured in terms of horizontal extent) is frequently related to the size of the forcing obstacle, but atmospheric conditions—particularly wind and vertical stability—also play an important role. Thus, orographically induced meso-scale circulations are partly due to forcing by meso-scale hills and partly due to atmospheric mechanisms that produce meso-scale circulations even from perturbations caused by much larger scale surface features. In this chapter we consider meso-scale waves resulting from airflow over mountains. The waves may conveniently be divided into those occurring over the mountain (mountain waves) and those occurring to the lee of the mountain (lee waves). The mountain wavelength is closely related to mountain size which may not, of course, be of meso-scale size. In contrast, lee wavelength is more a function of atmospheric characteristics which, in fact, generate classic meso-scale features, apparently regardless of the size of the forcing obstacle. Thus we see lee wavelengths of about 30 km downwind of the Rocky Mountains (Fig. 7(a)). In fact, as will become evident later, the waves are forced, not by the whole width of the Rocky Mountains, but by a ridge or two of the appropriate horizontal dimensions. Such lee waves have now been observed in both the troposphere and stratosphere (Nicholls, 1973a) and have important effects on both vertical flux of momentum and the existence of clear air turbulence. This chapter surveys our observational and theoretical knowledge of this type of wave. Detailed reviews of the whole problem of "airflow over mountains" appear in two Technical Notes of the World Meteorological Organization

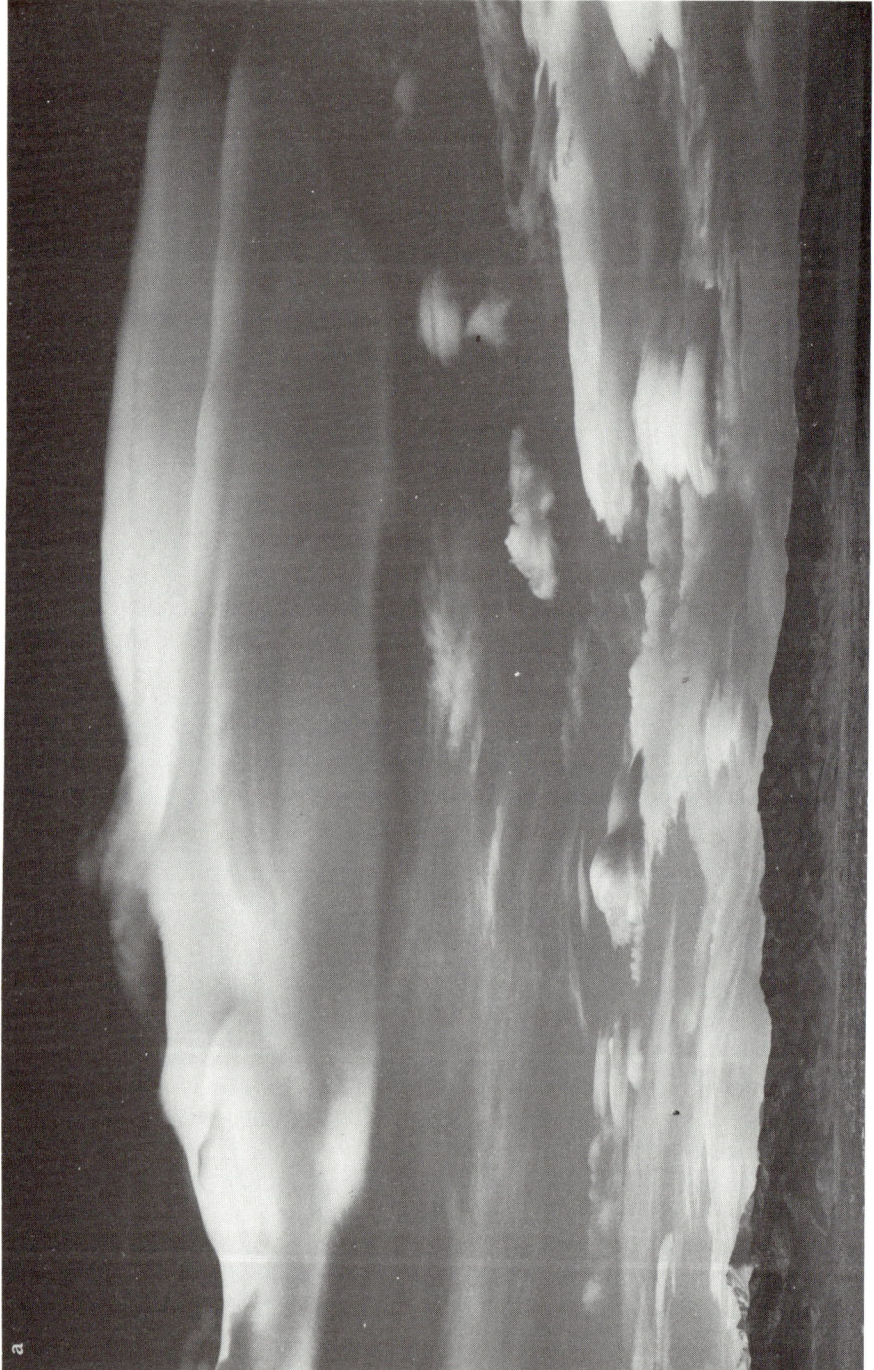

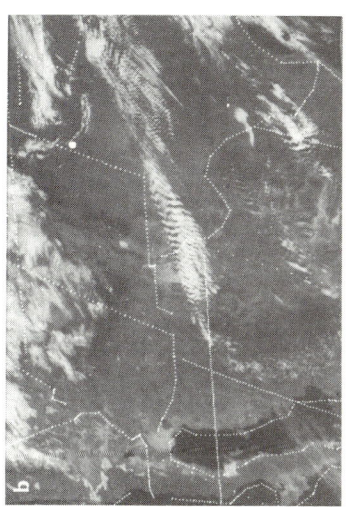

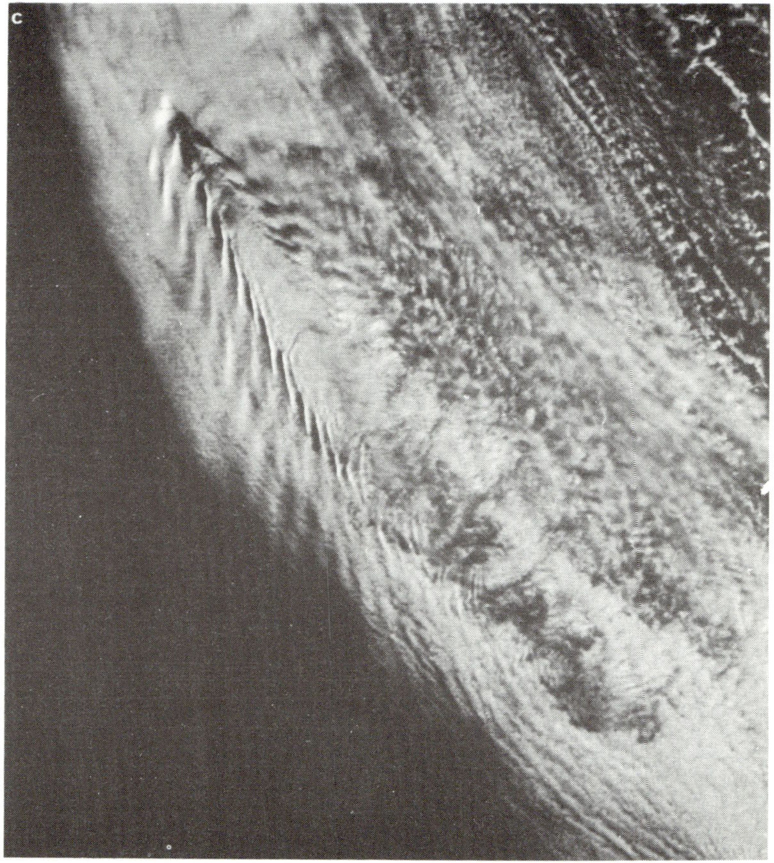

FIG. 7. (a) Ground photograph of clouds in lee waves downwind of the Rocky Mountains. (Courtesy of the National Centre for Atmospheric Research, Colorado, USA.) (b) Satellite photograph of clouds in a "two-dimensional" lee-wave train downwind of the Rocky Mountains over northern Mexico and western Texas. (Courtesy of the National Meteorological Centre, Maryland, USA.) (c) Satellite photograph of clouds in a "three-dimensional" lee-wave train downwind of Jan Mayen island. The island, about 50 km long, is marked by the bright patch at the head of the wave train. Distance from top to bottom of the picture is about 320 km. (Courtesy of Dundee University Electronics Laboratory, U.K.)

(Alaka, 1960; Nicholls, 1973a), in a paper by Corby (1954) and in a more general treatment of waves by Scorer (1978).

A. General mechanism of lee waves

The existence of waves to the lee of an obstacle is primarily due to the vertical oscillations of air induced by the obstacle. Thus, when air is forced to rise, restoring forces due to gravity generally come into play which tend to bring the airstream back to its original level. In fact, the air tends to overshoot this level and is then further subjected to forces which cause it to rise back to its original level. In other words particles of air experience a vertical oscillation. Clearly, if the air particles are also moving horizontally with the wind, they will move along a wave pattern in the vertical plane. The nature of the fundamental vertical oscillation was analysed by Brunt (1927). He showed that for simple vertical oscillations in the atmosphere the period for a particle displaced in the vertical is

$$\frac{2\pi}{\left[\frac{g}{T}(\Gamma-\gamma)\right]^{1/2}},$$

where γ is the lapse rate, which is positive when temperature (T) falls with height. In an isothermal atmosphere, $\gamma = 0$ and the period of oscillation is $2\pi/(g\Gamma/T)^{1/2}$ which, for $T = 300$ K, gives a period of rather less than 6 min. For inversions, the period of oscillation is less than this and as the lapse rate increases from zero up to the adiabatic value, the period increases indefinitely from 6 min upward. If an air particle moves horizontally with a velocity U (say) as it oscillates vertically, then the length of the wave form (λ) through which it passes will be

$$\lambda = \frac{2\pi U}{\left[\frac{g}{T}(\Gamma-\gamma)\right]^{1/2}}. \tag{15}$$

With typical values of $T = 250$ K, $\gamma = 0.006$ K m^{-1} and $U = 20$ m s^{-1}, $\lambda \simeq 10$ km, a frequently observed value in lee waves. λ is known as the "natural wavelength" of the airstream; that is, regardless of the cause of the oscillation, waves of that length would occur in the airstream. It is clear from equation (15) that this wavelength is a function of vertical stability and horizontal wind speed. Strong stability and low windspeeds favour short wavelengths whereas less stability and strong winds favour long wavelengths.

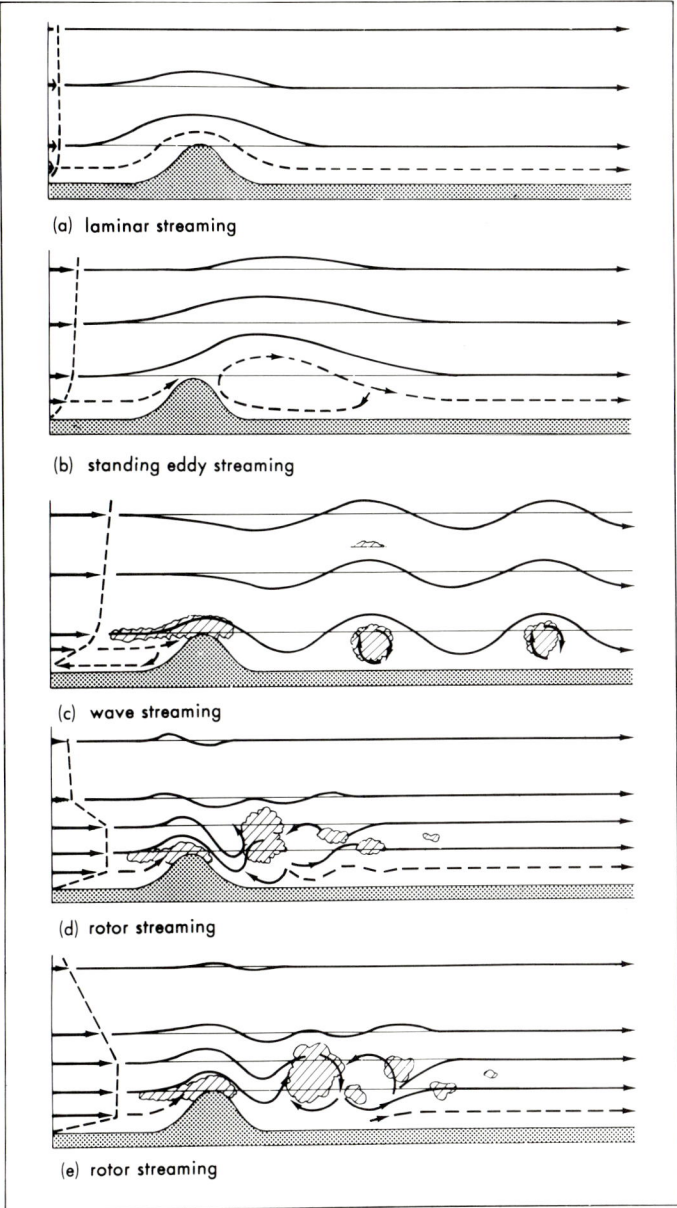

FIG. 8. Types of airflow over ridges: (a) laminar streaming; (b) standing eddy streaming; (c) wave streaming; (d) rotor streaming; (e) rotor streaming. Dashed line on left indicates vertical profile of horizontal wind speed. (After Förchgott, 1949.)

II. Observation

No doubt the often spectacular clouds associated with orographically induced waves have been noticed by laymen for centuries, but it is really only in the last half century that the waves have been seriously observed by scientists. Between 1920 and 1940 glider pilots such as Symons, Förchtgott and Küttner provided the bulk of our observational knowledge. In particular, as a result of substantial gliding experience in the European Alps, Küttner and Förchtgott were able to compile the first coherent descriptions of airflow over mountains and of wave flow in particular. Förchtgott (1949) went as far as classifying the airflow into four basic types (Fig. 8) and suggested that the differences between the types were a function of the vertical distribution of wind speed. In light winds the flow comprised a smooth shallow wave over the ridge. With somewhat stronger winds there was often evidence of a large semi-permanent eddy to the lee of the mountain above which the air flowed

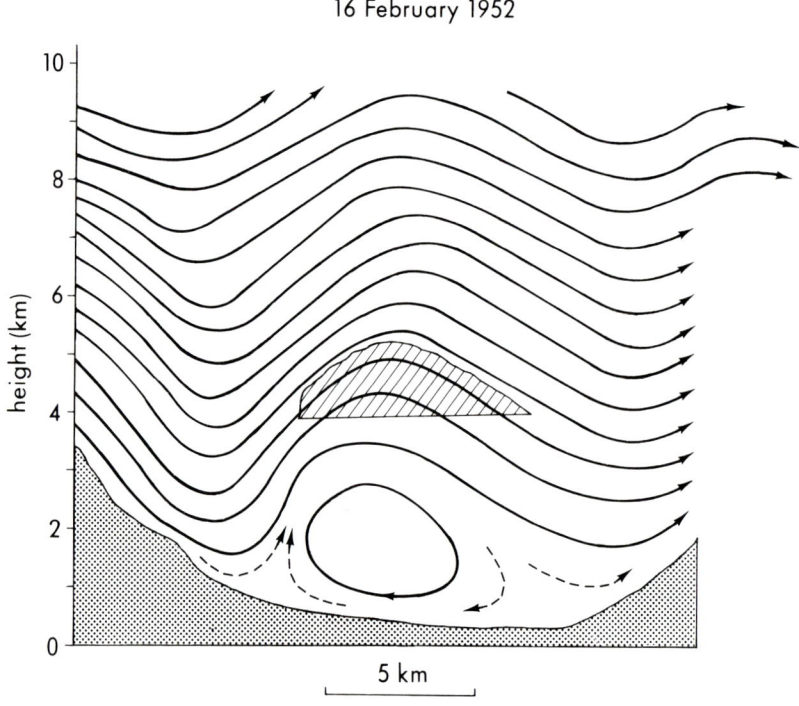

FIG. 9. Large lee wave, cloud and rotor to lee of Sierra Nevada, 16 February 1952. (After Holmboe and Klieforth, 1957.)

smoothly through a shallow wave. With stronger winds increasing with height a lee wave system developed downstream. The fourth category—rotor streaming—occurred in association with a strong wind maximum extending through a restricted vertical depth, comparable with the depth of the mountain.

Substantiation of Förchtgott's classification awaited further observations of mountain airflow by glider, powered aircraft, radar, lidar, constant level balloons and satellite (De, 1970) (Fig. 7(b)). But the capping cloud, or foehn wall, over the summit, the waves and the rotor were all confirmed by the contemporary work of Küttner (1938) and Manley (1945)—the former by glider, the latter by painstaking surface observational work in the Northern Pennines of England. As a result of the observational onslaught since 1950, we now know that lee waves occur in all parts of the world, from the Rocky Mountains (Lilly and Kennedy, 1973), Sierra Nevada (Holmboe and Klieforth, 1957) and Appalachians (Lindsay, 1962) of the United States to the Southern Alps of New Zealand (Hookings, 1968) and the Pennines of England (Aanensen, 1965). Nevertheless, despite work such as that by Yuyama (1972), a comprehensive climatology of their occurrence remains incomplete. We do, however, have sufficient data to provide a description of the waves in terms of their length, amplitude, vertical velocities, height, existence in trains and wakes and their duration. The bulk of these data apply to what may fairly be called "two-dimensional" waves, such as those forming to the lee of a long escarpment. But three-dimensional waves do, of course, exist, particularly in the lee of prominent islands (Fig. 7(c)) and other mountains (Abe, 1932, 1941).

A. Wave characteristics

The broad characteristics of lee waves have become clearer primarily as a result of observational programmes in California and Colorado in the United States, Alberta in Canada, the United Kingdom, France and Germany—particularly in the first two. In California the Sierra Wave Project of the early 1950s (Holmboe and Klieforth, 1957) provided the most comprehensive documentation of what were then recognized as the largest mountain waves in the world—the Bishop waves. Waves of similar size over the Rocky Mountains to the west of Boulder, Colorado were subject to scrutiny by the National Centre for Atmospheric Research in the late 1960s and, because better technology was available, this project was able to describe the airflow from the ground well into the stratosphere. Figure 9 shows the classic cross section of the flow in a strong wave over the Sierra Nevada on 16 February 1952, where the wave length was in accord with the distance between the Sierra in the west and the Ingo Mountains in the east. Figure 10

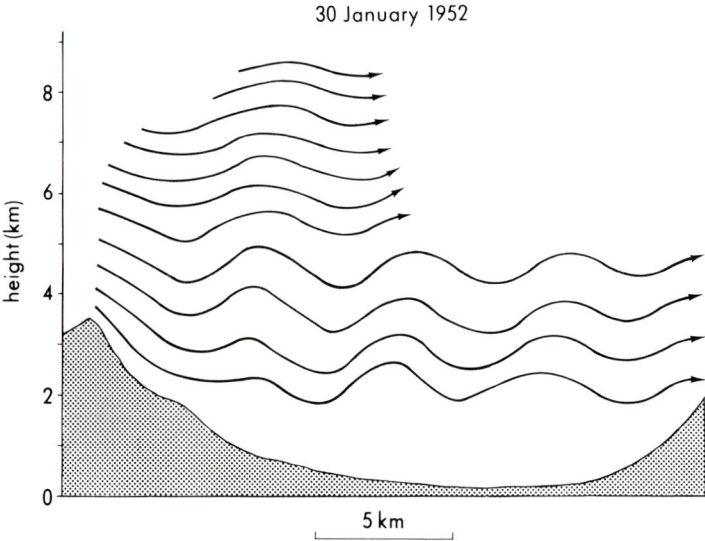

Fig 10. Small lee waves to lee of Sierra Nevada, 30 January 1952. (After Holmboe and Klieforth, 1957.)

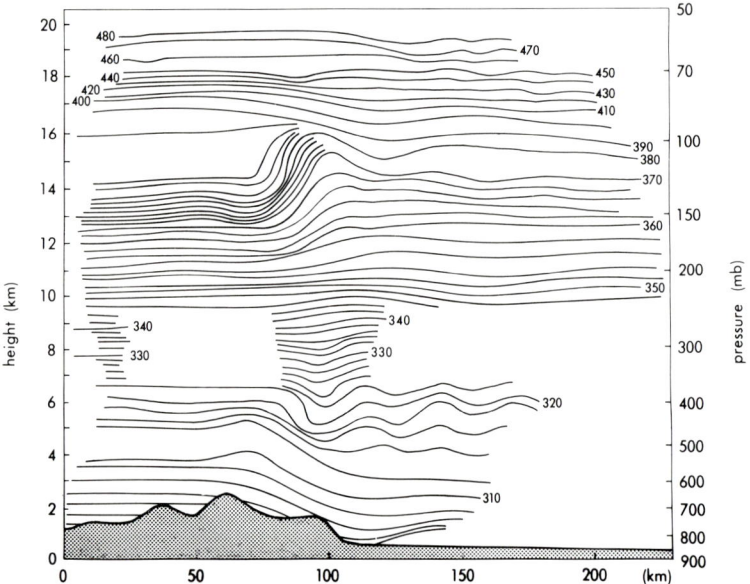

Fig. 11. Potential temperature cross-section (in kelvins) over Rocky Mountains near Boulder, Colorado, 17 February 1970. On the assumption of steady, adiabatic flow the isentropes may be considered as streamlines. (After Lilly and Kennedy, 1973.)

shows a different occasion on 30 January 1952 when several small waves were observed over the Owens Valley between the two mountain ranges. Figure 11, from Lilly and Kennedy (1973), shows flow over the Colorado Rockies, using isentropes as streamlines on the asumption of steady, adiabatic flow. This example clearly shows the trains of waves of different length in the troposphere, overlain by a massive wave between the 150 and 100 mb levels above the Front Range of the Rocky Mountains. On a separate occasion, the disturbance in the upper troposphere was so great that it took on the form of an hydraulic jump (Fig. 12), confirming the opinion of Vergeiner and Lilly (1970) that different nearly steady states of airflow over and to the lee of mountains can exist at different times. Cross-sections of the type shown here have been produced in virtually all serious investigations of lee-wave phenomena: they all have slight differences—but sufficient similarity to be typified by those used above.

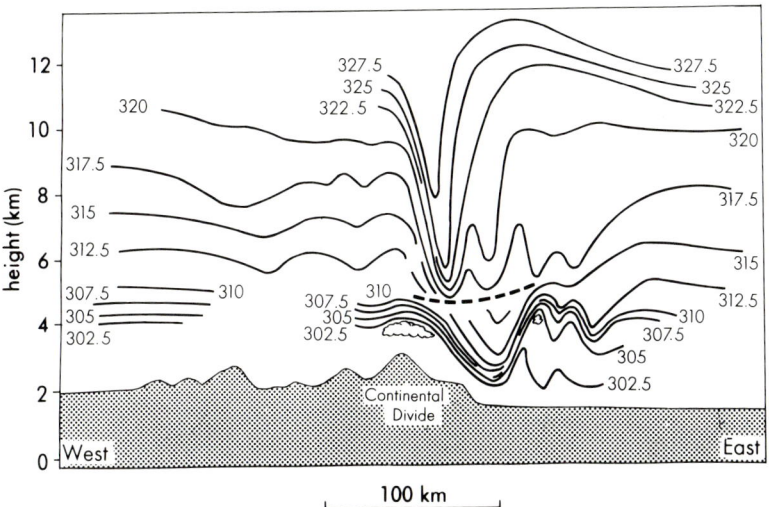

FIG. 12. Cross-section of potential temperature field (in kelvins) along an east–west line through Boulder, Colorado, 11 January 1972. For steady adiabatic flow the isentropes are good indicators of streamlines. Data above the heavy dashed line were gathered between 1700 and 2000 hours, while those below the line were gathered between 1330 and 1500 hours. (After Klemp and Lilly, 1975.)

Beneath the well-established wave shown in Fig. 9 lay a rotor in which the air at the base generally moved towards the mountain front. This is now a well-established phenomenon of lee-wave situations—particularly when the latter are well developed, as over the Sierra Nevada on 20 January 1952. Owing to the large vertical wind shear in the rotor, the characteristic roll cloud which often forms has the appearance of rotating about a horizontal

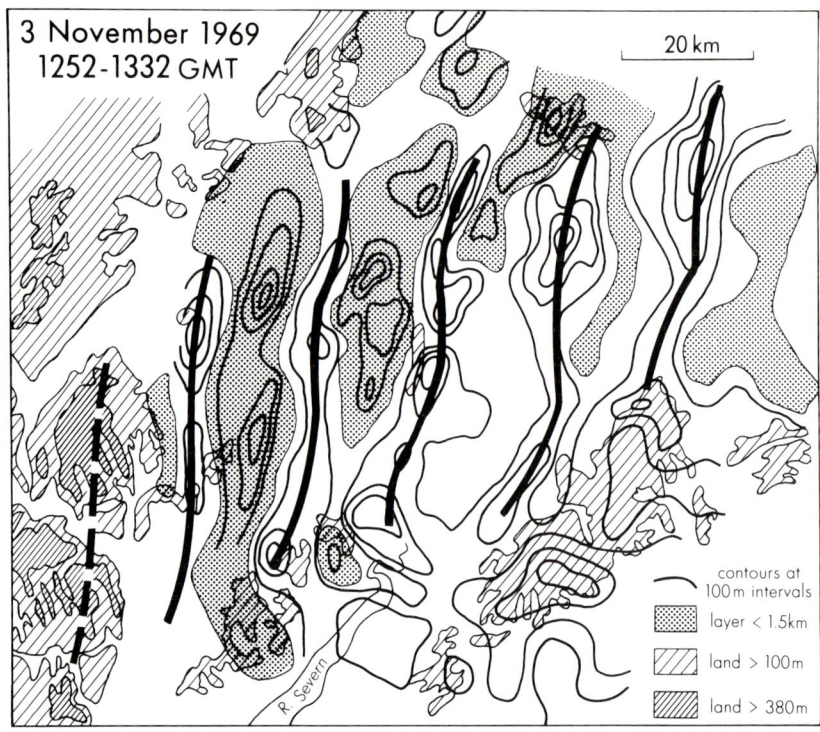

FIG. 13. Topography of a specified echo layer between 1252 and 1330 GMT on 3 November 1969, downwind of hills of southern Wales. Heavy lines indicate the positions of the wave crests. (After Starr and Browning, 1972.)

axis. The low level winds beneath rotors are much lighter than elsewhere, but violent turbulence frequently occurs in the vicinity of rotor clouds.

It is clear from the above diagrams that trains of lee waves exist. Satellite photographs show that clouds can extend for hundreds of kilometres parallel to wave-producing mountains (Fig. 7(b)) and up to 40 individual bands have been measured from the Andes to over 480 km downwind. In view of the complicated nature of most terrain (even "text-book" escarpments) the regularity of the plan view of lee-wave trains is quite surprising. Even at the small scale, the mountains of South Wales may generate a wave train of 110 km in length with a remarkable parallelism of crest and trough orientation (Fig. 13).

1. *Wavelength*

Individual lee waves are, of course, most sensibly described in terms of wavelength and wave amplitude. Table 3, compiled from several sources, shows a range of values from 1.8 km to nearly 70 km, but with the bulk of them lying between 5 and 20 km. Nearly all the projects found that wavelength varied with height, generally lengthening at higher levels (see Fig.

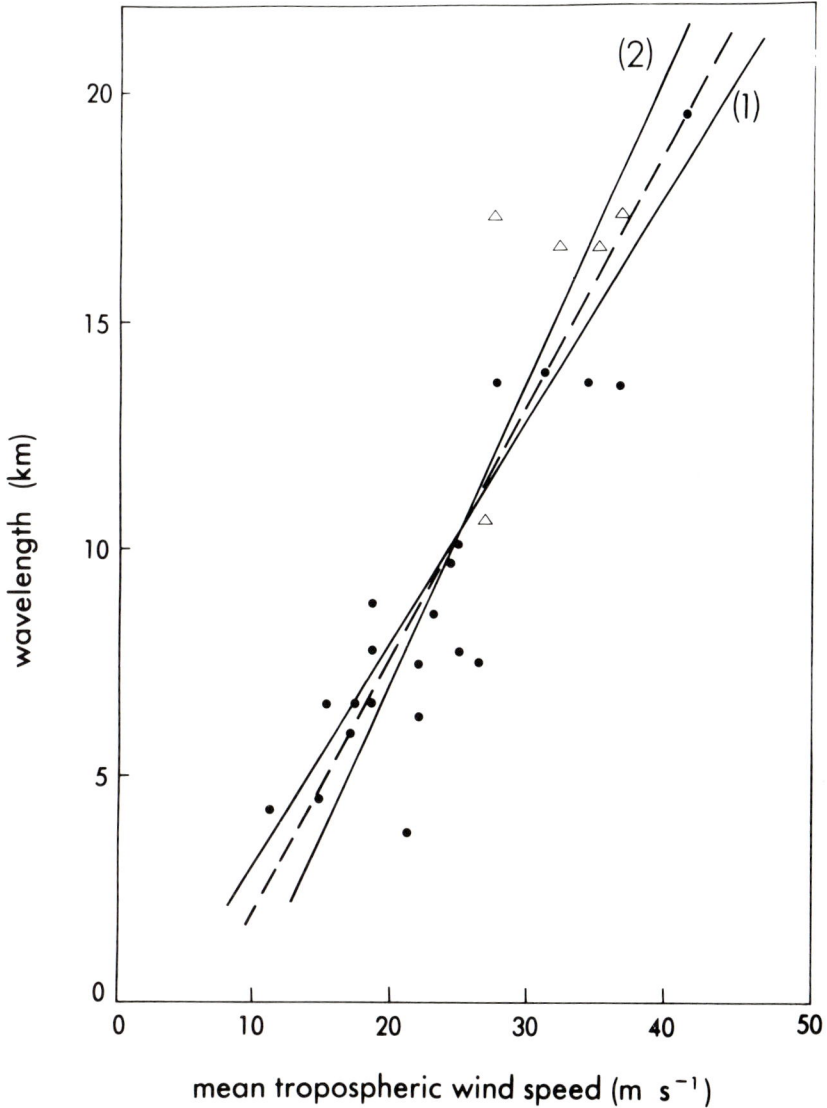

FIG. 14. Relation between mean tropospheric wind speed and wavelength of lee waves. Dots refer to data from Corby (1957a). Triangles refer to data from satellite photographs. Line 1 is the regression of wavelength on mean wind; line 2 is the regression of mean wind on wavelength. (After Anderson, 1966.)

Table 3 Wavelengths, amplitudes and vertical velocities in lee waves

Author	Wavelength (km)	Crest–trough amplitude (km)	Vertical velocity (m s^{-1})
Anderson (1966)	20	0.5	2.6
Bailey (1970)	24	—	2.0
Collis et al. (1968)	6	0.3	—
	17.5	0.3	—
	21.0	0.3	—
Colson (1952)	19	—	—
Corby (1957a)	10.7†	—	2.0
Foldvik (1962)	28	—	7
	31	—	2
	20	—	3.5
	7	—	c. 3
	15	—	2 to 3
Holmboe and Klieforth (1957)	8.8	0.6	+4.0 to −4.8‡
	16	1.1	+2.4
	14–18	—	±21.3
	4.4	0.3	+3.3 to −2.3
	8	0.6	+3.7 to −6.4
	18	2.1	+12.5 to −9.5
	9.2	0.4	+3.0 to −2.1
	28	1.2	+6.4
	7.6	0.9	+6.1 to −2.4
Holmes and Hage (1971)	45–56	0.3	—
	64–67	0.5–0.7	—
Küttner and Lilly (1968)	15	0.4	—
	16	2.0	—
	15	1.6	—
	11	2.0	—
	45	1.0	—
Lilly and Zipser (1972)	55	2.7	—
	13	6.0	—

† Average of 26 cases.
‡ Positive values upward; negative downward.

11). For example, Starr and Browning (1972) found wavelengths of about 16 km between heights of 5 and 7 km whereas between 2 and 5 km the wavelengths were typically 8–10 km. These differences are closely related to the wind speed of the layers in which the waves occur. This is shown in Fig. 14, which relates the wavelength of 26 lee-wave cases to the corresponding wind speed meaned over the layers contributing to the wavelengths assessments. The correlation coefficient is 0.91, which Corby (1957a) suggested was high enough to justify estimates of wavelength from mean tropospheric wind speed alone. In fact Fritz (1965) and Anderson (1966) also found this to be the case.

Table 3 (continued)

Author	Wavelength (km)	Crest–trough amplitude (km)	Vertical velocity (m s^{-1})
Manley (1945)	5–10	—	—
Nicholls (1973b)	10	0.2–0.5	—
Reynolds et al. (1968)	11.3	0.3	1.0
	9.8	0.6	2.9
	9.4	0.6	2.9
	11.3	1.1	5.0
Sarker and Calheiros (1974)	20.0	—	1.4
	18.2	—	0.8
	19.8	—	1.1
	20.1	—	2.1
	28.8	—	5.2
Starr and Browning (1972)	24	0.3	3.0
	25	0.8	2.6
Vergeiner and Lilly (1970)	6.8	—	1
	8.5	0.2	0.6
	10	0.4	1
	20	0.8	3
	20	1.1	3.5
	20	1.2	4
	10.4	0.8	5
	10.4	1.1	7
	11	0.6	3
	12	1.2	5.5
	12	1.0	4
	11.5	0.6	3.5
	14	0.7	3
	13.5	0.6	3
	17	0.5	2
Wooldridge and Lester (1969)	10.6§	—	—

§ Average of 14 cases.

2. Wave amplitude

Closely associated with wavelength is wave amplitude, here defined as the vertical distance between peak and trough of a streamline. Observed values are again shown in Table 3. They range from a few hundred metres to over 2 km. By far the most frequently observed maximum values for amplitude are 0.3–0.5 km. Similarly to wavelengths, amplitudes also vary with altitude, but not necessarily in any consistent relationship to wavelength. Under a given set of atmospheric conditions which establish the natural wavelength, the amplitude is greatest when the wavelength most nearly matches the shape of

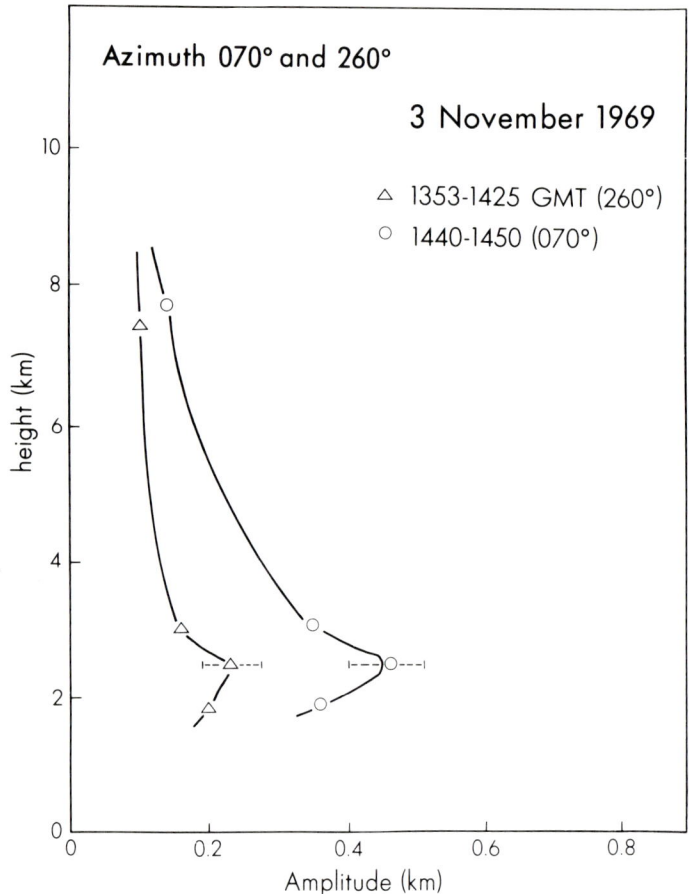

FIG. 15. Variation with height of the crest–trough amplitude of lee waves on 3 November 1969 downwind of hills of southern Wales. Data derived from radar data along azimuths 070° and 260°. Typical spread of amplitudes is shown by horizontal dashed lines. (After Starr and Browning, 1972.)

the obstacle. A most useful record of vertical variation of amplitude is provided by Starr and Browning (1972) for four cases downwind of the mountains of South Wales. Figure 15 shows maximum amplitudes of 0.3–0.6 km at an altitude of about 2 km, with values decreasing both above and below that height—the height of a temperature inversion, an important feature as shown earlier. But we should note from Fig. 12 that very large amplitudes have been observed in the stratosphere where the hydraulic jump of 11 January 1972 over the Colorado Rockies had a peak–trough distance of 6–7 km, the largest amplitude yet recorded.

The variation of the observed amplitudes and wavelengths in the vertical direction and with distance from the obstacle depends upon the summation of waves of different dimensions. Waves which are in phase tend to accentuate their characteristics (particularly their amplitudes) whereas waves out of phase tend to cancel out each other within short distances. The relative phase of waves and their relationships with the forcing obstacle are treated more fully by the theory outlined later.

3. *Vertical velocities*

Vertical velocities within lee waves are of course inextricably linked to wavelength and amplitude and such observations as exist are listed in Table 3. Values are usually of the order of 2–$6\,\text{m}\,\text{s}^{-1}$, but extremes of $15\,\text{m}\,\text{s}^{-1}$ have been claimed. In a study over small hills in eastern Scotland, Corby (1957a) provided a preliminary relationship between wavelength and maximum vertical velocity (Fig. 16) that suggested a maximum for that particular area for a wavelength of about 13 km, with smaller vertical velocities for both shorter and longer wavelengths. Corby further suggested that this was

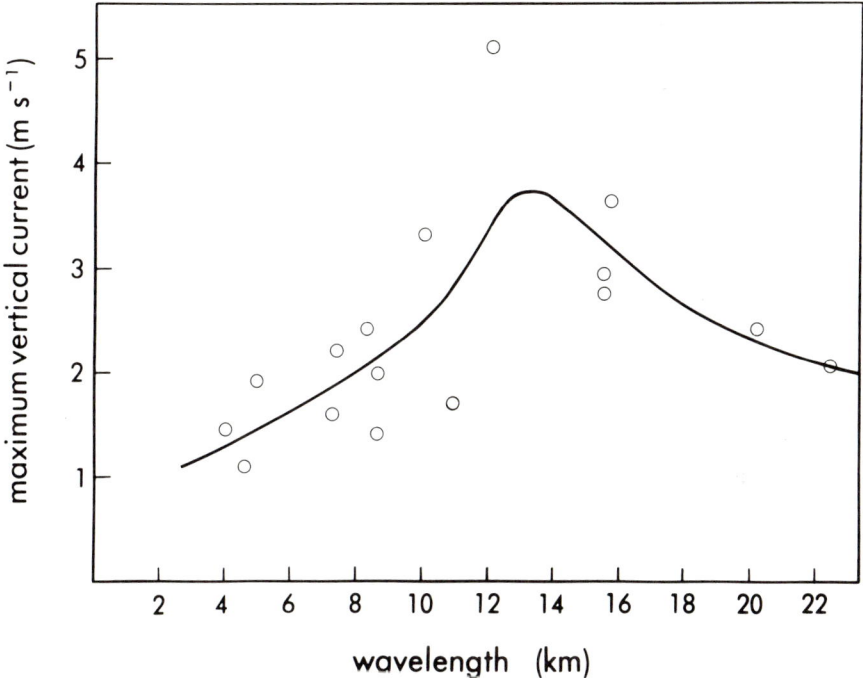

FIG. 16. Relation between maximum vertical current in and wavelength of lee waves. (After Corby, 1957a.)

probably due to a kind of resonance effect that occurred when the natural wavelength of the airstream matched the dominant wavelength component of the topography.

B. Atmospheric conditions suitable for waves

From the many observational studies mentioned above it soon became clear that, given a suitable obstacle, the occurrence of lee waves primarily depended upon two atmospheric characteristics; static stability and wind. Both factors were clearly evident for single cases observed in the Sierra project (Figs 9 and 10) but their overall importance to lee-wave formation was further stressed in the climatological study of Colson (1954). Table 4 shows the temperature differences for five atmospheric layers as observed at Santa Maria and Oakland (upwind of the wave area) from October 1951 to March 1952. Comparison of "wave days" with "all days" shows an increased instability on wave days from the surface to 850 mb, a slight increase in stability in the 850–700 mb layer and a definite increase in the stability in the 700–500 mb layer, which is the layer around the crest of the Sierras, the ridge that causes the Bishop wave. These observations and many others suggest that an essential requirement, at least for strong waves, is marked stability at levels where the air is disturbed by the mountain. We noted earlier that the maximum amplitude of lee waves is usually attained somewhere in or near the layers of maximum static stability (Starr and Browning, 1972).

Colson (1954) also showed (Fig. 17) that waves are favoured by strong winds and strong shear, clearly the case on 16 February 1952 (Fig. 9). The airstream should also blow more or less perpendicularly across the ridge and

Table 4 Vertical temperature distribution on wave days compared with that on all days. Data for period October 1951 to March 1952

	ΔT (°C)				
	SFC–850	850–700	700–500	500–300	300–200
Santa Maria					
All days	1.3	4.4	10.3	14.8	7.4
Wave days	2.1	3.5	8.6	15.0	8.8
Oakland					
All days	2.6	4.5	9.1	14.6	6.9
Wave days	4.7	4.4	8.2	14.4	6.4
Sth-Oak combined					
All days	2.0	4.5	9.7	14.7	7.2
Wave days	3.3	3.9	8.5	14.7	7.5

Source: Colson (1954).

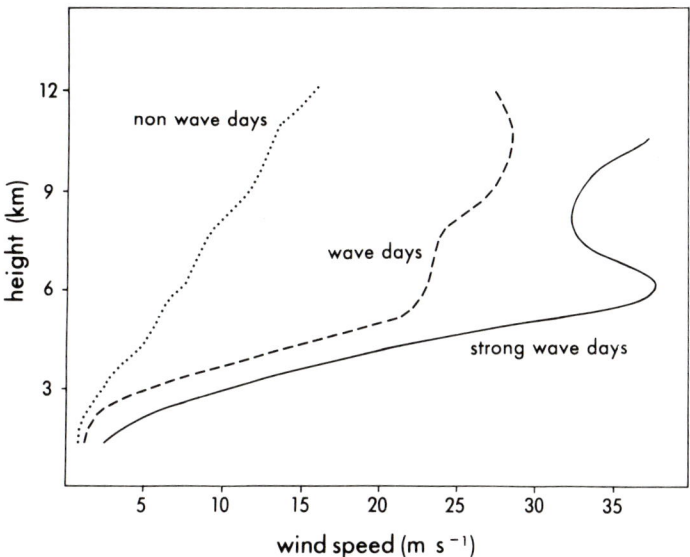

FIG. 17. Mean vertical wind shear on wave- and non-wave days. (After Colson, 1954.)

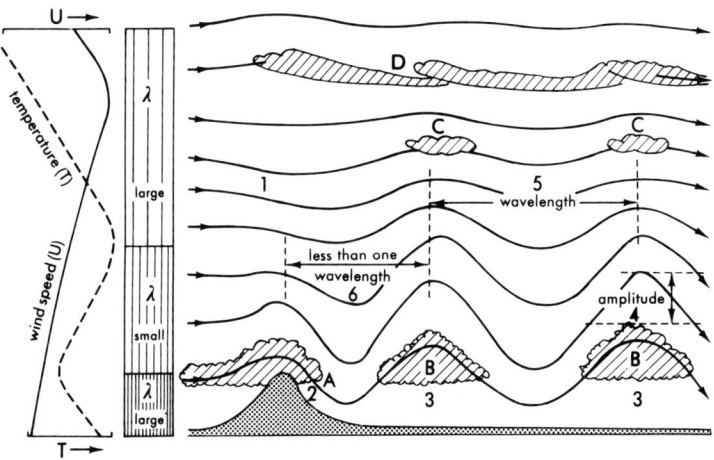

FIG. 18. Features of airflow across a long mountain range: (1) downdraught may occur at some levels to windward of ridge; (2) strong surface wind down lee slope; (3) variable surface wind; (4) maximum amplitude in stable layer; (5) order of wavelength, 5–30 km; (6) first wave crest usually less than one wavelength downstream of ridge. (A) Foehnwall; (B) roll cloud; (C) altocumulus lenticularis; (D) cirrus. λ denotes the natural wavelength determined by the airstream wind and temperature conditions. (After Wallington, 1960.)

this property should be maintained through a considerable depth of the atmosphere. Berenger and Gerbier (1957) have clearly shown that rotation and/or decrease of wind speed with height successfully suppress wave development.

Most of the observational characteristics of lee waves and associated vertical temperature and wind distributions are well summarized in Wallington's (1960) schematic diagram (Fig. 18).

III. Theory

Theoretical analysis of lee waves is but part of the general problem of wave occurrence in stratified fluids. Such analysis is necessarily rather mathematical and a full derivation would occupy more space than is available here. Quite comprehensive coverage of the theory of airflow over mountains appears in Alaka (1960), Corby (1954) and Scorer (1978). Both linear (small amplitude) and non-linear (large amplitude) models of orographically produced waves now exist, but by far the greater part of our understanding rests upon linear theory. It is only comparatively recently that non-linear theories have been developed (e.g. Clark and Peltier, 1977), but Smith (1977) was of the opinion that they had had little impact on the general problem. Consequently this section of the chapter concentrates largely upon linear models which, in essence, use perturbation methods (see Haurwitz, 1941; Panofsky, 1968) and which, as yet, have been more profitably applied to tropospheric than to stratospheric lee waves. The first works aimed at understanding the internal gravity waves produced by flow of stably stratified fluid over an obstacle were those of Lyra (1940, 1943) and Queney (1941, 1947, 1948) who used a linearized set of equations and boundary conditions. Scorer (1949) applied Queney's equation to separated layers characterized by different critical wavelengths. Following Scorer (1949), Corby and Sawyer (1958a), Sawyer (1960) and others solved the equations numerically in multi-layered models.

A. Small amplitude theory

Use of perturbation theory for the analysis of orographic waves requires certain assumptions about the flow. It should be steady, laminar, isentropic and non-viscous. With the exception of the last-mentioned, these assumptions do not accord well with observed atmospheric behaviour: nevertheless, they are necessary to make the problem tractable. Furthermore, both the obstacle and the resultant displacements of the airstream must be taken to be small compared with the corresponding undisturbed values. This enables the equations to be linearized since the products and squares of the disturbance quantities are then justifiably neglected. The condition is satisfied if the height of the mountain is small compared with its width.

1. *The work of Lyra and Queney*

Any examination of lee waves with the linearized equations rests to some degree on the efforts of Lyra and Queney. Lyra (1940) analysed the disturbance of a uniform stream due to a mountain ridge of rectangular section. His main result was that at higher levels the air underwent more than one oscillation. The oscillations increased in amplitude with height and decreased downwind and most of the disturbance to the airstream lay on the downwind side of the mountain. Lyra's results have been objected to on two counts: first that there was no upper limit to the disturbance and so every little hill should produce a disturbance in the stratosphere; and second, that the series of waves on the lee side vanished if a smooth mountain replaced the rectangular one.

Queney's studies formed part of a completely general theory of adiabatic perturbations in a stratified, rotating atmosphere. As such he modelled three types of motion that are caused directly by orography (Sawyer, 1959): (1) gravity oscillations that may extend upwards as far as the stratosphere; (2) oscillations about geostrophic equilibrium; (3) quasi-geostrophic motions.

Queney initially modelled flows over an infinitely long sinusoidal profile, not waves to the lee of obstacles. Later he did analyse the flow over an individual, smooth sided, two-dimensional ridge, but managed to produce only lee waves with wavelengths of hundreds of kilometres—a rather unrealistic result. In briefly reviewing Queney's work, Corby (1954) is an invaluable source. He noted:

> The detail of Queney's mathematics is elaborate and lengthy. For brevity... and...convenience...we may regard Queney's treatment as equivalent to the derivation of the wave equation represented by
>
> $$\left(1 - \frac{f^2}{k^2 U^2}\right)\frac{\partial^2 \psi}{\partial z^2} - \left(\frac{g}{c^2} + \beta\right)\frac{\partial \psi}{\partial z} + \left(\frac{g\beta}{U^2} - k^2\right)\psi = 0, \qquad (16)$$
>
> where U is the undisturbed, horizontal air velocity, β is a coefficient of stability defined as $(1/\theta)(\partial \theta/\partial z)$, ψ is a stream function and c is the speed of sound.
>
> This equation expresses the variation of the disturbance with height assuming that the variation in the horizontal is harmonic (Corby, 1954, p. 497).

Queney took the stability and wind to be constant with height and was then able to treat the coefficients of equation (16) as sensibly constant (for a given value of k) provided that the magnitude of the perturbation was not of the order of thousands of kilometres. His main results are shown in Fig. 19. Over terrain with short wavelength corrugations the solution of equation (16) gave wavy streamlines whose amplitude decreases exponentially upwards. No horizontal deformation occurs because the geostrophic forces have no

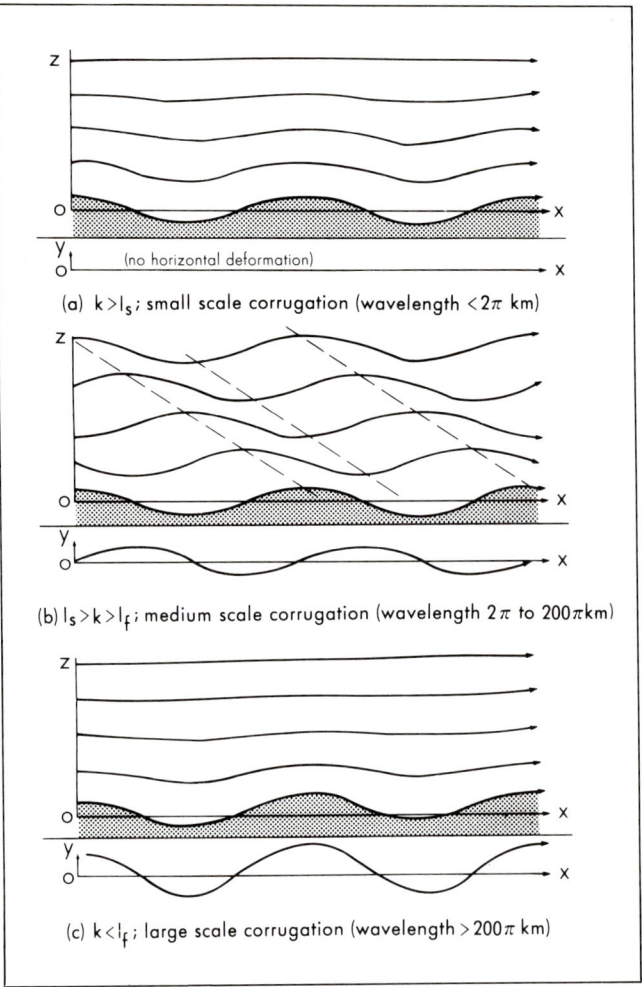

Fig. 19. Streamlines produced by an infinite sinusoidal ground profile. (After Queney, 1947.)

opportunity to affect the result. Thus gravity waves were produced due to the stability—but no lee waves occurred. At larger scales (Fig. 19(b), (c)) both gravity and coriolis effects become evident. When Queney extended his results to a single two-dimensional mountain of equivalent size to the corrugations in Fig. 19(a), (b), (c), he found that over the smallest hill (width about 100 m) there were no relevant restoring forces due to stability that might make some oscillatory motion possible, and the airstream merely executed a single symmetrical wave over the hill, dying away rapidly both

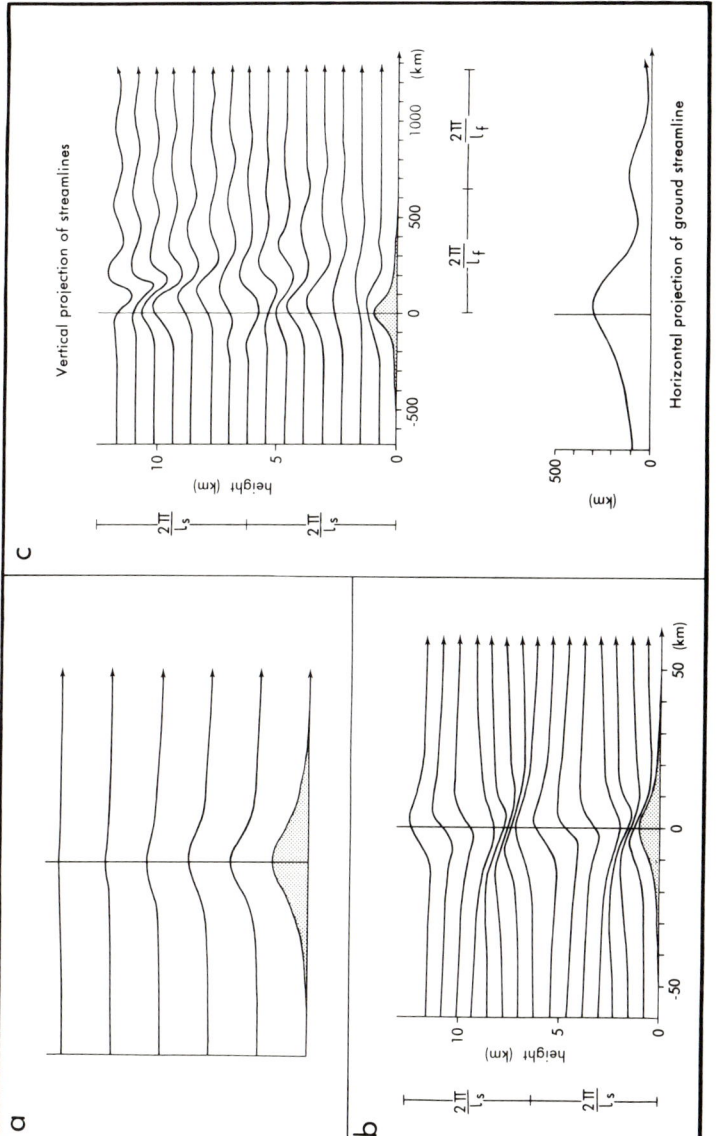

Fig. 20. (a) The flow over a very small ridge of width up to about 100 m; (b) the flow of a uniform airstream over a ridge having a width up to a few kilometres; (c) the flow of a uniform airstream over a ridge having a width of about 100 km. (After Queney, 1947.)

vertically and horizontally from the hill (Fig. 20(a)). Over the next larger hill (width about 15 km) the effects of stability, but not geostrophy, became evident and the flow was disturbed throughout its depth (Fig. 20(b)) as distinct from the last case in which the disturbance decreased rapidly upwards. Over the largest hill (width about 100 km) lee waves formed throughout the depth of the atmosphere and marked horizontal deformation of flow occurred (Fig. 20(c)). The horizontal deformation was much larger than the vertical and the horizontal wavelength was of the order of several hundred kilometres. The lee waves increased in amplitude upwards and persisted further downstream with greater height. In fact waves in this solution were primarily an oscillation about the geostrophic wind. The disturbance caused by the mountain upset the geostrophic balance between wind and pressure gradient and the result was a downstream oscillation for which geostrophic forces provided the restoring mechanism. Although such waves no doubt exist in the atmosphere, their mechanism cannot account for the normal lee waves that increase in amplitude up to some level and die away above and which have wavelengths of a few as opposed to hundreds of kilometres.

2. Scorer's work

Scorer (1949) recognized the inadequacies of the solutions obtained by Queney and Lyra and attempted to provide a more explanatory theory for lee waves. He considered frictionless, steady, laminar and isentropic flow but set out to provide for variations of lapse rate and wind shear in the vertical—two aspects ignored by Lyra and Queney. He ignored the Earth's rotation, a procedure shown by Sawyer (1959) to be quite valid in the present context. The perturbation equations for a steady, uniform airstream crossing a ridge with a velocity U in the undisturbed stream are

$$U\frac{\partial u'}{\partial x} - fv' = \frac{1}{\rho}\frac{\partial p'}{\partial x}, \qquad (17)$$

$$U\frac{\partial v'}{\partial x} + fu' = 0, \qquad (18)$$

where primes indicate perturbations from undisturbed values. Differentiation of equation (17) and substitution into equation (18) leads to

$$U\frac{\partial^2 u'}{\partial x^2} + \frac{f}{U}u' = \frac{1}{\rho}\frac{\partial^2 p'}{\partial x^2}. \qquad (19)$$

If the disturbance has a characteristic length, L, $\partial^2 u'/\partial x^2$ is of the order of u'/L^2 and the second term on the left-hand side of equation (19), which represents the effect of the Earth's rotation, is small compared with the first if

2 LEE WAVES

$$L^2 \ll \frac{U^2}{f^2} < \frac{U^2}{10f^2}.$$

With $U = 10\,\mathrm{m\,s^{-1}}$ and $f = 1.1 \times 10^{-4}\,\mathrm{s^{-1}}$ this requires $L < 30\,\mathrm{km}$. Consequently, as most lee wavelengths are of this order of size, the effects of the Earth's rotation may be safely ignored in their theoretical investigation.

In his classical paper, Scorer (1949) analysed the equations of motion and the continuity, adiabatic and perfect gas equations in perturbation form. Initially he considered the flow over a ground comprising continuous harmonic ridges. Early in the analysis he derived a relationship between the horizontal and vertical perturbation velocities as follows:

$$\frac{\partial u'}{\partial x} + \left(\frac{\partial}{\partial z} - \frac{g}{\gamma RT}\right) w' = 0, \qquad (20)$$

where γ is the ratio of specific heats at constant pressure and volume. Upon the introduction of a stream function ψ_0 and integration with respect to x, equation (20) became

$$u' = \frac{\partial \psi_0}{\partial z} - \frac{g}{\gamma RT} \psi_0. \qquad (21)$$

Substitution of equation (21) into the disturbance equations for pressure and vertical velocity and eliminating the latter resulted in the following:

$$\left\{1 - \frac{f^2}{U^2 k^2}\right\} \frac{\partial^2 \psi_0}{\partial z^2} + \left\{-\frac{g}{c^2} - \beta + \left(\frac{g}{c^2} + 2\beta\right) \frac{f^2}{U^2 k^2}\right\} \frac{\partial \psi_0}{\partial z}$$
$$+ \left\{\frac{g\beta}{U^2} - \frac{1}{U}\frac{\partial^2 U}{\partial z^2} - k^2\right\} \psi_0 = 0. \qquad (22)$$

When the coriolis effects are ignored, a procedure shown earlier to be valid, equation (22) reduces to

$$\frac{\partial^2 \psi_0}{\partial z^2} - \left\{\frac{g}{c^2} + \beta\right\} \frac{\partial \psi_0}{\partial z} + \left\{\frac{g\beta}{U^2} - \frac{1}{U}\frac{\partial^2 U}{\partial z^2} - k^2\right\} \psi_0 = 0. \qquad (23)$$

In turn this equation was further modified by Scorer and Sawyer (1960) elaborated this by noting that $g/c^2 + \beta = -\rho^{-1}(\partial \rho/\partial z)$ and by writing $\psi = (\rho_0/\rho)^{1/2} \psi_0$ where ρ_0 is density at height $z = 0$. Equation (23) then reduces to

$$\frac{\partial^2 \psi}{\partial z^2} + \left\{\frac{g\beta}{U^2} - \frac{1}{U}\frac{\partial^2 U}{\partial z^2} - k^2 - \frac{1}{2\rho}\frac{\partial^2 \rho}{\partial z^2} + \frac{1}{4\rho^2}\left(\frac{\partial \rho}{\partial z}\right)^2\right\} \psi = 0. \qquad (24)$$

The last two terms within the braces are of the order of $5 \times 10^{-3}\,\mathrm{km^{-2}}$ compared with values of the order of $1\,\mathrm{km^{-2}}$ for the first term. Consequently they are usually ignored and equation (24) reduces to

$$\frac{\partial^2 \psi}{\partial z^2} + (l^2 - k^2)\psi = 0, \qquad (25)$$

where

$$l^2 = \frac{g\beta}{U^2} - \frac{1}{U}\frac{\partial^2 U}{\partial z^2}.$$

Equation (25) expresses the variation of the disturbance with height assuming that the variation in the horizontal is harmonic. It is very similar to those employed by Queney: the essential differences are that the effect of the Earth's rotation is excluded and that there is an additional term in $\partial^2 U/\partial z^2$ representing the vertical rate of change of shear in U. As noted earlier Queney's and Lyra's analyses were constrained by a vertically uniform distribution of wind shear and stability. The important step forward made by Scorer (1949) was to allow both factors to vary with height. Even if we choose airstreams in which l^2 is constant with height, this still allows β and U to vary with height. We shall see later the importance of l^2.

The solution of equation (25) takes the form

$$\psi = A e^{\mu z} + B e^{-\mu z}, \qquad (26)$$

where $\mu = +(k^2 - l^2)^{1/2}$ and A and B are constants to be determined from boundary conditions. The vertical displacement of a streamline (ζ) will be given by

$$\zeta = \frac{\psi}{U} = \frac{1}{U}(A e^{\mu z} + B e^{-\mu z}). \qquad (27)$$

Two types of result are available from equation (27):

(1) If μ is real, i.e. if $k > l$, then $A = 0$, otherwise the disturbance increases indefinitely upward.

(2) If μ is imaginary, i.e. if $k < l$, then

$$\zeta = \text{Real}\left\{\frac{1}{U}(A e^{ivz} + B e^{-ivz})\right\}, \qquad (28)$$

where "Real" means "the real part of" and

$$v = +(l^2 - k^2)^{1/2}.$$

A major problem associated with this method is that both components e^{ivz} and e^{-ivz} are indeterminate as $z \to \infty$. Scorer argued that the only realistic choice was an upward progressive wave and therefore used the solution

$$\zeta = \text{Real}\left(\frac{B}{U}e^{-ivz}\right). \qquad (29)$$

For an infinite sinusoidal ground profile (represented by the real part of e^{ikx}) of wavelength $2\pi/k$, the streamline solutions are

$$k > l: \quad \zeta = \text{Real}\left\{\frac{U_0}{U_z} e^{ikx - \mu z}\right\} = \frac{U_0}{U_z} e^{-\mu z} \cos kx, \qquad (30)$$

$$k < l: \quad \zeta = \text{Real}\left\{\frac{U_0}{U_z} e^{i(kx - vz)}\right\} = \frac{U_0}{U_z} \cos(kx - vz), \qquad (31)$$

where $\mu = +(k^2 - l^2)^{1/2}$, $v = +(l^2 - k^2)^{1/2}$ and U_0 and U_z are the wind speeds in the undisturbed airstream at the ground level and level z respectively.

In applying these results to flow over and downstream of actual hills, Scorer followed Queney and used a profile represented by the Fourier integral:

$$\zeta(x, 0) = \text{Real}\left\{ah \int_0^\infty e^{ikx - ka} dk\right\}$$

$$= ah \int_0^\infty e^{-ka} \cos kx \, dk$$

$$= \frac{a^2 h}{a^2 + x^2}, \qquad (32)$$

where h is height of the "hill" and a is its half-width (the value of x for which the height is half the maximum). This bell-shaped "hill" is a synthesis of sinusoidal ground corrugations for all values of the wave number k, with the exponential factor e^{-ka} applied to each of them. An airstream having neither static stability (i.e. $\beta = 0$) nor shear (i.e. $U = $ constant) would have $l^2 = 0$ and the flow would be described by

$$\zeta = \frac{ah(a + z)}{(a + z)^2 + x^2}. \qquad (33)$$

This is simply flow forced over the obstacle where the disturbance dies out rapidly in the vertical and there are no leeward disturbances. If stability is put into the solution giving $l = $ constant (but not zero), then the disturbance is approximately given by

$$\zeta = \frac{U_0}{U_z} ah \frac{a \cos lz + x \sin lz}{(a^2 + x^2)}. \qquad (34)$$

The important difference between equations (33) and (34) is that the latter describes an airstream which is disturbed throughout its whole depth—but no lee waves appear in this solution.

Because the vertical velocity $w = \partial \psi / \partial x = ik\psi$, the wave equation (25) may be written as

$$\frac{\partial^2 w}{\partial z^2} + (l^2 - k^2)w = 0. \qquad (35)$$

If this equation is applied to flow over level ground to the lee of an obstacle, then the vertical velocity must be zero at the ground and tend to zero at infinite height. Such boundary conditions, in association with constant l, mean that w is everywhere zero and therefore no lee waves could occur in such an airstream. For realistic lee waves the vertical velocity must be finite at some levels, zero at the ground and tending to zero at great heights. Thus $\partial^2 w/\partial z^2$ must change sign at least once, and Scorer inferred that the value of l^2 must be less in some fairly deep upper layer than in a layer below. In fact he found that

$$l_L^2 - l_U^2 > \pi^2/4b^2,$$

where b is the depth of the lower layer, was a condition for waves to form over level ground in the lee of a hill. For mathematical convenience Scorer took the interface of the layers to be at $z = 0$ and the ground at $z = -b$. The streamlines over the ridge represented by equation (32) were given by

$$\zeta(z) = \text{Real} \int_0^\infty \frac{f_z(k)}{f_{-b}(k)} e^{ikx - ka} \, dk, \tag{36}$$

where

$$f_z(k) = \frac{1}{U_z} \left(\cosh \mu_1 z - \frac{\mu_2}{\mu_1} \sinh \mu_1 z \right) \tag{37}$$

and

$$\mu_1 = (k^2 - l_L^2)^{1/2}, \qquad \mu_2 = (k^2 - l_U)^{1/2}.$$

Solving this integral leads to the following equations for streamlines in the lower layer:

$$\zeta(z) = \text{Real} \left\{ ab \frac{f_z(0)}{f_{-b}(0)} \frac{(a+ix)}{(a^2+x^2)} + 2i\pi a h \frac{f_z(k)}{\partial f_{-b}(k)/\partial k} e^{ikx - ka} \right\} \quad \text{for} \quad x > 0 \tag{38}$$

and

$$\zeta(z) = \text{Real} \left\{ ab \frac{f_z(0)(a+ix)}{f_{-b}(0)(a^2+x^2)} \right\} \quad \text{for} \quad x < 0. \tag{39}$$

We see that the first term on the right-hand side of equations (38) and (39) describes the disturbance due to the ridge whereas the second term in equation (38) describes the lee waves downstream of the ridge. This term comprises several important effects which are analysed later.

3. Further analysis of factors affecting lee waves

(a) *Analytical results—two-dimensional models* Scorer's 1949 paper provided the foundation stone for several further penetrating analyses of lee waves by

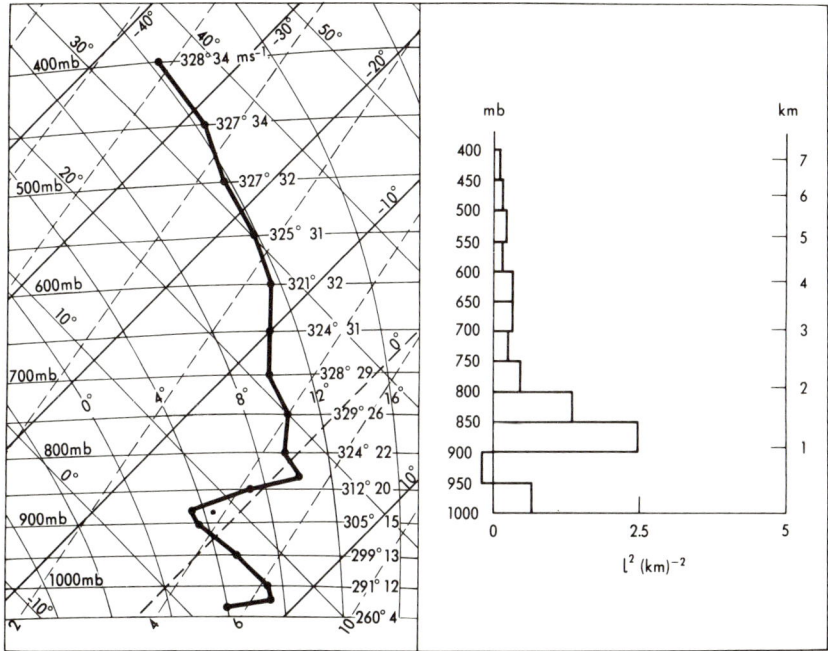

FIG. 21. Tephigram, upper winds and profile of l^2 on an occasion of lee-wave activity over Scotland, 11 March 1953. (After Corby, 1957b.)

himself and others. The importance of the parameter l^2 became increasingly evident as research progressed, so that by 1954 Scorer (1954) expressed clearly that the vertical profile of l^2 was critical to lee-wave formation and character. Familiarity with the two layer model confirmed his original suggestion that high values of l^2 at low levels were favourable to lee-wave formation. Observations also showed the validity of his ideas. Figure 21 illustrates the vertical sounding and distribution of l^2 on an occasion of strong wave development over the north-west of Great Britain. Marked stability in the 900–800 mb layer and winds in fairly constant direction, but increasing substantially in speed with height, gave large values of l^2 in the layer from 1 to 2 km. Waves were encountered in the stable layer with vertical velocities as high as $3.6 \, \text{m s}^{-1}$. In addition to allowing the prediction of the likely existence of waves from the vertical distribution of l^2, the theory also produced waves with a realistic wavelength, a maximum amplitude at some middle level and a dying away above that level.

It was also appreciated that there were many different l^2 profiles which could achieve very similar results. This is clearly shown in Fig. 22, based upon a two-layer model in which l^2 is constant in the upper layer and variable in the lower layer and an inversion of temperature is supposed to exist at the

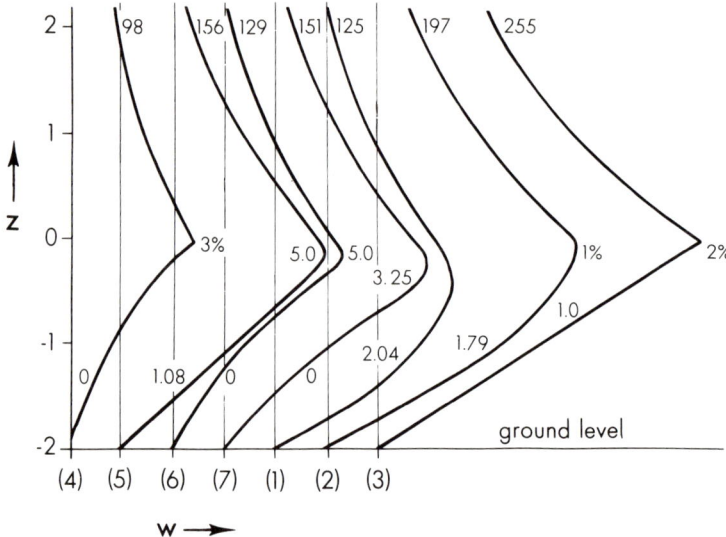

FIG. 22. Variation with height of vertical velocities in waves in the lee of the ridge described by equation (32) for seven different airstreams which all have the same lee wavelength of 2π km but different vertical profiles of l^2. The zero is displaced so as to separate the curves. In all cases $w = 0$ at the ground (at $z = -2$ km). The relative amplitude is indicated against the top of each curve and the number of the case against the bottom in parentheses. In all cases the value of l^2 for $z > 0$ was 0.75 km^{-2}. For $0 \geqslant z \geqslant -2$ km the value of l^2 is indicated against the curves. Where there is a discontinuity of density (cases 2, 3, 4) the percentage discontinuity is indicated. (After Scorer, 1954.)

interface. The wave amplitudes are quite similar so that airstreams with these l^2 profiles are, as Scorer puts it, "dynamically equivalent". This means that the variety of airstreams all possessing the same lee-wave length is infinite and that changing this wavelength by a factor of two or three will not obviously alter the nature of the flow. Realization of the tremendous capability of the atmosphere to produce different vertical profiles of l^2 and, within one type of l^2 profile, an infinite variety of thermal and wind distributions, was a salutary piece of insight which led Scorer (1954, p. 428) to conclude:

> The complexity of the mathematics prohibits the application of the formulae (i.e. the general theory) direct to actual cases, and in particular forecasting. Some possibilities have been illustrated here, and with an understanding of them it will probably be possible to explain the behaviour of actual airstreams; also it is now clear that the profile of the quantity l^2 ... is the feature that should be studied.

Scorer's plea for analysis of the effects of l^2 profiles was answered by Corby and Wallington (1956) in their investigation of the factors which affect lee-wave amplitude. Slight manipulation of the wave term of equation (38) allowed the flow pattern through the lee waves to be indicated by

$$\zeta(z) = -2\pi h a \, e^{-ka} \left(\frac{U_0}{U_z}\right) \frac{f_z(k)}{\partial f_{-b}(k)/\partial k} \sin kx, \qquad (40)$$

where U is horizontal wind speed and the suffix zero refers to ground level. This was broken down into three factors which have separate effects on lee-wave amplitude, viz.

$$h a \, e^{-ka}, \quad U_0/U_z \quad \text{and} \quad f_z(k)(\partial f_{-b}(k)/\partial k)^{-1}.$$

The first factor described the effect of mountain size and shape; the second the effect of the wind profile; and the third the effect of the l^2 profile.

The mountain effect is at a maximum when $a = k^{-1}$, which means that for a constant mountain height the lee-wave amplitude attains its greatest value when the mountain width is suitably adjusted to the natural wavelength of the

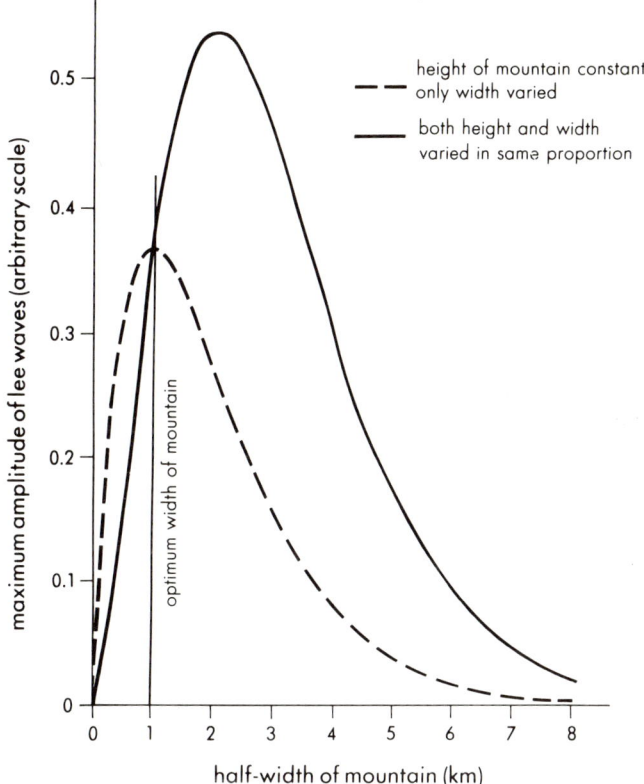

FIG. 23. Variation of lee-wave amplitude with size of mountain ($k = 1 \, \text{km}^{-1}$). (After Corby and Wallington, 1956.)

airstream. When $k = 1\,\text{km}^{-1}$, i.e. when the natural wavelength of the airstream is $2\pi\,\text{km}$, the amplitude falls off sharply for narrower or broader mountains (Fig. 23). This result, which agreed with intuitive preconceptions, was well summarized by Corby and Wallington (1956, p. 267): "We may liken this effect to resonance; the natural wavelength is a function of the l^2 profile, i.e. of the airstream's stability and wind profiles, but the lee-wave amplitude is likely to be small if the width of the mountain does not match the wavelength of the airstream". If both mountain height and width were varied together to retain the shape of the mountain but not its size, then amplitudes increased until $a = 2k^{-1}$, after which further increase in size produced a sharp decline in lee-wave amplitude (see Fig. 23). Thus, within the confines of linearized theory, for a given airstream the lee waves of largest amplitude are produced by mountains of appropriate width rather than of greatest height. Wallington (1958) also analysed the effects of mountain shape on lee-wave amplitudes, using numerical rather than analytical methods. By assigning different half-widths to so-called elementary ridges $ha^2/(a^2 + x^2)$ for $x \gtrless 0$, an asymmetrical

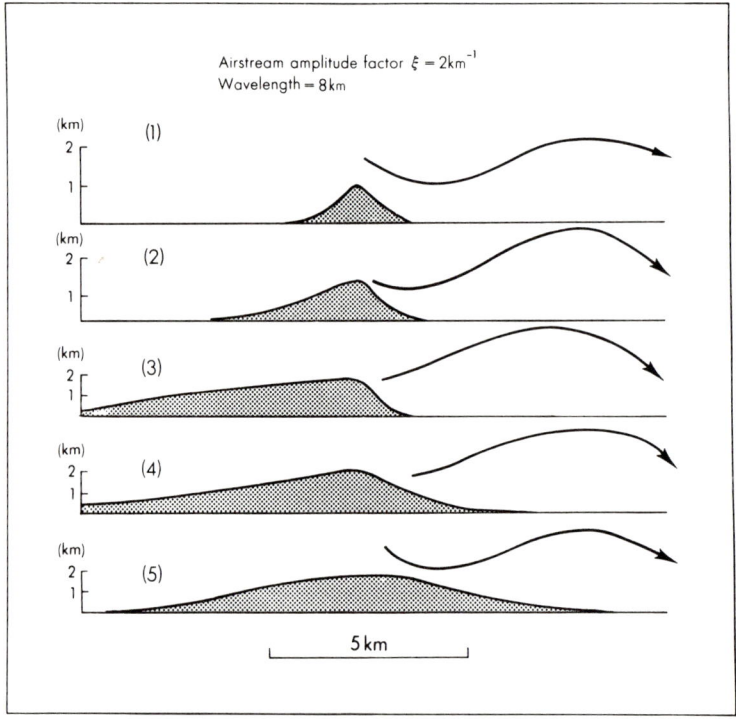

FIG. 24. Effect of mountain shape upon location and amplitude of lee waves. (After Wallington, 1958.)

ridge was arbitrarily but conveniently specified by summing a number of such elementary ridges. The results of the analysis, shown in Fig. 24, clearly indicated how asymmetry affected the location and amplitude of lee waves. Shape 2 brought the wave closer to the hill, but further lengthening of the upwind side (shape 3) brought the wave so far forward that the streamline soon turned upwards into the first lee-wave crest. Shape 4 had a similar effect but the breakaway occurred at a point further down the lee slope than in shape 3. Shape 5 was conducive to a local zone of strong surface winds.

The wind speed effect alone (U_1/U_z) would imply that a decrease of wind with height would favour lee waves of large amplitude. This partially contradicts the requirement that l^2 decrease with height, as such a decrease can be achieved by an increase of wind speed or by a decrease of stability with height, or by some combination of wind and stability distributions. Corby and Wallington concluded that, of two airstreams with favourable l^2 profiles, the one containing moderate or strong surface winds would be most likely to generate lee waves of marked amplitude.

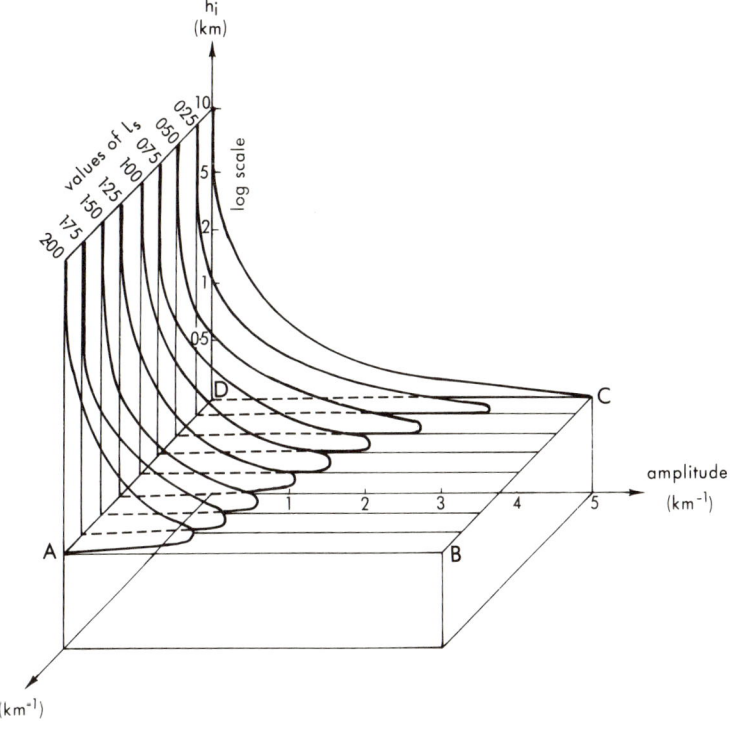

FIG. 25. Variation of lee-wave amplitude with airstream characteristics—L constant, see text. (After Corby and Wallington, 1956.)

Finally, the effect of the l^2 profile was tested by use of a model comprising (1) an adiabatic layer with $\partial^2 u/\partial z^2 = 0$ ($l^2 = 0$) from ground to a height h_a; (2) an intermediate layer in which $l^2 = l_i^2$ between levels h_a and $(h_a + h_i)$; (3) an upper layer in which $l^2 = l_s^2$ ($<l_i$) extending upwards from the level $(h_a + h_i)$. Using $L^2 = l_i^2 - l_s^2$ as a measure of the decrease of l^2 with height, Fig. 25 shows the variation of maximum lee-wave amplitude with l_s and h_i for a constant value of L^2 ($L = 5\,\text{km}^{-1}$). As Corby and Wallington point out, the result was that immediately above the plane ABCD (below which the condition for lee waves is not satisfied) a rapid change occurs from a condition in which there can be no lee waves to one in which they have large amplitudes. To quote them (Corby and Wallington, 1956, p. 271): "...it appears, that the optimum conditions, when the airstream is most sensitive, are very close to the critical conditions for waves, and that if conditions are made 'more favourable' for waves to occur the amplitude falls off sharply". Pursuing this theme, the effects of varying L^2 were analysed, holding l_s constant at $0.2\,\text{km}^{-1}$. Figure 26 shows the lee-wave amplitudes and lengths for $L = 2, 5$ and $10\,\text{km}^{-1}$. Once more airstreams appeared to be very sensitive to changes near the critical values of the parameters for waves to occur. Maximum amplitudes became potentially greater as L was made larger; and

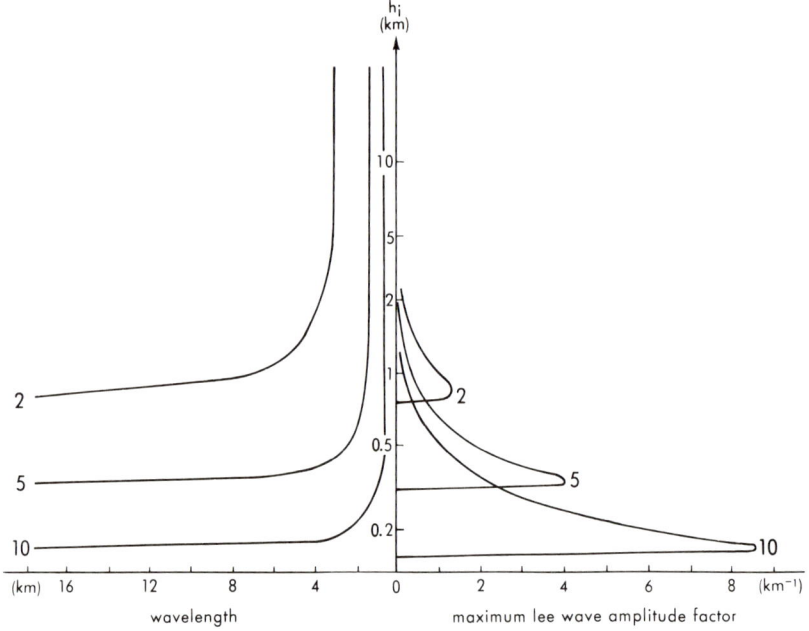

FIG. 26. Variation of lee-wave amplitude and wavelength with airstream characteristics (l_s = constant = $0.2\,\text{km}^{-1}$; values of L (2, 5 and $10\,\text{km}^{-1}$) are marked against the curves). (After Corby and Wallington, 1956.)

generally, greater amplitudes were possible if there was marked stability through a shallow layer rather than smaller stability through a deep layer.

The apparently steady increase in our understanding of the theory of lee waves in the 1950s hid an important difference of views on the nature of the upper boundary condition required to solve the lee-wave equation to give sensible results. Whereas Scorer (1958a,b) suggested that the problem is mathematically indeterminate, Palm (1958) and Corby and Sawyer (1958a,b) introduced devices which appeared to give lee-wave forms in close accord with reality. In particular, Corby and Sawyer (1958a) claimed that the familiar type of lee wave which has a maximum in the lower troposphere and which dies away above, is not at all sensitive to the upper boundary level condition. They acknowledged that lee waves in the stratosphere and upper troposphere could occur and that different boundary conditions had to be applied. Several subsequent studies, analytical, numerical and observational, tend to support the ideas of Corby and Sawyer.

Further understanding of wave-type airflow over and downstream of hills was still reliant upon intense investigation of the vertical variation of wind speed and stability—in effect Scorer's l^2 parameter. Two main approaches were employed. First, the well-tried analytical approach, but usually using three-layer models in which l^2 decreased exponentially with height in the lower and upper layers, the former representing the bulk of the troposphere and the latter the bulk of the stratosphere, the two separated by a layer extending from about 10 to 20 km within which l^2 increased with height; and second, the use of numerical methods to solve the wave equations, an approach which allowed the use of many levels, the consideration of three-dimensional flow and the analysis of large, meso-scale waves over mountain areas.

The exponential decay of l^2 with height of a form ($l^2 = l_0^2 e^{-\alpha z}$), where l_0^2 is the ground value of l^2, chosen constant for convenience, has been most widely used by Scandinavian scientists. Palm and Foldvik (1959) investigated mountain waves in the stratosphere and their relationships with tropospheric waves. Using the three-layer model outlined above, they found that realistic waves could exist in the stratosphere. With a mountain of height 700 m the maximum displacement at heights of 20–30 km was about 400 m and maximum vertical velocities were of the order of 1–3 m s^{-1}. In addition, where l^2 had a value at the ground which was at least 2.5 times as large as the minimum value, usually located at 7–10 km above the ground (as occurs with exponential distribution), the lower tropospheric wave motion was dependent on only wind and stability beneath the level of minimum of l^2, with no effects of stratospheric conditions. Within the lower tropospheric waves maximum vertical velocities of 6 m s^{-1} were found below a level of 5 km. Such a realistic picture was a gratifying result, particularly, as Palm and Foldvik noted, because it meant that, in future, the wave motion in the lower troposphere could, in fact, be obtained from a simple one-layer model, if it employed the exponential decay of l^2.

In a very similar, but independent investigation, Döös (1961) employed a one-layer (tropospheric) exponential model and calculated wavelengths and amplitudes for four occasions of lenticular cloud occurrence. Good agreement was found between calculated and observed wavelengths (as measured by distance between the clouds), particularly at the 2–3 km level. The amplitude calculations were less satisfactory and were worse at higher levels, probably because of the inaccuracy in the representation by the exponential function of the profile of l^2 at those levels. Later Döös (1962) compared his theoretical results with measurements of wavelength taken from a satellite photograph of clouds over and to the lee of the central Andes. The model predicted wavelengths within the lowest 2 km of the troposphere to within an accuracy of 2 km: but once again, amplitude calculations were difficult to verify.

One of the few analyses to employ both the analytical, exponential model and numerical methods also used satellite pictures of clouds over the Andes as its observational touchstone (Sarker and Calheiros, 1974). Table 5 shows the wavelengths measured from two types of satellite, and those calculated by the two different methods. Analytical results are seen to be slightly larger than the numerical ones which compare favourably with observed values, in spite of the inaccuracies involved in both computed and measured wavelengths.

Table 5 Lee wavelengths as observed by satellites and as computed by analytical and numerical models

Case no.	Observed wavelength (km)		Computed wavelength (km)	
	Heliosynchronous satellite	Geosynchronous satellite	Analytical	Numerical
1	20.0	18.0	21.0	19.1
2	18.2	17.0	20.0	18.0
3	19.8	—	—	20.4
4	20.1	19.2	—	20.3
5	28.8	24.5	32.0	31.4

Source: Sarker and Calheiros (1974).

(b) *Numerical results—two- and three-dimensional models* Perhaps the first, and certainly one of the most instructive, purely numerical analyses was performed by Sawyer (1960). Using Scorer's original equation, 17 levels at 1 km intervals, and a constant, specified value of l^2 above a height of 16 km, Sawyer calculated the vertical displacements of streamlines at each level for several types of airstream—idealized and real. A primary result was that

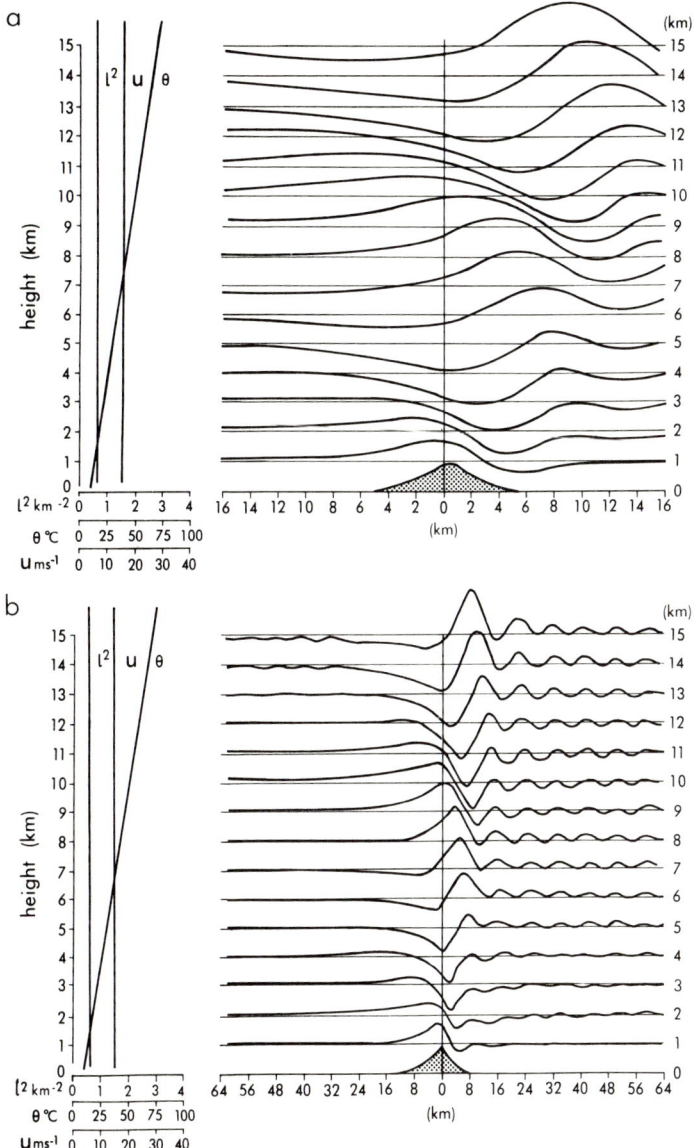

FIG. 27. Displacements of streamlines for an airstream for which l^2 is approximately independent of height. The left-hand section of the diagram shows the assumed potential temperature (θ), wind speed (u) and Scorer's parameter (l^2) as a function of height. The vertical displacement of the streamlines is plotted for each level on the same scale as the mountain profile at the bottom of each figure. In (b) the horizontal scale is one-quarter of that in (a)—otherwise the diagrams are similar. (After Sawyer, 1960.)

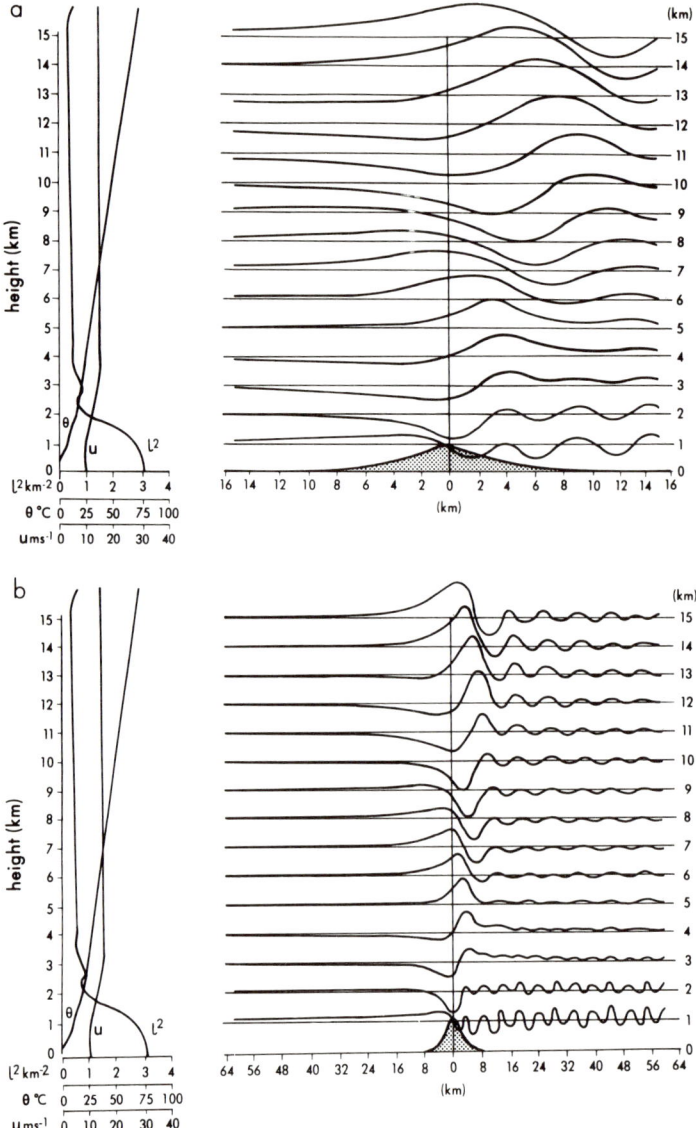

FIG. 28. Displacements of streamlines computed for an idealized airstream with a layer of high l^2 values near the ground. The left-hand section of the diagram shows the assumed potential temperature (θ), wind speed (u) and Scorer's parameter (l^2) as a function of height. The vertical displacement of the streamlines is plotted for each level on the same scale as the mountain profile at the bottom of each figure. In (b) the horizontal scale is one quarter that in (a)—otherwise the diagrams are similar. (After Sawyer, 1960).

changing the value of l^2 at heights above 16 km had little effect on the flow below that level, a conclusion in accord with the earlier analytical work of Corby and Sawyer (1958a) and Palm and Foldvik (1959). Sawyer went on to illustrate further the dependence of the computed streamlines upon the airstream characteristics, expressed in terms of the l^2 profile, and mountain width. With l^2 approximately independent of height the results of calculations for two mountain widths (about 10 and 20 km) are shown in Fig. 27. Over the smaller hill the flow was similar to that predicted by Queney for a uniform airstream. Over the larger hill lee waves do exist, being best developed above a height of 5 km. Sawyer (1960) claimed that there was no reason not to accept the solutions and that the waves over the larger hill were presumably similar to those computed by Queney (1947) for airstreams in which l^2 was strictly constant. When Sawyer inserted an l^2 profile with large values below a height of 2 km overlain by low values, realistic lee waves formed downwind of mountains of both sizes (Fig. 28). Finally, lee waves were also discovered if a high value of l^2 in the stratosphere overlay a deep layer of a substantially lower constant value of l^2. Figure 29 shows the flow resulting from such a configuration. As Sawyer (1960, p. 340) noted:

> The main effect of the introduction of a stable stratosphere has been to introduce an additional rapidly decaying lee-wave train of longer wavelength than that arising without the stratosphere. This is shown most clearly in the streamline for the 2 km level where the first two oscillations are clearly defined.

Sawyer (1962) showed later that such waves would be set up only by very long ridges, and would have only small amplitude in the lee of isolated hills. It is clear from Sawyer's study that, whilst a low level maximum of l^2 is favourable to lee-wave formation, as shown by Scorer, other distributions of l^2 as a function of height can also give rise to lee waves. Such lee waves maintain their amplitude up to the stratosphere, although they may decrease in amplitude downstream.

Numerical analyses such as those described above paved the way for more complicated treatments. Three-dimensional waves had been previously studied (e.g. Crapper, 1959, 1962; Wurtele, 1957) but most of the analyses had been restricted to airstreams in which wind speed, wind direction and stability were independent of height. In contrast, Scorer and Wilkinson (1956) discussed the character of waves which would form to the lee of an isolated hill in a two-layer airstream, with wind direction the same in both layers. With the aid of the computer, Sawyer (1962) was able to consider the forms of lee waves possible when the motion was three-dimensional and when the wind direction in the basic current varied with height.

Using a two-layer model similar to Scorer's, Sawyer established relationships between wave number — λ and μ respectively — (and thus wavelength) measured along (x direction) and normal (y direction) to the low level flow.

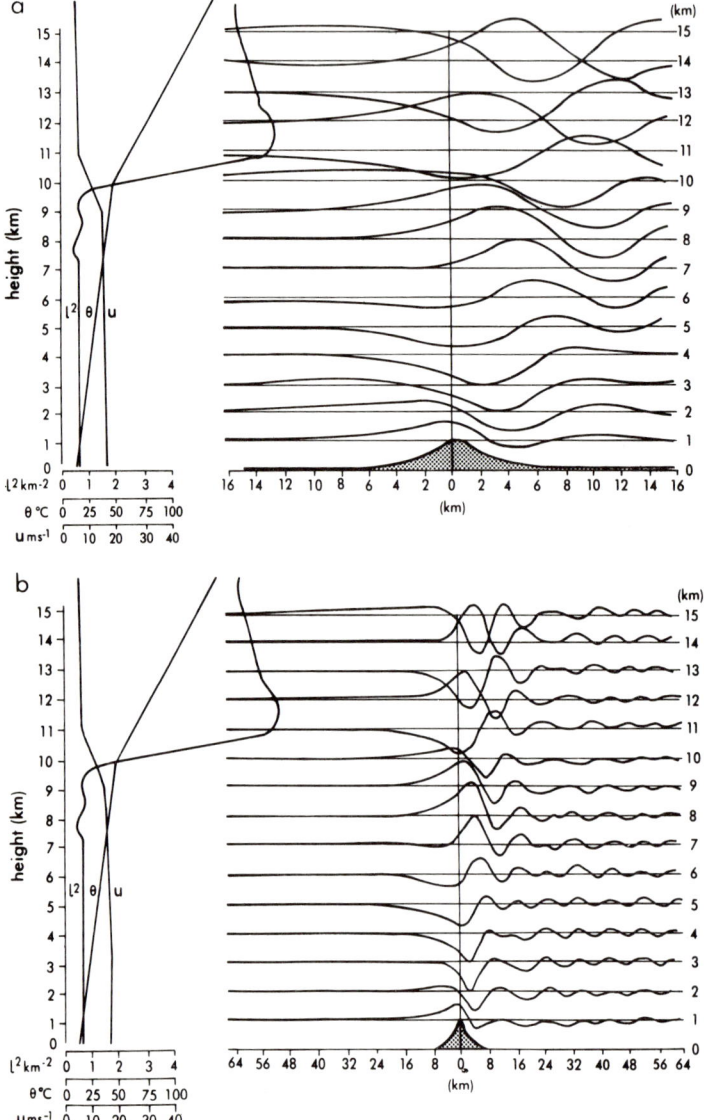

FIG. 29. Displacements of streamlines computed for an idealized airstream with approximately constant value of l^2 in the troposphere and substantially larger value of l^2 in the stratosphere. The left-hand section of the diagram shows the assumed potential temperature (θ), wind speed (u) and Scorer's parameter (l^2) as a function of height. The vertical displacement of the streamlines is plotted for each level on the same scale as the mountain profile at the bottom of each figure. In (b) the horizontal scale is one quarter that in (a)—otherwise the diagrams are similar. (After Sawyer, 1960.)

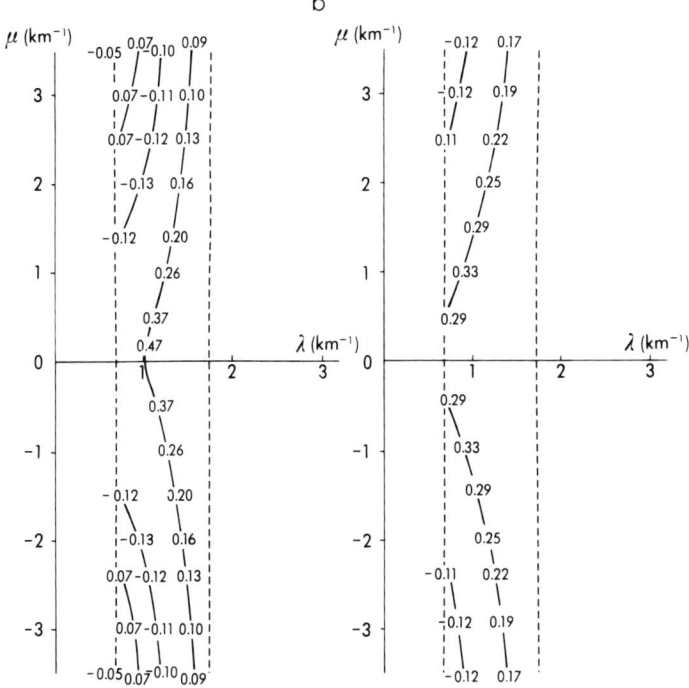

FIG. 30. Relation between λ and μ for lee waves; two-layer airstream, labelled 1 (upper) and 2 (lower). Undisturbed velocity components U, V; Scorer's parameter l^2; height of interface between layers h. Conditions in layers as follows:

$$l_1^2 = 0.5\,\text{km}^{-2}, \quad U_1 = 20\,\text{m s}^{-1}, \quad V_1 = 0;$$
$$l_2^2 = 3.0\,\text{km}^{-2}, \quad U_2 = 20\,\text{m s}^{-1}, \quad V_2 = 0.$$

(a) Case for $h = 1.5$ km; (b) case for $h = 0.9$ km. Numbers against curves represent the amplitude factor for the lowest level where it is a maximum. (After Sawyer, 1962.)

Figure 30 shows the relationship in two cases when the lower layer varies in depth and when there is no variation of wind direction with height. With this specification of wind all solutions are confined between $\lambda = l_1$ and $\lambda = l_2$, indicated by dashed lines on Fig. 30. Two-dimensional theory indicates one lee wavelength corresponding to $\lambda = 1.04$ km^{-1}, but other waves are possible in three dimensions. The amplitudes (shown by numbers on Fig. 30) are largest for the two-dimensional lee wave but drop only slowly as the wavelength in the y-direction increases. Whereas Scorer (1949) showed that two-dimensional waves could occur only if $l_2^2 - l_1^2 > \pi^2/4h^2$, where h is the depth of the lower layer, Sawyer showed that this condition no longer applied if wave motion in three dimensions was permitted and that lee waves were always

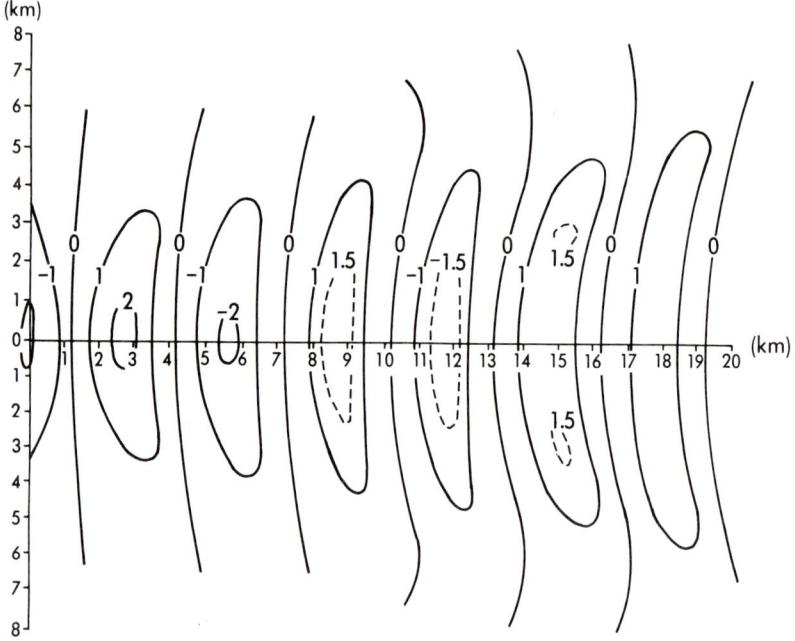

FIG. 31. Lee-wave component of vertical velocity at a height of 1.5 km for the two-layer airstream specified in Fig. 30. Quasi-circular hill with half-width of 2 km. Velocities in metres per second for a hill of 100 m height. (After Sawyer, 1962.)

possible in a two-layer airstream for some values of λ and μ. If the airstream that gave the results shown in Fig. 30(a) passed over a hill with the following shape:

$$b^2 H_0 e^{-(\lambda+\mu)b}, \quad \mu \geqslant 0,$$
$$b^2 H_0 e^{-(\lambda-\mu)b}, \quad \mu \leqslant 0,$$

where H_0 is the height of the hill and b is a width parameter, the resultant waves for $b = 2$ km and $H_0 = 100$ m would give a plan view as shown in Fig. 31. Changing the wind direction in the upper layer leads to the expected asymmetry in the relationship between λ and μ (Fig. 32) and in the plan view of the lee-wave component of the vertical velocity (Fig. 33).

A further major advance due to numerical techniques lay in the investigation of waves with lengths of the order of tens and hundreds of kilometres. As shown earlier, atmospheric circulations with such horizontal dimensions begin to be influenced by the coriolis force and this further complicates the analytical procedure. Risking the consequences of ignoring the coriolis effect, Farooqui and De (1974) used a two-dimensional model to calculate the flow over a small obstacle (half-width of 2 km), a large obstacle (half-width of 20 km) and the Assam Hills (200–300 km across). Their results in the latter

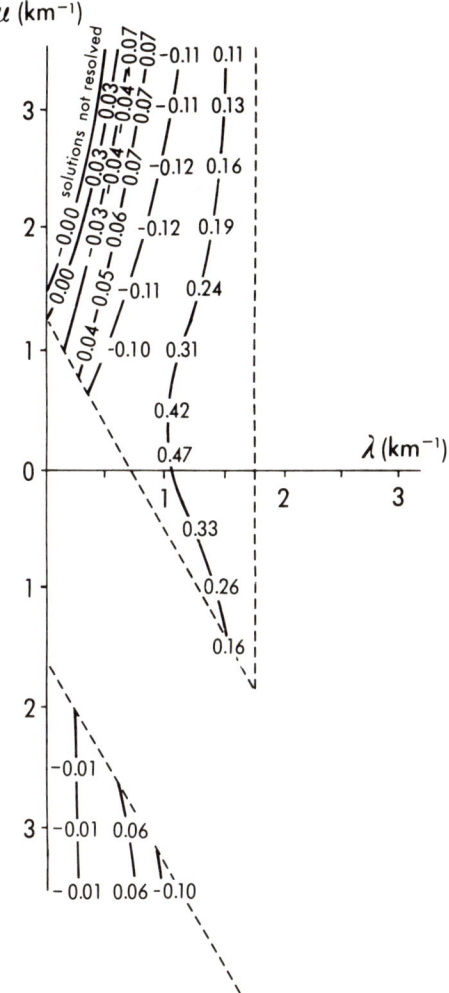

FIG. 32. Relation between λ and μ for lee waves; two-layer airstream, labelled 1 (upper) and 2 (lower). Undisturbed velocity components U, V; Scorer's parameter l^2; height of interface between layers h. Conditions in layers as follows:

$$l_1^2 = 0.5 \text{ km}^{-2}, \quad U_1 = 20 \text{ m s}^{-1}, \quad V_1 = 10 \text{ m s}^{-1};$$
$$l_2^2 = 3.0 \text{ km}^{-2}, \quad U_2 = 20 \text{ m s}^{-1}, \quad V_2 = 0;$$
$$h = 1.5 \text{ km}.$$

Numbers against curves represent amplitude factors. (After Sawyer, 1962.)

experiment show long waves (of length 20–40 km) and other large perturbations mainly between heights of 1 and 9 km. From 9 to 15 km perturbations are very small.

In contrast, Eliassen (1968) concentrated on the characteristics of air flow

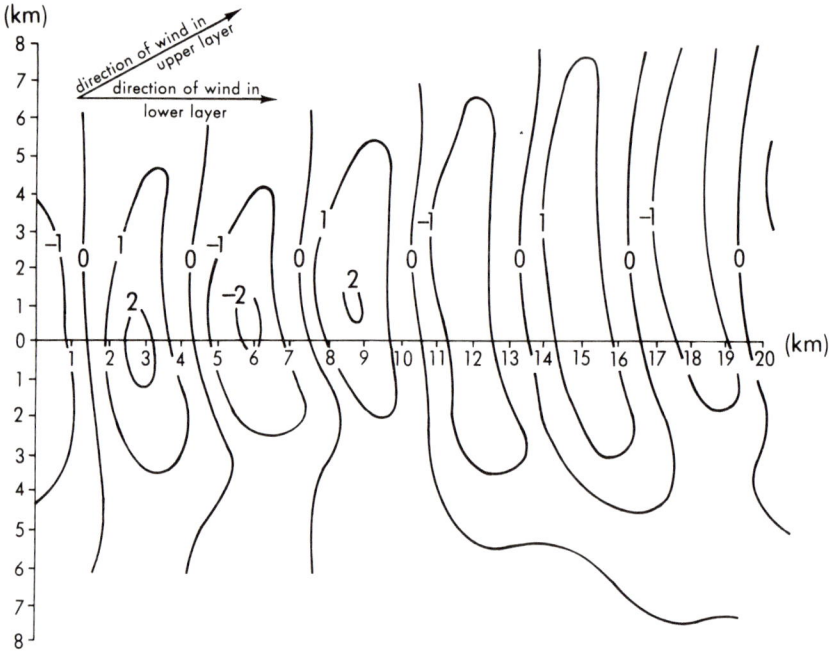

FIG. 33. Lee-wave components of vertical velocity at a height of 1.5 km for the two-layer airstream specified in Fig. 32. Quasi-circular hill with half-width of 2 km. (After Sawyer, 1962.)

over a mountain range of horizontal scale of the order of 100–1000 km and explicitly recognized the coriolis effect. The resultant waves are of the mixed gravity–inertia type that lie at the upper end of any spectrum of meso-scale motions. In a further example, Eliassen and Rekustad (1971) used a two-dimensional model of airflow over a mountain ridge with a width of about 400 km. In contrast to Sawyer's numerical solutions of steady state equations they provided numerical integration of time-dependent equations towards a steady state. The equations of motion employed in this analysis contained the coriolis parameter and were solved for the perturbation of isentropic surfaces in atmospheres of both strong and weak static stability (Fig. 34). The results steadied after 34.4 h and the flow over a mountain profile of height 890 m for both stability types is shown in Fig. 35. In strong stability, vertical wave amplitudes are moderate, with a maximum of about 500 m in the mid-troposphere. The tilt of the waves is strong, in marked contrast to the weak stability case. The main difference between the two cases is in the wave amplitude. In the weak stability case the strongest amplitude is found in mid-troposphere (4–5 km) in the second wave crest, located over the lee slope of the mountain ridge; its maximum value is about 1000 m. Figure 36 shows the horizontal projection of a number of streamlines (isentropes) at different

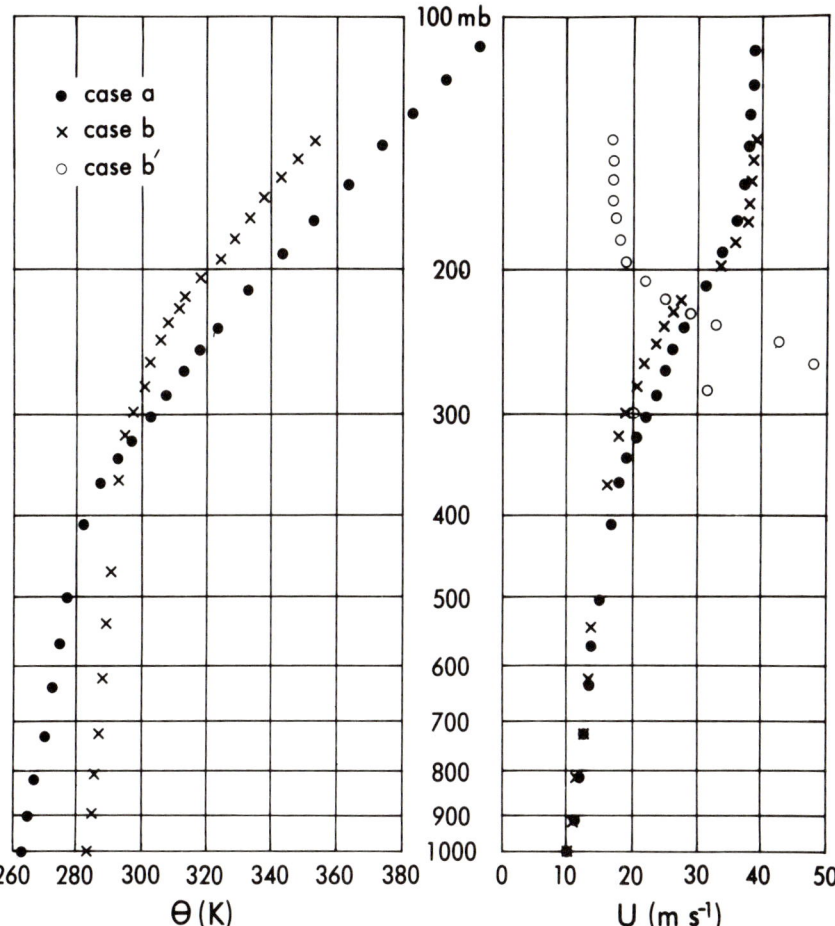

FIG. 34. Potential temperature (θ) and wind (U) profiles used by Eliassen and Rekustad. (a) Strong static stability; (b) weak static stability; (b′) weak static stability with a strong wind maximum near the tropopause. (After Eliassen and Rekustad, 1971.)

levels. At low levels the streamlines have cyclonic curvature on the windward side of the mountain, an anticyclonic bend over the highest part of the ridge, and again a cyclonic bend on the lee side reflecting vortex stretching and compression. At higher levels, there is just an anticyclonic bend over the ridge.

B. Large amplitude theory

"It is unfortunate from the theoretician's point of view that large-amplitude waves for which linear theory is invalid play such an important role in the

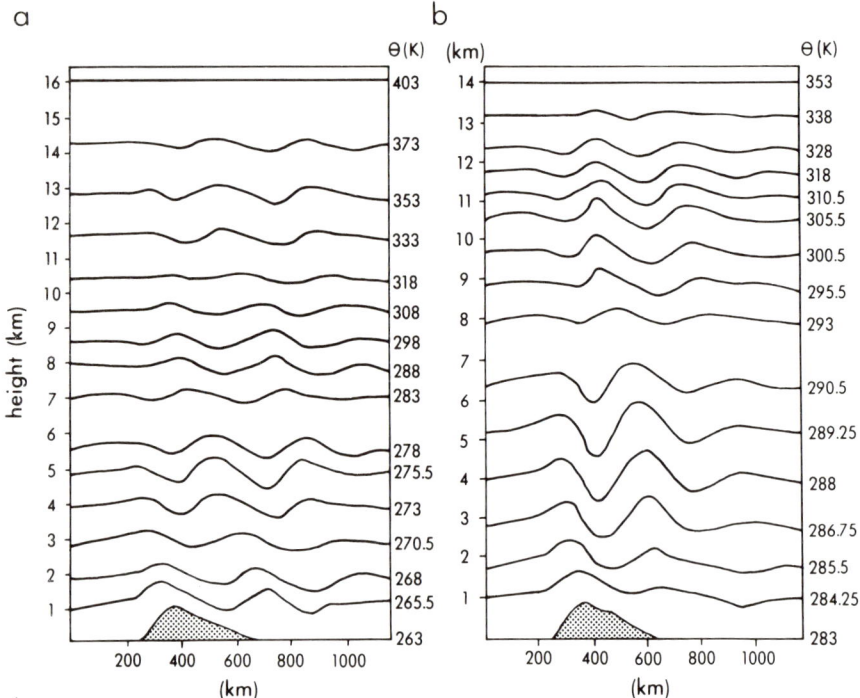

FIG. 35. Vertical cross-sections along the flow over a mountain of height 890 m after a period of 34.4 h: (a) strong static stability; (b) weak static stability. (After Eliassen and Rekustad, 1971.)

atmosphere" (Smith, 1977, p. 1634). But despite the difficulties the theoreticians have made a start on the development of large amplitude theory. Much of the work, involving two-dimensional flow represented by both linearized and unlinearized equations and three-dimensional flow, is directly applicable only to liquid motions over barriers. Atmospheric flows are differentiated from many of the studied fluid flows by their compressibility, markedly variable upstream conditions and non-confinement between rigid planes. Consequently their analysis is more difficult. Nevertheless, such flows are important to the formation of rotors and hydraulic jumps and as such have attracted the attention of meterologists. In particular rotors were investigated by Queney (1955) and Scorer and Klieforth (1959). Although their results were not directly applicable to the forecasting of large amplitude waves, Nicholls (1973a) noted that they were important since they demonstrated that rotors could occur as a solution to equations governing real atmospheric flow and that it was unnecessary to resort to analogies with hydraulic jump theory to explain their presence. This

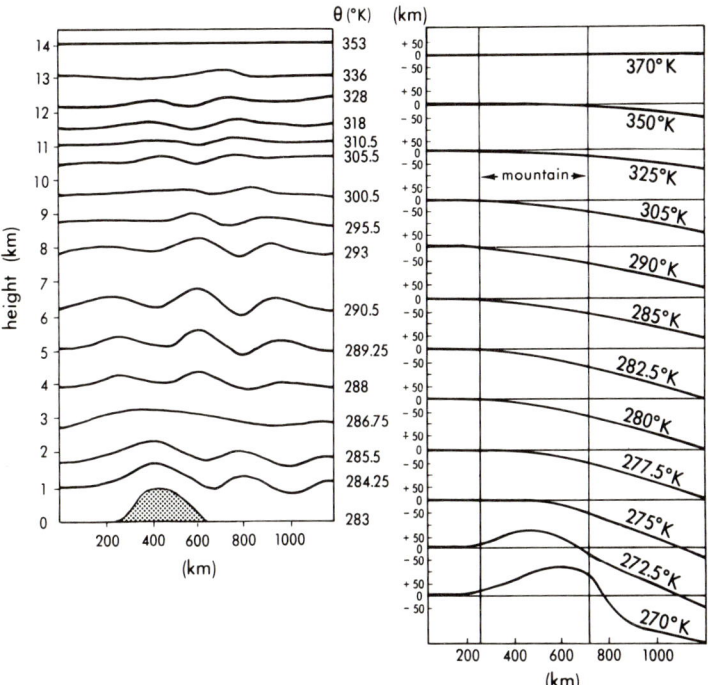

FIG. 36. Streamlines in horizontal projection at different levels for the case of strong static stability illustrated in Fig. 35(a). (After Eliassen and Rekustad, 1971.)

latter idea was primarily due to Küttner (1958) and is more fully developed in Chapter 3.

IV. Hardware models

In common with other branches of physics, meteorology has employed experimental methods to complement observational and theoretical investigations. Simulation techniques have been applied at the micro-, meso- and macro-scales, providing notable insights particularly into cloud physics and the behaviour of rotating fluids at the planetary scale. Any simulation of fluid flows must ensure the dynamical similarity between the model and reality. This is achieved by use of the appropriate dimensionless parameters listed in Table 2 (see Chapter 1). Because of the large range of scale of atmospheric phenomena, any attempt to simulate one particular type must choose a key dimensionless parameter which is believed to control the behaviour of that type. To quote Hidy (1967, p. 145): "The trick in building 'successful' laboratory analogues is to delineate the key parameters as defined by proper characteristic properties like velocity and length. Some of the 'models' are found by accident, and others develop rationally, for example, from the theory of fluid dynamics".

In his analytical approach to the problem, Scorer was not unaware of modelling techniques. However, with the aid of dimensional analysis, he argued that wind tunnel simulation was scarcely possible at the time of writing — the early 1950s. Dynamical similarity for a model 3×10^{-5} times full size would require a wind-tunnel flow speed as low as $4\,\mathrm{cm\,s^{-1}}$, too low for good experimental control. We see later that this has now been achieved. But, contemporaneous with the analytical work of the early 1950s, hardware modelling was being practised by Long (1953a) in the USA. Long was interested in the behaviour of the flow of stratified fluids and soon realized the relevance of his work to atmospheric flow over hills. In a theoretical paper (Long, 1953b) he derived a non-linear equation describing a stratified but incompressible fluid flow between upper and lower rigid boundaries. Whereas earlier authors had used perturbation theory to analyse finite amplitude disturbances, Long's equation accounted for the variation of a disturbance without the restriction of small perturbations. The non-linearity of the equation was eliminated by considering the one type of flow in which the density gradient with height was linear and also that ρU^2 was constant, where ρ is density and U is the velocity.

In a subsequent paper Long (1954) supported his theoretical results with a model of stratified flow comprising two fluids of different density. Three non-dimensional numbers were used to ensure dynamical similarity, as follows:

$$\mathrm{Fr}_i = \frac{U}{\{g(\Delta\rho/\bar{\rho})H\}^{1/2}}, \qquad R_0 = \frac{h_0}{H}, \qquad \beta = \frac{b}{H},$$

where Fr_i is called the internal Froude number, $\bar{\rho}$ is mean density of the fluids, $\Delta\rho$ is the density difference between the layers, H is the total depth of the two fluids, h_0 is the initial height of the lower layer upstream and b is the maximum height of the obstacle.

In the absence of any variation in h_0, b and H, variations in flow configuration are due to variations in U and $\bar{\rho}$, that is variation in the Froude number. Long ran the apparatus with $R_0 < \frac{1}{2}$ and $R_0 > \frac{1}{2}$, varying the Froude number in each case. When $R_0 < \frac{1}{2}$ and for any moderate values of β the flow varied as follows: for small Froude numbers (about 0.050), the fluids flowed over the obstacle with little or no turbulence; as the Froude number increased a hydraulic jump occurred to the lee of the barrier; and at maximum Froude number values (0.50 and greater), the interface between the fluids swelled symmetrically over the obstacle. When $R_0 > \frac{1}{2}$ and the obstacle was small compared to the depth of the lower layer, weak lee waves appeared at low speeds, increasing in amplitude as the approach velocity of the fluid increased. Long's main conclusion from this experiment was that unless the tropospheric inversion is quite high and the barrier quite small, the main phenomena of interest were not lee waves but rather shock waves resembling hydraulic jumps, surges or bores. This occurred because the pressure distribution in the model was closely hydrostatic except in the vicinity of the jump. In a third paper Long (1955) reported on experiments in which continuous density gradients were employed. With Froude numbers ranging from 0.250 to 0.143 long waves formed downstream of the obstacle, but Long did not consider them to be classical lee waves. At other Froude number values, complicated combinations of waves and eddies formed to the lee of the obstacle.

The three papers summarized above had comparatively little to say about lee waves in their own right. Later in the 1950s Long (1959) used a model of the Sierra Nevada with the explicit intention of simulating the Bishop wave with the aid of insight gained in the previous experiments. Geometric similarity and equality of the internal Froude number ensured similarity in the flow patterns. As the geometry was unchanged, the flows varied with the Froude numbers. In single fluid flows, a Froude number of 0.240 provided a close analogy with the single massive wave found over the Owens Valley on 16 February 1952 (Fig. 9). Situations with three waves were simulated with Fr_i values between 0.05 and 0.09. These experiments suggested that the combination of a realistic model orography and a flow with the appropriate Froude number could indeed simulate lee-wave situations.

We noted earlier that wind-tunnel simulation of lee waves (as opposed to water and salt solution simulation) was deemed unworthy of effort in the 1950s, despite the earlier efforts of Abe (1942). After 1960 the innovation of large wind-tunnel facilities (see, for example, Yamada and Meroney, 1974) and ingenious micro-scale sensors meant that the challenge of this problem could be met. The study by Lin and Binder (1967) exemplified the kind of work being done. Within the wind tunnel at Colorado State University, lee waves were simulated in a flow where the density stratification was produced by heating the air and cooling the lower boundary. The flow comprised two layers of which the lower one (18 cm deep) had large stability. Velocities were as low as 6 cm s^{-1}. Froude numbers, as defined by Long, were calculated on the basis of the height, stability and average velocity in the lower layer. Two bell-shaped model mountains of the same height (11 cm) but of different horizontal width were used.

Figure 37 shows that both the wavelength and amplitude increase with the Froude number, but that the amplitude decreases with height, a result of the decreasing static stability. The first wave crests lay about $0.8-0.85\lambda$ downstream of the mountain top, which compares well with the 0.75λ (where λ is the wavelength) displacement predicted by theory. Within the airflow in the tunnel, isotherms were coincident with streamlines: this provided an ingenious way of drawing streamlines from the measured temperatures. Figure 38 clearly shows the simulated lee waves in flows with Froude

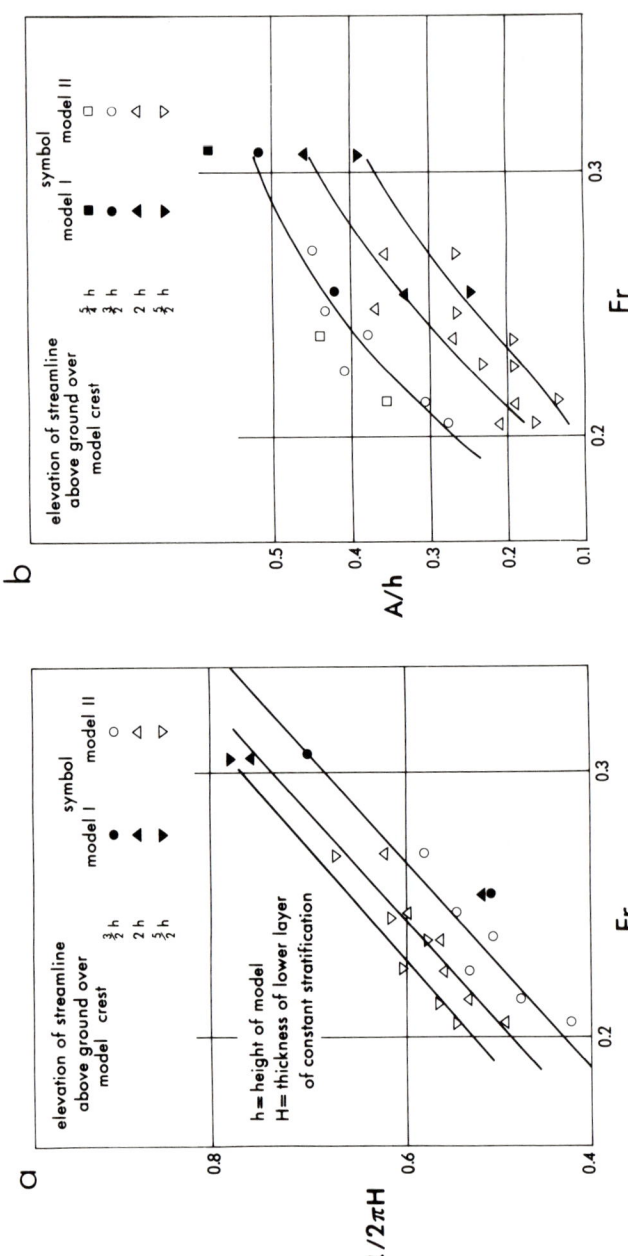

FIG. 37. (a) Lee wavelengths at different elevations as a function of Froude number. (b) Lee-wave amplitudes at different elevations as a function of Froude number. (After Lin and Binder, 1967.)

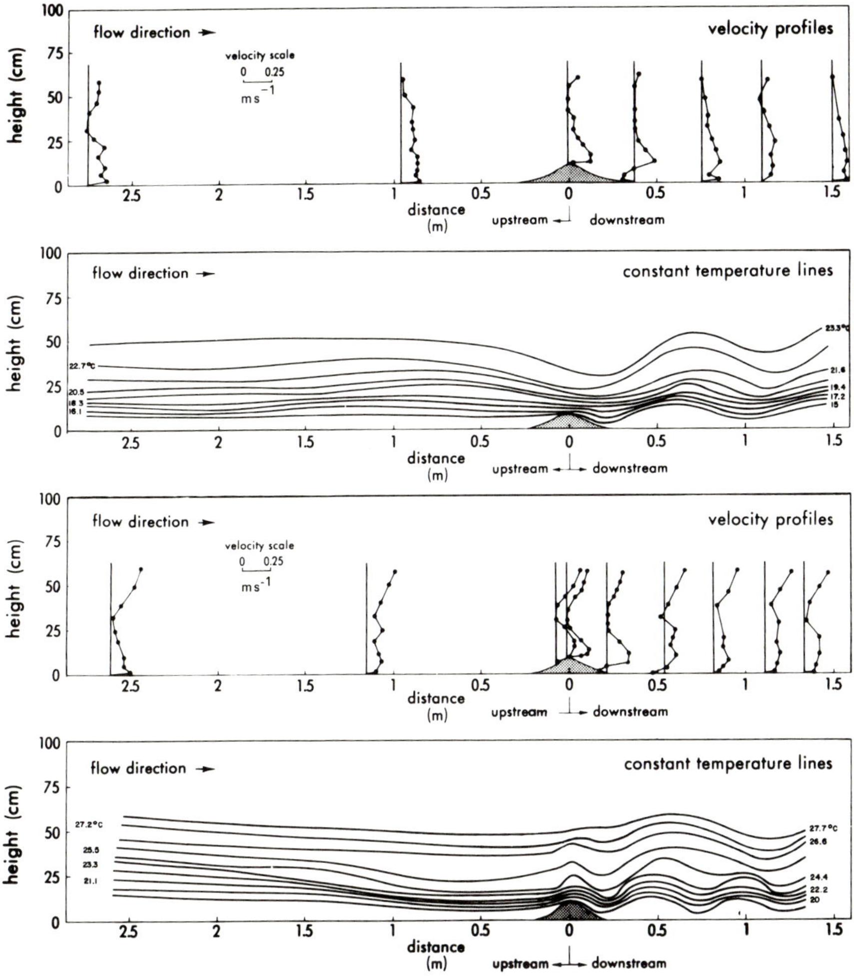

FIG. 38. Constant temperature lines and velocity profiles. Top two boxes: $U = 0.084 \, \text{m s}^{-1}$, $H = 0.182 \, \text{m}$; Fr = 0.238. Bottom two boxes: $U = 0.076 \, \text{m s}^{-1}$; $H = 0.182 \, \text{m}$; Fr = 0.214. (After Lin and Binder, 1967.)

numbers ranging from 0.238 to 0.214. Such experiments as these suggest that there are real potentialities in wind-tunnel modelling of lee-wave flow for both academic and practical purposes.

V. Momentum transfer by lee waves

The waves and associated features described in this chapter result from the disturbances of stably stratified airstreams as they pass over orography.

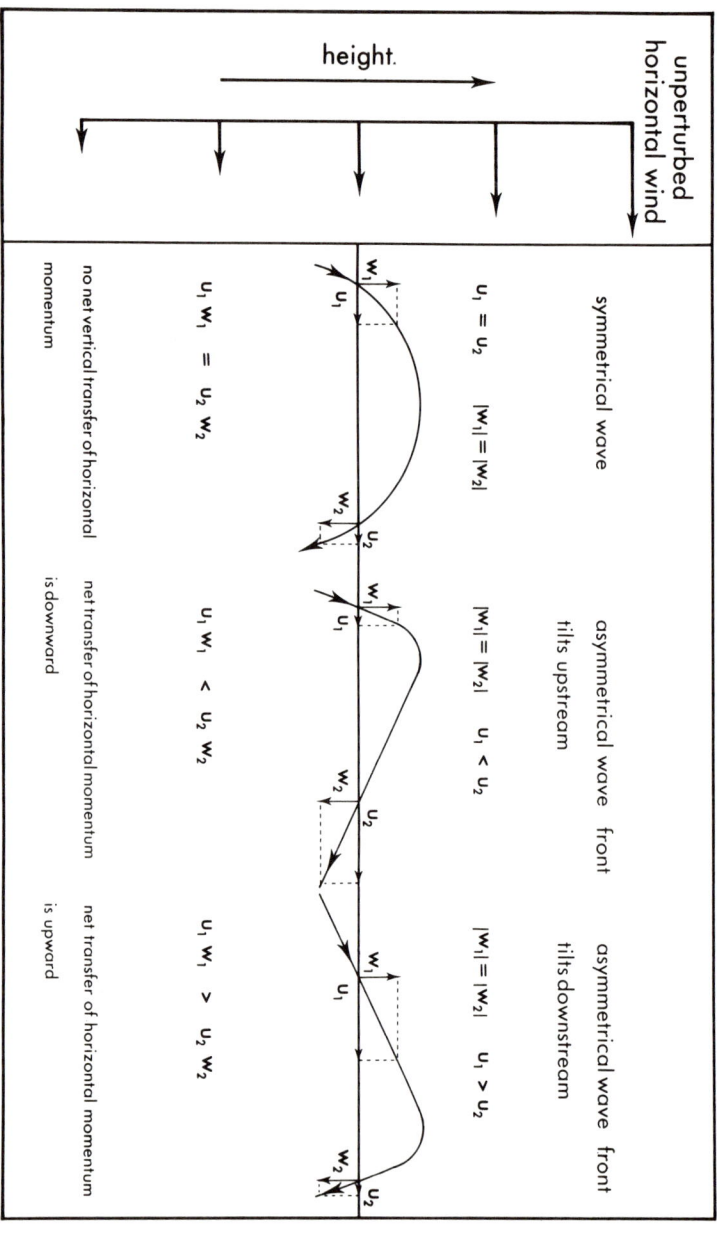

Fig. 39. Vertical transfer of horizontal momentum by waves. (a) Symmetrical wave; (b) asymmetrical wave—front tilts upstream; (c) asymmetrical wave—front tilts downstream.

Frequently the wave front tilts upwind in sympathy with asymmetrical wave forms: such configurations have been observed by Bradbury (1972). In such a situation the upward transfer of horizontal momentum in the upwind part of a wave is generally less than the downward transfer in the downwind part of a wave (Fig. 39(b)). Consequently the net effect of a wave is to achieve a downward flux of momentum within the airstream. A suite of waves in the vertical will have the same effect. This transfer is associated with pressure being systematically higher on the upwind slopes than on the downwind ones, resulting in a net force being exerted on the ground. Conversely a drag—known as wave drag—acts on the airstream. Sawyer (1959) was of the

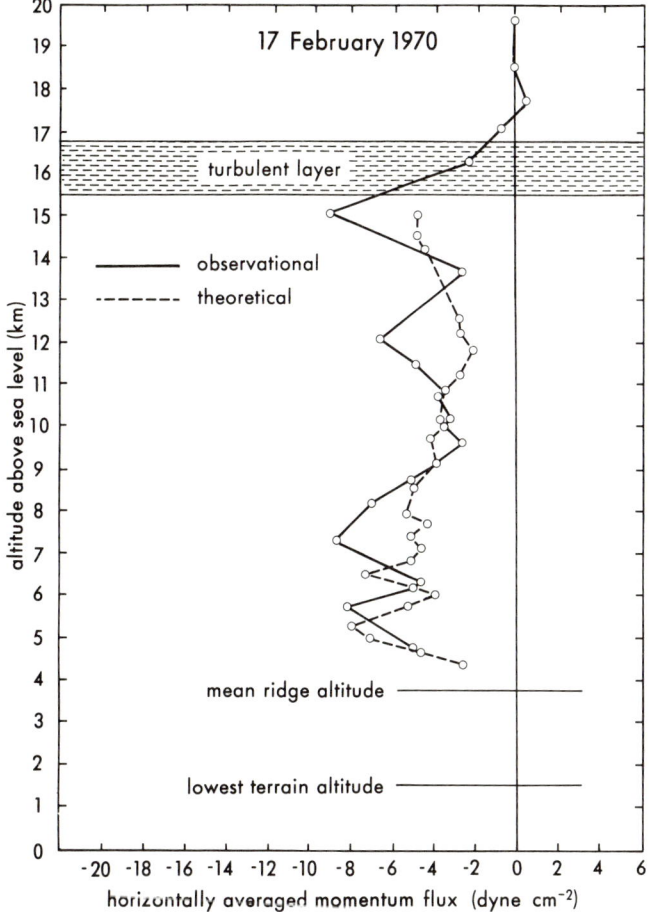

FIG. 40. Vertical flux of momentum due to waves over the Colorado Rocky Mountains. Negative values indicate downward flux. (After Lilly and Kennedy, 1973.)

opinion that this wave drag on the atmosphere may be a significant fraction of the direct frictional drag due to turbulence induced by surface roughness and estimated its value to be 7.6 dyne cm^{-2}. Bretherton (1969) calculated the wave drag in a 19 m s^{-1} wind over North Wales, giving values of 4 dyne cm^{-2}, of which 3 dyne cm^{-2} probably acted in the atmosphere above 20 km. Sawyer's preliminary estimates received support from the detailed analysis by Lilly and Kennedy (1973) of well-developed waves over the Colorado Rockies on 17 February 1970. Figure 40 shows downward fluxes of 4–6 dyne cm^{-2} over most of the troposphere with maxima of 8–10 dyne cm^{-2} near the surface and near 200 mb.

In contrast to the above results, Wooldridge (1970) found upward momentum fluxes of several tens of dynes per square centimetre in the lower tropospheric layers associated with intense gravity wave activity over the southern Rocky Mountains. By close examination of radiosonde, pibal and satellite data Wooldridge concluded that the wave front was tilted downwind, a configuration which would indeed lead to upward momentum flux, in a counter-gradient sense (Fig. 39(c)). Clearly, as Lilly (1972) suggests, far more work needs to be done to clarify the role of gravity waves in general, and lee waves in particular, in the energy and momentum budgets of the atmosphere.

References

Aanensen, C. J. M. (1965). Gales in Yorkshire in February 1962. *Geophysical Memoirs*, **108**, Meteorological Office, London.

Abe, M. (1932). The formation of cloud by the obstruction of Mount Fuji. *Geophys. Mag.*, **6**, 1–10.

Abe, M. (1941). Mountain clouds, their forms and connected current. Part II. *Bull. central met. Obs. Japan*, **7**, 93–145.

Abe, M. (1942). An attempt to make visible the mountain air current. *J. met. Soc. Japan*, **20**, 69–76.

Alaka, M. A. (ed.) (1960). The airflow over mountains, *Technical Note*, **34**, World Meteorological Organization.

Anderson, I. I. (1966). Mountain lee wave activity over Tasmania on 17 April, 1966. *Aust. Met. Mag.*, **14**, 119–131.

Bailey, M. (1970). Mountain lee-wave incidents in Scotland. *Met. Mag.*, **99**, 110–117.

Berenger, M. and Gerbier, N. (1957). Les movements ondulatoires à Saint Auban-sur-Durance. In Première campagne d'étude à des mesures, January 1956, *Monographe met. nat.*, no. 4.

Bradbury, T. A. M. (1972). Tilted or asymmetric mountain waves. *Met. Mag.*, **101**, 85–90.

Bretherton, F. P. (1969). Momentum transport by gravity waves. *Q. Jl R. met. Soc.*, **95**, 213–243.

Brunt, D. (1927). The period of simple vertical oscillations in the atmosphere. *Q. Jl R. met. Soc.*, **53**, 30–31.

Clark, T. L. and Peltier, W. R. (1977). On the evolution and stability of finite-amplitude mountain waves. *J. atmos. Sci.*, **34**, 1715–1730.

Collis, R. T. H., Fernald, F. G. and Alder, J. E. (1968). Lidar observations of Sierra-wave conditions. *J. appl. Met.*, **7**, 227–233.
Colson, de Ver (1954). Meteorological problems in forecasting mountain waves. *Bull. Am. met.'Soc.*, **35**, 363–371.
Colson, de Ver (1952). Results of double-theodolite observations at Bishop, California, in connection with the 'Bishop-wave' phenomena. *Bull. Am. met. Soc.*, **33**, 107–116.
Corby, G. C. (1954). The airflow over mountains; a review of the state of current knowledge. *Q. Jl R. met. Soc.*, **80**, 491–521.
Corby, G. A. (1957a). A preliminary study of atmospheric waves using radiosonde data. *Q. Jl R. met. Soc.*, **83**, 49–60.
Corby, G. A. (1957b). Airflow over mountains. *Meteorological Report*, **18**, Meteorological Office, London.
Corby, G. A. and Sawyer, J. S. (1958a). The airflow over a ridge—the effects of the upper boundary and high level conditions. *Q. Jl R. met. Soc.*, **84**, 25–37.
Corby, G. A. and Sawyer, J. S. (1958b). Airflow over mountains: indeterminacy of solution. *Q. Jl R. met. Soc.*, **84**, 284–285.
Corby, G. A. and Wallington, C. E. (1956). Airflow over mountains: the lee wave amplitude. *Q. Jl R. met. Soc.*, **82**, 266–274.
Crapper, G. D. (1959). A three-dimensional solution for waves in the lee of mountains. *J. Fluid Mech.*, **6**, 51–76.
Crapper, G. D. (1962). Waves in the lee of a mountain with elliptical contours. *Phil. Trans. R. Soc. A*, **254**, 601–623.
De, U. S. (1970). Lee waves as evidenced by satellite cloud pictures. *Indian J. Met. Geophys.*, **21**, 637–647.
Döös, B. R. (1961). A mountain wave theory including the effect on the vertical variation of wind and stability. *Tellus*, **13**, 305–319.
Döös, B. R. (1962). A theoretical analysis of lee wave clouds observed by Tiros I. *Tellus*, **14**, 301–309.
Eliassen, A. (1968). On meso-scale mountain waves in the rotating earth. *Geofys. Publ*, **27**, (6).
Eliassen, A. and Rekustad, J.-E. (1971). A numerical study of meso-scale mountain waves. *Geofys. Publ.*, **28**, (3).
Farooqui, S. M. T. and De, U. S. (1974). A numerical study of the mountain wave problem. *Pageophysics*, **112**, 289–300.
Foldvik, A. (1962). Two-dimensional mountain waves—a method for the rapid computation of lee wavelengths and vertical velocities. *Q. Jl R. met. Soc.*, **88**, 271–285.
Förchtgott, J. (1949). Wave currents on the leeward side of mountain crests. *Meteorologicke Zpravy*, **3**, 49–51 (as on translation in Library, British Meteorological Office).
Fritz, S. (1965). The significance of mountain lee waves as seen from satellite pictures. *J. appl. Met.*, **4**, 31–37.
Haurwitz, B. (1941). *Dynamic Meteorology*, McGraw-Hill, New York.
Hidy, G. M. (1967). Adventures in atmospheric simulation. *Bull. Am. met. Soc.*, **48**, 143–161.
Holmboe, J. and Klieforth, H. (1957). Investigations of mountain lee waves and airflow over the Sierra Nevada. Final Report, Contract no. AF19(604)-728, University of California.
Holmes, R. M. and Hage, K. D. (1971). Airborne observations of three Chinook-type situations in southern Alberta. *J. appl. Met.*, **10**, 1138–1153.

Hookings, G. A. (1968). The lee wave systems of New Zealand. *Schweiz. Aero-Rev.*, **43**, 200–206.
Klemp, J. B. and Lilly, D. K. (1975). The dynamics of wave-induced downslope winds. *J. Atmos. Sci.*, **32**, 320–329.
Küttner, J. P. (1938). Moazagotl und Föhnwelle. *Beitr. Phys. freien Atmos.*, **25**, 79–114.
Küttner, J. P. (1958). The rotor flow in the lee of mountains. *Schweiz. Aero-Rev.*, **33**, 208–215.
Küttner, J. P. and Lilly, D. K. (1968). Lee waves in the Colorado Rockies. *Weatherwise*, **21**, 180–185.
Lilly, D. K. (1972). Wave momentum flux—a GARP problem. *Bull. Am. met. Soc.*, **53**, 17–24.
Lilly, D. K. and Kennedy, P. J. (1973). Observations of a stationary mountain wave and its associated momentum flux and energy dispersion. *J. atmos. Sci.*, **30**, 1135–1152.
Lilly, D. K. and Zipser, E. J. (1972). The Front Range windstorm of 11 January 1972. *Weatherwise*, **25**, 56–63.
Lin, J. T. and Binder, G. J. (1967). Simulation of mountain lee waves in a wind tunnel. US Army Research Grant DA-AMC-28-043-65-G20, Fluid Dynamics and Diffusion Laboratory, Colorado State University, Fort Collins.
Lindsay, C. V. (1962). Mountain waves in the Appalachians. *Mon. Weath. Rev.*, **90**, 271–276.
Long, R. R. (1953a). A laboratory model resembling the "Bishop-wave" phenomenon. *Bull. Am. met. Soc.*, **34**, 205–211.
Long, R. R. (1953b). Some aspects of the flow of stratified fluids. I: A theoretical investigation. *Tellus*, **5**, 42–58.
Long, R. R. (1954). Some aspects of the flow of stratified fluids. II: Experiments with a two fluid system. *Tellus*, **6**, 97–115.
Long, R. R. (1955). Some aspects of the flow of stratified fluids. III: Continuous density gradients. *Tellus*, **7**, 341–357.
Long, R. R. (1959). Laboratory model of air flow over the Sierra Nevada mountains. In *The Atmosphere and Sea in Motion*, Rockefeller Institute Press, New York, pp. 372–380.
Lyra, G. (1940). Uber den Einfluss von Bodenerhebungen auf die Strömung einer stabil geschichteten Atmosphäre. *Beitr. Phys. freien Atmos.*, **26**, 197–206.
Lyra, G. (1943). Theorie de stationaren Leewallenstroming in freier Atmosphäre. *Z. angew. Math. Mech.*, **23**, 1–28.
Manley, G. (1945). The helm wind of Crossfell, 1937–1939. *Q. Jl R. met. Soc.*, **71**, 197–220.
Nicholls, J. M. (1973a). The airflow over mountains: Research 1958–1972, *Technical Note*, **127**, World Meteorological Organization.
Nicholls, J. M. (1973b). Aircraft measurements of disturbed airflow over mountains. *Weather*, **28**, 141–153.
Palm, E. (1958). Airflow over mountains: indeterminacy of solution. *Q. Jl R. met. Soc.*, **84**, 464–465.
Palm, E. and Foldvik, A. (1959). Contribution to the theory of two-dimensional mountain waves. *Geofys. Publr*, **21**, (6).
Panofsky, H. A. (1968). *Introduction to Dynamic Meteorology*, Pennsylvania State University, University Park, Pennsylvania.
Queney, P. (1941). Ondes de gravite produites dans un courant aerien par une petite chaine de montagnes. *C. r. hebd. Séanc. Acad. Sci. Paris*, **213**, 588–591.
Queney, P. (1947). Theory of perturbations in stratified currents with application to

air flow over mountain barriers. University of Chicago, Department of Meteorology, Miscellaneous Report no. 23.

Queney, P. (1948). The problem of air flow over mountains: a summary of theoretical studies. *Bull. Am. met. Soc.*, **29**, 16–26.

Queney, P. (1955). Rotor phenomena in the lee of mountains. *Tellus*, **7**, 367–371.

Reynolds, R. D., Lamberth, R. L. and Wurtele, M. G. (1968). Investigation of a complex mountain wave situation. *J. appl. Met.*, **7**, 353–358.

Sarker, R. P. and Calheiros, R. V. (1974). Theoretical analysis of lee waves over the Andes as seen by satellite pictures. *Pageophysics*, **112**, 301–319.

Sawyer, J. S. (1959). The introduction of the effects of topography into methods of numerical forecasting. *Q. Jl R. met. Soc.*, **85**, 31–43.

Sawyer, J. S. (1960). Numerical calculation of the displacements of a stratified airstream crossing a ridge of small height. *Q. Jl R. met. Soc.*, **86**, 326–345.

Sawyer, J. S. (1962). Gravity waves in the atmosphere as a three-dimensional problem. *Q. Jl R. met. Soc.*, **88**, 412–425.

Scorer, R. S. (1949). Theory of waves in lee of mountains. *Q. Jl R. met. Soc.*, **75**, 41–56.

Scorer, R. S. (1954). Theory of airflow over mountains: III—airstream characteristics. *Q. Jl R. met. Soc.*, **80**, 417–428.

Scorer, R. S. (1958a). Airflow over mountains: indeterminacy of solution. *Q. Jl R. met. Soc.*, **84**, 182–183.

Scorer, R. S. (1958b). Reply to letter by E. Palm entitled "Airflow over mountains: indeterminacy of solution". *Q. Jl R. met. Soc.*, **84**, 465–466.

Scorer, R. S. (1978). *Environmental Aerodynamics*, Ellis Horwood Ltd, Chichester.

Scorer, R. S. and Klieforth, H. (1959). Theory of mountain waves of large amplitude. *Q. Jl R. met. Soc.*, **85**, 131–143.

Scorer, R. S. and Wilkinson, M. (1956). Waves in the lee of an isolated hill. *Q. Jl R. met. Soc.*, **82**, 419–427.

Smith, R. B. (1977). The steepening of hydrostatic mountain waves. *J. atmos. Sci.*, **34**, 1634–1654.

Starr, J. R. and Browning, K. A. (1972). Observations of lee waves by high power-radar. *Q. Jl R. met. Soc.*, **98**, 73–85.

Vergeiner, I. and Lilly, D. K. (1970). The dynamic structure of lee wave flow as obtained from balloon and airplane observations. *Mon. Weath. Rev.*, **98**, 220–232.

Wallington, C. E. (1958). A numerical study of the topographical factor in lee waves. *Q. Jl R. met. Soc.*, **84**, 428–433.

Wallington, C. E. (1960). Introduction to lee waves in atmosphere. *Weather*, **15**, 269–276.

Wooldridge, G. L. (1970). Vertical momentum transport over mountainous terrain. Atmospheric Science Paper no. 164, Department of Atmospheric Science, Colorado State University, Fort Collins, Colorado.

Wooldridge, G. and Lester, P. E. (1969). Detailed observations of mountain lee waves and a comparison with theory. Atmospheric Science Paper no. 138, Department of Atmospheric Science, Colorado State University, Fort Collins, Colorado.

Wurtele, M. G. (1957). The three-dimensional lee wave. *Beitr. Phys. freien Atmos.*, **29**, 242–252.

Yamada, T. and Meroney, R. N. (1974). A wind-tunnel facility for simulating mountain and heat island gravity waves. *Boundary Layer Met.*, **7**, 65–80.

Yuyama, Y. (1972). A climatological study of orographic wave clouds on and leeside of Mt Fuji (in Japanese). *J. met. Res.*, **24**, 415–418.

3

Downslope Winds

I. Introduction

The occurrence of downslope winds and their effects on settlements on mountain fringes have excited the curiosity of both meteorologist and layman alike for well over a century. In Europe particularly, a huge literature, mostly in German, describes many individual cases of downslope winds and speculates on their origin. The English-language literature is not as voluminous, but recent work in Colorado, USA, has provided not only excellent observations of these winds but also a preliminary theory of their origin.

In common with lee waves, downslope winds clearly exist primarily because of the obstacle-effect of the mountain. Whereas the lee waves are dependent upon but not spatially coincident with the obstacle that causes them, the downslope winds, by definition, must remain close to the mountain. It would seem reasonable to argue that, for such winds to be considered as meso-scale systems, they must form over an obstacle of the appropriate horizontal dimensions. This is in contrast to lee waves whose wavelengths are frequently of the meso-scale regardless of the size of the forcing obstacle. There is, however, increasing evidence that downslope winds and lee waves are due to similar mechanisms and if this is the case (to be discussed later) downslope winds may be classified as meso-scale phenomena on dynamical criteria rather than on the grounds of their chance formation over a meso-scale orographic feature. In reality there is little doubt that many of these winds frequently form part of a meso-scale system. Figure 41 clearly shows that downslope winds at Boulder, Colorado are part of a system with characteristic horizontal dimension of 40 km. It also illustrates how strong these winds may be with mean speeds of 20 m s^{-1} and peak gusts of 36 m s^{-1} in Boulder at the foot of the eastern Rocky Mountain slope. A gust of

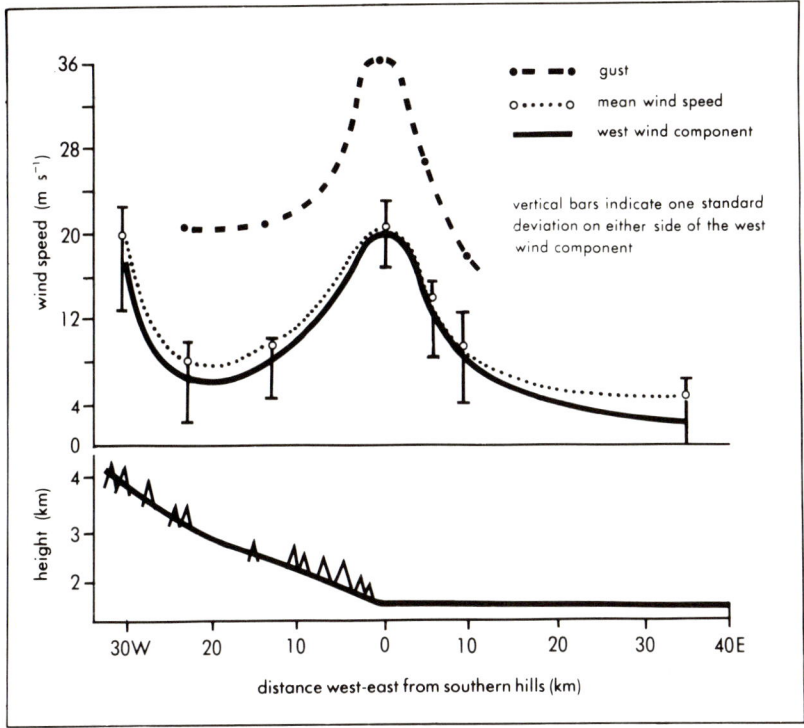

FIG. 41. Mean surface winds for the hour with highest west wind component at the eastern edge of the Rocky Mountains near Boulder, Colorado. The mean was calculated from hourly wind speeds in 20 storms. (After Brinkmann, 1974.)

54 m s^{-1} was noted in the great storm of 11 January 1972 at Boulder (Lilly and Zipser, 1972).

Despite the strength of most downslope winds, investigation throughout the last century has concentrated upon their temperatures and humidities. This is not surprising because surface-bound scientists of the late nineteenth century were confronted with most striking changes in both temperature and humidity at the onset of downslope winds. In Alpine Europe, a warm, dry downslope wind had been known since Roman times. Brinkmann (1971) suggested that the German word "*foehn*", now widely recognized as the generic term for such winds, may derive from the Latin word "*favonius*", which means west wind. As Brinkmann (1971, p. 230) said:

> To the Romans the most important characteristic of the "favonius" was not its direction but its warmth. As they reached the northern slopes of the Alps,... and experienced a warm wind, the Romans believed this to be the "favonius" from the Mediterranean. Thus the name came to be applied to the warm, dry wind descending the northern Alpine slopes.

In contrast, the eastern coastlands of the Adriatic Sea frequently experience a cold, dry downslope wind known as the *"bora"*. This word, meaning "north" in Greek has now become the generally accepted generic term for this type of wind. In the case of both "foehn"- and "bora"-type winds, different localities have different names for their local winds. Some of them appear later in this chapter: more comprehensive lists are provided by Becker (1948) and Schmitt (1930).

The apparently simple subdivision of downslope winds into warm and cold types belies a confusing situation in the diagnosis of foehn and bora. In fact no clear agreement exists on whether these winds are, or should be, generically or genetically classified. There are elements of the chicken and egg problem here. Until the type of wind is clearly defined and observed it cannot be explained. Yet the explanation of the wind would be, and has been used as (albeit not always successfully), a most desirable basis for the definition of the wind. The difficulties and confusions of the problem were well summarized by Brinkmann (1971) who noted (p. 235) that "...in a number of studies foehn wind cases have been selected without satisfactory definition or explanation". Much, if not all, of her discussion was as readily applicable to the definition of the bora as it was to the foehn, with the proviso of course that the changes of temperature were in opposite senses in the two winds.

According to Brinkmann (1971) the most common approach to the identification of foehn occurrences is the "three-criteria" definition as used by Conrad (1936), Osmond (1941) and Obenland (1956). In this definition the foehn is characterized by a simultaneous and abrupt change in temperature and relative humidity, together with a surface wind from the direction of the mountain range. Brinkmann noted that this method is less satisfactory during the warm season when temperatures are generally higher and that it has been found to include foehn-like winds produced by other weather phenomena such as a change in air mass. As noted above, wind speed has largely been ignored on the grounds that the wind may be of any strength, while in other cases the foehn is defined particularly as a strong wind. Schuetz and Steinhauser (1955) used potential temperature to identify downward-moving air but they ignored wind characteristics and, as a result, found that all wind directions were possible during the foehn and that speeds varied from 0 to 20 m s^{-1}. Such confusion over definition clearly bodes ill for climatological studies, a point emphasized by Frey (1957). Brinkmann well illustrated how different definitions lead to different results. In studies of foehn frequency in the Alps, covering approximately the same period, Ungeheuer (1952) found the maximum to occur during the period April to August while Obenland (1956) found a December to April maximum. Also, Ungeheuer found a mean frequency of 114 foehn winds per year while Obenland's figure is only 22. In Brinkmann's (1970) own analysis of the chinook (North American name for foehn wind to the east of the Rocky Mountains) in Alberta, she used two criteria: the wind at upper levels had to have a strong component normal to the mountain range; a "foehn nose" had to exist. This latter feature is a deformation of the isobars on a synoptic surface chart with a ridge of high pressure on the windward side and a trough of low pressure to the leeward of the mountain range. A number of periods with winds from the directions west-north-west, west and west-south-west (the preferred directions for the chinook at Calgary) were classified as either chinook or non-chinook winds on the basis of these criteria. However, discriminant analysis of the five surface parameters, wind speed, temperature, relative humidity, vapour pressure, station pressure, revealed that 20 per cent of the cases were mis-classified. If only wind speed, temperature and relative humidity were used, the discriminant function mis-classified almost 50 per cent of the periods. In Brinkmann's (1970, p. 276) own words: "It is therefore concluded that it is not possible to differentiate between chinooks and non-chinooks on the basis of surface criteria alone". The analysis may, however, have revealed weaknesses in the original criteria rather than the combination of surface criteria, but Brinkmann makes no comment on this point.

It may seem surprising that after knowing of the existence of foehns for two millennia and after a century of fairly close examination, particularly in Europe, we are still uncertain as to what they (and bora) really are. Perhaps the current uncertainty arises from the asking of more probing questions and a deeper appreciation of the complexity of this particular atmospheric feature. Recent studies (to be outlined later) have emphasized the dynamics of the downslope winds at the expense of their temperature and humidity. Before considering this "unified" approach, salient observational studies of warm and cold downslope winds are reviewed.

II. Observation

Lack of precise definitions of "foehn"- and "bora"-type winds has not hindered the production of a multitude of observational studies within the last 100 years. The bibliographies compiled by Stepanova (1951) and Yoshino (1972) provide scores of examples of individual cases in many parts of the world, but particularly in Alpine Europe, Yugoslavia, southern Russia and Japan. Climatological studies are fewer and comparisons from place to place are very difficult due to the use of different criteria. The following section outlines the main characteristics of downslope winds and, notwithstanding the problems indicated earlier, attempts a climatological sketch of their occurrence.

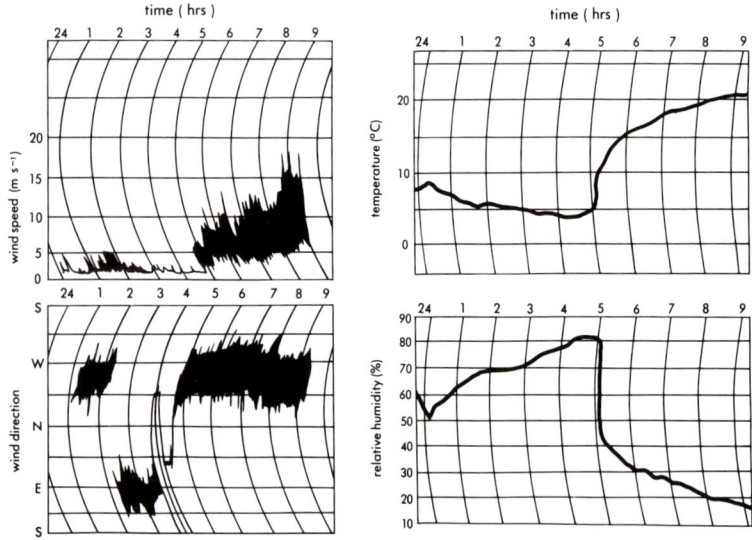

FIG. 42. Abrupt start of a foehn on 26 May 1963 at Obihiro, Japan, shown by changes in wind speed and direction, temperature and humidity. (After Arakawa, 1969.)

A. Surface characteristics

1. *Warm winds—foehn*

Foehn winds have been responsible for some of the most dramatic changes in temperature, humidity and wind speed ever observed. Even if we discount as not a true foehn (for reasons discussed later) the famous Black Hills case of 15–16 January 1943, when temperatures changed from $-20\,°C$ to $+7\,°C$ within only 2 min (Hamann, 1943), several spectacular changes caused by foehns are on record. For example on 19 December 1933 at Havre, Montana, temperatures rose almost instantly from $-23\,°C$ to $-8\,°C$ due to a chinook which hit "like a shot from a cannon" (Glenn, 1961) and temperatures continued upward more slowly for the next 24 h reaching $+7\,°C$ (Math, 1934). Closely associated with temperature rises of this kind are sharp falls (of up to 50 per cent) in relative, but not usually absolute, humidity. The rapidity of these temperature and humidity changes is closely associated with the frequently sudden onslaught of the foehn wind. Typical changes in wind speed from less than $5\,m\,s^{-1}$ to over $45\,m\,s^{-1}$ in a matter of seconds have been recorded (Julian and Julian, 1969). Once the foehn has set in it tends to be very gusty but maintains a fairly steady direction. Figure 42 illustrates the main features outlined above. The wind starts in a matter of minutes and takes on a fairly constant west-north-westerly direction; temperature rises

10°C within 1 h; and relative humidity drops dramatically by over 50 per cent in less than 30 min. Yet strong, warm downslope winds are not always associated with marked changes in temperature and humidity. In what he called "an unusual chinook case", Riehl (1971) noted that, in contrast with a spectacular wind record, temperature and relative humidity remained almost constant. No satisfactory explanation was given for these observations.

Not all foehn winds "hit like a cannon". Many build up over a period of hours and consequently are less readily recognized until after the event (Brinkmann and Ashwell, 1968). Figure 43 shows a typical sequence of foehn behaviour over a period of 5 days, summarizing most of the surface characteristics. In particular the humidity changes stand out clearly. It is interesting that the author of the diagram recognizes the difficulties of definition by noting that "periods when *suspected* chinooks occurred are shown by brackets" (Beran, 1967, p. 869; present author's italics).

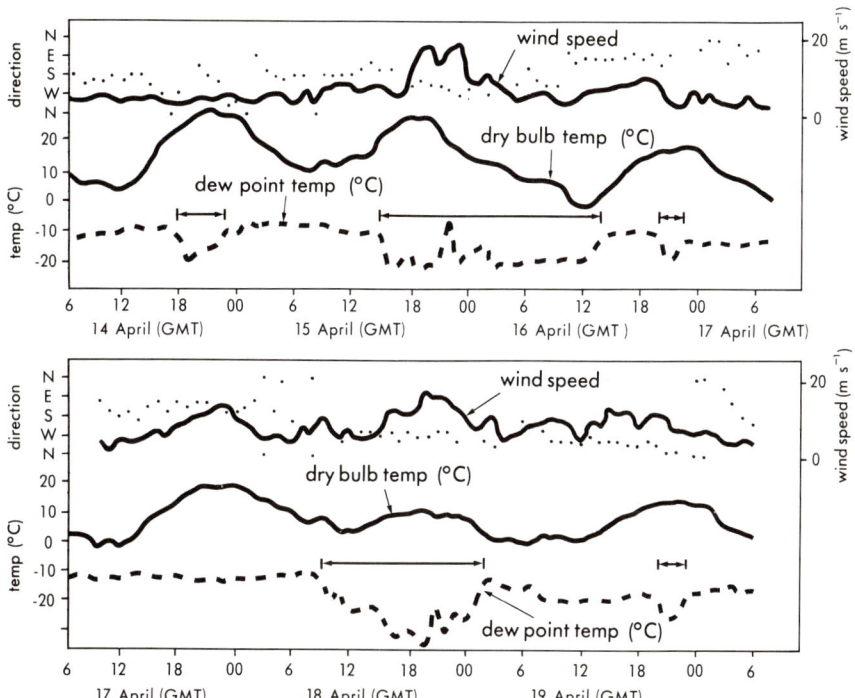

FIG. 43. Gradual start of foehn conditions from 14 to 19 April 1963 at Denver, Colorado, shown by changes in wind speed and direction (the latter shown by dots), dry and dew point temperatures. Periods when suspected chinooks occurred are shown by the brackets between the dry bulb and dew point temperature curves. (After Beran, 1967.)

2. Cold winds—bora

Descriptions of the bora-type of downslope wind concentrate upon temperatures and wind speeds. For example Mazelle (1907) noted temperatures dropping to $-12.8\,°C$ at Trieste on 23 January 1907, being accompanied by north-easterly winds (the bora from the nearby mountains) of over $28\,\mathrm{m\,s^{-1}}$ for 8 h. In one of the few observational studies using especially collected rather than routine data, Yoshimura et al. (1976) found wind gusts ranging from 9 to $32\,\mathrm{m\,s^{-1}}$ and temperature falls between 0.5 and $10.3\,°C\,h^{-1}$ in boras on the Yugoslav coast. No measurements of humidity were made. Figure 44 illustrates a temperature drop of $10\,°C$ in just over 2 h due to the sudden arrival of east-north-east winds with speeds of about $10\,\mathrm{m\,s^{-1}}$. As in the case of foehn winds, not all boras start so abruptly. Yoshimura et al. noted that many boras develop gradually over several hours as the synoptic scale wind changes to the appropriate direction.

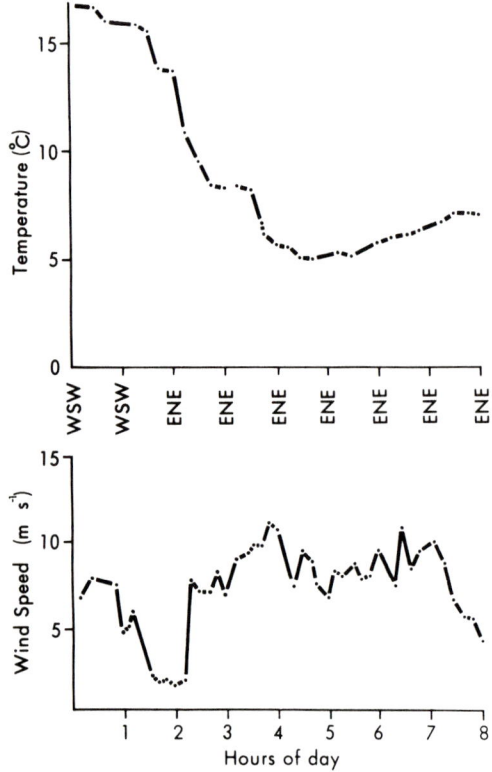

FIG. 44. Start of bora on 12 November 1972 on the Adriatic coast shown by changes in wind speed and direction and temperature. (After Yoshimura et al., 1976.)

B. Climatology

Despite the difficulties of definition outlined in the introduction, climatological studies of downslope winds do exist. As far as can be determined all the studies are of either warm or cold winds rather than downslope winds as a whole. The few papers concerned with downslope winds, both warm and cold (e.g. Brinkmann, 1974; Arakawa, 1969), are not, despite one of the titles, essentially climatological in nature.

1. Warm winds—foehn

Two of the more notable climatological studies of foehns are those concerned with Europe and North America by Flohn (1941) and Riehl (1974) respectively. Riehl used the following criteria to identify the foehns at Fort Collins, Colorado: the surface wind direction must be between 250° and 290° as the mountains are due west of Fort Collins; a sharp onset of high wind speeds must be identifiable; and the strongest winds must attain at least $15\,\mathrm{m\,s^{-1}}$ in winter and $11\,\mathrm{m\,s^{-1}}$ in summer. There was no mention of temperature and humidity changes in the criteria despite the use of the word chinook in the title of the paper. Yet it is clear from the text that temperature and humidity changes have been implicitly incorporated in the analysis.

Riehl analysed foehns within the period 1964–71 and normalized his frequencies to a 10-year basis. On average, about 10 foehns occurred each year, seven of them between October and March (Table 6). Using the criterion that a foehn ceased when the average wind speed dropped to $7\,\mathrm{m\,s^{-1}}$, Riehl compiled Table 7 which shows foehn duration. The longest durations, up to 16 h, occurred in winter but the seasonal pattern was quite irregular. Duration from 4 to 8 h was most common. It is worth noting here that Riehl excluded from his analysis those occasions when moderately high winds from the west continued for 2–3 days under general anticyclonic flow aloft over the entire western mountain area of USA. Such winds have a desiccating effect similar to that of pronounced foehns and in summer may lead to periods with extreme maximum temperature. We see later that similar discrimination was not employed in the majority of climatological studies of bora-type winds. Peak gusts in the Colorado foehn (Table 8) most frequently ranged from 16 to $19\,\mathrm{m\,s^{-1}}$, particularly in winter, while gusts in the range 20–$29\,\mathrm{m\,s^{-1}}$ occurred with only slightly less frequency.

2. Cold winds—bora

Much of our recent knowledge on cold downslope winds results from the work of M. M. Yoshino (1976a) and his group in western Yugoslavia. Unfortunately the studies are weakened by lack of consideration of the problems of definition outlined earlier in this chapter. In virtually all the

Table 6 Frequency of warm winds per year

| | \multicolumn{12}{c|}{Month} |
	Jan	Feb	Mar	Apr	May	Jun	Jul	Aug	Sep	Oct	Nov	Dec
Colorado—chinook[†] Frequency	2.4	0.9	1.2	0.8	1.0	0.3	0.3	0.4	0.3	0.4	0.9	1.7
California—Newhall storms[‡] Frequency	4.8	3.4	3.1	2.0	0.6	0.1	0.1	0.1	0.4	1.7	3.6	4.3

[†] *Source*: Riehl (1974). Period of observations: January 1964 to April 1971.
[‡] *Source*: Koutnik (1968). Period of observations: 1957–66.

Table 7 Duration of chinooks in 2 h intervals

	Jan	Feb	Mar	Apr	May	Jun	Jul	Aug	Sep	Oct	Nov	Dec	Total	Per cent
0–2	1		1	1	1			1					5	7
2–4	3			2	1	1	1	1		1	2	5	18	23
4–6	2	1	5		2	1			1		1	2	14	18
6–8	4	2	2	2	2		1			1	1	5	19	25
8–10	1	2		1	1	1		1	1		2		10	13
10–12	4	1	1	1									7	9
12–14	2		1										3	4
14–16	1												1	1

Source: Riehl (1974). Period of observations: January 1964 to April 1971.

Table 8 Frequency distribution of peak gusts in chinooks

Speed range (m s^{-1})	Month												Total	Per cent
	Jan	Feb	Mar	Apr	May	Jun	Jul	Aug	Sep	Oct	Nov	Dec		
11–15	5	1			1	1			1				7	9
16–19	2	1	5	3	3		2	2			2	2	21	28
20–24	7	1	2	1	3	1				2		3	17	23
25–28			1					1	1	1	2	5	17	22
29–33	3	2		2								2	10	13
34–37			1										1	1
38–43	1	1									1		3	4

Source: Riehl (1974). Period of observations: January 1964 to April 1971. Note that speed values have been converted from miles per hour.

Table 9 Monthly mean frequencies of bora-days

Place	Month												Year
	Jan	Feb	Mar	Apr	May	Jun	Jul	Aug	Sep	Oct	Nov	Dec	
Adriatic coast†	11.0	9.5	9.6	7.7	7.3	6.8	6.0	7.0	6.3	10.5	9.5	10.6	101.8
Trieste‡	12.5	12.0	12.8	10.0	9.7	7.6	8.3	10.2	9.8	12.6	13.5	13.2	132.2
Ajdovscina‡	10.2	7.4	6.4	8.2	7.9	5.2	6.3	9.1	7.0	11.5	11.4	10.4	101.0

† *Source:* Tamiya (1972). Period of observations: 1956–65. Bora-days identified from surface pressure charts *Täglicher Wetterbericht* published by Deutschern Wetterdienst, Offenbach.
‡ *Source:* Yoshimura (1972). Period of observations: 1956–65. Bora-days defined as days when wind direction is from north to east with the peak gust exceeding 10 m s^{-1}.

contributions the bora is simply identified as a wind from the east-north-east, sometimes with the additional requirement of a gust of over $10\,\mathrm{m\,s^{-1}}$; this wind, in turn, is occasionally identified, not from instruments, but from synoptic scale pressure maps of south-eastern Europe. All but one of the chapters are based upon routinely collected meteorological data at places such as Trieste and Split, rather than data collected in especially designed programmes. One result of using routine data is that the "bora-day" rather than each bora occasion is the basis of frequency counts. A bora-day is a day on which one or more boras blew.

Monthly frequencies of bora-days have been calculated according to two definitions (Table 9) both of which reveal the same broad seasonal variation. In winter about 10 bora-days occur each month whereas in summer the figure falls to about six. The agreement between the figures derived according to different criteria is gratifying because, as Tamiya (1972, p. 53) admitted: "Strictly speaking, the bora-day, determined by the present criterion (i.e. from pressure charts—BWA) should be called a 'potential' bora-day, because bora must be detected by the meteorological elements observed at the local stations on the Adriatic coast". Despite these deficiencies, the Japanese results are in qualitative agreement with earlier European studies, for example those by Grober (1948), Jedina (1892) and Potapov (1961) and they provide useful quantitative estimates of bora frequency.

In common with foehn winds, bora-type winds have high peak gusts, such as $41.7\,\mathrm{m\,s^{-1}}$ in Trieste (Yoshino, 1976b), frequently result in relative humidities of less than 40 per cent, tend to be more frequent at night-time and usually last for 12–20 h (Yoshino, 1969).

C. Synoptic background of downslope winds

The classification of meteorological phenomena according to their association with particular synoptic scale circulations may be either a means to an end or the end itself (viz. Barry and Perry, 1973). As a means to an end it often provides the first clues to the mechanism of the circulation in question. Consequently it is not very surprising that the first steps of many attempts to explain the occurrences of both warm and cold downslope winds involved the methods of synoptic climatology. Classifications have been made throughout most of this century, many involving quite numerous subdivisions. Notable contributions include those of Ficker and Rudder (1943) on the foehn and Paradiz (1957), Poje (1962) and Yoshino (1971) on the bora. A little scrutiny reveals many similarities in the essentials of the classifications for the two types of wind. The important conclusion emerging from these studies is that the existence of both foehn and bora are encouraged by, first, a pressure gradient along the mountain axis, thus giving a gradient wind normal to the axis and, second, a stable layer at roughly the same height as the mountain. It is only within the last decade that these close links between warm and cold downslope winds have been explicitly studied.

Brinkmann's (1974) analysis of strong, downslope winds at Boulder, Colorado, included both warm and cold winds. She clearly demonstrated many similarities between the two types of wind and identified only one significant difference—namely the windward and leeward surface pressure tendencies that created the pressure gradient that, in turn, drives the downslope wind. A windward/leeward pressure gradient of 8 mb, common to both warm and cold winds, was generated by different processes. In the case of warm winds, pressure dropped first and fell twice as quickly in the lee as to the windward of the mountain. In the case of cold winds pressure rose first and continued to rise twice as quickly to the windward as to the leeward of the mountain. These interesting differences were not explained by Brinkmann but she did show that the resulting pressure gradient was adequate to accelerate the downslope winds (both warm and cold) to speeds of over 30 m s^{-1}. Similarly, a pressure difference of 5 mb between Boulder (at the foot

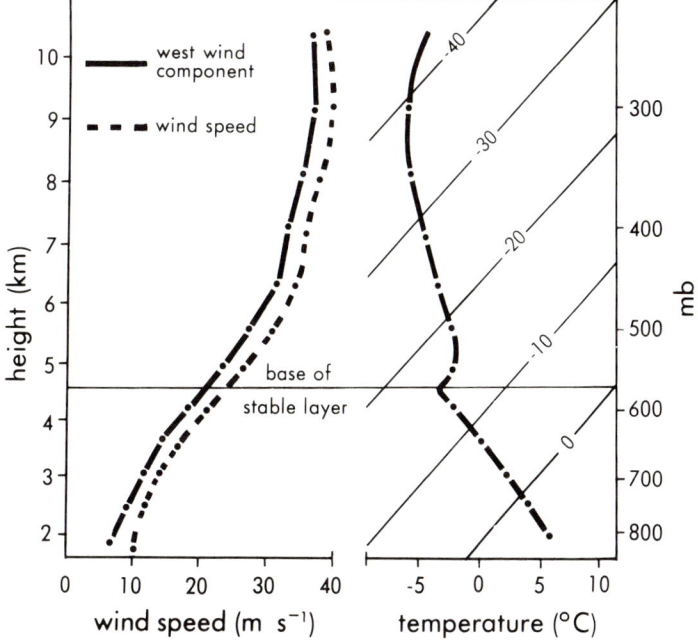

FIG. 45. Composite upwind sounding for a time close to the beginning of downslope wind storms in Boulder, Colorado. (After Brinkmann, 1974.)

of the slope) and Denver (about 20 km east of Boulder on the plains), increasing from Boulder to Denver, was adequate to decrease wind speeds from over $30 \, \text{m s}^{-1}$ to about $10 \, \text{m s}^{-1}$, a frequently observed value in downslope wind situations.

Thermal and wind characteristics of airstreams that gave downslope winds at Boulder are shown in Fig. 45. To avoid excessive smoothing resulting from straight averaging of the 24 soundings used, a composite temperature and wind speed profile was produced using as a reference point the base of a stable layer which appeared in all the soundings and was considered by Brinkmann to mark the upper limit of wind-storm air. No justification was given for this assumption about the upper limit, but it detracts little from the analysis and allowed for the first time an explicit demonstration of the relationships between strong downslope winds, atmospheric stability and vertical distribution of wind speed. The analysis also revealed that mean gusts at the surface during wind storms in Boulder ($36 \, \text{m s}^{-1}$) were much higher than the mean wind up to the 400 mb level in the undisturbed flow. The strong surface winds were therefore not immediately attributable to strong winds aloft—for example, a low level jet descending to the surface.

D. Concluding remarks

It is now clear that warm and cold downslope winds are similar in the following respects:

(1) They frequently start suddenly, are strong, gusty and fairly constant in direction.

(2) They are associated with low relative humidities but little change or even a slight increase in absolute humidities.

(3) Their diurnal and monthly frequencies reveal nocturnal and winter maxima.

(4) The airstreams in which the winds occur usually contain a stable layer at a critical height and move in a direction normal to the mountains.

Thus the only important difference between the winds is the temperatures they bring to the foot of mountain slopes. These temperatures are clearly "warm" or "cold" only relative to the temperatures already existing at the bottom of the slope prior to the wind. Thus foehns are warm relative to cold, continental air typical of winters in central North America and Alpine Europe. On the other hand, boras are cold relative to warm, frequently maritime air such as found in winter on the Adriatic coast. It is thus quite possible that, as Brinkmann (1974, p. 601) concludes, "... warm and downslope wind storms (foehns and boras) are generated by similar mechanisms". This idea is examined in the next section.

III. Theory

Explanations of some facets of downslope winds have a century-old pedigree. In particular the warmth of the foehn wind was partially explained as early as 1866 and it is fair to say that throughout the following century the majority of foehn studies have not fundamentally added to our understanding. In brief, the problems of downslope winds are twofold:

(1) What is the mechanism of the downward flow?
(2) What is the cause of the relative warmth or cold of the wind?

The answer to the second question lies partially in the concluding paragraph of the observational section. It is now clear that the relative warmth or coldness of the downslope wind is as much a function of the original temperatures of places prior to the occurrence of the wind as it is to the temperature of the downward moving air. It is possible that downslope winds experience one mechanism that causes temperature to change and yet still arrive at the slope foot as either foehn or bora. Hann (1866) realized that the temperature change would be an increase. In his explanation of the foehn he originally suggested the "thermodynamic theory", which is as follows. Moist air rises over the mountain, is cooled and in all probability gives cloud and precipitation over the mountain top. The condensation process releases latent heat and thus warms up the air. Consequently, upon descent in the lee of the mountain, the air is warmer, level for level, than it was on the windward side. In all probability the downward moving air would also frequently be warmer than the air it "replaced" at the foot of the slope, and thus it would be recognized as a foehn. Nearly 20 years later Hann (1885) realized that condensation was not necessary for the heating to occur. If air sinks, it is warmed by adiabatic compression and consequently dry, upper level air is a potential source for the foehn wind. This mechanism was further clarified by Bilwiller (1899), who showed that there is no essential difference between the mountain foehn and the warm, sinking air of anticyclones. Figure 46 illustrates these two mechanisms which account for the warmth of the foehn. We should note that they may also occur in a bora, a point frequently made by Yoshino (1976a), but that the warming is inadequate to transform the wind sufficiently for it not to be felt as a bora on warm coastal strips.

In addition to these two basic mechanisms, some authors (e.g. Glenn, 1961; Beran, 1967) suggest two other factors important to foehn winds, in particular the chinook. The first of these is, in reality, a change of air mass when warm Pacific air comprising the "foehn" displaces cold Arctic air which frequently lies to the east of the Rocky Mountains. This was the primary reason for the very dramatic temperature changes in the Black Hills in 1943 noted earlier. The second factor involves the mixing by the onslaught of the chinook of

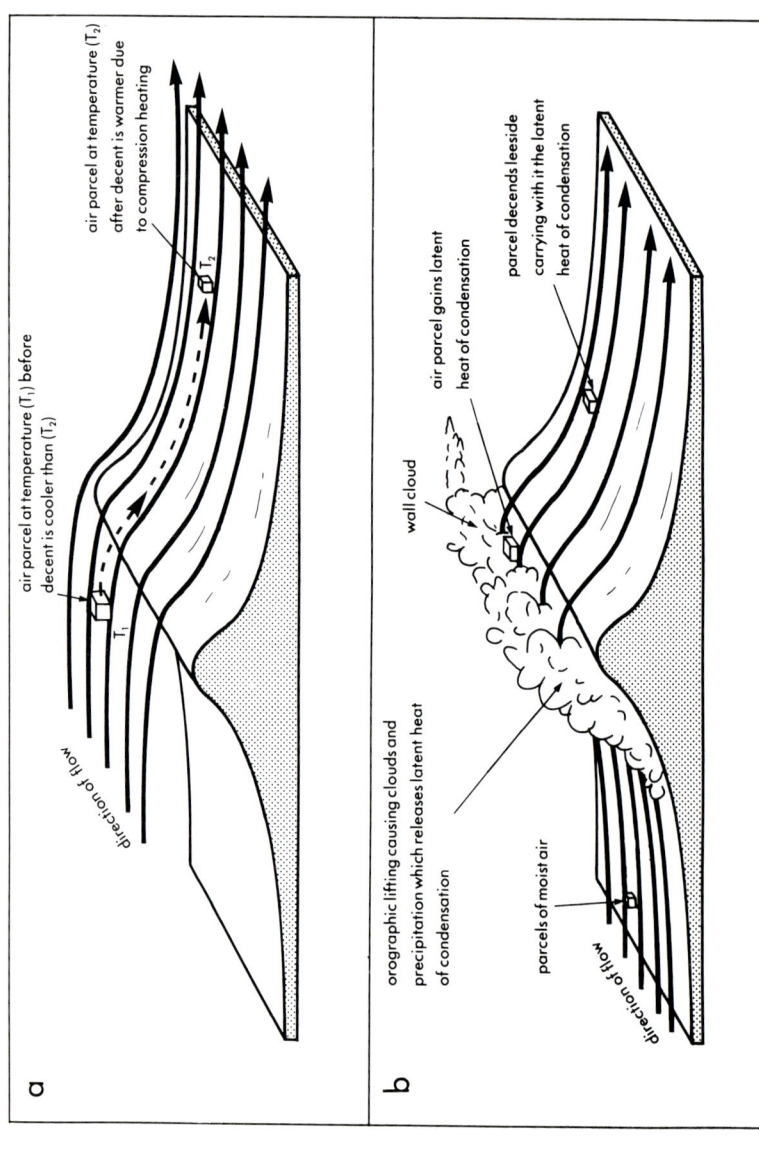

FIG. 46. Two possible mechanisms of a foehn. (a) Cross-section through a ridge showing descent of an air parcel on the leeward side. The temperature of the parcel at T_2 is greater than the temperature of the air parcel at T_1 due to compressional heating caused by the descent. (b) Cross-section through a ridge showing moist air being lifted on the windward side. This lifting causes condensation and the release of latent heat. The air parcel then descends the leeward side and heat gained from the condensation process is realized as warming at stations on the leeward slope. (After Beran, 1967.)

leeward air which is cooling by nocturnal radiation. The mixing reduces the rate of cooling. Whereas both factors may operate at times they are not fundamental to the downslope wind mechanism.

Hann's thermodynamic theories provide satisfactory answers to the second question. The answers to the first question may even yet not have been found. Over the last half century four main theories for downslope winds have been suggested: they are reviewed below.

Implicit in Hann's studies and more explicitly presented by Ficker (1905) is the idea that the descent of foehn air is a simple consequence of withdrawal of cold surface air from the lee valleys, since the mass of the mountains prevented replacement from anywhere except aloft. Crude as this argument appears there may be a large element of truth in the idea of "blocking", even if the reasons given for it were rather vague. Over 50 years later the same idea was suggested by Scorer and Klieforth (1959) as an incidental facet of an investigation of rotors. Implicit in their analysis were two mechanisms that could account for downslope winds. First, they derived a solution to the following equation:

$$\frac{\partial^2 \zeta}{\partial z^2} + 2S \frac{\partial \zeta}{\partial z} + (l^2 - k^2)\zeta = 0, \qquad (41)$$

where ζ is the displacement of a streamline from its undisturbed position; $S = (1/U)(\partial U/\partial z)$, where U is the undisturbed horizontal wind speed in the x direction; $l^2 = g\beta/U^2$, where $\beta = (1/\theta)(\partial\theta/\partial z)$. Note that the wind shear term (numerically small) was omitted in this version of l^2.

The solution is of the form

$$\zeta = \zeta_1(x)\cos lz + \text{"lee wave" terms}, \qquad (42)$$

where the mountain shape is given by

$$\zeta = \zeta_1(x) \quad \text{at} \quad z = 0.$$

This solution was then used to define both airstreams and mountain shapes that would form lee waves of large amplitudes. As a result of this analysis two types of flow that could account for downslope winds emerged. The first involved blocking of the lower layers and the second the manifestation of the large amplitude wave as the downslope wind.

Scorer and Klieforth showed that in a uniform airstream without lee waves the general solution would be

$$\zeta = \zeta_1(x)\cos lz + f(x)\sin lz, \qquad (43)$$

where either $f(x) = 0$ or $f(x) = \bar{\zeta}_1(x)$ and where $\bar{\zeta}_1$ is a function derived from ζ_1 by the application of a boundary condition.

With any of the solutions of equation (43) there are values of z which will give nodal surfaces (i.e. $\zeta = 0$) within the flow. If a mountain is higher than this nodal surface, theoretically there will be no flow over the mountain. With solutions to equation (42), i.e. in an airstream with lee waves, the maximum vertical distance between successive nodal surfaces is π/l; therefore, flow will be impossible in the layer below the top of a mountain with height of at least π/l. With the aid of Long's (1955) results, Scorer and Klieforth suggested that one of two things could happen when an airstream was confronted by a mountain with height greater than π/l. In view of the apparent misunderstanding in some articles of what Scorer and Klieforth actually wrote, we quote the following (Scorer and Klieforth, 1959, p. 137):

> Long found that the steam was modified either upstream by the passage of a wave upstream from the mountain, reducing the value of l, or there was a hydraulic jump modifying the stream permanently on the lee side; this happens if there are no waves which can travel upstream to modify the oncoming flow because the stream is too rapid. Equally the flow might not occur in the manner supposed: for instance, the lower layers might be blocked and become stationary with the upper layers descending to the surface on the lee side. Consequently we cannot suppose that any airstream will flow across any mountain—still less that its flow can be represented by the present means even when the equations are linear. Whether any of these things happen in the atmosphere has not been investigated; but the obvious suggestion is that foehn winds will occur if an airstream reaches a mountain, say on the arrival of a cold front, whose height exceeds π/l.

Scorer and Klieforth went on to identify airstreams that will form lee waves of large amplitude and to determine the form of the mountain ridge that will excite such waves. Whilst the main aim of the paper was to present a theory of flow which could generate rotors, the resultant flow gave strong downslope winds to the lee of the mountain. This was one of the first demonstrations that such winds could be due to large amplitude lee waves.

Scorer and Klieforth's suggestion that foehns could be caused by the blocking of low level inversion layers and associated large amplitude lee waves was tested by Beran (1967) east of the Rocky Mountains and by Lockwood (1962) in the UK. In both cases a close relationship existed between mountain height and foehn wind occurrence, given an airstream normal to the mountain. Figure 47 shows a particularly good example. Although Beran clearly considered blocking and large amplitude lee waves as parts of the same mechanism, in testing the hypothesis by the sole use of the blocking criterion, he found several cases where blocking did not lead to foehns when it (being representative of large amplitude lee waves) would have been expected to do so. Whilst recognizing the blocking effect Lockwood suggested that large amplitude lee waves probably aided the downslope winds, implying a separate mechanism to the blocking. This idea is more in accord with the original theory of Scorer and Klieforth than that used by Beran. Of the two mechanisms inherent in Scorer and Klieforth's theory, the important one for downslope winds is the large amplitude lee wave. Within

3 DOWNSLOPE WINDS

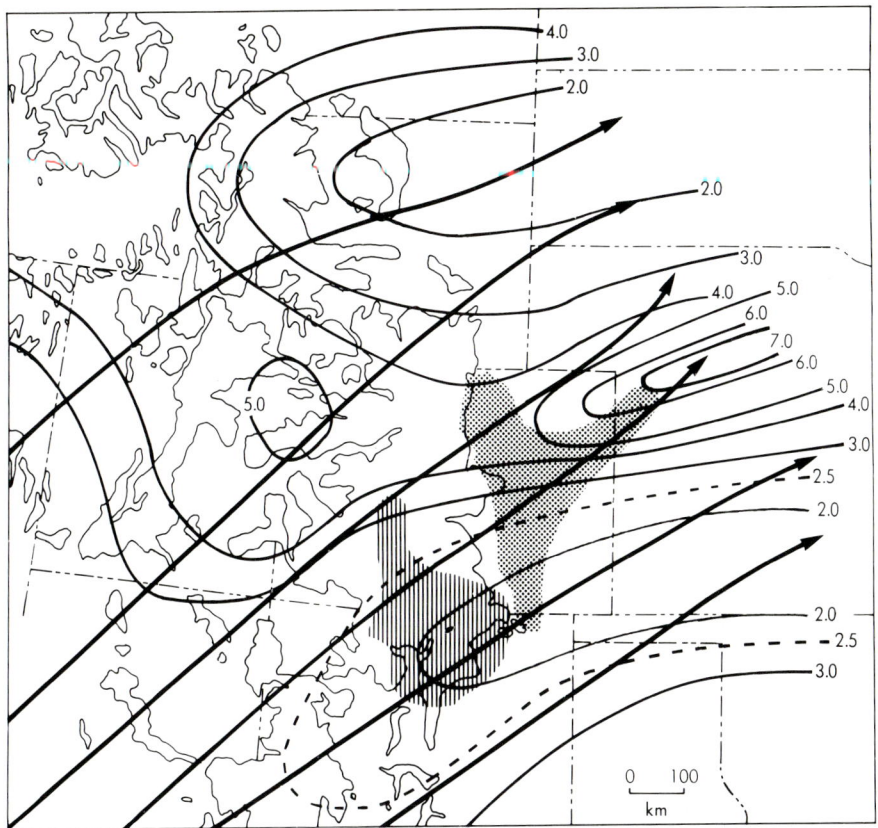

FIG. 47. Field of $(\pi/l+z)$ (solid lines, in kilometres), where z is the height of each station, for 1200 GMT 15 April 1963 superimposed on a map of terrain contours of the Colorado Rockies, showing land over 2100 m. The area where the mountains are higher than the $(\pi/l+z)$ field is line shaded. Heavy lines are the 700 mb streamline pattern. Dotted portion shows the area affected by a chinook at 1800 GMT 15 April. (After Beran, 1967.)

an airstream giving such waves, and therefore, potentially giving foehn winds, blocking would occur only if the mountain height were greater than π/l.

The third main theory for downslope winds, after Ficker's ideas on blocking and the large amplitude lee-wave theory, was the hydraulic jump theory. First put forward by Küttner (1939), it has subsequently received support from Schweitzer (1953), Küttner (1958) himself, Houghton and Kasahara (1968) and Arakawa (1968, 1969). Using the observation that downslope winds are frequently associated with an inversion just above the mountain top, Arakawa manipulated the equations of motion into a generalized form which described the behaviour of shallow water. He therefore expected that atmospheric phenomena under an inversion would be similar to those in shallow water. On this basis, the theory of hydraulic jumps could be applied to downslope winds. A complete theory based on the time-dependent, non-linear shallow water equations has been worked out by Houghton and Kasahara (1968) for a one-layer system and by Houghton and Isaacson (1969) for a two-layer system.

Texts on hydraulics show that the velocity and depth of water flow in a channel are closely related. The discharge has a maximum when the fluid has a "critical height" of h_0 and the velocity reaches the critical value (v_0)

$$v_0 = (gh_0)^{1/2}. \tag{44}$$

If the Froude number $\text{Fr} = v_0/(gh_0)^{1/2} = 1$, then critical flow is said to occur. If $\text{Fr} > 1$, supercritical (rapid) flow occurs; if $\text{Fr} < 1$, subcritical (tranquil) flow occurs. In a channel with a sloping floor, gravity may accelerate the stream from critical to supercritical flow prior to an eventual return to subcritical flow downstream, where the slope disappears.

The atmosphere provides a fairly close analogy to the above situation. Figure 48 schematically illustrates a "reservoir" of air to the left of the mountain crest. The equivalent of the free surface of the water is considered to be an inversion in the air mass. As the inversion is the interface between a colder and warmer airmass the density on top of the flow is not zero but only slightly reduced by an amount dependent on the temperature difference across the inversion. The gravitational force acting on the interface is reduced to the modified gravity:

$$\varepsilon = \frac{\Delta\theta}{\theta} g, \tag{45}$$

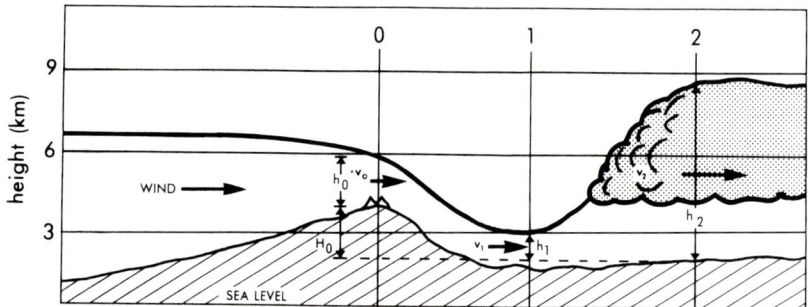

FIG. 48. Schematic diagram of hydraulic air flow over a mountain range. H_0 is the height of the mountain above the valley floor. h_0 is the height of the streamline at an inversion (considered to be equivalent to a free water surface) above the mountain crest. h_1 and h_2 are the heights of the same streamline at sections 1 and 2. v_0, v_1, v_2 are the velocities at sections 0, 1 and 2. (After Küttner, 1958.)

which is not a constant and may be 50 times smaller than gravity. Thus the critical velocity over the mountain crest becomes

$$v_0 = (\varepsilon h_0)^{1/2}. \tag{46}$$

As air flows from section 0 to section 1 in Fig. 48 application of Bernouilli's energy equation to the top streamline gives

$$\varepsilon(H_0 + h_0 - h_1) = \tfrac{1}{2}(v_1^2 - v_0^2) \tag{47}$$

where H_0 is the height of the mountain above the valley floor, h_0 is the height of the streamline above the mountain crest, h_1 is the height of the streamline above the leeward valley floor, v_0 and v_1 are the velocities at sections 0 and 1 and ε is as defined in equation (45).

With a constant discharge $Q = vh$, where h is the height of the top streamline above the ground and v the critical velocity defined by equation (46), we arrive at the following equation for the depths of the flow at the two cross-sections:

$$\left(\frac{h_0}{h_1}\right)^3 - \left(2\frac{H_0}{h_0} + 3\right)\frac{h_0}{h_1} + 2 = 0. \tag{48}$$

Solution of equation (48) shows that the flow depth shrinks by about 50 per cent and the velocity doubles if $h_0 = 0.4H_0$. If $H_0 = 1500$ m (typical of the Rocky Mountains) this means that $h_0 = 600$ m, a figure quite in keeping with observed values of inversion height above a mountain top. Combined with a strong inversion this situation leads to strong downslope winds as gravity would cause the critical flow to become "supercritical" or "shooting". Küttner was of the opinion that this is the basic mechanism of the foehn and bora.

Further downstream, at section 2, the supercritical flow is decelerated by friction and the undisturbed air over the non-sloping land and thus tends towards subcritical flow. As a result the flow in the valley raises its depth from h_1 to h_2 (Fig. 48). If this happens waves can travel upstream in the subcritical area near h_2, while in the supercritical area near h_1 they cannot. The wave front therefore steepens until it breaks between sections 1 and 2. In this way the hydraulic jump is formed and becomes stationary over ground.

Attractive as the hydraulic jump theory is, both for its comparative simplicity and its apparent success at closely modelling real situations, Klemp and Lilly (1975, 1978) were of the opinion that many downslope winds are due to a different mechanism. They also suggested that Scorer and Klieforth's theory, which is a possible explanation in terms of rather short (less than 20 km wavelength), quasi-periodic, large amplitude lee waves, is inadequate to explain downslope winds in Colorado. These winds, and similar ones in Alberta (Lester, 1976; Lester and MacPherson, 1977), are associated with hydrostatic waves of horizontal wavelength of order 50–100 km, are essentially

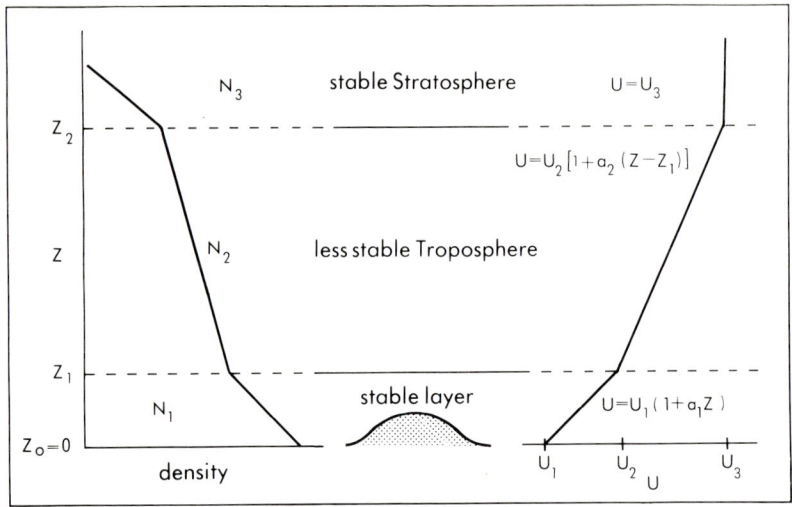

FIG. 49. Density and velocity profiles in a three-layer incompressible atmosphere with constant stability and linear shear in each layer. The three layers represent an idealized atmosphere having a low-level stable region (layer 1), a less stable troposphere (layer 2) and a stable stratosphere (layer 3). In an incompressible atmosphere the stability is defined as $N^2 = -g(\partial(\ln \rho)/\partial z)$ while in the compressible analogue it is given by $N^2 = g(\partial(\ln \theta)/\partial z)$. (After Klemp and Lilly, 1975.)

forced by the contours of the mountain and propagate freely in the vertical (Lilly, 1978). Trapped resonant waves (such as the short wavelength lee waves) do not appear to play an important role.

Klemp and Lilly's (1975, 1978) model is the fourth main theory for downslope winds reviewed here. They first obtained analytical solutions for linear multilayer models which were then used to identify the upstream atmospheric conditions responsible for the generation of intense surface winds. This was followed by the derivation of a linear numerical wave model which was run with real data from a particularly strong downslope wind situation.

The analytical model (Fig. 49) comprised an incompressible atmosphere in which there was a low level, stable layer, a less stable troposphere and a stable stratosphere. The model was linear, steady state and two-dimensional. To investigate the atmospheric conditions capable of producing a strong response, Klemp and Lilly obtained solutions for flow over a single Fourier component of a mountain contour. Using the Boussinesq approximation, the linear, hydrostatic, steady-state equation for the vertical velocity (w) becomes

$$\frac{\partial^2 w}{\partial z^2} + l^2 w = 0. \tag{49}$$

where

$$l^2 = -\frac{g}{U^2}\frac{\partial(\ln\rho)}{\partial z} = \frac{N^2}{U^2}$$

in an atmosphere with linear vertical shear, U is the undisturbed horizontal wind speed and N is the Brunt–Väisälä frequency. In an incompressible atmosphere the stability is defined as $N^2 = -g(\partial(\ln\rho)/\partial z)$. With appropriate boundary conditions Klemp and Lilly provided solutions of equation (49) for each layer for flow over a single Fourier component. They then used this solution for w in the continuity equation to find the maximum perturbation wind speed at the surface—in effect the downslope wind. This took the form

$$|U(0)| = \frac{1}{kU_1}|U_1\mu_1 b_1 + \tfrac{1}{2}N_1\,\mathrm{Ri}^{-1/2}a_1|, \qquad (50)$$

where U_1 is the horizontal wind speed in the lowest layer, N_1 is the stability in the lowest layer and Ri is the mean Richardson number in the lowest layer ($=N_1^2/\alpha_1^2 U_1^2$, where α represents the linear wind shear factor; see Fig. 49). μ_1, a_1 and b_1 are complicated functions of k, U_1, N_1 and Ri in all layers, together with the phase shift of the waves throughout the depth of each layer. Due to the complexity of equation (50), Klemp and Lilly considered the case with weak wind shear. The assumption that Richardson number values were much greater than unity reduced equation (50) to

$$|U(0)| = N_1 H A, \qquad A = \left(\frac{X+Y}{X-Y}\right)^{1/2}, \qquad (51)$$

where H is the height of the mountain and X and Y are both functions of the stabilities in all layers and the phase shifts in layers 1 and 2. In a one-layer atmosphere having constant properties N_1 and U_1, the maximum perturbation of the horizontal surface velocity is $N_1 H$. Consequently, Klemp and Lilly considered A to be an amplification factor multiplying the velocity that would be attained in a single-layer system having $N = N_1$. The strongest surface winds would occur with maximum A and this occurs when the phase shifts in layers 1 and 2 are equal to $\pi/2$. This means that maximum amplification of the surface velocity in a three-layer system occurs when the lower two layers each have a thickness equalling one-quarter of the vertical wavelength in the respective layer. As the vertical wavelength is shorter in the more stable layer, this optimal structure is consistent with the presence of a relatively thin, low level inversion with a thicker, less stable troposphere above.

On the insertion of the following realistic values,

$$N_1 = 1.6 \times 10^{-2}\,\mathrm{s}^{-1}, \qquad U_1 = 15\,\mathrm{m\,s}^{-1},$$
$$N_2 = 0.9 \times 10^{-2}\,\mathrm{s}^{-1}, \qquad U_2 = 25\,\mathrm{m\,s}^{-1},$$
$$N_3 = 2.0 \times 10^{-2}\,\mathrm{s}^{-1}, \qquad U_3 = 45\,\mathrm{m\,s}^{-1},$$

the computed layer depths for optimal response were 1920 m for the stable layer and 5940 m for the upper troposphere, giving a tropopause height of 7860 m above the lowest layer of air which actually passes over the mountain. The computed amplification factor $A = 4.0$ and the predicted maximum surface wind speed for a 500 m sinusoidal mountain (1000 m ridge to trough) was 32 m s^{-1}. As Klemp and Lilly (1975, p. 327) noted: "Characteristics of this optimal structure are consistent with...observations...".

Precision about the mechanics of this amplification process is difficult. Klemp and Lilly claimed that the amplification could be interpreted in terms of partial wave reflections which occur in a multilayer medium having differing propagation characteristics. When the low level, stable layer and the less stable region above have optimum thicknesses, partial reflections of the wave motion reinforce the wave in the lower part of the atmosphere and can produce a wave amplitude several times that of the mountain height.

In addition to showing the relationship between layer depth and vertical wavelength, Klemp and Lilly assessed the sensitivity of the surface windspeed to the height of the inversion layer. Figure 50 shows that the maximum perturbation surface velocity is 40 m s^{-1}, occurring with an inversion layer depth of 1.57 km. The wind speed falls off quite rapidly as the thickness of the stable layer deviates from this optimal value. In summary, Klemp and Lilly's analytical theory indicated that fairly long waves can become very intense if

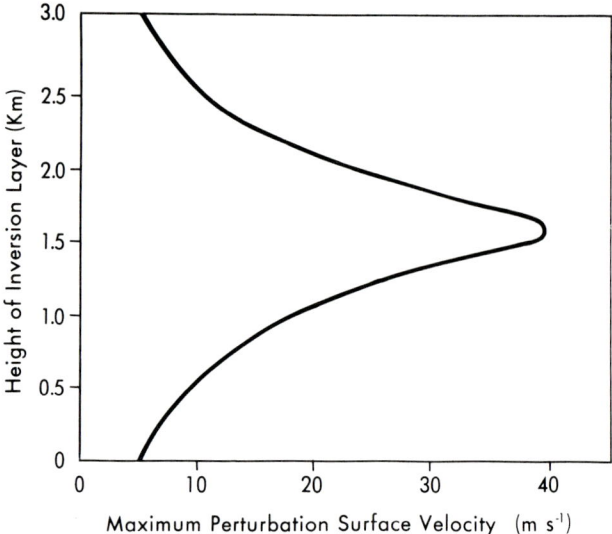

FIG. 50. Variation of the maximum perturbation surface velocity with changes in the thickness or height of the low level, stable layer. Line represents the three-layer atmosphere with an inversion layer of constant stability. (After Klemp and Lilly, 1975.)

an inversion is present near mountain-top level in the upstream environment and if the stability and wind profiles are such that the waves approximately reverse phase between the surface and the tropopause.

The modification of the theory to provide the capability of simulating real atmospheric conditions also allowed an assessment of the effect of mountain shape on the downslope winds. Figure 51 shows that a "ramp" shaped mountain produced maximum surface winds of 50 per cent greater than an isolated, symmetrical mountain. This ramp-shaped mountain is roughly analogous to the situation west of Boulder, Colorado, where many severe downslope winds have occurred.

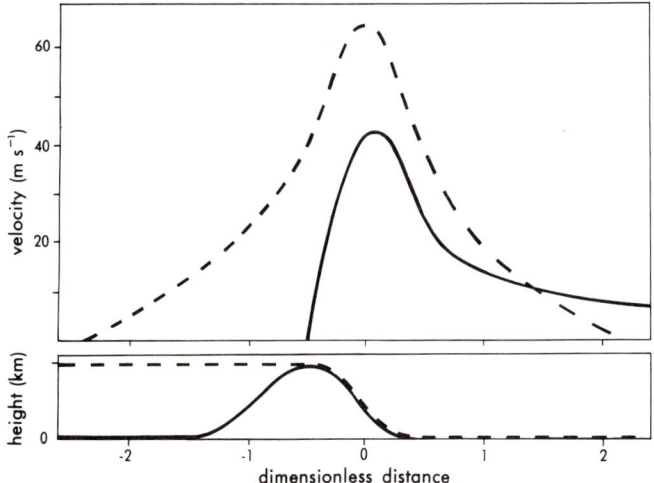

FIG. 51. Perturbation surface velocity profiles for two different mountain shapes. The solid line is a mountain contour represented by one cycle of a sine wave; the dashed line a ramp-shaped mountain with height gradually dropping off to zero for upstream. (After Klemp and Lilly, 1975.)

Klemp and Lilly concluded their investigation of the dynamics of wave-induced downslope winds with a comparison of their results with those produced by hydraulic jump models. In these numerical experiments the mountain height was 0.5 km and the undisturbed atmosphere was characterized by a constant mean wind of $20\,\mathrm{m\,s^{-1}}$ and a 10 K increase in potential temperature across the inversion. The Froude number was set at unity, which then required the top of the inversion to be at 1.17 km. Under these conditions integration of the shallow-water equations produced both upstream and downstream propagating jumps (Fig. 52). In comparison Klemp and Lilly obtained non-linear solutions for the same two-level structure with a non-zero stability above the inversion surface. These

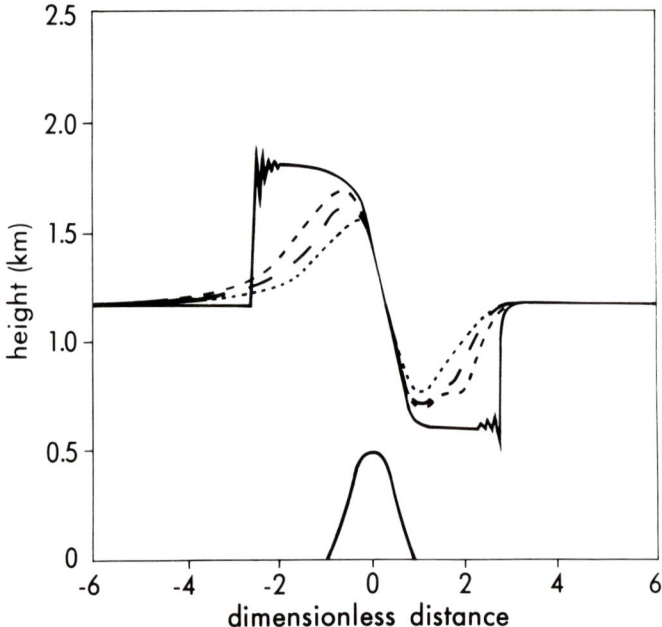

Fig. 52. Position of the inversion surface for differing stabilities above the inversion as obtained from non-linear numerical solutions. In all cases the wind speed was 20 m s^{-1}, the Froude number equal to unity and there was a 10 K increase in potential temperature across the inversion interface with neutral stability below. The temperature lapse rates (in kelvins per kilometre) above the inversion were as follows: solid (neutral stability), 9.8; light dashed, 9; heavy, long dashed, 8; heavy short dashed, 6. (After Klemp and Lilly, 1975.)

solutions were generated by a non-linear wave model using various subadiabatic lapse rates above the inversion, viz. 9, 8 and 6 K km^{-1}. Figure 52 clearly shows that as the stability above the inversion increases, the solution begins to deviate significantly from the jump solution.

Klemp and Lilly stressed that their results did not imply that hydraulic jumps do not occur in the atmosphere, nor that strong wave and downslope winds do not show some aspects of jump behaviour. Their words provide a suitable conclusion to this section.

> Rather, it is felt that hydraulic jump analyses require idealized simplifications of the atmosphere which may produce artificial or misleading results. The analysis presented here emphasizes the sensitivity of the nature of the wave motion to the detailed structure of the atmosphere and indicates that care must be taken to include its essential features in modelling the mountain wave phenomenon. In particular, our theoretical and observational results emphasize the importance of the entire troposphere and lower stratosphere and the vertical propagation of wave energy in producing strong wave response (Klemp and Lilly, 1975, p. 338).

IV. Hardware models

Laboratory experiments of downslope winds have not been conducted in isolation: they have usually been a part of a more general investigation of airflows over obstacles. Some results of these experiments are of relevance to lee waves and these have been reviewed in Chapter 2. In the current context the focus of interest is the flow over the leeward slope of the obstacle rather than waves to its lee. Such a particular aim was rarely the objective or the achievement of either the earlier workers such as Abe (1932, 1942) or the later modellers such as Suzuki and Yabuki (1956) and Long (1953, 1954). In these later studies Long clearly simulated several aspects of airflow over mountains with the aid of a moving obstacle in a channel containing fluid comprising three layers of different density. Using an internal Froude number,

$$\text{Fr}_i = U \bigg/ \left(g \frac{\Delta \rho}{\bar{\rho}} D\right)^{1/2},$$

where U is the speed of the current, $\bar{\rho}$ is the mean density of the three fluids, $\Delta \rho$ is the density difference between the top and bottom layers and D is the total depth of the fluids, Long simulated four types of flow for values of Fr_i equal to 0.13, 0.14, 0.17 and 0.19. In the last three runs a hydraulic jump was created, its amplitude increasing with the values of Fr_i. Long (1953) tentatively suggested that such jumps may be the explanation for the "Bishop wave" phenomenon in California but felt then, as we know now, that the jump fails to explain all the relevant facets of the "Bishop wave". Long's (1954) later paper confirmed the frequent occurrence of the hydraulic jump, but also managed to simulate lee waves, and apparently more important for downslope winds, an absolutely supercritical flow over the obstacle. In this latter situation flow closely followed the shape of the obstacle and contained no other waves or jumps. We should note, however, that this required a wind speed of 92 m s^{-1} and so is a type of flow unlikely to be found over large areas and for any length of time in the real atmosphere. In all the runs the ratio h_0/D (where h_0 is the initial height of the lower layer upstream) was equal to 0.33, and the ratio b/D (where b is the maximum height of the obstacle) ranged from 0.067 for the lee wave case to 0.205 for subcritical jump and supercritical flows. The variation in the last three flow types was due to Fr_i values to 0.048, 0.220 and 0.583 respectively.

In common with other branches of meteorology the introduction of the computer and associated numerical modelling techniques has probably prevented a proliferation of hardware experiments. Valuable as Long's experiments were to our general understanding of airflow over mountains, a more particular knowledge of downslope winds no doubt awaits more complicated non-linear numerical models.

References

Abe, M. (1932). The formation of cloud by the obstruction of Mount Fuji. *Geophys. Mag.*, **6**, 1–10.
Abe, M. (1942). An attempt to make visible the mountain air current. *J. met. Soc. Japan*, **20**, 69–76.
Arakawa, S. (1968). A proposed mechanism of fall winds and Dashikaze. *Met. Geophys.*, **19**, 69–99.
Arakawa, S. (1969). Climatological and dynamical studies on the local strong winds, mainly in Hokkaido, Japan. *Geophys. Mag.*, **34**, 349–425.
Barry, R. E. and Perry, A. H. (1973). *Synoptic Climatology; Methods and Applications*, Methuen, London.
Becker, R. (1948). Die Winde der Erde mit Eigennamen. *Wetter Klima*, **1**, 358–371.
Beran, D. W. (1967). Large amplitude lee waves and chinook winds. *J. appl. Met.*, **6**, 865–877.
Bilwiller, R. (1899). Uber verschiedene Entstehungsarten und Erscheinungsformen des Föhns. *Met. Z.*, **16**, 204–215.
Brinkmann, W. A. R. (1970). The Chinook at Calgary (Canada). *Arch. Met. Geophys. Bioklim. B*, **18**, 269–278.
Brinkmann, W. A. R. (1971). What is a foehn? *Weather*, **26**, 230–239.
Brinkmann, W. A. R. (1974). Strong downslope winds at Boulder, Colorado. *Mon. Weath. Rev.*, **102**, 592–602.
Brinkmann, W. and Ashwell, I. Y. (1968). The structure and movement of the chinook in Alberta. *Atmosphere*, **6**, 1–10.
Conrad, V. (1936). Die klimatologischen Elemente und ihre Abhängigkeit von terrestrischen Einflüssen. In *Handbuch der Klimatologie* (W. Köppen and R. Greiger, eds), Gebrüder Bornträger, Berlin.
Ficker, H. von (1905). Innsbrucker Föhnstudien I. *Denkschr. Wien Akad. Wissenschaften*, **78**, 83–163.
Ficker, H. von and Rudder, B. de (1943). *Föhn und Fohnwirkungen der gegenwartige Stand der Frage*, Leipzig, Akademische Verlags.
Flohn, H. (1941). Häufigkeit, Andauer und Eigenschaften des "freien Föhns". *Beit. Phys. freien Atmos.*, **27**, 110–124.
Frey, K. (1957). Zur Diagnose des Foehns. *Met.-Rund.*, **10**, 181–185.
Glenn, C. L. (1961). The Chinook. *Weatherwise*, **14**, 175–182.
Grober, K. W. (1948). Aerologische Beobachtungen über Bora und Scirocco in Sibenik an der dalmatinischen Küste. *Z. Met.*, **2**, 145–148.
Hamann, R. R. (1943). The remarkable temperature fluctuations in the Black Hills Region. *Mon. Weath. Rev.*, **71**, 29–32.
Hann, J. V. (1866). Zur Frage über den Ursprung des Föhn. *Z. öst. ges. Met.*, **1**, 257–263.
Hann, J. (1885). Einige Bemerkungen zur Entwicklungs-Geschichte der Ansichten über den Ursprung des Fohn. *Met. Z.*, **2**, 393–399.
Houghton, D. D. and Isaacson, E. (1968). Mountain winds. *Stud. numer. Anal.*, **2**, 21–52.
Houghton, D. D. and Kasahara, A. (1968). Non-linear shallow fluid flow over an isolated ridge. *Commun. pure appl. Math.*, **21**, 1–23.
Jedina, R. (1892). Die Teildepression des Mittelmeeres und die Borastürme Trieste. *Met. Z.*, **9**, 344–345.
Julian, L. T. and Julian, P. R. (1969). Boulder's winds. *Weatherwise*, **22**, 108–109.
Klemp, J. B. and Lilly, D. K. (1975). The dynamics of wave-induced downslope winds. *J. atmos. Sci.*, **32**, 320–339.

Klemp, J. B. and Lilly, D. K. (1978). Numerical simulation of hydrostatic mountain waves. *J. atmos. Sci.*, **35**, 78–107.
Koutnik, W. (1968). Newhall winds of the San Fernando Valley. *Weatherwise*, **21**, 186–189.
Küttner, J. (1939). Moazagotl und Föhnwelle. *Beit. Phys. freien Atmos.*, **25**, 79–114.
Küttner, J. (1958). The rotor flow in the lee of mountains. *Schweiz. Aero.-Rev.*, **33**, 208–215.
Lester, P. F. (1976). Evidence for long lee waves in southern Alberta. *Atmosphere*, **14**, 28–36.
Lester, P. F. and MacPherson, J. I. (1977). Waves and turbulence in the vicinity of a chinook arch cloud. *Mon. Weath. Rev.*, **105**, 1447–1457.
Lilly, D. K. (1978). A severe downslope windstorm and aircraft turbulence event induced by a mountain wave. *J. atmos. Sci.*, **35**, 59–77.
Lilly, D. K. and Zipser, E. J. (1972). The Front Range windstorm of 11 January 1972. *Weatherwise*, **25**, 56–63.
Lockwood, J. G. (1962). Occurrence of föhn winds in the British Isles. *Met. Mag.*, **91**, 57–65.
Long, R. R. (1953). A laboratory model resembling the "Bishop wave" phenomenon. *Bull. Am. met. Soc.* **34**, 205–211.
Long, R. R. (1954). Some aspects of the flow of stratified fluids. II. Experiments with a two-fluid system. *Tellus*, **6**, 97–115.
Long, R. R. (1955). Some aspects of the flow of stratified fluids. III. Continuous density gradients. *Tellus*, **7**, 341–357.
Math, F. A. (1934). A battle of the Chinook at Havre, Montana. *Mon. Weath. Rev.*, **62**, 54–56.
Mazelle, E. (1907). Kälteeinbruch und Bora in Trieste, Januar 1907. *Met. Z.*, **24**, 171–173.
Obenland, E. (1956). Untersuchungen zur Foehnstatistik des Oberallgäus. *Ber. dt. Wetterdienstes* no. 23.
Osmond, H. W. (1941). The Chinook wind east of the Canadian Rockies. *Can. J. Res.*, **19**, 57–66.
Paradiz, B. (1957). Burja v Sloveniji. *Hidrometeorološki Zavod Ir Slovenije 10 Let Hidrometeorološke Sluzke*, Ljubljana, 147–172.
Poje, D. (1962). Ein Beitrag zur Aerologie der Bora über der Adria. *VI. Int. Tagung F. Alpine Met. Bled. Jugoslawien 14–16 Sept 1960.* Fed. Hydro-Met. Inst. d. F. Yugoslawien, Beograd, pp. 371–383.
Potapov, N. S. (1961). Bora na iuzhnom beregu Kryma. *Priroda*, **11**, 81–82.
Riehl, H. (1971). An unusual chinook case. *Weather*, **26**, 241–246.
Riehl, H. (1974). On the climatology and mechanisms of Colorado Chinook winds. *Met. Inst. Bonn. met. Abh.*, **17**, 493–504.
Schmitt, W. (1930). Föhn Erscheinungen und Föhngebiete. *Deutscher und Osterreichischer Alpenuerein, Wissenschaft Liche Veröffentlichungen* no. 8.
Schuetz, J. and Steinhauser, F. (1955). Neue Foehnuntersuchungen aus dem Sonnenblick. *Arch. Met. Geophys. Bioklim. B*, **6**, 207–224.
Schweitzer, H. (1953). Versuch einer Erklärung des Föhns als Luftströmung mit überkritischer Geschwindigkeit. *Arch. Met. Geophys. Bioklim. A*, **5**, 350–371.
Scorer, R. S. and Klieforth, H. (1959). Theory of mountain waves of large amplitude. *Q. Jl R. met. Soc.*, **85**, 131–143.
Stepanova, N. A. (1951). Selective annotated bibliography on special winds. *Met. Abstr. Bibliogr.*, **2**, 586–629.
Suzuki, S. and Yabuki, K. (1956). The airflow crossing over the mountain range. *Geophys. Mag.*, **27**, 273–291.

Tamiya, H. (1972). Chronology of pressure patterns with bora on the Adriatic Coast. *Climatol. Notes*, **10**, 52–63.

Ungeheuer, H. (1952). Zur statistik des Foehns im Voral—pengebiet. *Ber. dt. Wetter dienstes US-Zone*, no. 38, 117–120.

Yoshimura, M. (1972). Chronology of the bora-day at Ajdovscina and Trieste. *Climatol. Notes*, **10**, 64–78.

Yoshimura, M., Nakamura, K. and Yoshino, M. M. (1976). Local climatological observation of Bora in the Senj region on the Croation Coast. In *Local Wind Bora* (M. M. Yoshino, ed.), Tokyo University Press, Tokyo, pp. 21–40.

Yoshino, M. M. (1969). Synoptic and local climatological study on bora in Yugoslavia. *Geogr. Rev. Japan*, **42**, 747–761.

Yoshino, M. M. (1971). Die Bora in Jugoslawien; Eine Synoptisch-klimatologische Betrachtung. *Annln Met.*, **5**, 117–121.

Yoshino, M. M. (1972). An annotated bibliography on bora. *Climatol. Notes*, **10**, 1–22.

Yoshino, M. M. (ed.) (1976a). *Local Wind Bora*, University of Tokyo Press, Tokyo.

Yoshino, M. M. (1976b). Bora in Trieste, Italy. In *Local Wind Bora* (M. M. Yoshino, ed.), Tokyo University Press, Tokyo, pp. 127–134.

4

Circulations in Wakes

I. Introduction

Orography affects airflow in a multitude of ways. In addition to the generation of such features as lee waves and downslope winds, upland areas frequently cause easily identified wakes. Huschke (1959, p. 617) defines a wake as follows: "The region of turbulence immediately to the rear of a solid body in motion relative to a fluid". He goes on to note that, "under certain conditions a series of vortices may form in the wake and extend downstream; such a vortex train in a turbulent wake is called a vortex street". In addition to such vortices, meso-scale low pressure areas, not necessarily having cyclonic circulation, have also been observed in wakes, taken in a broad sense, and as such they are, of course, particular types of the familiar "lee lows". Much of our knowledge of lee lows is derived from study of synoptic scale circulations such as those that quite frequently form to the lee of the Rocky Mountains and the European Alps, particularly over the Gulf of Genoa. Observation and analysis of much smaller lows are very sparse in the literature, but some coverage is attempted in this chapter. In contrast to these meso-scale lee lows, lee vortices, usually in the form of vortex streets, have been quite fully investigated over the past two decades (Chopra, 1972, 1973). Virtually all the observational evidence came from satellite photographs of stratiform cloud distribution to the lee of oceanic islands. It appears that their explanation lies in the classical von Kármán theory of some 70 years' standing. These aspects are reviewed below.

II. Observation

A. Meso-scale lee lows

Aanensen (1965) has provided an excellent analysis of meso-scale lows to the lee of the Pennines in the United Kingdom. Figure 53 shows three such features within a trough of low pressure extending north–south on the east side of the Pennines. A further small low pressure centre lay over Flintshire to the east of the high ground in North Wales. In association with the low pressure to the lee of the Pennines was a complementary small centre of high pressure to the west of the Pennines. Aanensen concluded that the lee trough

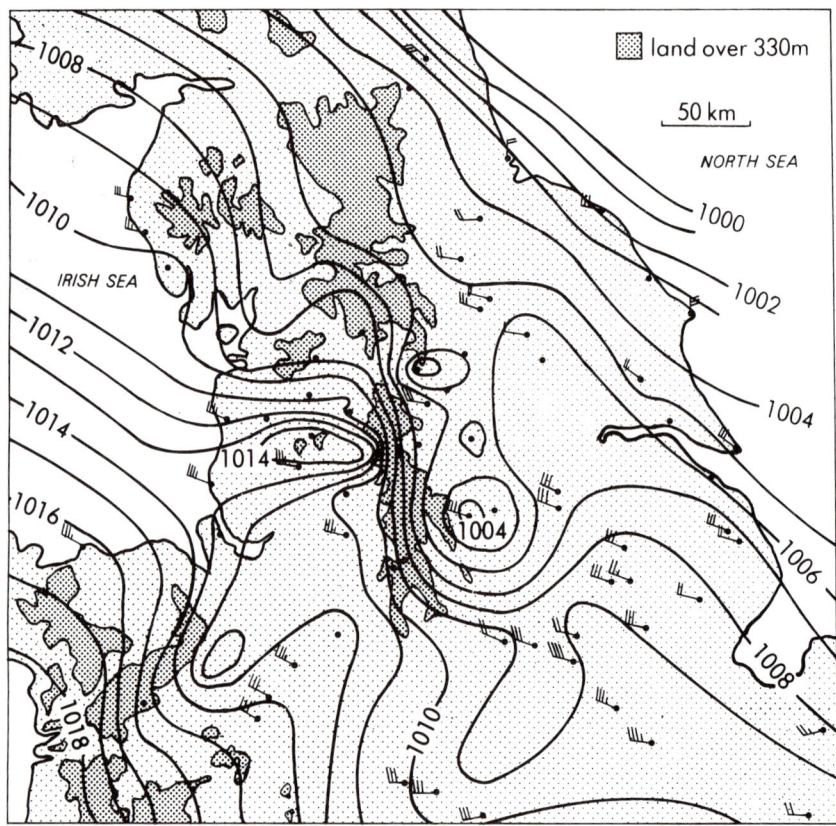

Fig. 53. Surface pressure at 0900 GMT, 16 February 1962, in northern England. Leeward lows and windward high are clearly visible on eastern and western sides respectively of the Pennine hills. (After Aanensen, 1965.)

persisted for as long as the very strong winds of that day lasted (about 24 h) but that within the trough some of the "meso lows" occasionally broke away from the trough and moved downstream, to be replaced by another meso low for the process to repeat itself. It is clear from Aanensen's evidence that the wind and pressure fields in these meso lows were not in balance. Within the lows there is evidence of neither down- nor along-gradient flow, but the latter was to be expected as the systems were too small to experience coriolis effects. In addition, the horizontal pressure gradient between the upwind "meso-high" and the lee "meso-lows" was of the order of 1 mb per 3 km. For such a pressure gradient, the equilibrium geostrophic wind would have to be about $180\,\mathrm{m\,s}^{-1}$ from a northerly direction. Clearly, as winds of this strength were not experienced, geostrophic balance was not reached, probably, in

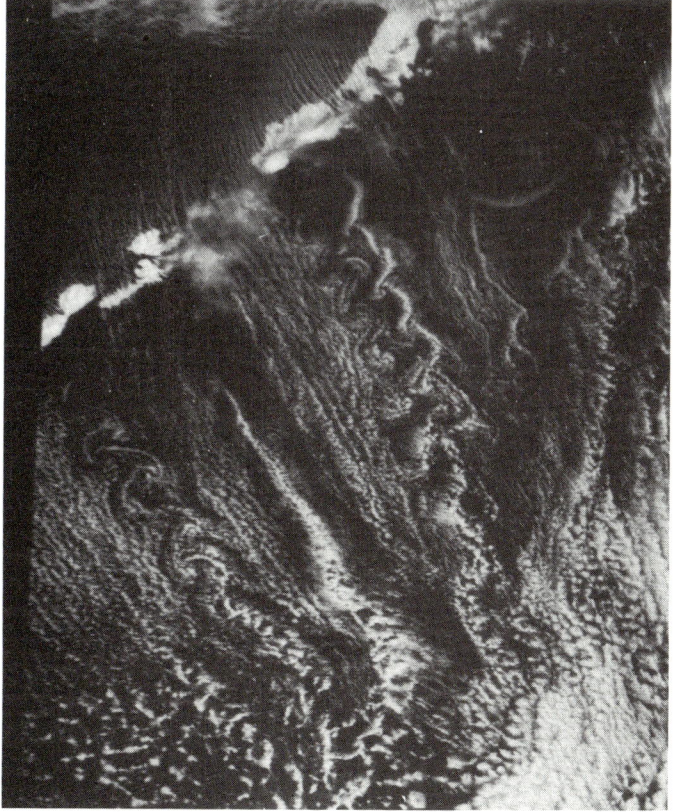

FIG. 54. Lee vortices revealed on a portion of a NOAA 4 satellite VHRR (visible) image taken at 2006 GMT, 5 April 1976, over the north-eastern Pacific Ocean. The Alaskan Peninsula and eastern Aleutian Islands appear in the upper left-hand corner. (Courtesy of MacDonald, Dettwiler and Associates Ltd.)

Aanensen's opinion, because individual air particles remained under the influence of this intense pressure gradient for only a very short time and were therefore moving under the influence of the large scale general pressure gradient. These interesting results suggest that further examination is desirable, but lack of suitable data is a severe hindrance.

B. Meso-scale lee vortices

One of the first rewards of the observational capacities of meteorological satellites was the regular surveillance of cloud patterns over hitherto poorly observed oceanic areas. Many fascinating new patterns emerged, one of the most striking being the vortex streets downwind of small frequently isolated islands (Fig. 54). Although Bowley *et al.* (1962) provided early satellite evidence of wakes under low level inversions, it was probably Hubert and Krueger (1962) who first noticed that such wakes often comprised quite definite vortices. It was not long before the increasing quantity and quality of satellite photographs encouraged a fuller examination of these features by Chopra and Hubert (1964, 1965), Lyons and Fujita (1968), Tsuchiya (1969) and Zimmerman (1969). The vortex patterns consist of two roughly parallel rows of vortices such that a vortex in one row is situated across the mid-point of the two adjacent vortices in the other row (Fig. 55(a)). When first shed the vortices are comparable in diameter to those of the islands, usually about 40 km. As they move downstream they are spaced 50–100 km apart longitudinally (distance a in Fig. 55(a)) and 30–50 km laterally (distance h in Fig. 55(a)). They tend to increase in diameter as they move downwind and may persist to form a wake 100 km wide and 400–600 km long. A pair of vortices is shed by the island about once every 8 h and may last as long as 30 h. All the vortices in one row have similar circulation, but in a sense opposite to that in the other row. In his study of vortices to the lee of Cheju Island Tsuchiya (1969) found that their displacement speed was about three-quarters that of the general airstream speed. Tangential velocities within the vortices were about 3 m s^{-1}, or about one-third of the undisturbed flow speed and vorticity was $2.5 \times 10^{-4} \text{ s}^{-1}$. Table 10 summarizes the major characteristics of vortices shed by Madeira and several of the Aleutian and Canary Islands.

Atmospheric conditions associated with the vortex streets are typically as follows: strato-cumulus clouds identifying the vortex exist just below a low level inversion from 450 to 2000 m above the ocean; the tops of the islands are well above the inversion layer; the general winds are steady, about 10 m s^{-1}; the vortices tend to occur in areas with nearly straight surface isobars over a large area. The height of the inversion is particularly important because it means that air goes around the obstacle more readily than over it. Clear areas

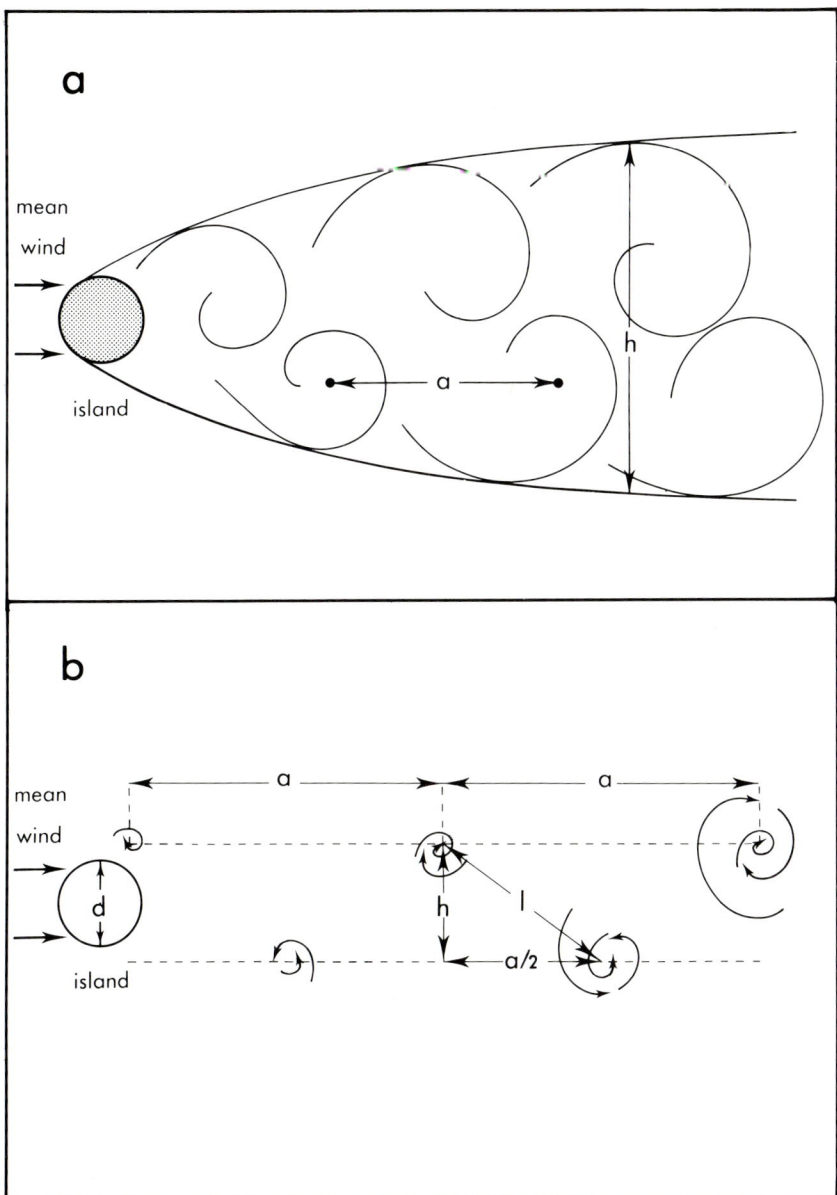

FIG. 55. (a) Schematic diagram of atmospheric vortices to the lee of an isolated island. The vortices diverge as they become larger further downstream in the wake. (b) Kármán vortex street produced by an obstacle of diameter d in two-dimensional flow. (After Chopra and Hubert, 1975.)

Table 10 Observed vortex characteristics

Location	Height (m)	Effective width (km)	Longitudinal spacing (km)	Distance between rows (km)	Total length of street (km)	Number of vortices	Undisturbed wind speed (m s^{-1})
Pavlof Volcano (Alaskan Peninsula)[†]	2518	18	65	25	370	2	10
Shishaldin Volcano (Unimak Island)[†]	2857	32	75	35	630	7	10
Pogromni Volcano (Unimak Island)[†]	2002	13	50	30	370	6	10
Mount Vsevidof (Unimak Island)[†]	2109	17	75	35	390	5	10
Kiska Island[†]	1216	18	85	25	475	3	10
Tenerife Island[‡]	3720	40	122	48	—	—	7
Gran Canaria Island[‡]	1950	20	132	51	—	—	7
Madeira Island[‡]	1860	40	217	85	—	—	7

[†] *Source:* Thomson et al. (1977).
[‡] *Source:* Zimmerman (1969).

in the cloudy wakes were attributed by Lyons and Fujita (1968) to the downward entrainment of drier air from the inversion.

III. Theory

A. Meso-scale lee lows

Lee lows and troughs in general, that is, regardless of size, tend to be explained in terms of the conservation of vorticity. With suitable modification the equation of vorticity conservation for a column of air of depth D may be written as

$$\frac{d}{dt}\left(\frac{f+\zeta}{D}\right) = 0,$$

where ζ is the relative vorticity about the vertical. This equation may profitably be applied to a consideration of flow over and to the lee of mountains. To illustrate the point we may assume a large scale, straight, uniform, westerly current of zero relative vorticity impinging upon a north–south mountain range. In order to simplify the discussion we also assume that any vorticity that may appear will do so as curvative rather than as shear. Furthermore, we assume that the vertical perturbation of the air as it moves over the mountain decreases in magnitude with elevation. Thus there will be some elevation above which no perceptible effect of the mountain occurs. Although this assumption means gross simplification of the problem, it serves to elucidate one of the essential features of the application of the vorticity theorem to flow over orography.

Figure 56 illustrates the behaviour of the current outlined above. On the upwind side of the mountain there are no changes in either depth or latitude to cause the airstream to curve. When the air begins to rise over the mountain the depth begins to decrease, causing anticyclonic curvature to appear. The depth continues to decrease until the top of the barrier has been reached. At this point the current has its maximum anticyclonic curvature because there it has minimum depth. During the descent on the leeward side the depth gradually increases again while the vorticity returns to zero. If the divergence effect alone is considered, the current should reach the leeward plain with the same relative vorticity it had to begin with, namely zero. It would thus be a straight, uniform current, but the mountain's influence would have turned it from a westerly to a north-westerly current. However, the current is travelling towards decreasing latitude during the descent and thus it arrives at the foot of the barrier with some cyclonic curvature due to the latitudinal variation of the coriolis parameter. Once past the mountain the air is controlled solely by the latitude effect since depth no longer changes.

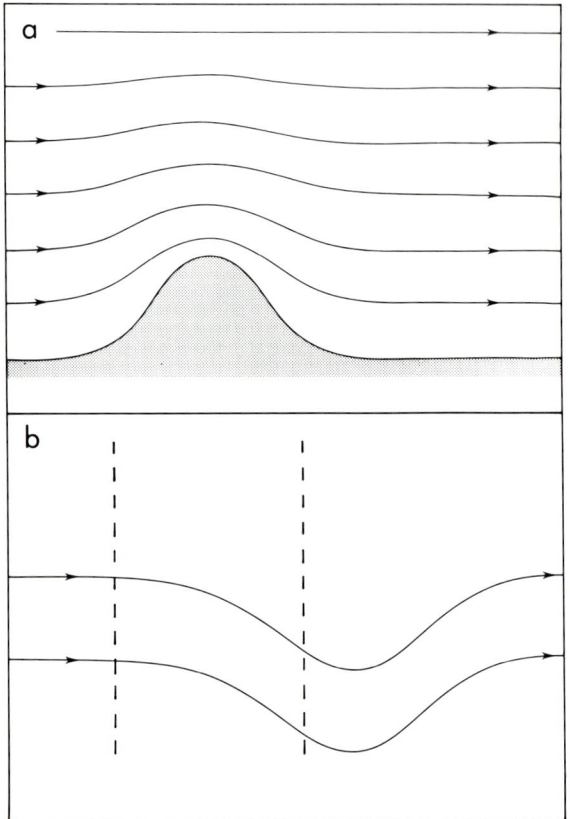

FIG. 56. Vertical cross-section (a) and plan view (b) of streamlines of an initially zonal current crossing a large mountain range in the northern hemisphere. (After Hess, 1959.)

Such a mechanism clearly operates on mountain ranges of both substantial height and latitudinal extent. It no doubt accounts for lee troughs of the type described in the first part of the chapter. Its relevance to the "meso-lows" and "meso-highs" is less clear. We noted earlier that the flow in these systems was not in geostrophic balance and consequently could not behave in accord with the simplified vorticity equation outlined above. In addition we noted that the "meso lows" appeared to detach themselves from the trough and move downstream. Such behaviour is akin to that of the shedding of lee vortices covered in Section I.B, but it is probably dangerous to pursue that comparison in the absence of dynamical understanding.

In common with synoptic scale pressure systems the origins of the meso lows and highs no doubt lie in ageostrophic motions. On the upwind side of the obstacle a low level velocity and mass convergence must occur due to deceleration of the flow, resulting in the meso-scale high pressure noted earlier. On the downwind side of the obstacle, accelerating flow would lead to low level velocity and mass divergence resulting in the meso lows. It is interesting to contrast this process with that outlined by Bjerknes and Holmboe (1944) for the origin of synoptic scale depressions. In that case it is high level divergence in the jet stream which is the primary cause of surface low pressure: high level convergence in the upwind parts of the Rossby waves results in surface high pressures.

B. Meso-scale lee vortices

It did not take scientists long, after first sight of satellite photographs such as Fig. 54, to see the striking resemblance of the quasi-two-dimensional atmospheric cloud patterns to the vortices produced in the wakes of bluff bodies in the laboratory. The formation of vortex streets associated with periodical shedding of vorticity into a wake was first described by Strouhal in 1878. But it was not until von Kármán's (1911) theoretical formulation that the subject received active study in the laboratory. The term "Kármán vortex street" has subsequently been adopted for these patterns despite the fact that von Kármán's work originally dealt with only the idealized and readily parameterized arrangement of point vortices generated by initially two-dimensional potential flow incident upon a parallel cylinder (Fig. 55(b)).

Chopra and Hubert (1965, p. 652) were among the first "to examine the properties of (atmospheric) eddies in the light of their resemblance to the classical pattern, taking advantage of the relationship between the pattern geometry and various parameters derived from drag theory". They noted that a vortex street develops behind an obstacle when the flow in the wake region does not mix with that in the surrounding region. The transfer of momentum between the wake and the basic flow region is minimized, thereby contributing to the long lifetime and persistence of the street downstream. Within the wake vorticity is generated in the boundary layer near the vertical sides of the obstacle, is added to the vortex pair shed at points of separation in the flow and is diffused from the vortex layers into the main body of the fluid. In suitable conditions the rates of generation and diffusion of vorticity are equal and the equilibrium is maintained through an appropriate elongation of the vortices along the direction of the flow. Chopra and Hubert (1965) were also of the opinion that the viscosity that diffuses momentum horizontally plays two roles in vortex streets: first, it leads to the formation of the boundary layer in which vorticity is generated and the vortex pairs are formed, the vortices being shed alternately near each edge of the obstacle; second, the strength of the eddies is spread horizontally by the mechanism of horizontal diffusion. The size of an eddy or vortex increases and its strength near the centre decreases as it propagates downstream until its region of influence overlaps and counteracts that of a neighbouring vortex of opposite circulation. At this stage the vortex street destroys itself.

Laboratory analysis of Kármán vortex streets in the wakes of cylinders has provided several descriptive parameters, including the ratio h/a, the Reynolds (Re) and Strouhal (St) numbers, the Lin parameter (β), the eddy viscosity (K^M), the propagation speed (u_e) and the period of vortex formation (T).

According to von Kármán's theoretical formulation, $h/a = 0.2805$ for neutrally stable dimensions. In actual laboratory situations, however, this ratio is dependent upon the shape of the obstacle, the characteristics of the flow and the distance along the wake and is observed to lie in the range $0.28 < h/a < 0.52$. If the Reynolds number Re $= Ud/v$, where U is the undisturbed airflow speed and d is the diameter of the obstacle, in laboratory flow falls within the approximate range $40 \lesssim$ Re $\lesssim 200$, the vortex street behind a parallel cylinder is laminar and stable. For geophysical scale motions, determination of the Reynolds number Re $= Ud/K^M$ requires the independent measure of the eddy viscosity K^M. Thomson et al. (1977) suggest that since observed values of K^M are found to vary over many orders of magnitude, the usual procedure in the case of meso-scale vortex streets is to obtain an indirect estimate of K^M via the Strouhal number St $= jd/U$, and Lin's (1959) parameter $\beta =$ St/Re $= vj/U^2$. This latter ratio is independent of the size of the obstacle and β is thus an inherent property of the vortex street. In laboratory experiments with circular cylinders, the values of β range from 10^{-3} to 2.5×10^{-3} whenever a stable vortex street is clearly discernible. From the above, $K^M = \beta U^2/j$, where each item on the right-hand side is known. Experimental values of eddy viscosity have been found to range from about $10\,\text{m}^2\,\text{s}^{-1}$ to $10^5\,\text{m}^2\,\text{s}^{-1}$ (Heffter, 1965). For Reynolds numbers in the range $40 \lesssim$ Re $\lesssim 200$ the Strouhal number for laboratory models is $0.12 \lesssim$ St $\lesssim 0.19$.

The ratio of the vortex propagation speed (u_e) to that of the undisturbed air speed (U) was found in laboratory experiments (Birkhoff and Zarantonello, 1957) to be 0.85. The period of vortex formation (T) is given by $T = a/u_e$ and the rate of shedding (j) by $j = 1/T$.

In their analyses of atmospheric meso-scale vortex streets, Chopra and Hubert (1965), Zimmerman (1969), Tsuchiya (1969) and Thomson et al. (1977) calculated the values of all the above parameters, with the exception of β. In the cases of the propagation speed and period of pair formation they all assumed that the former was 0.75 of the undisturbed airspeed. This assumption in turn affected the formation period. Table 11 shows their values for various islands and in all cases the calculations were a good estimation of the actual values, strongly suggesting that, at least in the case of these two attributes, the atmospheric vortices were behaving in accord with that of Kármán vortices. Strong supporting evidence comes from the remainder of

Table 11 Calculated values that characterize atmospheric meso-scale vortex streets and associated flow

Location	h/a	Propagation speed ($m\,s^{-1}$)	Period of vortex pair formation ($=1/j$, where j is frequency of formation) (h)	Eddy viscosity ($m^2\,s^{-1}$)	Reynolds number	Strouhal number
Pavlof Volcano[†]	0.38	7.5	2.4	1.5×10^3	120	0.21
Shishaldin Volcano[†]	0.47	7.5	2.8	1.8×10^3	183	0.32
Pogromni Volcano[†]	0.60	7.5	1.9	1.2×10^3	112	0.19
Mount Vsevidof[†]	0.47	7.5	2.8	1.8×10^3	97	0.17
Kiska Island[†]	0.30	7.5	3.1	1.9×10^3	100	0.16
Tenerife Island[‡]	0.39	5	6.8	1.2×10^3	210	0.21
Gran Canaria Island[‡]	0.39	5	7.3	1.3×10^3	170	0.17
Madeira Island[‡]	0.39	5	12.0	2.2×10^3	150	0.15
Cheju Island, Korea[§]	0.33	7	4.4	1.8×10^3	180	0.20

[†] *Source*: Thomson *et al.* (1977).
[‡] *Source*: Zimmerman (1969).
[§] *Source*: Tsuchiya (1969).

Table 11. All but one of the values of h/a lie within the Kármán range specified earlier. Similarly, the values of eddy viscosity, Reynolds and Strouhal numbers virtually all lie in the range predicted by Kármán theory. Calculation of further parameters, such as drag coefficient, circulation strength and rate of energy dissipation per unit mass, all support the above results. In summary, the bulk of the available evidence strongly suggests that meso-scale vortex streets are the atmospheric analogue of Kármán vortex streets that develop in the wake of bluff bodies in laboratory experiments.

IV. Hardware models

Hardware modelling of meso-scale circulations in wakes has concentrated upon lee vortices. It may seem somewhat ironic that atmospheric vortex streets should be the subject of hardware modelling as much of our theoretical knowledge is already derived from laboratory experiments of Kármán vortices around rough cylinders. But it is precisely because this previous laboratory work was not specifically geared to atmospheric situations that renewed efforts are being made. Barnett's (1972) study clearly illustrates the application of wind-tunnel techniques to the simulation of atmospheric vortex streets. His approach was to determine as many non-dimensional parameters as practical which describe the atmospheric conditions for vortex-street occurrence and which could be made numerically equal in a wind-tunnel flow. Geometrical similarity between a typical isolated island and the wind-tunnel model was also a factor.

Six simulation criteria were chosen: Reynolds number; Strouhal number; a low layer of air next to the Earth's surface with a moderately large negative

Table 12 Comparison of values of simulation criteria between atmosphere and wind tunnel for vortex streets

Simulation criterion	Atmosphere	Wind tunnel
Reynolds number	177.0	96.0
Lowest-layer Richardson number	−4.2	−0.1
Middle-layer Richardson number	+2.6	+0.07
Top-layer Richardson number	+9.0	+1.4
Strouhal number (observed from vortex shedding)	0.18	0.14
Physical model	Isolated island: diameter ≃ 40 km, peak > 2 km	Uniform cylinder: diameter = 0.95 cm, vertical

Source: Barnett (1972).

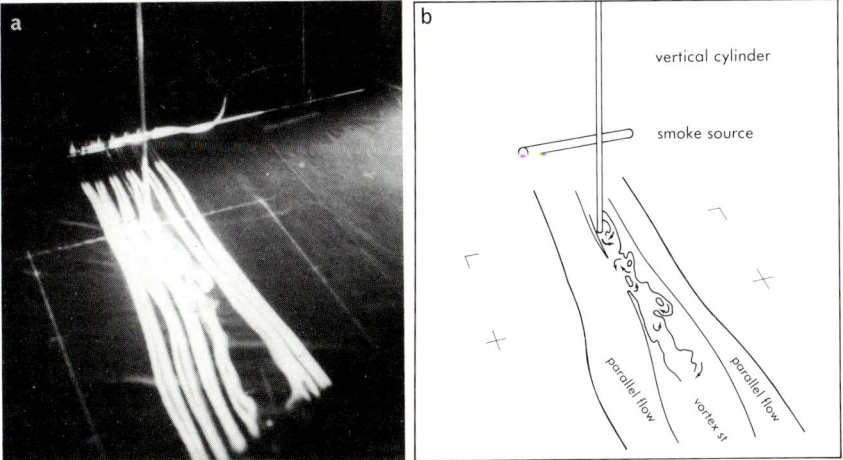

FIG. 57. (a) Photograph of vortex street in the lee of a vertical cylinder in a thermally stratified wind-tunnel flow. (b) Major features of Fig. 57(a). The cylinder is 0.95 cm in diameter. (After Barnett, 1972.)

Richardson number; a middle layer of air with a small positive Richardson number and the vortex street near its top; a top layer of air with a larger positive Richardson number; and a physical model of the island which was geometrically similar to the real thing and extended vertically upward through all three layers.

By clever use of the wind tunnel, Barnett produced a flow creditably comparable to that observed in the atmosphere. Table 12 shows the values of the six simulation criteria in both model and atmosphere. The model Strouhal number is slightly less than in the atmosphere and the Reynolds number is just over half the average value taken by Barnett from earlier observational studies. As regards the Richardson numbers, Barnett (1972, p. 439) noted that the values in the three layers "compare well between the atmosphere and the tunnel, in a qualitative sense". Figure 57 shows the result of one of the simulations and clearly illustrates a marked degree of success. This study showed that vortices could be produced in a flow containing fairly realistic vertical gradients of temperature and horizontal speed: these conditions are not applied in classical laboratory experiments of Kármán vortex streets. As such Barnett's experiment was that much nearer to reality.

References

Aanensen, C. J. M. (1965). Gales in Yorkshire in February 1962. *Geophysical Memoirs*, **108**, Meteorological Office, London.

Barnett, K. M. (1972). A wind-tunnel experiment concerning atmospheric vortex streets. *Boundary Layer Met.*, **2**, 427–443.

Birkhoff, G. and Zarantonello, E. H. (1957). *Jets, Wakes and Cavities*, New York, Academic Press.

Bjerknes, J. and Holmboe, J. (1944). On the theory of cyclones. *J. Met.*, **1**, 1–22.

Bowley, C. J., Glaser, A. H., Newcomb, R. J. and Wexler, R. (1962). Satellite observations of wake formation beneath an inversion. *J. atmos. Sci.*, **19**, 52–55.

Chopra, K. P. (1972). Velocity field in vortices leeward of islands. *J. atmos. Sci.*, **29**, 396–399.

Chopra, K. P. (1973). Atmospheric and oceanic flow problems introduced by islands. *Adv. Geophys.*, **16**, 297–421.

Chopra, K. and Hubert, L. F. (1964). Kármán vortex-streets in the earth's atmosphere. *Nature, Lond.*, **203**, 1341–1343.

Chopra, K. P. and Hubert, L. F. (1965). Meso-scale eddies in wake of islands. *J. atmos. Sci.*, **22**, 652–657.

Heffter, G. L. (1965). The variation of horizontal diffusion parameters and travel periods of one hour or longer. *J. appl. Met.*, **4**, 153–156.

Hess, S. L. (1959). *Introduction to Theoretical Meteorology*, Henry Holt and Co., New York.

Hubert, L. F. and Krueger, A. F. (1962). Satellite pictures of meso-scale eddies. *Mon. Weath. Rev.*, **90**, 457–463.

Huschke, R. E. (ed.) (1959). *Glossary of Meteorology*, American Meteorological Society, Boston, Mass.

Lin, C. C. (1959). On periodically oscillating wakes in the Oseen approximation. In *Studies in Fluid Mechanics Presented to R. von Mises*, Academic Press, New York, pp. 170–176.

Lyons, W. A. and Fujita, T. (1968). Meso-scale motions in oceanic stratus as revealed by satellite data. *Mon. Weath. Rev.*, **96**, 304–314.

Strouhal, V. (1878). Über eine besondere Art der Tonerregung. *Annln Phys. Chem.*, New Series **5**, 216–251.

Thomson, R. E., Gower, J. F. R. and Bowker, N. W. (1977). Vortex streets in the wake of the Aleutian Islands. *Mon. Weath. Rev.*, **105**, 873–884.

Tsuchiya, K. (1969). The clouds with the shape of Kármán vortex street in the wake of Cheju Island, Korea. *J. met. Soc. Japan*, **47**, 457–465.

von Kármán, Th. (1911). Über den Mechanismus des Widerstandes, den ein bewegter Körper in einer Flüssigkeit erfährt, Göttinger Nachrichten. *Math. Phys. Kl.*, **4**, 509–517.

Zimmerman, L. I. (1969). Atmospheric wake phenomena near the Canary Islands. *J. appl. Met.*, **8**, 896–907.

Part II

Topographically induced circulations

B. Thermally induced circulations

5
Sea/Land Breeze Circulation

I. Introduction

In coastal or lake-side areas we frequently observe in the course of a day both onshore and offshore winds. Those blowing from the sea/lake in daytime are known as sea/lake breezes and those blowing from the land at night-time as land breezes. Although long recognized by fishermen as their passport to the fishing grounds at night and a safe return home the following morning (Neumann and Partsch, 1885; Neumann, 1973), these breezes were only rarely instrumentally observed (e.g. Sherman, 1880) before the twentieth century. Throughout the last five decades fairly steady progress has been made in our understanding of the sea/land breeze. Most of the first observational studies concentrated on monitoring the surface characteristics of the winds, but by the 1930s an appreciation of the upper-air circulations appeared in the literature—primarily due to the increasingly available pilot balloon observational techniques. These observational studies were soon followed by analytical theoretical treatments in the 1940s and early 1950s. In turn, numerical theoretical studies began to appear in the late 1950s and they continue to do so.

A. The mechanism of the sea and land breeze

The broad outline of the sea-breeze mechanism has been known for many decades. Both Bjerknes' circulation theorem and the observations mentioned above pointed towards a thermally direct, diurnally reversible circulation in the vertical plane, the bottom limb of which is alternately the sea and the land breeze. Qualitatively, the mechanism is as follows (Fig. 58). In a calm, or near

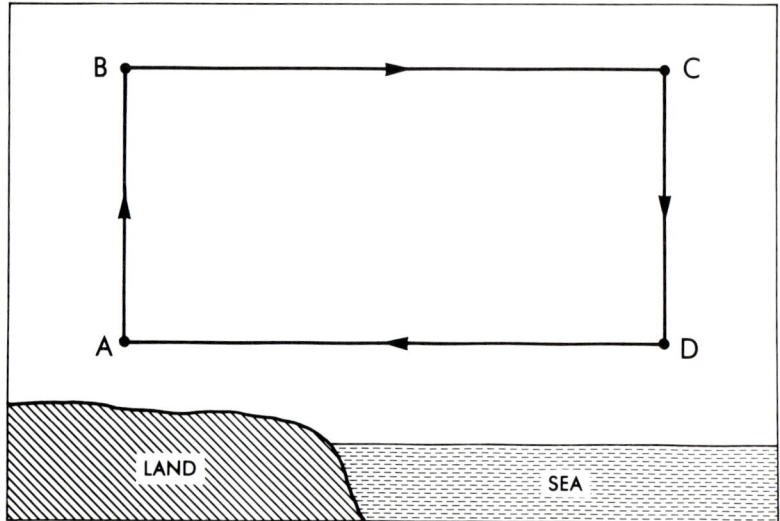

Fig. 58. Schematic circulation in a mature sea breeze.

calm, clear atmosphere, solar radiation heats up a land surface more rapidly than a water surface, causing horizontal temperature gradients of about 1 K per 20 km (Hsu, 1967). In turn the air over the land heats up and thus expands more rapidly than that over the water. Because of hydrostatic conditions, the vertical gradient of pressure is greater in the cooler air over the water than in the warmer air over the land. This means that, at a given constant level above both land and water, pressure is higher over the land than over the water. This pressure gradient (of the order of 1 mb in 50 km) produces a slight flow of air from land (B) to sea (C). Convergence near C leads to an increase of pressure so that subsidence occurs from C to D in response to the departure from hydrostatic equilibrium and flow from D to A develops because of the hydrostatic pressure gradient between D and A. This is the sea breeze. Simultaneously the divergence near B leads to a decrease of pressure there, so that flow from A to B develops in response to the departure from hydrostatic equilibrium in the vertical AB. The land breeze responds to this mechanism in reverse. At night the land cools more rapidly than the sea and a comparatively shallow layer of overlying air takes on the same thermal regime. Thus pressure is relatively high at "upper" levels over the sea and low at the same level over the land. The upper level air flow in response to this pressure gradient leads to convergence over the land, which in turn leads to an increase in pressure. Air subsides in response to the departure from hydrostatic equilibrium and flow from land to sea develops at low levels in response to the horizontal pressure gradient. To complete the circulation air rises over the sea. The land breeze circulation is not as intense as the sea breeze either in velocity or height of development, since there exists no lower heat source to carry the circulation to greater heights as with the sea breeze.

II. Observation

The clear response of the sea/lake breezes to thermally induced pressure gradients means that they are comparatively easy to identify in the observational record—particularly if the gradient wind is very light. The onset of the sea breeze is usually marked by an increase in wind speed, a fall in temperature and a rise in humidity (Fig. 59(a)). If there is a gradient wind, then the onset of the sea breeze may be slightly more difficult to detect. An onshore gradient wind masks the sea breeze whereas an offshore wind either prevents the breeze altogether or sharpens its leading edge into a "sea-breeze front". In the discussion that follows the "essence" of the sea breeze is described in the context of a zero gradient wind. The relationships between sea breeze and gradient wind are reviewed later.

In contrast to the sea/lake breeze, the land breeze barely registers on purpose-built, let alone conventional meteorological instruments. In one of the very few observational studies of the land breeze *per se*, Feit (1969) used sensitive autographic instruments together with mobile towers with sensors at four levels up to 28 m to investigate the onset and structure of the breeze. Figure 59(b) shows the onset of the breeze at 2400 hours in an otherwise calm atmosphere—the land breeze blowing from the north over the Texas coast. The directional trace suggests that the breeze blew all night. There was no noticeable temperature change at the onset of the breeze.

A. Climatology

A direct thermally driven circulation such as the sea/land breeze would be expected to occur more frequently and with greater regularity in the tropics than in the middle and high latitudes. Such is indeed the case and in fact it is comparatively easy to construct a climatology of sea and land breezes in the tropics. A similar exercise is difficult for the higher latitudes.

Many climatological studies of tropical sea/land breezes have been undertaken in India. Most authors point out that the south-west monsoon gives consistent sea breezes (in the sense that the wind blows from the sea) on the west coast of peninsular India whereas the north-east monsoon acts in similar fashion on the east coast. The climatological analyses of the sea/land breezes have thus been restricted to those seasons in which the breezes clearly blow in the opposite direction to the main monsoonal drift. In the following tables, the seasonal reversal of the monsoon is reflected in the seasonal

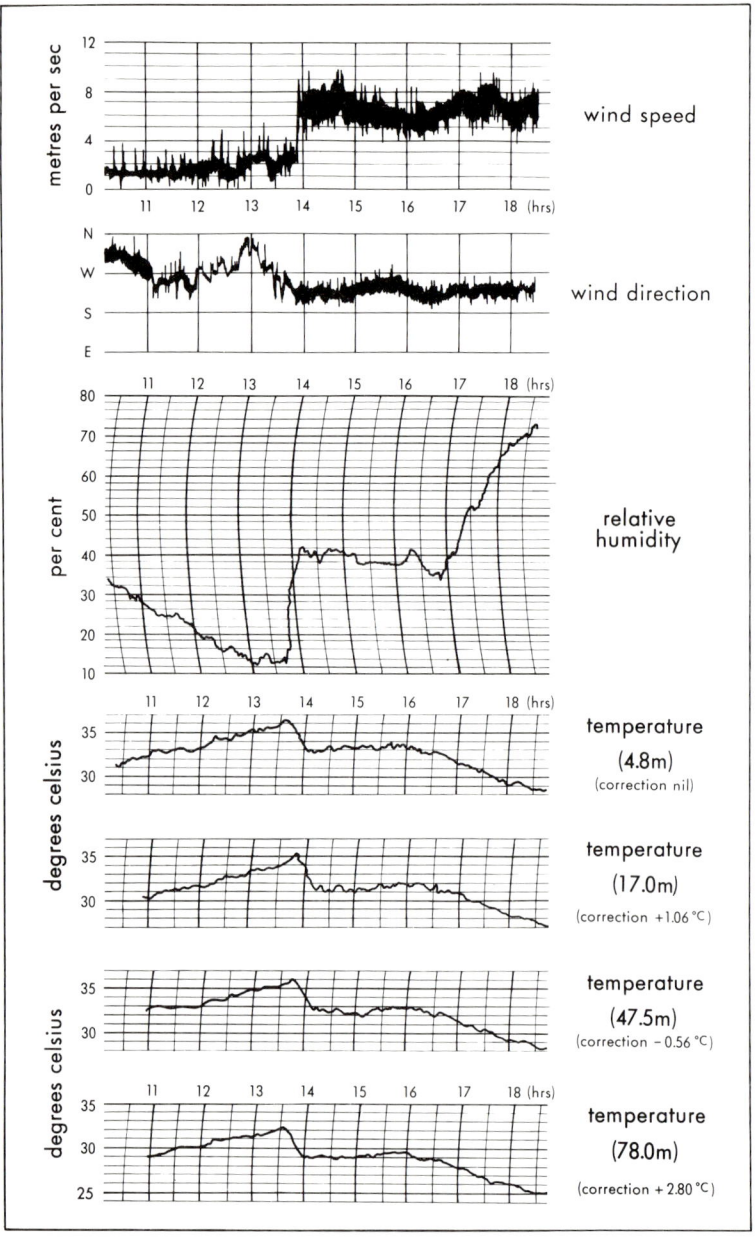

FIG. 59. (a) Onset of a sea breeze on 15 April 1930 at Karachi as shown by changes in wind speed (in metres per second) and direction, relative humidity (in per cent) and temperatures (in degrees Celsius) at four heights above the ground. (After Ramdas, 1931.)

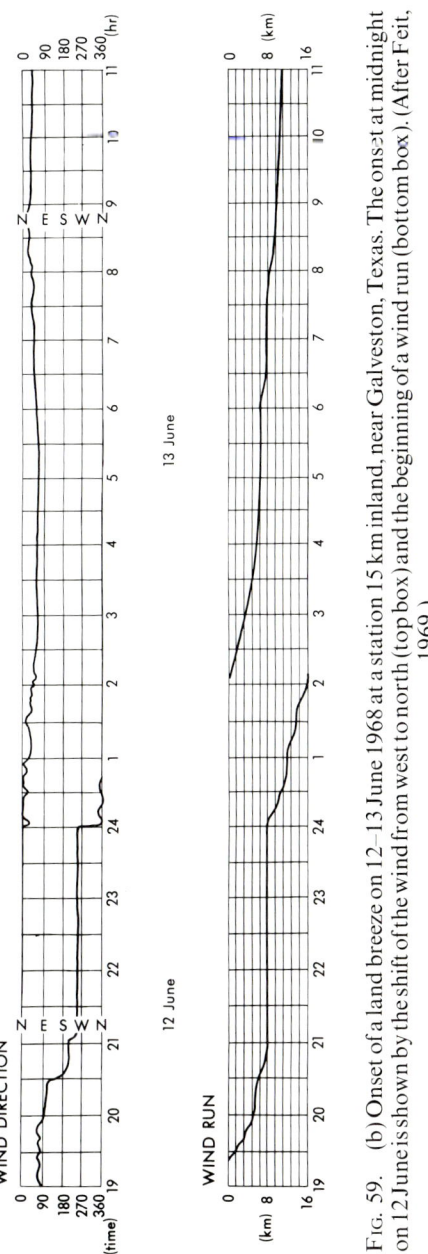

FIG. 59. (b) Onset of a land breeze on 12–13 June 1968 at a station 15 km inland, near Galveston, Texas. The onset at midnight on 12 June is shown by the shift of the wind from west to north (top box) and the beginning of a wind run (bottom box). (After Feit, 1969.)

Table 13 Monthly mean number of days with sea breezes

Location and period of observations	Month											
	Jan	Feb	Mar	Apr	May	Jun	Jul	Aug	Sep	Oct	Nov	Dec
(a) Tropics†												
Bombay (1961–64)	28	27	30	26	10	Considered to be daily or continuous				23	28	27
Thumba (1964–66)	29	24	27	16	Considered to be daily or continuous						12	21
Karachi (1929–30)	9	11	9.5	8	Considered to be daily or continuous					8	9	13
Madras (1938–40)	Considered to be daily or continuous					27	25	21	24	27	29	29
(b) Mid-latitudes												
Chicago (1966–68)	0	0	0	0	10	11	14	10	0	0	0	0
Lake Constance (1913–20)	0	4.0	4.7	5.0	6.7	5.0	6.4	9.3	7.9	3.3	2.0	0
Porton, UK (40 km inland) (1956–61)	0	0	1.0	1.1	1.3	1.0	1.5	1.0	0.7	0.2	0	0
Worthy Down, UK (65 km inland) (1927–32)	0	0	1.3	0.7	2.0	2.0	1.2	1.5	1.3	0	0	0

Sources: Bombay, Dekate (1968); Thumba, Narayanan (1967); Karachi, Ramdas (1931); Madras, Roy (1940); Chicago, Lyons (1972); Lake Constance, Kopfmüller (1922); Porton, Elliott (1964); Worthy Down, Peters (1938).

† Bombay, Karachi and Thumba are on the west coast and Madras is on the east coast of peninsular India. Reversal of the monsoon accounts for the different seasonal incidence of the definitely identifiable sea breeze.

Table 14(a) Times of onset of the sea breeze at three locations in India and Pakistan: average number of days

Local time	Location	Jan	Feb	Mar	Apr	May	Jun	Jul	Aug	Sep	Oct	Nov	Dec
Before 1000	B	0.3	0.5	1.3	7.5	5.3	—	—	—	—	0.3	0.0	0.0
Before 1100	M	—	—	—	—	2	1	0	1	1	6	—	—
1000–1059	B	2.0	1.7	8.3	11.7	3.7	—	—	—	—	3.0	0.3	1.0
1100–1159	B	5.0	9.5	14.5	5.7	0.7	—	—	—	—	6.3	1.7	2.7
	M	—	—	—	—	8	1	0	3	2	5	—	—
1200–1259	B	8.5	9.3	4.3	0.5	0.3	—	—	—	—	2.0	6.3	4.5
	M	—	—	—	—	9.5	2.0	2.5	3.5	3.0	8.5	—	—
	K	0.0	0.5	1.5	—	—	—	—	—	—	1.0	0.0	0.0
1300–1359	B	5.3	3.7	1.3	0.0	0.3	—	—	—	—	4.7	10.7	6.7
	M	—	—	—	—	6.0	5.5	3.5	4.5	5.5	1.5	—	—
	K	0.0	0.5	0.5	—	—	—	—	—	—	1.5	0.0	0.5
1400–1459	B	4.7	1.5	0.0	0.0	0.0	—	—	—	—	4.3	5.5	8.0
	M	—	—	—	—	2.0	7.0	5.0	3.5	4.0	2.5	—	—
	K	0.7	2.0	2.5	—	—	—	—	—	—	2.5	2.0	0.5
1500–1559	B	1.5	0.7	0.0	0.0	0.0	—	—	—	—	0.7	2.3	3.0
	M	—	—	—	—	1.5	5.5	4.5	1.0	2.5	1.0	—	—
	K	1.3	3.5	2.5	—	—	—	—	—	—	0.0	3.5	5.0
1600–1659	B	0.7	0.0	0.0	0.0	0.0	—	—	—	—	2.0	1.3	0.3
	M	—	—	—	—	0.0	1.5	4.0	3.5	4.0	1.0	—	—
	K	2.3	2.5	1.5	—	—	—	—	—	—	1.0	1.5	3.5
1700–1759	B	0.0	0.0	0.0	0.0	0.0	—	—	—	—	0.7	0.3	0.5
	M	—	—	—	—	1.0	1.5	2.5	1.5	1.0	0.0	—	—
	K	3.0	0.5	0.5	—	—	—	—	—	—	0.5	1.5	2.5
1800 and later	M	—	—	—	—	0.0	2.0	2.5	0.5	1.0	1.0	—	—
	K	1.7	1.5	0.5	—	—	—	—	—	—	0.5	0.5	1.0

B = Bombay. *Source:* Dekate (1968). Period of observations: 1961–64.
M = Madras. *Source:* Roy (1940). Period of observations: 1938–40.
K = Karachi. *Source:* Ramdas (1931). Period of observations: 1929–30.
Dash indicates onset not observed because breeze continuous.

Table 14(b) Times of onset of the sea breeze at Worthy Down†: actual number of days

GMT	Jan	Feb	Mar	Apr	May	Jun	Jul	Aug	Sep	Oct	Nov	Dec
1200–1259	0	0	0	0	1	0	0	0	0	0	0	0
1300–1359	0	0	1	0	1	1	0	0	0	0	0	0
1400–1459	0	0	0	0	0	0	1	0	0	0	0	0
1500–1559	0	0	2	0	2	0	2	0	0	0	0	0
1600–1659	0	0	0	1	2	3	0	4	3	0	0	0
1700–1759	0	0	2	1	2	2	2	1	2	0	0	0
1800–1859	0	0	2	1	1	2	2	1	1	0	0	0
1900–1959	0	0	1	0	1	2	0	1	2	0	0	0
2000–2059	0	0	0	0	0	1	0	0	0	0	0	0
2100–2159	0	0	0	0	0	0	0	1	0	0	0	0

† Worthy Down is 65 km inland. *Source*: Peters (1938). Period of observations: 1927–32.

Table 14(c) Time of onset of the sea breeze at Porton†: total number of days in 6 years

	1400–1459	1500–1559	1600–1659	1700–1759	1800–1859	1900–1959	2000–2059
Total no. of days in 6 years	1	9	13	12	5	6	1

† Porton is 40 km inland. *Source*: Elliott (1964). Period of observations: 1956–61.

5 SEA/LAND BREEZE CIRCULATION

incidence of sea/land breezes at various stations in India (Bombay, Calcutta, Karachi and Madras). Table 13(a) shows that tropical sea breezes occurred on at least two-thirds of the days, even in the "non-continuous" season. In contrast, the lake breeze at Chicago (Table 13(b)) occurred only in the summer months and then on only about one-third of the total possible days. At Porton and Worthy Down the frequencies were even lower. The high frequencies for Eskmeals probably owe something to the definition of the sea breeze used by Brittain (1978). Any surface onshore wind with a component normal to the coastline stronger than the equivalent component of the "free stream wind" was considered to be a sea breeze. Tables 14(a), (b) and (c) show that within the day the tropical sea breeze set in most frequently between 1200 and 1400 hours, but that the range of onset times extended from before 1000 to later than 1800 hours. In mid-latitudes Lyons (1972) found that the summer Chicago lake breeze started most frequently between 0800 and 0900 hours, with a range from 0600 to 1800 hours. At Athens in summer the sea breeze started at about 0800 and ended at 2100 hours. In winter the times were 1030 and 1900 hours respectively (Zambakas, 1973). At Kinloss (Scotland), Gill (1968) noted onset times of 1200 hours in March and October with a progressively earlier onset in the intervening months culminating in 0700 hours in June and July. At Porton, 40 km inland, onset times were correspondingly later. Table 15(a) reveals that the sea breezes at Bombay tended to last for 7–8 h, but with a range extending from less than 3 h to more than 10. Gill's (1968) results suggest that March and October breezes lasted for 2–4 h, but at the peak of their development in June and July they lasted for about 10 h.

Table 15(a) Duration of the sea breeze at Bombay: average number of days

Duration (h)	Month							
	Oct	Nov	Dec	Jan	Feb	Mar	Apr	May
<3.0	2.0	0.5	—	2.5	1.5	1.0	—	—
3.0–4.5	—	1.5	3.5	5.0	2.0	1.0	—	—
4.5–6.0	1.5	3.0	4.0	7.5	6.0	2.0	1.0	—
6.0–7.5	3.5	8.5	8.5	3.5	3.0	3.0	0.5	—
7.5–8.0	6.0	12.0	7.5	3.5	5.5	6.5	1.5	—
9.0–10.5	6.0	2.0	3.0	4.5	6.0	8.0	6.0	—
10.5–12.0	—	—	0.5	1.5	4.5	4.0	6.5	—
>12.0	1.5	—	—	—	—	3.0	10.0	11.0
Mean duration	7.48	7.18	6.58	6.27	7.47	8.36	11.16	>12

Source: Dekate (1968). Period of observations: 1963–64.
Dash indicates no occurrence.

Table 15(b) Duration of the sea breeze at Worthy Down: actual number of days

Duration (h)	Month							Total
	Mar	Apr	May	Jun	Jul	Aug	Sep	
<1	1	0	1	1	0	1	2	6
1–1.9	1	0	2	0	0	2	4	9
2–2.9	3	1	1	5	2	2	1	15
3–3.9	2	2	2	2	2	3	0	13
4–4.9	0	1	2	1	1	1	1	7
5–5.9	0	0	1	1	0	0	0	2
6–6.9	0	0	0	1	2	0	0	3
7–7.9	0	0	2	0	0	0	0	2
8–8.9	1	0	0	0	0	0	0	1

Source: Peters (1938). Period of observations: 1927–32.

Equivalent data on land breezes are more scarce, but Dekate (1968) and Sen Gupta and Chakravortty (1947) provided some useful information. Table 16(a) clearly reveals a marked seasonality in the Bombay land breeze, being non-existent from June to August but occurring nearly every night in November. In contrast, at Calcutta (Table 16(a)) land breezes occurred on less than a third of the possible occasions, being most frequent in December. South African observations (Table 16(b)) also showed a wide range of values, the major common feature being a winter maximum of occurrence. Tables 17(a) and (b) show the fairly large range of times of onset of the land breeze. At Bombay (Table 17(a)) most of the breezes start between 2300 and 0300 hours whereas at Calcutta (Table 17(b)) the range is from 2400 to 0500 hours.

B. Mean surface characteristics of the tropical sea/land breeze

At the onset of the sea breeze, the wind speed and direction usually change. Indeed, it is often the change in direction, or the appearance of a definite direction in otherwise calm conditions, which helps to determine the existence of the breeze. Table 18(a) indicates an average speed of the sea breeze at the time of onset at Bombay of 4.4–4.7 m s^{-1}, the highest speeds occurring, albeit infrequently, in October and November. The Madras results (Table 18(b)) are in reasonable accord with those at Bombay. The monthly mean temperature fall ranges from 1.1 °C in April and October to 4.4 °C in July at Madras (Table 19(a)). The frequency analysis in Table 19(b) illustrates the range of values of the temperature fall on a monthly basis at Bombay and Madras. Clearly the former has far more small temperature falls than the latter, possibly a result of local site conditions affecting the representivity of the observations. Rao

Table 16(a) Monthly mean number of days with land breezes

Location	Jan	Feb	Mar	Apr	May	Jun	Jul	Aug	Sep	Oct	Nov	Dec
Bombay†	26.0	20.5	22.0	17.5	6.5	0.0	0.0	0.0	3.3	19.5	27.3	26.3
Calcutta‡	10	9	6	2	0	0	0	0	0	3	8	12

† *Source*: Dekate (1968). Period of observations: 1961–64.
‡ *Source*: Sen Gupta and Chakravortty (1947). Period of observations: 1935–40.

Table 16(b) Percentage frequency of days with land breezes

Location	Jan	Feb	Mar	Apr	May	Jun	Jul	Aug	Sep	Oct	Nov	Dec
Durban†	2.7	3.2	4.0	7.4	15.6	20.9	19.2	11.2	4.8	3.8	3.0	4.4
Bluff Signal Station, South Africa‡	23.0	36.0	26.0	40.0	71.0	67.0	50.0	55.0	10.0	29.0	20.0	13.0

† *Source*: South African Weather Bureau (1960). Period of observations: 1946–55.
‡ *Source*: Jewell (1964). Period of observations: 1963.

Table 17(a) Times of onset of the land breeze at Bombay—monthly mean number of days

	Sunset–2059	2100–2259	2300–0059	0100–0259	0300–0459	0500–sunrise
Sep	—	0.7	0.5	1.3	0.7	—
Oct	0.5	3.7	6.5	4.7	1.3	1.7
Nov	1.0	8.7	8.0	5.0	2.5	2.0
Dec	0.5	5.7	8.0	6.7	3.3	2.0
Jan	2.0	2.3	6.3	4.7	4.0	4.7
Feb	1.3	4.3	5.0	3.7	2.3	4.0
Mar	—	1.5	4.7	5.7	6.5	3.5
Apr	0.3	—	2.3	6.0	5.5	4.3
May	—	—	—	1.7	3.0	1.7

Source: Dekate (1968). Period of observations: 1961–64.

(1955) shows that the magnitude of the fall depends on the time of onset of the breeze within the day (Table 19(c)). Late afternoon breezes cause falls of over 4.4 °C, no doubt because air temperatures over the land are at their highest between 1600 and 1800 hours, thus giving maximum contrast between sea and land air temperatures.

The incursion of sea air causes an increase in the absolute and relative humidities at coastal stations. Tables 20(a), (b) and (c) provide frequency analyses of the increases in relative humidities associated with the sea breezes at Bombay, Karachi and Madras. The range of rises is from 5 to 30 per cent with average values at Bombay of about 10–12 per cent. Table 20(d) shows

Table 17(b) Times of onset of the land breeze at Calcutta—monthly mean number of days

	2200–2259	2300–2359	2400–0059	0100–0159	0200–0259	0300–0359	0400–0459	0500–0559	0600–0659	Time of sunrise
Oct	1	0	0	2	6	3	3	2	0	0557
Nov	1	0	7	3	12	9	7	2	0	0613
Dec	0	5	8	10	21	6	7	2	1	0632
Jan	0	3	7	5	9	12	10	5	0	0643
Feb	1	1	3	6	7	12	9	4	1	0633
Mar	0	3	4	5	6	5	3	3	0	0609
Apr	2	0	0	1	3	4	1	1	0	0540
Total	5	12	29	32	64	51	40	19	2	
% frequency	2	5	11	13	25	20	16	7	1	

Source: Sen Gupta and Chakravortty (1947). Period of observations: 1935–40.

5 SEA/LAND BREEZE CIRCULATION

Table 18(a) Frequencies of speed of sea breeze at time of onset at Bombay

Speed		Month							
km h^{-1}	m s^{-1}	Jan	Feb	Mar	Apr	May	Oct	Nov	Dec
<5	<1.4	0.0	0.0	0.0	0.0	0.0	0.0	0.0	0.0
6–10	1.7–2.8	2.3	2.5	0.7	0.0	0.5	1.0	0.5	2.0
11–15	3.1–4.2	12.5	6.7	5.5	3.5	2.7	5.5	6.7	10.7
16–20	4.4–5.6	12.7	15.0	19.5	18.3	4.0	15.3	18.7	13.0
21–25	5.8–6.9	0.5	2.5	3.7	3.7	3.0	1.0	1.5	0.7
26–30	7.2–8.3	0.0	0.0	0.0	0.0	0.0	0.3	0.3	0.0
Average speed (km h^{-1})		15.0	16.3	17.1	18.0	17.7	16.9	16.9	15.4

Source: Dekate (1968). Period of observations: 1961–64.
No figures are given for periods when the sea breeze is considered to be continuous.

that vapour pressure increases of over 8 mb were typical of sea breezes which start any time between 1200 and 1800 hours.

Once again data on the land breeze are scarce. As expected, most of them have speeds less than $2\,\mathrm{m\,s^{-1}}$ (Table 21(a)) and only small impact on temperature and humidity at a given station (Tables 21(b) and 21(c)). It is surprising that the temperatures rose on most occasions at the onset of the land breeze, a feature also observed by Feit (1969). He tentatively suggested that adiabatic compression in the subsiding, cool land air could be the cause of this feature. The breezes at Calcutta (Sen Gupta and Chakravortty, 1947) lasted for less than 2 h, nearly half of them for less than 1 h (Table 22). This is in apparent contradiction to Feit's (1969) observations which showed that the

Table 18(b) Frequencies of speed of sea breeze at time of onset at Madras

Speed		Month						
mi h^{-1}	m s^{-1}	Apr	May	Jun	Jul	Aug	Sep	Oct
0–4	0–1.8	18	12	11	10	17	2	8
5–9	2.2–4.0	4	5	6	9	11	14	2
10–14	4.5–6.3	9	21	30	12	13	12	4
15+	6.7+	1	11	4	5	0	1	0
All speeds		32	49	51	36	41	29	14

Source: Rao (1955). Period of observations: 1945–46.
No figures are given for periods when the sea breeze is considered to be continuous.

Table 18(c) Total frequency of speeds of sea breezes at Karachi

Speed		Number of occasions
$mi\,h^{-1}$	$m\,s^{-1}$	
⩽5	⩽2.2	8
6	2.7	2
7	3.1	7
8	3.6	10
9	4.0	17
10	4.5	27
11	4.9	14
12	5.4	17
13	5.8	12
14	6.3	5
15	6.7	7
16	7.2	5
17	7.6	2
18	8.0	2

Source: Ramdas (1931). Period of observations: 1929–30.

breeze frequently lasts for 4–6 h. However, Feit did find definite surges of the land breeze (of up to $5\,m\,s^{-1}$) with durations of 0.5–2 h and in the light of the conventional instrumentation used by Sen Gupta and Chakravortty, it is reasonable to conclude that they observed only the strongest parts of each land breeze.

C. The sea/land breeze circulation

It is clear from the qualitative outline of the breeze mechanism that the resultant circulation is essentially a thermally driven over-turning in the

Table 19(a) Temperature fall at onset of the sea breeze: average temperature fall at Madras

	Month							
	Apr	May	Jun	Jul	Aug	Sep	Oct	Year
Temperature fall (°C)	1.1	2.8	3.9	4.6	3.5	2.8	1.1	2.8

Source: Rao (1955). Period of observations: 1945–46.
No figures are given for periods when the sea breeze is considered to be continuous.

Table 19(b) Temperature fall at onset of the sea breeze: monthly frequencies of temperature fall at Bombay and Madras

Temperature fall		Month											
°C	°F	Jan	Feb	Mar	Apr	May	Jun	Jul	Aug	Sep	Oct	Nov	Dec
Bombay													
0.0–0.9		14.0	14.3	19.5	19.7	9.5					13.5	11.7	11.3
1.0–1.9		11.3	8.7	9.0	4.7	0.5					8.7	13.5	12.3
2.0–2.9		2.5	3.3	1.0	0.7	0.0					0.5	2.3	2.7
3.0–3.9		0.3	0.7	0.0	0.0	0.0					0.3	0.0	0.5
Madras													
	<1.0					4	2	2	5	3	11.5		
	1.0–1.9					4.5	3.5	2.5	6.5	8	9		
	2.0–2.9					5	2.5	9	3	6	0.5		
	3.0–3.9					5.5	4.5	2	2	4	2		
	4.0–4.9					4.5	2.5	4	2	1	0		
	5.0–5.9					2	2.5	3.5	0	1	0		
	6.0–6.9					3	4	1	0.5	0.5	0		
	>7.0					2	3	0.5	2.5	0	0		

Sources: Bombay, Dekate (1968), period of observations, 1961–64; Madras, Roy (1940), period of observations, 1938–40. No figures are given for periods when the sea breeze is considered to be continuous.

vertical. Any description of the sea/land breeze phenomenon should therefore concentrate on this vertical structure and in the following paragraphs the important components are outlined. Attention is purposely restricted to the typical, fully developed circulation, whilst not forgetting that the circulation

Table 19(c) Diurnal variation of temperature fall at onset of sea breeze at Madras

	Time				
	1000–1159	1200–1359	1400–1559	1600–1759	1800–2000
Mean fall (°C)	2.6	3.0	3.7	4.7	3.5

Source: Rao (1955). Period of observations: 1945–46.

Table 19(d) Frequency of temperature fall at different heights in sea breezes at Karachi

	Number of occasions of temperature fall of			
	°F: 0.5–1.4	1.5–2.4	2.5–3.4	⩾3.5
Height (m)	°C: 0.3–0.8	0.8–1.3	1.4–1.9	⩾2.0
4.9	35	28	5	2
17.1	22	28	8	4
47.6	23	23	12	2

Source: Ramdas (1931). Period of observations: 1929–30.

Table 20(a) Humidity changes at the onset of the sea breeze: monthly frequency (days) of relative humidity rises (in per cent) at Bombay

Relative humidity rise (%)	Month							
	Jan	Feb	Mar	Apr	May	Oct	Nov	Dec
⩽5	5.3	2.0	3.3	11.3	6.5	5.5	1.3	3.7
6–10	13.7	10.3	12.7	9.0	2.7	6.5	12.0	12.0
11–15	3.3	4.7	3.7	3.0	0.7	2.0	4.3	4.3
16–20	3.5	7.7	5.5	0.5	0.3	4.5	6.7	3.5
21–25	0.7	1.7	2.7	0.7	—	2.0	1.5	1.3
26–30	0.5	0.7	0.7	0.5	—	1.7	1.0	1.0
⩾31	1.0	0.5	9.7	0.5	—	0.7	1.0	0.7
Average	10.7	13.2	15.7	7.8	5.1	12.9	13.6	11.7

Source: Dekate (1968). Period of observations: 1961–64.

Table 20(b) Humidity changes at the onset of the sea breeze: frequency of occasions of relative humidity increases at Karachi

	Relative humidity rise (%)						
	≤5	6–9	10–14	15–19	20–24	25–29	≥30
Number of occasions	16	24	19	15	6	8	3

Source: Ramdas (1931). Period of observations: 1929–30.

Table 20(c) Humidity changes at the onset of the sea breeze: monthly frequency (in days) of relative humidity rises (in per cent) at Madras

Relative humidity rise (%)	Month					
	May	Jun	Jul	Aug	Sep	Oct
5	2.5	3.0	1.5	1.5	1.5	6.5
5–9	4.0	4.5	4.0	7.5	5.5	6.0
10–14	9.0	2.0	3.5	5.5	7.0	7.5
15–19	7.0	7.0	4.5	3.5	6.0	2.0
20–24	4.0	5.5	8.5	1.0	3.5	1.0
25–29	3.5	3.0	2.5	0.5	1.0	0.0
30	0.5	0.5	0.5	2.0	0.0	0.0

Source: Roy (1940). Period of observations: 1938–40.

Table 20(d) Humidity changes at the onset of the sea breeze: diurnal changes in vapour pressure at Madras

	Time of day				
	1000–1159	1200–1359	1400–1559	1600–1759	1800–2000
Vapour pressure rise (mb)	7.2	8.1	8.4	8.3	6.9

Source: Rao (1955). Period of observations: 1945–46.

Table 21(a) Speed of land breeze at Calcutta

	Speed			
mi h^{-1}:	<5	5–9	10–14	>14
m s^{-1}:	<2.2	2.2–4.0	4.5–6.3	>6.3
No. of occasions	196	50	8	0
Percentage frequency	77	20	3	0

Source: Sen Gupta and Chakravortty (1947). Period of observations: 1935–40.

Table 21(b) Temperature change at onset of land breeze at Calcutta

	Temperature change						
°F: °C:	< +1 < 0.7	1–1.9 0.7–1.0	2–2.9 1.1–1.6	3–3.9 1.7–2.2	4–4.9 2.3–2.7	⩾ 5 ⩾ 2.8	0 or fall
No. of occasions	81	103	30	10	2	0	28
Percentage frequency	32	40	12	4	1	0	11

Source: Sen Gupta and Chakravortty (1947). Period of observations: 1935–40.

Table 21(c) Relative humidity change at onset of land breeze at Calcutta

	Relative humidity change (%)					
	−1 to −4	−5 to −9	−10 to −14	−15 to −19	−20 to −24	−25 or more	0 or rise
No. of occasions	82	82	43	20	10	9	8
Percentage frequency	32	32	17	8	4	4	3

Source: Sen Gupta and Chakravortty (1974). Period of observations: 1935–40.

evolves through a definite life cycle within the duration of about 8 h. Clearly the development and dissipation stages of the circulation will exhibit less marked characteristics than the fully developed stage.

It is convenient to describe the circulation in terms of its

Table 22 Monthly mean frequency of duration of land breeze

Month	Time (min)						
	20	20–39	40–59	60–79	80–99	100–119	⩾120
Oct	6	4	3	1	3	0	0
Nov	6	9	3	7	5	3	8
Dec	10	19	12	10	4	4	1
Jan	15	9	7	5	6	4	5
Feb	7	12	12	6	3	2	2
Mar	3	8	4	5	4	4	1
Apr	3	4	1	2	1	0	1
Total	50	65	42	36	26	17	18
Frequency (%)	20	25	17	14	10	7	7

Source: Sen Gupta and Chakravortty (1947). Period of observations: 1935–40.

"limbs"—particularly the lower one which comprises the sea/land breeze and the upper, "return" current. Their depth and horizontal extent determine the overall size of the circulation and are closely related to its intensity, i.e. the wind speeds within it.

Table 23 Depth of sea and land breeze

Source	Location	Depth	Comments
Sea breeze			
Air Ministry (1943)	British Isles	150–300 m	
Craig *et al* (1945)	Massachusetts	120 m	
de Felice and Gasne-Tabbagh (1971)	Brittany	700 m	
Dixit and Nicholson (1964)	Bombay	<1 km	
Findlater (1963)	UK	1.5 km	A quoted rather than observed figure
Frizzola and Fisher (1963)	Long Island	<1 km	
Hatcher and Sawyer (1947)	Madras	170–200 m	
Hsu (1970)	Texas	670 m	
Johnson and O'Brien (1973)	Oregon	1 km	
Keen and Lyons (1978)	Lake Michigan	500–800 m	
Kimble *et al.* (1946)	Tropics	1–1.2 km	
Lyons (1972)	Lake Michigan	500 m	
Moroz (1967)	Lake Michigan	500 m	
Narayanan (1967)	Thumba, India	0.8–1 km	
Pedgley (1958)	Ismailia	1 km	
Peters (1938)	Worthy Down, UK	300–600 m	
Preston-Whyte (1969)	Natal	600–900 m	
Ramanadham and Subbaramayya (1965)	Visakhapatnam, India	600 m	
Ramanathan (1931)	Poona	1.2 km	
Ramdas (1931)	Karachi	0.5–1.5 km	
Roy (1940)	Madras	0.5–1 km	
Schroeder *et al.* (1967)	California	0.5–2 km	
Sherman (1880)		150–200 m	No location specified
Smith (1974)	East Anglia, UK	1.2 km	
Sutcliffe (1937)	Felixstowe, UK	500–600 m	
van Bemmelen (1922)	Batavia	1–1.2 km	
Wexler (1946)	Danzig	200 m	Derived from other sources
	Tropics	1–2 km	
Land breeze			
Meyer (1971)	Atlanta City	<90 m	

1. Depth of sea and land breezes

As indicated earlier, most of the early instrumental observations of the sea/land breeze were restricted to the near-surface layers. Estimates of the depth of the breezes relied upon such information as could be gleaned from, for example, smoke plumes. The classic paper by van Bemmelen (1922) was almost invariably referred to in pre- and immediately post-War papers (e.g.

Table 24 Height and depth of the return current

Source	Location	Depth (D) and/or height of base (H)	Comments
Craig et al. (1945)	Massachusetts Bay		Only the existence noted—no depths and speed
de Felice and Gasne-Tabbagh (1971)	Brittany	$H = 700$ m	
Dixit and Nicholson (1964)	Bombay	$H = 3$ km	
Eddy (1966)	Texas	$D = 0.9$–2.4 km; $H = 1.2$–1.5 km	
Fisher (1960)	Massachusetts	$H = 750$ m	
Frizzola and Fisher (1963)	Long Island	$H = 670$ m	"The return flow was difficult to detect…"
Hsu (1970)	Texas	$D = 2$ km; $H = 670$ m	
Johnson and O'Brien (1973)	Oregon		Only the existence noted
Keen and Lyons (1978)	Lake Michigan	$D = 1.1$ km; $H = 500$–800 km	
Lyons (1972)	Lake Michigan	$D = 1$ km; $H = 500$ m	
Moroz (1967)	Lake Michigan	$D = 2$ km; $H = 500$ m	
Pedgley (1958)	Ismailia	$D = 1$ km; $H = 900$ m	
Ramanadham and Subraramayya (1965)	Visakhapatnam, India	$D = 2$ km; $H = 600$ m	
Ramanathan (1931)	Poona	$D = 1.5$–2 km; $H = 2$ km	
Ramdas (1931)	Karachi	$H = 1$ km	
Sherman (1880)		$D = 240$–400 m	No location specified
van Bemmelen (1922)	Batavia	$D = 2.8$ km; $H = 1.2$ km	

Kimble, 1946; Wexler, 1946), presumably because it was one of the few good observational studies of the sea/land breeze up to that time. Fortunately the last quarter-century has seen a substantial increase in the number of observational studies which concentrated on elucidating the vertical structure. Table 23 summarizes the depths of sea and land breezes as measured in the mid-latitudes and in the tropics. Values for sea breezes range from just over 100 m to 1 km. Meyer (1971) and Feit (1969) recorded land breezes of 100 m in depth. Wexler (1946) quoted values of 1–2 km in the

Table 25 Values and height of maximum velocities in sea-breeze and return current

Source	Location	Speed ($m\,s^{-1}$)	Height at which speed occurred (m)
(a) *Sea breeze*			
Air Ministry (1943)	Lossiemouth		60–150
	Mt Batten	4.5–6.3	150–240
	Felixstowe		<150
	Wick	4.5–5.4	<100
de Felice and Gasne-Tabbagh (1971)	Brittany	3	<700
Dixit and Nicholson (1964)	Bombay	10.3	
Hsu (1970)	Texas	2.2	
Johnson and O'Brien (1973)	Oregon	7.2	
Keen and Lyons (1978)	Lake Michigan	4.1	150
Kimble *et al.* (1946)	Tropics	7.7	
Kopfmuller (1922)	Lake Constance	2.3	
Lyons (1972)	Lake Michigan	6	
Mizumi and Kakuta (1974)	Japan	8–10	100
Moroz (1967)	Lake Michigan	4–7	250
Narayanan (1967)	Thumba, India	11.3	
Pedgley (1958)	Ismailia	5	
Preston-Whyte (1969)	Natal	4	180–240
Ramanadham and Subbaramayya (1965)	Visakhapatnam, India	5–7	200
van Bemmelen (1922)	Batavia	7.2	
Wexler (1946)	Tropics	7	
	Europe	2	
(b) *Return current*			
Dixit and Nicholson (1964)	Bombay	7.7	
Eddy (1966)	Texas	4	
Fisher (1960)	Massachusetts	2	
Koschmeider (1941)	Danzig	4	
Lyons (1972)	Lake Michigan	4	
Ramdas (1931)	Karachi	1–1.3	
van Bemmelen (1922)	Batavia	3	

tropics but gave no supporting evidence or reference. Despite the great variation in depths, there is a suggestion in Table 23 that tropical breezes are deeper than those in mid-latitudes.

2. Return current

The return current predicted by theory has not always been observed, particularly in early mid-latitude investigations. For example, it was not evident in Sutcliffe's (1937) observations at Felixstowe, England and even Fisher (1960) and Frizzola and Fisher (1963) in their carefully set-up experiments found that "the return flow was difficult to detect..." (Frizzola and Fisher, 1963, p. 738), particularly if there was a gradient wind. Perhaps one of the reasons for the lack of observation of the return current is that instruments (mainly pilot balloons) were simply not sent high enough: it was the sea breeze *per se* that claimed attention, not the overlying return current. Table 24 shows observations that indicate that the current does exist with depths ranging from less than half to nearly 3 km, but it should be noted that some of the observations were made a century ago.

Table 25(a) shows several measurements of the maximum velocity of the sea breeze, that is, the velocity at a height of about 100 m away from the friction layer and at the time of strongest development of the breeze. These values are free of any influence by ambient winds. As expected by continuity requirements the speeds of the return flow, which is approximately twice the depth of the sea breeze, are about half those of the sea breeze (Table 25(b)). The shallow, nocturnal land breeze is considerably weaker than the sea breeze.

3. Horizontal extent

For obvious observational reasons, the penetration inland of the sea breeze is the better documented part of this circulation. Assuming a symmetrical circulation, its total "horizontal" extent would be twice the distance of inland penetration. In mid-latitudes, sea breezes tend to penetrate 20–50 km (Table 26), but in the tropics, distances of 300 km and over have been both quoted (Wexler, 1946, p. 281) and directly claimed (Clarke, 1955).

(a) *Coriolis effects* Whereas most of the description of the sea/land breeze in the late nineteenth century literature dealt primarily with the reversal of the sea breeze circulation in a single vertical plane perpendicular to the coastline, temporal variation of the circulation in the horizontal was recognized. As early as 1801 Capper (1801) noted that winds at Malabar rotated through 360° in the course of a day, following the sun. Similar "rotation" of the winds was apparently observed in the nineteenth century by Prestel (1864), Loomis (1871), Woeikof (1875), Taylor (1877), Ringe (1882) and Davis *et al.*

Table 26 Horizontal extent of sea and land breezes

Source	Location	Inland penetration (km)	Seaward extension (km)
Sea breeze			
Air Ministry (1943)	British Isles	48–64	19–32
Blanford (1889)	India	110	
Clarke (1955)	Australia	250–330	
Dixit and Nicholson (1964)	Bombay	>80	120–150
Fegusson (1971)	Harrogate, UK	113–129	
Findlater (1963)	British Isles	55	
Hsu (1970)	Texas	50–65	
Johnson and O'Brien (1973)	Oregon	>60	
Keen and Lyons (1978)	Lake Michigan	25	
Kimble et al. (1946)	Tropics	80	80
Marshall (1950)	UK	160	
Moroz (1967)	Lake Michigan	25–30	
Pedgley (1958)	Ismailia	110	
Preston-Whyte (1969)	Natal	65	
Ramakrishnan and Jambunathan (1958)	Madras	>120	
Ramanathan (1931)	Poona	110	
Schroeder et al. (1967)	Various quoted	Up to 300	
Smith (1974)	UK	160	
Wexler (1946)	Tropics		200–300
	Massachusetts	40	
	California	100	
	UK	10–15	
Land breeze			
Meyer (1971)	North-eastern USA		22–26

(1890)—all quoted in the most useful bibliographies by Jehn (1973) and Baralt and Brown (1965). This "rotation" manifests itself in a veering (in the northern hemisphere) and backing (in the southern hemisphere) of the winds that comprise the circulation such that, late in their lifetimes, the sea and land breezes blow parallel to rather than across the coastline. Despite the nineteenth century observations outlined above, and more recent confirmatory evidence by Sutcliffe (1937), it was the theoretical analysis by Haurwitz (1947) (to be examined in more detail later) which prompted the majority of definitive observations of the phenomenon. Most results are displayed as either a hodograph of winds at one level or a cross-section of the wind components across and parallel to the coast (Fig. 60). Whereas, in a careful study off Rhode Island, Fisher (1960) only inferred an effect from changes in the gradient wind, most observational experiments, such as those

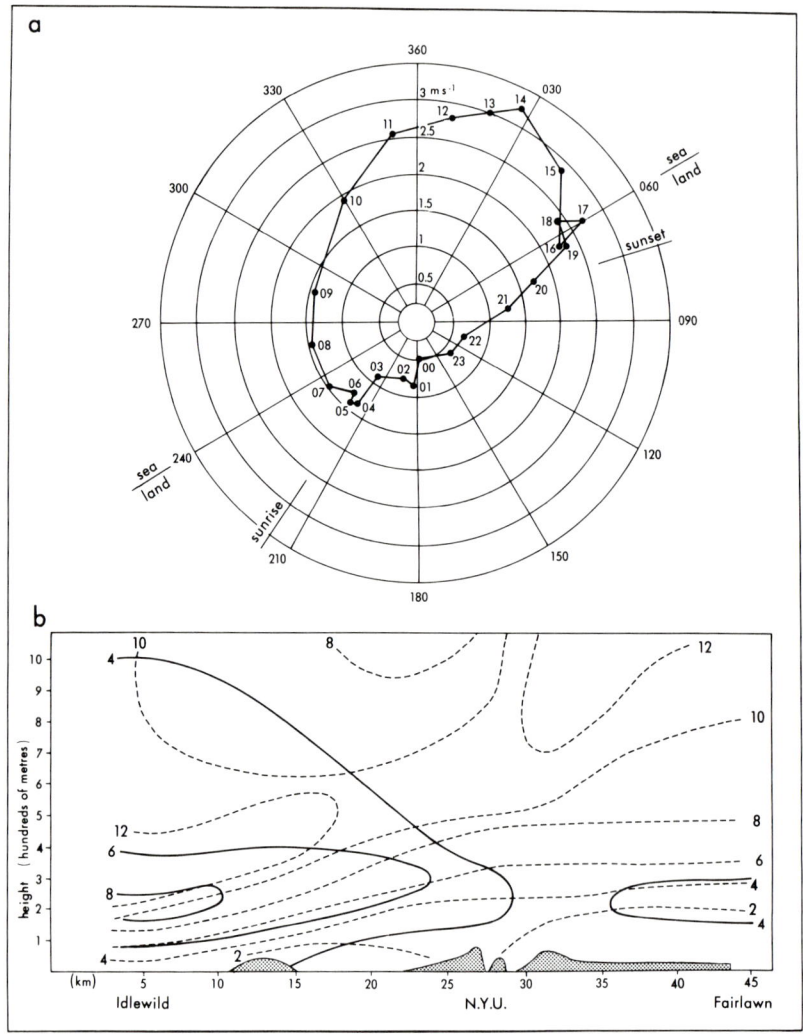

FIG. 60. (a) Mean hourly wind vectors for sea breeze days in May 1958–62 at Kinloss, Scotland. In the diagram the centre is taken as a station circle and the wind vectors are directed from the plotted point towards the centre. Numbers near dots indicate time of day in hours. (After McCaffery, 1966).

(b) Sea breeze at 2000 hours on 2 June 1960 over New York. Solid lines show component of breeze across shoreline (u), positive values toward land (right). Dashed lines show component of breeze parallel to the shoreline (v), positive values out of page. Values are in metres per second. (After Frizzola and Fisher, 1963.)

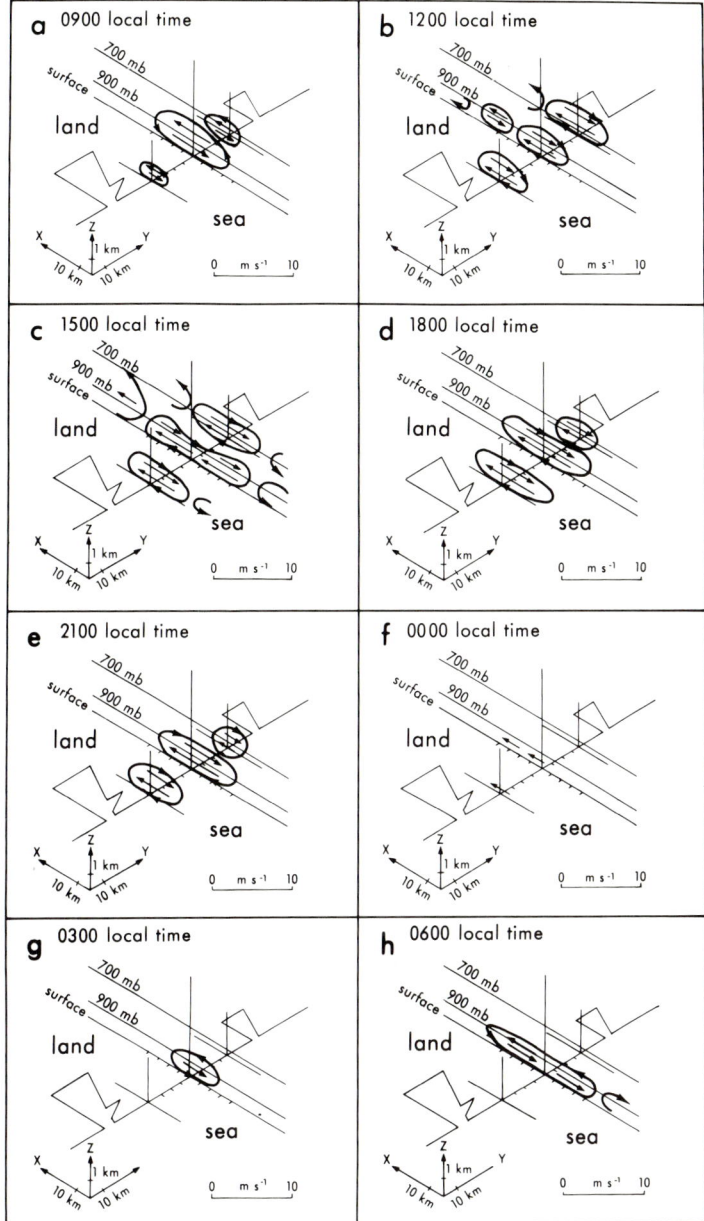

FIG. 61. Synthesized empirical model of sea/land-breeze circulation over the Texas coast. (After Hsu, 1970.)

by Staley (1957), Frizzola and Fisher (1963), Hsu (1970) and Moroz (1967), clearly observed this veering of the breezes. Haurwitz's (1947) theoretical analysis attributes the turning to the coriolis effect, thus making the sea/land breeze circulation one of the few meso-scale circulations that may be affected by the turning of the Earth.

4. *Life cycle of the sea/land breeze circulation*

We noted earlier that the sea/land breeze undergoes diurnal variation. Having reviewed the structure of the circulation at its greatest intensity, an appropriate summary is provided by considering the life cycle of the circulation. Fortunately the comprehensive study by Hsu (1970) outlines the life cycle in the context of the Texas Gulf–coast circulation. Hsu examined the circulations along and across 40 km of flat coastline and summarized his findings in Fig. 61.

The diagram shows schematically the typical vertical circulation relative to the coastline between Galveston Bay and Sabine Lake. As the convective condensation level is approximately at 900 mb, which is about 1 km, and since the 700 mb level (about 3 km) is near the top of the sea breeze circulations, these two isobaric surfaces are included. The height or location of each arrow represents the position of the maximum speed (in metres per second) of the land or sea breeze. It is interesting that some of the "return currents" appear to be as strong as their lower level counterparts.

At 0900 hours the air temperature over the land is still lower than that over the sea and thus the land breeze is still blowing. By noon, the land as a whole is warmer than the surrounding waters. A baroclinic field exists, resulting in a direct circulation from near shore to about 20 km inland. As the land breeze still prevails at this time at 20 km inland, a low level convergence is established. Scattered cumuli with a base at about 1 km tend to form a line.

At 1500 hours the sea breeze is in the fully developed stage, since the air temperature difference between land and water reaches the maximum near noon. The sea breeze has pushed further inland and the fully developed convergence line at the landward limit of the breeze may cause showers along a line roughly parallel with the coastline. Mass convergence in the upper return flow and the associated pressure gradient in the vertical leads to subsidence of air over the sea. In the next 6 h the sea breeze still prevails but is of reduced intensity. By midnight, land air has cooled sufficiently to eradicate all sea breeze pressure differences and the air is consequently virtually calm. Three hours later, the land air is cooler than the sea air, resulting in the generation of the land breeze to a distance of 20 km offshore. By 0600 hours the horizontal temperature gradient from land to sea is at its maximum and the land breeze is accordingly stronger, extending 30 km seaward. This breeze continues to blow into the midmorning hours when the circulation cycle starts again.

D. Factors affecting sea/land breeze circulation

The typical sea/land breeze circulation as analysed above is seldom completely free of effect from other factors—both atmospheric and non-atmospheric. Of the atmospheric factors, the gradient wind and the stability are the most important.

1. *Gradient wind*

Comparatively few observational studies of sea/land breezes are undertaken when strong gradient winds exist for the obvious reason that they frequently "swamp" the sea/land breeze circulation. The studies that do exist have considered gradient wind effects dependent upon its direction relative to the coastline, that is on-, parallel to or offshore.

In onshore gradient winds, the horizontal temperature gradient from land to sea is hindered in its development and consequently so is the associated horizontal pressure gradient. In turn this reduces the opportunity for the development of a sea breeze. For example, in south-easterly winds over southern Britain and inland temperatures of 30 °C no breeze developed because cool sea air warmed up rapidly within a few kilometres of the coast (Findlater, 1963). Similarly, it would probably prevent the existence of a land breeze because of the general weakness of the latter. Kimble *et al.* (1946) maintained that land breezes in the tropics are prevented with an onshore gradient wind of $3.5-5.5 \, \text{m s}^{-1}$. There is the possibility that a very light onshore wind would not destroy a land breeze, rather that it would sharpen temperature gradients at the leading edge of the breeze, yet retreat before it. Such a mechanism operates in the sea breeze as outlined below.

Sea breezes are, of course, observed when the gradient wind is onshore. Sutcliffe (1937) found that the sea breeze added $1 \, \text{m s}^{-1}$ to the speed of the gradient wind up to a height of 300 m. Pedgley (1958) claimed that sea breezes had a very marked leading edge in onshore winds, a feature usually found only in offshore winds as shown below. Frizzola and Fisher (1963) also found that in light onshore winds the sea breeze developed early in the day (0800 hours local time) and moved 20–30 km inland by mid-afternoon. Full penetration was 40–50 km. The maximum vertical extent under light gradient conditions was 900 m with maximum speeds of $7-10 \, \text{m s}^{-1}$ found below 330 m. Temperature and dew point changes were gradual although local wind changes could be abrupt.

Whereas Wexler (1946) was of the opinion that sea breezes rarely develop in gradient winds parallel to the coast (in either sense, i.e. land to right or left), both Dixit and Nicholson (1964) and Frizzola and Fisher (1963) found that such flow placed little hindrance on the development of the breezes. The effect of an offshore wind on the sea breeze has received more attention than the other two gradient wind directions considered above. Perhaps the earliest comprehensive study was that by Koschmeider (1936, 1941) who suggested that the offshore wind had the effect of pushing out to sea the thermal (and therefore pressure) gradient necessary for the sea breeze. Thus the sea breeze begins several kilometres out to sea and advances landward, often not reaching the coast until mid-afternoon. Such behaviour was subsequently observed by Kimble et al. (1946), Wexler (1946), Fisher (1960) and Frizzola and Fisher (1963). The observations near New York (Frizzola and Fisher, 1963) showed that with an offshore wind the sea breeze formed later in the day was shallower, did not extend as far inland, had much lower maximum velocities and retreated much earlier than a breeze forming in a calm atmosphere. If the offshore gradient wind was about $8-10 \text{ m s}^{-1}$ and the land–sea temperature contrast was less than $6\,°C$, then no sea breeze formed. At another mid-latitude site Watts (1955) showed that sea breezes would not form even in calm if the land–sea temperature difference were less than $1\,°C$ and that the critical gradient wind speed for prevention of the breeze rose to about 8 m s^{-1} even when the excess of land temperature was as high as $11.2\,°C$. Kimble et al. (1946) and the Air Ministry (1943) both suggested that an offshore wind of over 8 m s^{-1} would prevent the formation of a sea breeze. Figure 62 clearly shows the difference in the structure of the sea breeze between that forming in a light onshore wind and that forming in an offshore wind.

A major result of Koschmeider's work was the identification of the sea breeze front in offshore winds. This front was so called because it bore many scaled down similarities in temperature, humidity, pressure and wind changes to a synoptic scale cold front. The front is primarily due to the tightening of gradients because of the convergence between the offshore wind and the sea breeze. Whilst the sea breeze front has not been consistently observed in offshore winds (e.g. Fisher, 1960), sufficient observations have accumulated to allow Schroeder et al. (1967) to recognize three types of front. The first type is the classical one mentioned above in which there is a sharp temperature fall, humidity rise and sharp wind velocity change. Such fronts have been observed not only by Koschmeider but also by Wallington (1959, 1961, 1965), Simpson (1964), Frizzola and Fisher (1963) and Simpson et al. (1977). The second type of front is that of a wind-shift line as reported by Fosberg and Schroeder (1966). This type of front is a modified air mass type, that is it is modified thermally but not kinematically. The third type is that of the cool change where the front is characterized by a sustained cooling and humidity rise and no wind shear line.

The detailed nature of sea breeze fronts has been elucidated by glider flights and radars (Simpson, 1967; Simpson et al., 1977). The most intense part of the front is usually 100–250 m wide, quite turbulent and has uplift of about $1-2 \text{ m s}^{-1}$ (Wallington, 1959, 1961, 1965). Simpson (1964) found that the

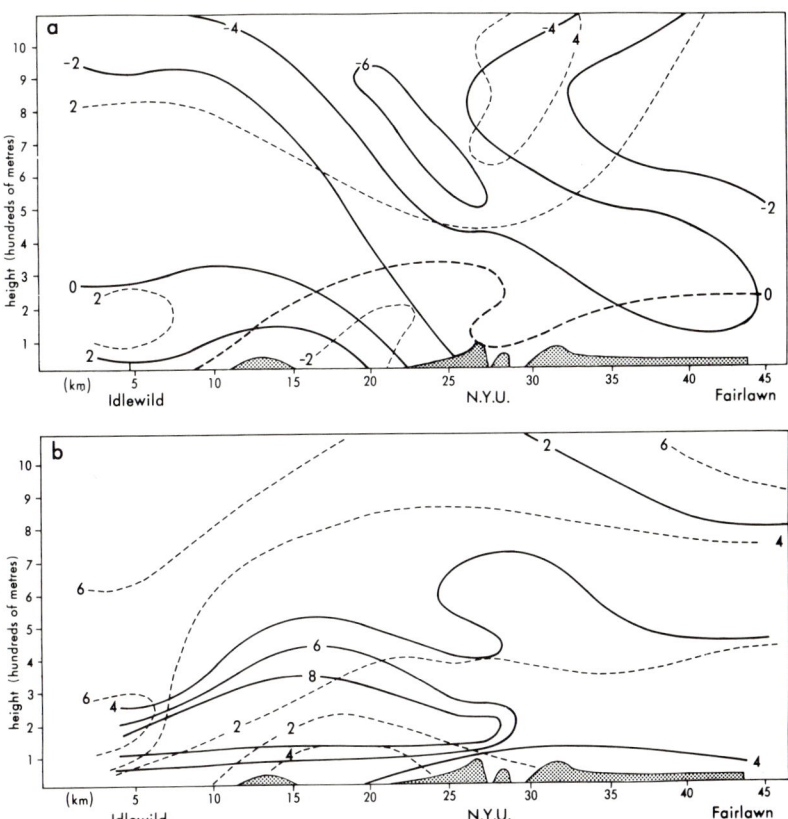

FIG. 62. (a) Sea breeze at 1700 hours on 7 June 1960 over New York. The gradient wind was from right to left in the diagram and the sea breeze was consequently shallow and weak. Solid lines represent the component of the wind across the coastline, positive toward the land (right) and dashed lines represent the component of the wind, parallel to the coast, positive out of the page. Values in metres per second.

(b) Sea breeze at 1700 hours on 2 June 1960 over New York. The gradient wind was very light. The sea breeze was deeper and stronger than the case in (a). Solid and dashed lines as in (a). (After Frizzola and Fisher, 1963.)

fronts moved at about $1-1.5 \text{ m s}^{-1}$ near the coast and speeded up as they moved inland. In their later study Simpson *et al.* (1977) found that fronts moving inland from the southern coast of England had a speed of $2-8 \text{ m s}^{-1}$ at a distance of 60 km inland and 3.5 m s^{-1} at a distance of 80 km inland. In contrast Wallington (1959) found that the fronts moved at $1.5-2.0 \text{ m s}^{-1}$ near the coast and $0.3-0.6 \text{ m s}^{-1}$ inland, moving in a series of pulsations rather than with a steady movement. As there appears to be a friction head in these pulsations, Wallington's results accord with theoretical postulates on cool

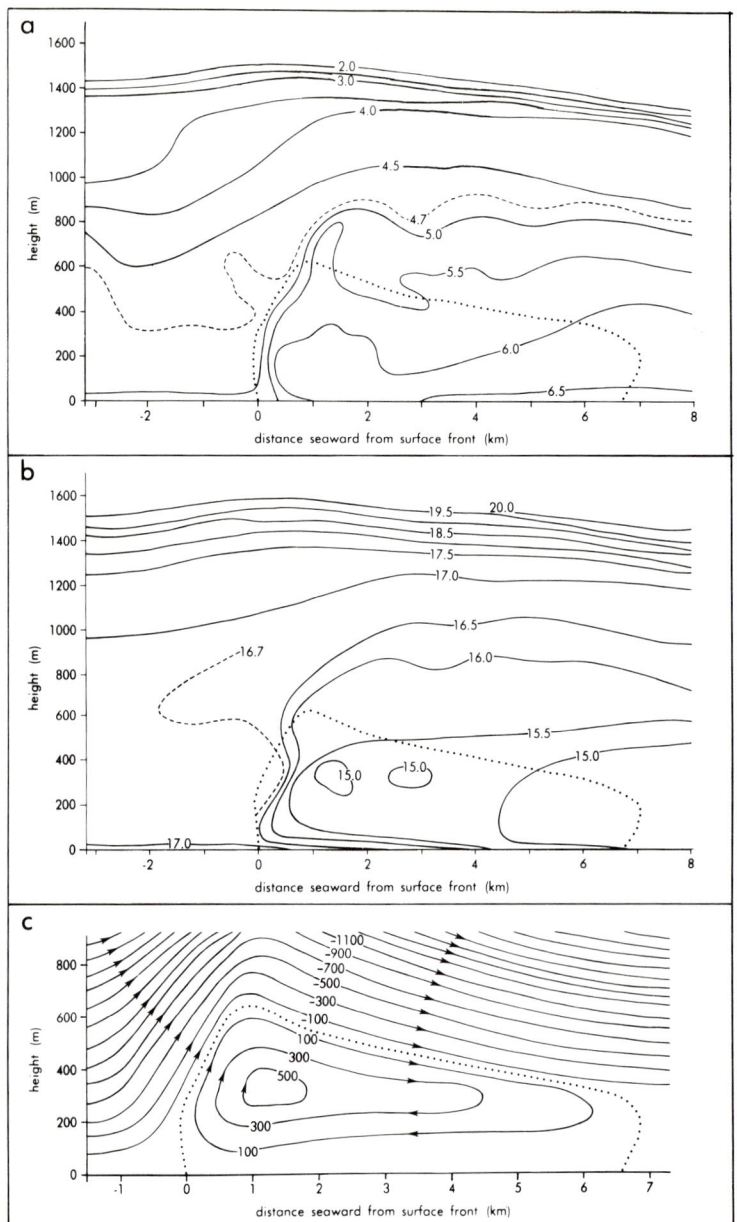

FIG. 63. Characteristics of a sea breeze front on 14 June 1973 over southern England. (a) Humidity field at 1700 GMT. Isopleths of mixing ratio in grams per kilogram. The dotted line shows the extent of the closed circulation of the cut-off vortex. (b) Potential temperature (in degrees Celsius) for same case as (a). (c) Streamlines of flow relative to the front. The contour interval is 200 m³ s⁻¹ per metre thickness. The zero streamline is dotted and marks the boundary of the cut-off circulation. (After Simpson et al., 1977.)

gravity currents. Most other observations of the cross-sectional slope of the front agree with Wallington's suggestions. Near Danzig Koschmeider (1941) found the leading edge of sea air to be very steep while a few kilometres to the rear (i.e. seaward) the slopes were quite small. At the leading edge slopes were 1 : 14 whereas 5 km to the rear slopes were 1 : 100. Wallington's (1965) later observations supported these results and his own earlier suggestions. Whichever is the case as regards the speed of the sea breeze front, it is clear from observations that, at some stage in its life, the sea breeze air itself has a higher speed than that of the front. Continuity suggests that the sea air should rise and this is the uplift and associated mixing in the front which glider pilots have used for many years. Near the narrow zone of uplift there is a sinking of air on the seaward side of the sharp frontal zone. More details of the wind, temperature and humidity structure of the sea breeze front and its environs have been provided by Simpson et al. (1977). Figure 63(a), (b) and (c) clearly reveals the sharp increase in mixing ratio and an accompanying decrease in temperature at the onset of the sea breeze front. Figure 63(c) also shows that the air behind the front moves within a cut-off circulation of about 7 km horizontal extent. The forward end of this circulation was about 600 m high in the case studied by Simpson et al. (1977), falling seaward to a height of about 300 m. Simpson et al. (1977) went on to compare the flow near the sea breeze front to that in a gravity current head. Figure 64, a schematic representation of the latter type of flow, bears a striking similarity to the sea breeze flow. The cut-off circulation is clear in Fig. 63(c), and in Fig. 63(a) there is a hint of the

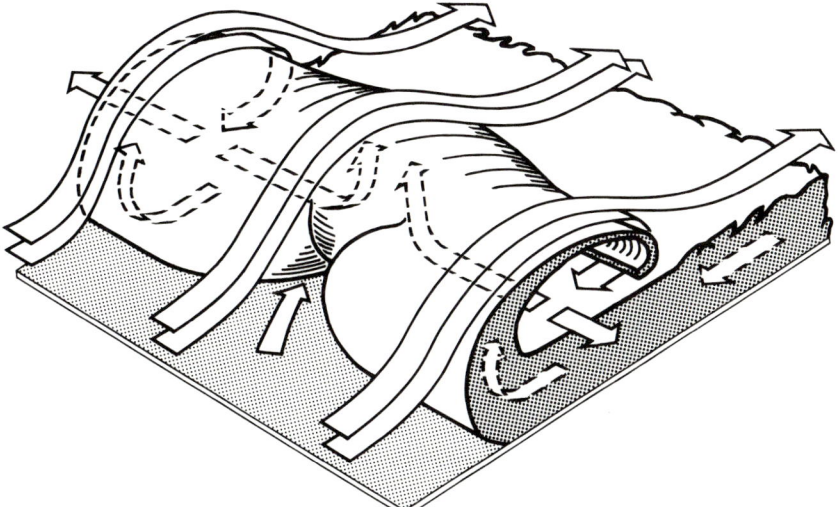

FIG. 64. Schematic representation of the flow at a gravity current head. (After Simpson et al., 1977.)

incorporation of drier air at heights of 500–700 m at distances of 1–3 km seaward of the surface front.

On a slightly larger scale the sea breeze front is associated with a convergence of low level winds. Lyons (1972) noted a zone 1–2 km wide in which updraughts of over 1 m s^{-1} and convergences of over $3 \times 10^{-3} \text{ s}^{-1}$ were measured. Findlater (1964) found the convergence zone to be about 3 km wide with upward velocities of up to 1.5 m s^{-1} on the landward side and downward velocities of $1.0–2.0 \text{ m s}^{-1}$ on the seaward side. Wallington's (1965) results were in good agreement with Findlater's. Strong convergences, such as these, led to the growth of cumulus clouds along the line of the sea breeze front and Findlater (1964) has shown the value of meso-scale analysis of sea breeze convergence zones for forecasting the occurrence of cumulus activity and even thunderstrom occurrence. Figure 65 shows the very close coincidence between the location of storm centres and the overlapping convergence zones on 12 June 1963.

2. *Stability*

Few observations of the relationship between sea breezes and vertical stability exist in the literature but accumulated experience of meteorologists based at operational coastal stations shows that a fairly clear relationship does exist. Wexler (1946) noted that the most favourable time for sea breeze penetration is the time of greatest vertical instability. He was of the opinion that if the land air remains stable the sea breeze may not even reach the shore. The later work of Johnson and O'Brien (1973), Pearce (1968) and Brittain (1978) confirmed Wexler's views. In a stable atmosphere, the upper stable layers act as a strong damping mechanism on the vertical circulation of the sea breeze. In contrast, less stable or unstable air encourages an extension of the circulation both horizontally and vertically, and an increase in intensity (Patrinos and Kistler, 1977).

3. *Effects of topography*

Topography may influence the sea breeze circulation in two main ways: through the form of the land and through the vegetation cover. Hills near a coastline may accentuate the sea breeze if equatorward—facing slopes heat up more than surrounding flat land. Kimble *et al.* (1946) noted that sea breezes were much better developed along the hilly west coast of Sumatra than along the low lying east coast south-east of Palembang. Once the sea breeze blows, hills and valleys may profoundly affect its direction. For instance, studies in Washington and Oregon (Staley, 1957; Olsson *et al.*, 1973) clearly revealed the channelling effect of gaps in the coastal range. The fairly low, strong marine inversion in that part of the world favours such steering effects, forcing the breeze around hills rather than over them.

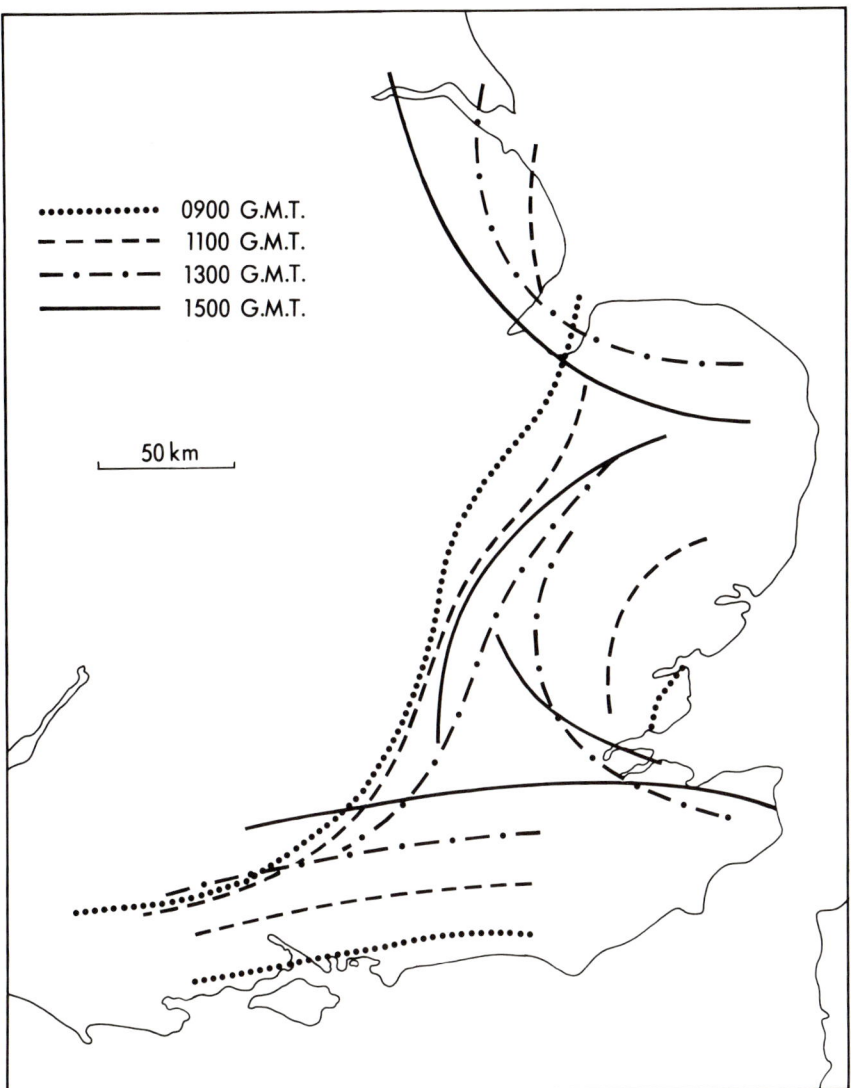

FIG. 65. Successive positions of sea breeze and inland convergence zones on 12 June 1963 over England. (After Findlater, 1964.)

If the forces working for and against sea breezes are nearly equal, the character of the vegetation cover may become important. Dry, barren coastlines heat up more quickly than moist, well-vegetated ones given the same radiative input of energy. Consequently more heat becomes available to initiate the sea breeze circulation.

III. Theory

Despite the apparent simplicity of the sea/land breeze mechanism, particularly when compared with, for instance, lee-wave mechanics, theoretical studies of the sea breeze are rare in the literature prior to 1945. Yet over the past 35 years most of the major problems posed by the sea breeze circulation have been tackled, if not fully solved. In particular the effects on the circulation of the following have been investigated: land–sea temperature contrast; the coriolis force; friction; vertical variations in eddy diffusivity; vertical stability; the geostrophic wind; and topography. In the latter case both idealized and real topographies have been studied. Theoretical analysis of these features was initially analytical but after 1955 numerical modelling became the main method of investigation. The bulk of the numerical models are two-dimensional and time dependent, but three-dimensional models do exist. This section reviews the theoretical investigation by both analytical and numerical techniques of the several effects listed above.

A. Analytical results

Arguably the first attempt at a quantitative theory of sea/land breezes lies once more in the classical paper by Jeffreys (1922). Within his threefold classification of winds (see Chapter 1) Jeffreys recognized an antitriptic wind as one where the pressure gradient force is primarily balanced by frictional forces. With the aid of a simple dimensional analysis Jeffreys concluded that sea/land breezes exemplify one type of antitriptic wind. For a land/sea temperature difference of 20 °C over a distance of 30 km the theory predicted a surface wind of about 8 m s^{-1} with zero wind at 150 m altitude and a return current above the sea breeze. In view of the many approximations made in the analysis and its comparative brevity and simplicity, there is much validity in Jeffreys' (1922, p. 42) claim that "the theory gives a good first approximation to the truth".

In his 1922 paper, Jeffreys considered winds as a whole: indeed the main object of the paper was to investigate the possibility of classifying winds within the context of their dynamics. Consequently Jeffreys' consideration of the sea/land breeze was of necessity somewhat superficial. About a decade later the sea/land breeze circulation was analysed in more detail by Kobayasi and Sasaki (1932) and Arakawa and Utsugi (1937). These papers laid the foundation for all subsequent work in that they explicitly recognized and attempted to model the dynamical processes involved in sea/land breeze generation—i.e. that the land/sea difference causes horizontal temperature gradients that cause horizontal density gradients that in turn are related to horizontal pressure gradients. The breezes blow of course in response to the pressure gradients. The central theoretical problem in these early stages was to understand the link between an assumed temperature distribution which accorded with land/sea thermal differences and the pressure and motion fields that form in response to it. Later in the development of the theory we see that the temperature field itself is predicted from the distribution of land and water. Prior to the 1950s the investigation of the thermodynamics of the sea/land breeze was restricted to analysis of the linearized equations of motion, yet fundamental insights were gained, particularly between 1947 and 1950. The advent of computers in the 1950s meant that the non-linear terms of the equations of motion could be accommodated by numerical methods and sea breeze modelling progressed apace.

In the 15 years after World War II five important analytical papers appeared in the literature (Haurwitz, 1947, 1959; Schmidt, 1947; Pierson, 1950; Defant, 1950). All five were primarily concerned with explaining the effects of temperature, friction and the coriolis force on the character of the sea/land breeze. Two of the papers, those by Haurwitz (1947) and Schmidt (1947) reach very similar conclusions by different routes. The authors had identified the same important problems and clearly must have been working on them at the same time: their manuscripts were received for publication within three days of each other and the papers were published together in the same issue of the *Journal of Meteorology*.

Haurwitz (1947, p. 1) clearly stated that "friction and the deflecting force of the earth's rotation are the specific factors discussed in this paper". He analysed the effect of friction with the aid of Bjerknes' circulation theorem (see Hess, 1959, Chapter 16). As applied to this problem it tells us that in the absence of friction, sea breeze intensity (as measured in terms of circulation, C would increase as long as the land-air temperature (T_a) is higher than the sea-air temperature (T_b), i.e. $dC/dt \neq 0$ when $T_a - T_b > 0$, where C is the circulation. Haurwitz showed that, owing to the effect of friction, maximum sea breeze intensity occurs not when $T_a - T_b$ has decreased to zero, but earlier while the land is still warmer than the sea. A specific positive temperature difference is required just to overcome the frictional force. As soon as $T_a - T_b$ has fallen below this critical value, the sea breeze starts to slow down.

Haurwitz considered a two-dimensional vertical circulation with the x-axis perpendicular to the coastline. Coriolis forces were ignored in the analysis of friction as they do not affect the argument. The frictional force was assumed to be opposite to and proportional to the wind velocity (the Guldberg–Mohn friction formula).

The equations of motion as used by Haurwitz were

$$\frac{du}{dt} + k_f u = -\frac{1}{\rho}\frac{\partial p}{\partial x}, \tag{52}$$

$$\frac{dw}{dt} + k_f w = -\frac{1}{\rho}\frac{\partial p}{\partial z} - g, \tag{53}$$

where k_f is a constant that expresses the intensity of the frictional force. In reality du/dt and dw/dt were treated as $\partial u/\partial t$ and $\partial w/\partial t$, i.e. the equations were linearized. If equations (52) and (53) are multiplied by dx and dz, respectively, and added, the resulting equation may be integrated around a closed curve in the x, z plane to give the rate of change of circulation C:

$$\frac{dC}{dt} = \oint\left(\frac{du}{dt}dx + \frac{dw}{dt}dz\right) = -\oint\frac{dp}{\rho} - \oint g\,dz - k_f C. \tag{54}$$

The second integral on the right-hand side vanishes since g is a single-valued function of z. Integration of the first term on the right-hand side is performed around a quadrilateral in the vertical plane, bounded on upper and lower side by pressure surfaces (p_1 and p_0) and on its vertical sides by a T_a isotherm over land and a T_b isotherm over the sea (see Fig. 66). Equation (54) can then be written as

$$\frac{dC}{dt} = R(T_a - T_b)\ln\frac{p_0}{p_1} - k_f C, \tag{55}$$

where p_0 and p_1 are the lower and upper pressures respectively. If the length of the path of integration is L and the acceleration along L is $d\bar{V}/dt$, then $dC/dt = L(d\bar{V}/dt)$ and $C = \bar{V}L$, where $\bar{V}$ is the average velocity. Therefore equation (55) may be rewritten as

$$\frac{d\bar{V}}{dt} + k_f \bar{V} = (T_a - T_b)\frac{R}{L}\ln\frac{p_0}{p_1}. \tag{56}$$

Haurwitz then assumed that the temperature difference, which is a periodic function of time, may be expressed as follows:

$$\frac{R}{L}\ln\frac{p_0}{p_1}(T_a - T_b) = A\cos\Omega t. \tag{57}$$

This assumption implies that the time is reckoned from the instant when $T_a - T_b$ reaches its maximum. The introduction of this substitution into equation (56) allows integration to give

$$\bar{V} = \text{Const } e^{-k_f t} + A(k_f^2 + \Omega^2)^{-1}(\Omega\sin\Omega t + k_f\cos\Omega t). \tag{58}$$

Haurwitz then put $\chi = \tan^{-1}(\Omega/k_f)$ and the expression for $\bar{V}$ became

$$\bar{V} = \text{Const } e^{-k_f t} + A(k_f^2 + \Omega^2)^{-1/2}\cos(\Omega t - \chi). \tag{59}$$

The arbitrary constant in equation (58) was assumed by Haurwitz to be zero because in the absence of a temperature difference ($A = 0$) the wind should be zero. But the temperature difference and wind ($\bar{V}$) are zero at time $t = \pi/2\Omega$, in

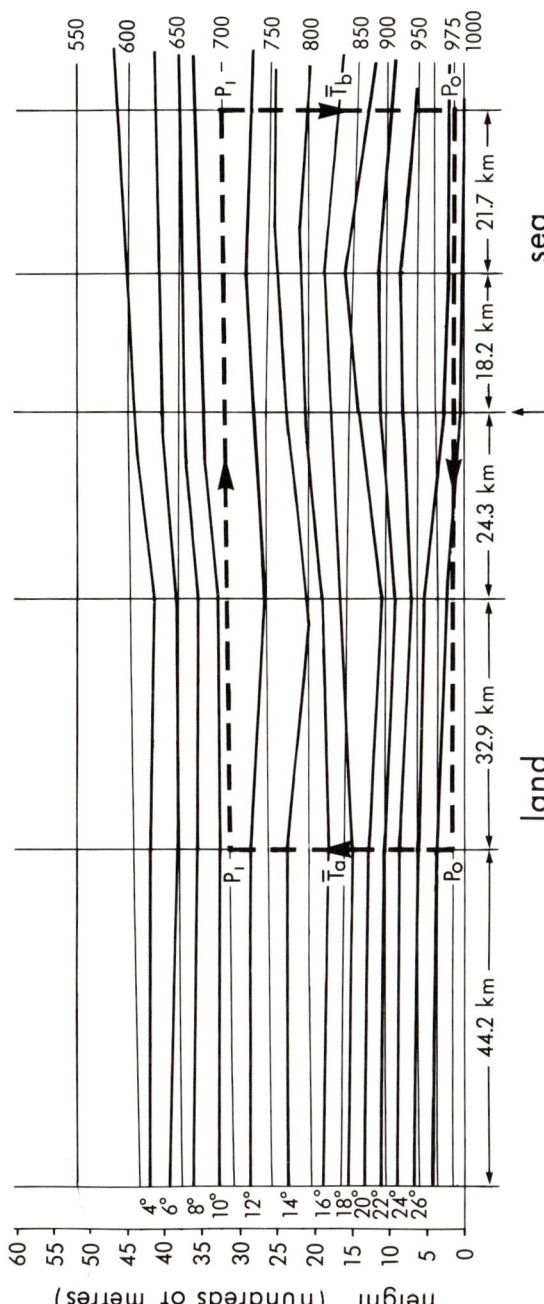

Fig. 66. Baroclinic and barotropic fields associated with a sea breeze over the Texas coast. Thin lines are isobaric surfaces (in millibars); thick lines are isotherms (in degrees Celsius); heavy dashed line is the path of integration for the sea breeze circulation. (After Hsu, 1970.)

which case the constant is not zero; but as t increases the $e^{-k_f t}$ term becomes decreasingly important.

Equation (59) shows the effects of friction on the sea breeze. The phase angle χ shows how much time elapses between the occurrence of maximum temperature difference and the occurrence of maximum sea breeze intensity. If $k_f = 0$, i.e. there is no friction, then the phase angle is 90° or equivalent to a lag of 6 h. This means from equation (59) that the maximum sea breeze occurs when $t = \pi/2\Omega$ which in turn, from equation (57), is when the temperature difference between land and water decreases to zero. With increasing friction, the phase angle decreases, i.e. the time difference between the maxima of the sea breeze and of the temperature difference decreases. Haurwitz used the few observations available to him to estimate values of k_f (from 2×10^{-5} to $8 \times 10^{-5} \, \text{s}^{-1}$) and showed that the maximum sea breeze should occur about 3 h after the maximum temperature difference between land and water. He recognized that in reality the maximum sea breeze occurs at approximately the same time as the maximum temperature difference and claimed that the discrepancy was probably due to the crude measure of frictional effects employed in the study together with the omission of advective, non-linear effects. Nevertheless, the effect of friction had been clearly demonstrated. Hsu (1970) used Haurwitz's analysis to assess the effect of friction on the Texas sea breeze (Fig. 66). With $p_0 = 1000 \, \text{mb}$, $p_1 = 700 \, \text{mb}$, $L = 200 \, \text{km}$, $T_a - T_b = 5°C$, $(R/L) \ln(p/p_1)$ takes a value of $0.05 \, \text{cm s}^{-2} \, \text{K}^{-1}$ and $A = 0.25 \, \text{cm s}^{-2}$. Substitution of these figures into equation (58) with $k_f = 2 \times 10^{-5} \, \text{s}^{-1}$ gave a mean speed of the sea breeze perpendicular to the Texas coast of $8.8 \, \text{m s}^{-1}$. This was in agreement with Hsu's observations of the horizontal limbs of the sea breeze circulation. Complementary to Hsu's work. Feit (1969) applied Haurwitz's model to the Texas coast land breeze. With $p_0 = 1015 \, \text{mb}$, $p_1 = 1000 \, \text{mb}$, $L = 80 \, \text{km}$, $T_a - T_b = 8°C$ and $A = 4.37 \times 10^{-2} \, \text{cm s}^{-2}$ the mean land breeze speed was $5.8 \, \text{m s}^{-1}$. Measurements taken during the land breeze showed that the maximum winds were about $4-5 \, \text{m s}^{-1}$ at a height of 28 m: mean speeds were of course less.

In his analysis of the sea breeze, Jeffreys ignored the effects of the coriolis force, claiming that the circulation was too small to be affected by the Earth's rotation. In fact, observation in mid-latitudes has shown that as the sea breeze circulation grows horizontally from the coast throughout the day, the sea breeze does turn to blow approximately parallel to the coast—usually by late afternoon. This means that air particles within the circulation must be subject to the coriolis acceleration for several hours. Bearing in mind typical sea breeze velocities of $5-10 \, \text{m s}^{-1}$ over a horizontal range of about 100 km an air particle would lie in the sea breeze part of the sea breeze circulation for only about 3 h, probably not sufficient to account for the full veering effect on the wind. However, if the air particles rose at the sea breeze front and formed part of the return current they would still be under the influence of the coriolis force within the sea breeze circulation and thus encourage further veering of both the surface sea breeze and the upper return current. Such veering of both currents was observed by Fisher (1960). The above argument implies that the sea breeze circulation is largely a closed one, a conclusion which differs from Wexler's (1946, p. 273) comment that:

> It is not to be considered that the same air continually rotates within a closed circulation. The downward motion over the sea is completed through a gradual subsidence of the sea air as it spreads inland. The upward motion over the land is partially the sea air and partially the land air. As the sea breeze advances inland new air is continually brought into circulation.

However, it is significant that Wexler's paper appeared one year before Haurwitz's analysis of coriolis effects and contains virtually no mention of such effects. Perhaps surprisingly there appear to be no direct observations to

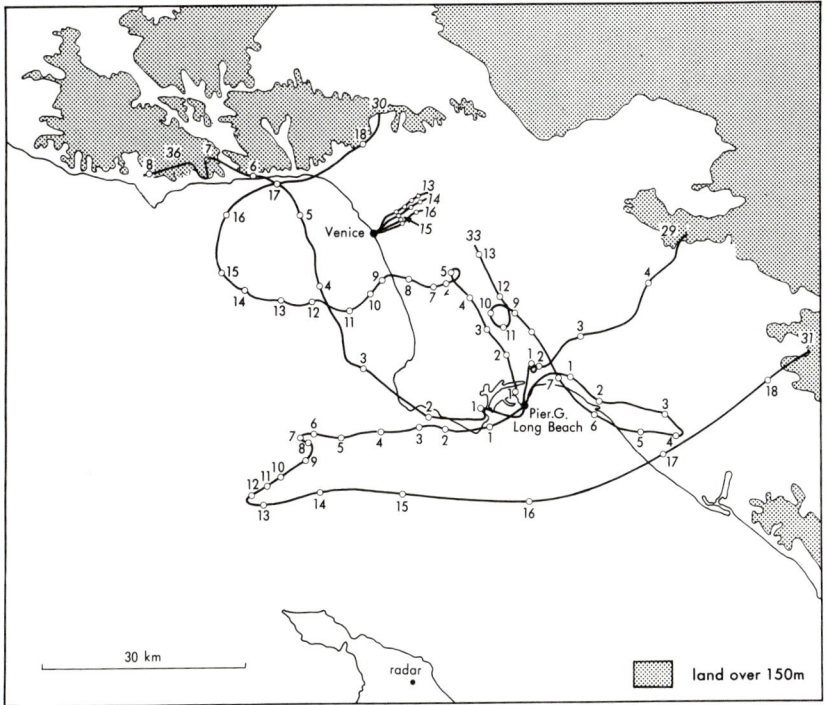

FIG. 67. Series of tetroon trajectories over Los Angeles basin. Those originating at Long Beach show the land and sea breezes; those originating at Venice (positions at 6 min intervals) show the veering with time of the sea breeze. In the Long Beach cases the number of hours after tetroon release are indicated next to the small circles; the tetroon flight number is at the end of the trajectory. (After Pack and Angell, 1963.)

test the validity of Wexler's statement. The nearest approach lies in the use of tetroons, ballons which allow the two-dimensional trajectories of small volumes of air to be followed. Unfortunately the tetroons are constrained to float at one pressure level but, despite this, Pack and Angell (1963) found that a typical single tetroon found itself in both land and sea breeze trajectories over a period of about 18 h (Fig. 67).

In a later paper, Angell and Pack (1965) showed that the tetroon trajectory direction itself veered with time, at a rate nearly in accord with the pendulum day period. Here, at last, was direct observation of a particle of air being accelerated by the coriolis force. Trajectory direction after $1\frac{1}{4}$ h was about 12° different from that at the beginning of the period. This means that at latitude 40°N (where Angell and Pack made their observations) any air particles initially moving perpendicular to a coastline in response to a thermally induced pressure gradient would take $7\frac{1}{2}$ times longer, i.e. about 9 h, to turn through 90° and blow parallel to the coast. At 30°N the appropriate pendulum day period is 12 h and Hsu (1970) found that the sea breeze in Texas as recorded at the beach turned through 90° in 12 h to blow parallel to the coast by 2100 hours local time. Ten miles out to sea the breeze blew perpendicular to the coast at 1300 hours local time and parallel by 0400 hours local time, an interval of 15 h. These results agree with the conclusions of Mizuma and Kakuta (1974) who claim that tetroons flown at 200 m altitude reveal "a closed circulation system of air flow in the vicinity of the coast" (Mizuma and Kakuta, 1974, p. 426). It is clear that the coriolis force affects the sea breeze direction as long as the air remains within the sea breeze circulation. If air in middle latitudes originated from well outside the sea breeze circulation at sea, took part in the circulation for (say) 6 h and then left the circulation inland, it is difficult to see how the circulation could turn through the 90° to give a wind parallel to the coast. If we assume no gradient wind, then it is the sea breeze circulation alone which causes the local operation of the coriolis effect. Air not in the circulation will not be subject to the effect and therefore will not turn parallel to the coast. As we observe that the breeze does indeed veer this must mean that in latitudes 30–60° at least it must be affected by the coriolis force for about 6–12 h. A typical sea breeze velocity of 8 m s^{-1} would move a particle of air over a horizontal distance of about 300 km in 10 h. As most sea breezes usually extend no more than 50 km landward and seaward, if the air particle continues to be affected by the coriolis force it must remain within the sea breeze circulation. Consequently the evidence suggests that the sea breeze circulation is largely closed and that it is this very closure which lends significance to coriolis effects.

By arguing that they are of the same order of magnitude as the coriolis terms, which he had ignored, Jeffreys chose also to ignore the local derivatives of the wind. Haurwitz argued, first qualitatively and then quantitively, that both the local derivatives and the coriolis terms must be included in a realistic dynamical theory of the sea breeze. He set up a simple model in which spatial changes and changes in air compressibilities were ignored. The equations of motion were as follows:

$$\frac{du}{dt} - fv + k_f u = P_x - F(t), \qquad (60)$$

$$\frac{dv}{dt} + fu + k_f v = P_y, \qquad (61)$$

where x is the horizontal axis positive from land to water, y is the horizontal axis parallel to the shore, P_x and P_y are components of large scale pressure gradient force and $F(t)$ is the periodic force caused by the variable temperature difference between land and water. It was chosen to be $A/\pi + \frac{1}{2} A \cos \Omega t$, where

$$A = \frac{1}{\rho} \frac{\partial p_0}{\partial x} = \frac{gz}{T} \frac{\partial T}{\partial x},$$

showing the relationship between horizontal temperature and pressure gradients and p_0 is the surface pressure. Haurwitz chose this particular form of $F(t)$ for the following reasons: a simple harmonic function did not adequately fit the observations at his disposal; the sea breeze itself moderates the diurnal rise of air temperature over land; while the temperature curve in day-time follows a harmonic function quite well under undisturbed conditions, nocturnal cooling follows a different regime, with the minimum temperature not occurring 12 h after the maximum temperature, but later, at sunrise; finally, observations at Boston showed that as a rule the air over land did not become as much colder than the air over water at night as it became warmer in day-time. Substituting $w = u + iv$ and $P_z = P_x + iP_y$, equations (60) and (61) were manipulated into

$$\frac{dw}{dt} + (if + k_f)w = P_z - F(t). \qquad (62)$$

This was integrated to give

$$w = \frac{P_z}{if + k_f} - \frac{A}{\pi} \frac{1}{if + k_f} - \frac{A\Omega \sin \Omega t + (if + k_f) \cos \Omega t}{2 \quad (if + k_f)^2 + \Omega^2} + C e^{-(if + k_f)t}. \qquad (63)$$

In the fourth term C is an arbitrary integration constant. Since the velocity w should vanish if both parts of the pressure gradient force are zero, namely $P_z = 0$ and $A = 0$, it follows that $C = 0$. The first term represents the motion caused by the constant part of the pressure gradient force. The second and third terms represent the effect of the pressure gradient which is due to the temperature difference between land and water. Equation (63) separates into real and imaginary parts to give

$$u = \frac{k_f P_x + f P_y}{f^2 + k_f^2} - \frac{A}{\pi} \frac{k_f}{f^2 + k_f^2} - M \sin(\Omega t + \chi), \qquad (64)$$

where, in this case,

$$\tan \chi = \frac{k_f(k_f^2 + \Omega^2 + f^2)}{\Omega(k_f^2 + \Omega^2 - f^2)}$$

and

$$M = \frac{A}{2} \left(\frac{k_f^2 + \Omega^2}{(k_f^2 + \Omega^2 - f^2)^2 + 4f^2 k_f^2} \right)^{1/2},$$

and also

$$v = \frac{-f P_x - k_f P_y}{f^2 + k_f^2} + \frac{A}{\pi} \frac{f}{f^2 + k_f^2} - N \cos(\Omega t + \psi), \qquad (65)$$

where

$$\tan \psi = \frac{2 k_f \Omega}{\Omega^2 - f^2 - k_f^2}$$

and

$$N = \frac{Af}{2} [(k_f^2 + \Omega^2 - f^2)^2 + 4f^2 k_f^2]^{-1/2}.$$

Haurwitz himself admitted that equations (64) and (65) are difficult to discuss in general terms because of the auxiliary constants. He proceeded by taking a numerical example in which $k_f = 0.58 \times 10^{-4} \, \text{s}^{-1}$, $f = 10^{-4} \, \text{s}^{-1}$, $\Omega = 0.7 \times 10^{-4} \, \text{s}^{-1}$, $A = 0.048 \, \text{cm s}^{-2}$. This gives the following values: $\chi = 95° \, 13'$, $\psi = 116° \, 41'$, $M = 192 \, \text{cm s}^{-1}$, $N = 204 \, \text{cm s}^{-1}$ and

$$\frac{k_f P_x - f P_y}{f^2 + k_f^2} = -0.436 v_g + 0.753 u_g,$$

$$-\frac{f P_x - k_f P_y}{f^2 + k_f^2} = 0.753 u_g + 0.436 v_g,$$

where u_g and v_g are the components of the geostrophic wind,

$$\frac{A}{\pi} \frac{k_f}{f^2 + k_f^2} = 66.5 \, \text{cm s}^{-1}; \quad \frac{A}{\pi} \frac{f}{f^2 + k_f^2} = 115 \, \text{cm s}^{-1}.$$

Figure 68 shows typical hodographs of winds resulting from this calculation when the geostrophic wind was taken as $5 \, \text{m s}^{-1}$. When Haurwitz varied the value of the coefficient of friction, hodograph shapes changed; with an increase of k_f to $10^{-4} \, \text{s}^{-1}$ the hodographs took on a much flatter elliptical

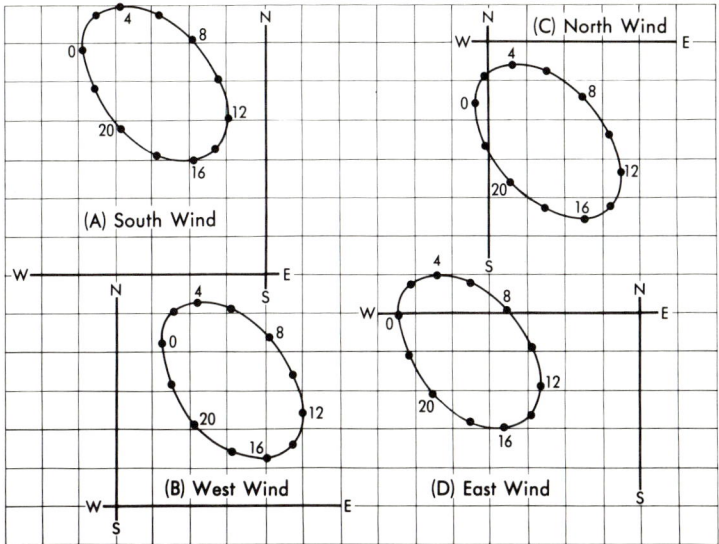

FIG. 68. Hodographs of theoretical sea breeze for different directions of the general geostrophic wind of speed $5 \, \text{m s}^{-1}$. One side of a square corresponds to $1 \, \text{m s}^{-1}$. The dots mark the end points of vectors from the origin at the intersection of the axes and the numbers represent the duration (in hours) from the time of maximum difference between the air temperatures over land and water. In all cases land lies to the left and water to the right of the N–S line. (After Haurwitz, 1947.)

shape. Haurwitz's theoretical hodographs bore a strong resemblance to those available from observations at the time. Subsequent observations have confirmed his theory, despite its rudimentary nature compared with present-day models. Haurwitz was clearly aware of the shortcomings of his theory, mentioning in particular his omission of advective effects, the variation in the vertical, the effect of eddy viscosity and atmospheric stability. Most of these features were included in later models as outlined below.

Schmidt's (1947) theory of sea breezes also showed the effects on the sea breeze circulation of both frictional and the coriolis force—but using a different method to that employed by Haurwitz. Whereas Haurwitz specified reasonable values for the diurnal temperature changes that ultimately drive the sea breeze, considered the atmosphere to be incompressible and ignored variations in the vertical, Schmidt included air compressibility, considered vertical variations and specified the temperature distribution in the horizontal, the vertical and through time. In a further attempt to clarify the coriolis effects Schmidt analysed the sea/land breeze circulation in both equatorial regions and mid-latitudes. In common with Haurwitz he represented frictional effects by a force proportional to the wind speed (the

Guldberg–Mohn friction formula) but added a consideration of vertical variation of frictional effects.

For sea/land breezes in equatorial areas, where coriolis effects were ignored and attention was concentrated upon frictional effects, Schmidt logically took as his starting point the temperature distribution, described as follows:

$$T_z = T_0 - \bar{\gamma}z + \tfrac{1}{2}\tau_0 \, e^{-az} \sin \Omega t [1 + \sin (\pi x/\lambda)], \qquad (66)$$

where T_z is the temperature at a level z, T_0 is the temperature at the surface, $\bar{\gamma}$ is the mean lapse rate of temperature, τ_0 is the amplitude of the daily radiative temperature wave inland well away from the coast, a is a parameter representing the decrease of the temperature amplitude with height, x is the horizontal distance reckoned from the coastline, positive inland and $\tfrac{1}{2}\lambda$ is the distance from the coastline to the line where the daily temperature wave reaches its full amplitude τ_0. This equation gives an approximate description of the temperature distribution near a coastline, including variation with x for $x > 0$. Haurwitz did not include this variation. Equation (66) neglects advective effects and assumes an exponential decrease of the temperature amplitude with height, both shortcomings clearly recognized by Schmidt but retained for their simplifying effect on the problem. Following his own earlier analysis (Schmidt, 1946), Schmidt related the air density change due to the temperature change in the following way:

$$\frac{\partial \rho}{\partial t} = -\frac{\pi}{T} \delta_0 \, e^{-(a+b)z} \{(1-(a+b)z\} \cos \Omega t \left(1 + \sin \frac{\pi x}{\lambda}\right), \qquad (67)$$

where δ_0 is the amplitude of the daily density range and b is a parameter representing the decrease of density with height. Pressure changes were considered by Schmidt to be partly due to the density changes resulting from the temperature changes, but more importantly due to the divergence and convergence of the compressible air. Schmidt claimed that the influence of divergence on pressure changes decreases with increasing height approximately according to an exponential law (Schmidt, 1946) and thus represented the pressure change at a level z as follows:

$$\frac{\partial p}{\partial t} = -\int_0^z g \frac{\partial \rho}{\partial t} dz - e^{-sz} \int_0^\infty g \frac{\partial (\rho u)}{\partial x} dz, \qquad (68)$$

where $1/s$ indicates the level at which the influence of divergence is reduced to $1/e$ of its value at the ground. The horizontal motion which results from the horizontal pressure gradients caused by pressure changes described above is governed by the following more familiar equation:

$$\frac{\partial u}{\partial t} = -\frac{1}{\rho}\frac{\partial p}{\partial x} - k_f u, \qquad (69)$$

5 SEA/LAND BREEZE CIRCULATION

where k_f is a function of height. $\partial u/\partial t$ is used instead of du/dt because advective effects are ignored in this "elementary theory", to use Schmidt's own description. Combination of equations (68) and (69) gives the following equation for the horizontal velocity component due to the land/sea breeze circulation:

$$\frac{\partial^2 (\rho u)}{\partial t^2} = g \int_0^z \frac{\partial^2 \rho}{\partial t\, \partial x} dz + g\, e^{-sz} \int_0^\infty \frac{\partial^2 (\rho u)}{\partial x^2} dz - k_f \frac{\partial (\rho u)}{\partial t}. \tag{70}$$

By ignoring $\partial \rho/\partial t$ with respect to $\partial u/\partial t$ in subsequent manipulation, equation (70) reduces to

$$\frac{\partial^2 u}{\partial t^2} = -\frac{\pi^2 g \delta_0 \alpha_0}{\lambda T} \cos \Omega t \, \frac{\cos \pi x}{\lambda}$$

$$\times \left(z\, e^{\{h-(a+b)\}z} - \frac{s}{(a+b)^2} e^{(h-s)z} \right) - k_f \frac{\partial u}{\partial t}, \tag{71}$$

where $\alpha_0\, e^{hz}$ is written for $1/\rho$. The solution to equation (71) is of the form

$$u = u_m \sin(\Omega t + \beta) \cos(\pi x/\lambda), \tag{72}$$

where $\beta = \tan^{-1}(-\Omega/k_f)$ and u_m is a function of z given by

$$u_m = -\frac{\pi^2 g \delta_0 \alpha_0/(\lambda T)}{\Omega k_f \cos \beta - \Omega^2 \sin \beta} \left(z\, e^{\{h-(a+b)\}z} - \frac{s}{(a+b)^2} e^{(h-s)z} \right). \tag{73}$$

Solutions are available from equations (72) and (73) on the insertion of suitable values for the constants. Schmidt took the following values: $k_f = (k_f)_0\, e^{-rz}$ where $(k_f)_0 = 1.5 \times 10^{-4}\, \text{s}^{-1}$ and $r = 10^{-3}\, \text{m}^{-1}$; a varied from 1×10^{-3} to $2 \times 10^{-3}\, \text{m}^{-1}$; b was equal to about $10^{-4}\, \text{m}^{-1}$; and s was estimated to be $10^{-3}\, \text{m}^{-1}$. Using these values Schmidt found good agreement between the vertical distribution of the theoretical sea breeze at its maximum development and the observations at Batavia: the agreement for land breezes was not so good (Fig. 69).

By applying the same kind of reasoning Schmidt elucidated the coriolis effects on mid-latitude sea/land breeze circulations. In this case the equations of motion were

$$\frac{du}{dt} = fv - \frac{1}{\rho} \frac{\partial p}{\partial x} - k_f u, \tag{74}$$

$$\frac{dv}{dt} = -fu - k_f v, \tag{75}$$

where x is perpendicular to the coast, positive towards inland and y is along the coast.

It was assumed that there was no component of the induced pressure

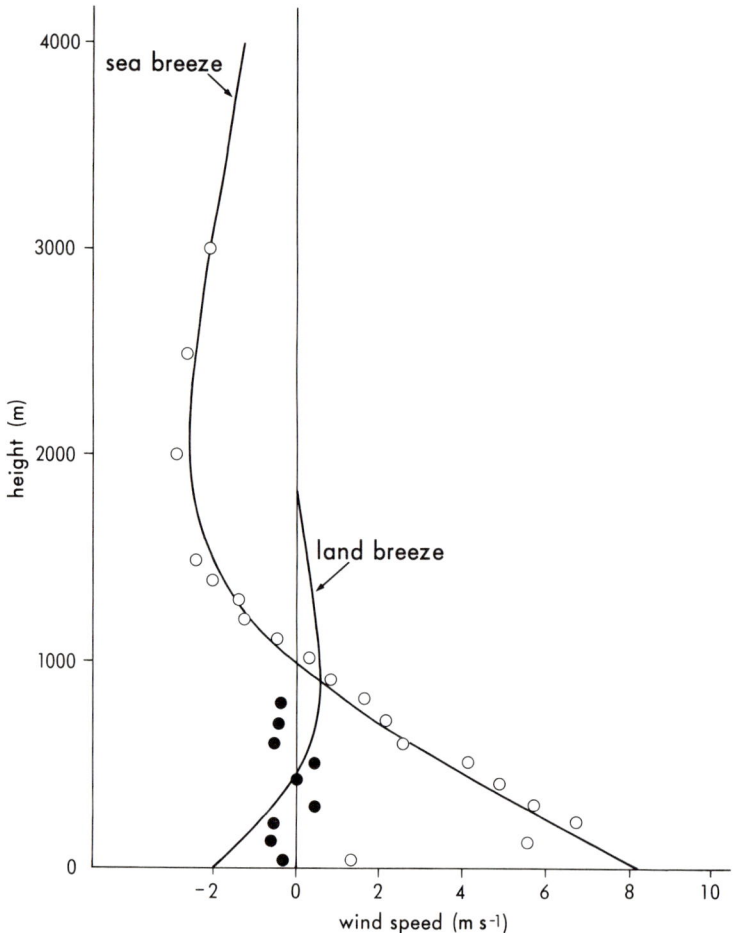

FIG. 69. Curves showing the variation with height of theoretical maximum sea and land breezes. Open and closed circles give values for sea and land breezes respectively as observed at Batavia. (After Schmidt, 1947.)

gradient force in the y direction. Also $\partial v/\partial t$ was substituted for dv/dt as the circulation in the immediate neighbourhood of the coast was considered.

The solutions of equations (74) and (75) are of the following form:

$$u = u(z)\cos(\pi x/\lambda)\sin(\Omega t + \beta), \tag{76}$$

$$v = v(z)\cos(\pi x/\lambda)\sin(\Omega t + \beta + \beta'), \tag{77}$$

where $u(z)$ and $v(z)$ are velocities which are functions of height. The second and third terms in both equations are the modifications on $u(z)$ and $v(z)$ with distance from the coast and with time respectively. Schmidt demonstrated that

$$\tan \beta' = -\Omega/k_f,$$

$$v(z) = Au(z)/\Omega f,$$

$$A = -\Omega f^2/(\Omega^2 + k_f^2)^{1/2}.$$

Further manipulation of the equations showed that

$$\tan \beta = -\frac{2\pi - (AT/2\pi)\sin \beta'}{k_f T - (AT/2\pi)\cos \beta'}$$

and

$$u(z) = -\frac{\pi g \delta_0 \alpha_0}{\lambda} \left(z e^{\{h-(a+b)\}z} - \frac{s}{(a+b)^2} e^{(h-s)z} \right)$$
$$\times \left(2k_f \cos \beta - 2\Omega \sin \beta + \frac{A}{\pi}(\sin \beta \sin \beta' - \cos \beta \cos \beta') \right)^{-1}. \quad (78)$$

Consequently the velocity distribution within a coastal region where the land/sea breeze circulation has developed was given by

$$u = u(z)\sin(\Omega t + \beta)\cos\frac{\pi x}{\lambda}, \quad (79)$$

$$v = \frac{Au(z)}{\Omega f}\sin(\Omega t + \beta + \beta')\cos\frac{\pi x}{\lambda}, \quad (80)$$

where $u(z)$ is given by equation (78) and $Au(z)/\Omega f = v(z)$. This analysis clearly showed the coriolis effects in giving a v component, notably absent in the analysis of the low latitude case. Whereas the vertical distribution of the speed of the land/sea breeze is of the same character as that occurring in tropical regions, equations (79) and (80) show that the wind varies continually in direction. With the substitution of realistic values for the constants a veering sea breeze which accorded well with observations was predicted.

In broad strategy Pierson's (1950) study was identical to Schmidt's. The pressure gradient force that drives the land/sea breeze circulation was derived from the diurnal temperature changes across a land–water boundary. This pressure gradient force was then combined with coriolis and frictional effects in the equations of horizontal motion to predict the variation of the circulation in space and time. The main contributions of Pierson's study were in his more realistic modelling of the vertical variation of the amplitude of daily temperature changes with the aid of eddy diffusion and the representation of frictional forces in the equations of motion by the terms $K_z^M(\partial^2 u/\partial z^2)$ and $K_z^M(\partial^2 v/\partial z^2)$. The latter gave rise to an Ekman spiral, a realistic feature not simulated by either Haurwitz or Schmidt.

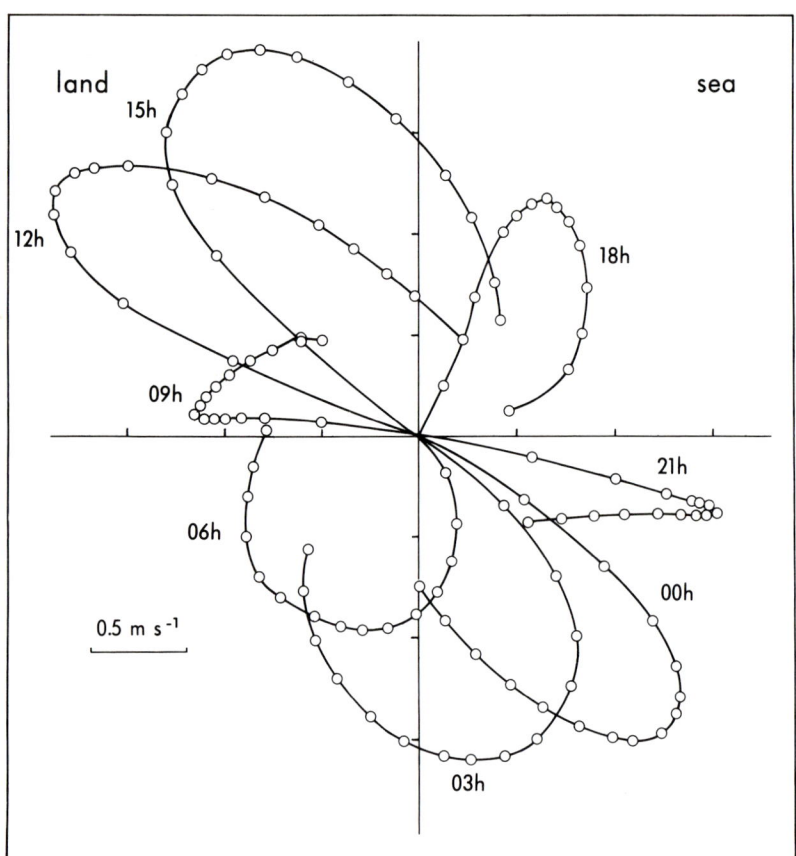

FIG. 70. Hodographs showing variation of winds with both height and time in a theoretical sea/land breeze circulation. On each curve (labelled with appropriate time) the open circles indicate the end points of vectors from the origin and are 50 m apart up to a height of 500 m, then 100 m apart to a height of 1 km. (After Pierson, 1950.)

Pierson computed both hodographs and trajectories for conditions typical of Boston. Figure 70 shows how the sea/land breeze veered with height at a particular time and also how the whole Ekman spiral itself veered in response to the coriolis effects. The theory also, not surprisingly, predicted that, in maturity, there should be a return current aloft. For example, at 1500 hours the wind becomes offshore above a height of about 640 m (Fig. 70). Pierson's calculated trajectories suggested that, despite the omission of vertical velocities, an air particle "would tend to return nearly to its starting point after twenty four hours and would thus be expected to remain under the influence of the applied forces for a greater length of time and produce a condition more nearly like the non-transient state" (Pierson, 1950, p. 22). This is consistent with the earlier argument that the sea breeze circulation is probably largely a closed one. In addition to the general mid-latitude case of the land/sea breeze Pierson simulated the circulations at Boston and Batavia in particular because observations were available from these stations. He found very good agreement between theory and observation at Boston for a theoretical situation where the sea breeze (*not* the whole vertical circulation) was 300 m deep. The model also allowed the calculation of the vertical circulation at Batavia to compare with van Bemellen's (1922) observations. Figure 71 shows the theoretical time–height section calculated with an eddy viscosity K_z^M equal to $6.7 \times 10^4 \, \text{cm}^2 \, \text{s}^{-1}$. Although the predicted offshore winds at night at the surface and the onshore winds aloft were too strong, the overall agreement with observations is impressive.

The fourth important post-war analytical study of the land/sea breeze was by Defant (1950) who took a completely different view of the problem from the earlier authors. Defant treated the land/sea breeze as a single circulation cell as in convection theory and explicitly considered both vertical and horizontal heat transfer together with frictional and coriolis effects. This inclusion of more realistic heating functions (albeit for transfer in the fluid) was the major contribution of Defant's theory.

The temperature distribution across the coastline was simulated by

$$\theta = M \, e^{i\Omega t} \sin lx, \tag{81}$$

where M is the amplitude of the temperature variation, l determines the

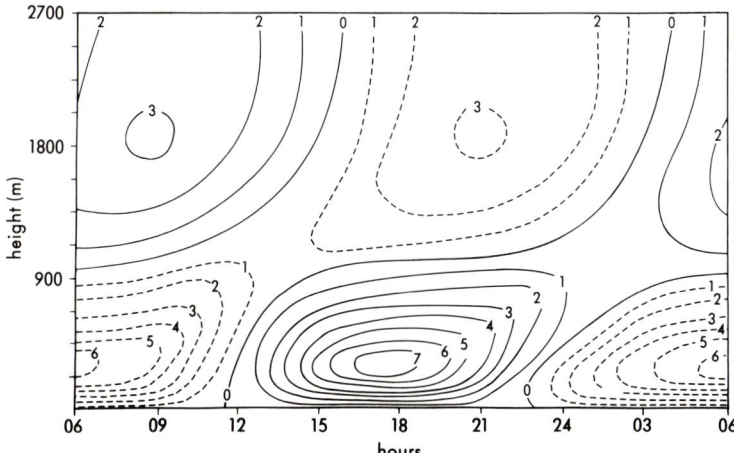

FIG. 71. Time height cross-section for a theoretical sea breeze at Batavia. Velocities in metres per second; solid lines onshore and dashed lines offshore. (After Pierson, 1950.)

length of the circulation cell ($L/2$) by the relation $L/2 = \pi/l$ and x is the direction normal to the coast, positive toward inland. The equations of motion used in the analysis were

$$\frac{\partial u}{\partial t} = fv - k_f u - \frac{1}{\rho}\frac{\partial \bar{\omega}}{\partial x},$$

$$\frac{\partial v}{\partial t} = -fu - k_f v, \qquad (82)$$

$$\frac{\partial w}{\partial t} = -k_f w - \frac{1}{\rho}\frac{\partial \bar{\omega}}{\partial z} + \gamma\theta,$$

where $\bar{\omega}$ is related to pressure (p) by the following relationship:

$$p = \bar{\omega} - \gamma\rho_0 \int \theta\, dz,$$

where ρ_0 is density in a homogeneous atmosphere; $\gamma = g\alpha$, where α is the coefficient of thermal expansion. In addition to these equations Defant treated the air as incompressible and used the continuity equation in the form

$$\frac{\partial u}{\partial x} + \frac{\partial w}{\partial z} = 0 \qquad (83)$$

and the heat transfer equation:

$$\frac{\partial \theta}{\partial t} + \frac{\partial \theta}{\partial z} w = K^H \left(\frac{\partial^2 \theta}{\partial x^2} + \frac{\partial^2 \theta}{\partial z^2}\right). \qquad (84)$$

All these equations are very similar to those used by Arakawa and Utsugi (1937) with two differences. Defant included the coriolis term in this equation whereas Arakawa and Utsugi did not; and Defant used the Guldberg–Mohn frictional formula in contrast to the $K^M \nabla^2 u$ and $K^M \nabla^2 v$ terms used by the Japanese workers. Nevertheless, the results of both studies show an agreement to be expected from the similarity in analyses. In turn, the Japanese Yamashita (1953) also closely followed the earlier methods of his countrymen with similar results.

The solutions of equations (82), (83) and (84) are of the form

$$u = u(z)e^{i\Omega t}\cos lx,$$
$$v = v(z)e^{i\Omega t}\cos lx,$$
$$w = w(z)e^{i\Omega t}\sin lx,$$
$$\bar{\omega} = \bar{\omega}(z)e^{i\Omega t}\sin lx,$$
$$\theta = \theta(z)e^{i\Omega t}\sin lx.$$

On the insertion of boundary conditions that $z = 0$, $w = 0$ and $\theta = M$ and that for large values of z the circulation becomes negligibly small, the solutions are

$$u = \frac{rM}{(a^2-b^2)l}(a e^{+az} + b e^{-bz})e^{i\Omega t}\cos lx,$$

$$v = -\frac{f}{i\Omega + k_f}u,$$

$$w = -\frac{rM}{(b^2-a^2)}(e^{+az} - e^{-bz})e^{i\Omega t}\sin lx,$$

$$\theta = M\left\{e^{-bz} + \frac{b^2-s}{b^2-a^2}(e^{+ax} - e^{-bx})\right\}e^{i\Omega t}\sin lx,$$

where a, b, r and s are functions of the constants l, Ω, k_f, f, K^H, γ.

Defant's results highlighted the effects of frictional and coriolis forces on u, v and w for a given temperature amplitude. Working on the basis of maximum land–water temperature difference at 1200 hours, the theory showed that, in the absence of a coriolis force, the depth of the sea breeze increased with friction from 320 m to about 500 m. This apparently paradoxical result occurred possibly because the friction increased the efficiency of vertical turbulent heat flux although in the model vertical heat flux and friction were independent mechanisms. The effect of friction on the velocity is qualitatively as expected, reducing the values from about 5.5 m s^{-1} to 1.7 m s^{-1} per unit of temperature amplitude (usually 1 °C). Arakawa and Utsugi (1937) had previously calculated that for a horizontal land–sea temperature difference of 1 °C the maximum sea breeze velocity was 1.6 m s^{-1} and that for a temperature difference of 10 °C the sea breeze velocity is about 16 m s^{-1}. Defant found a velocity of 10 m s^{-1} for a temperature difference of 5 °C. The introduction of a realistic coriolis parameter for latitude 45° of 1.03×10^{-4} s^{-1} did not radically alter these frictional effects, but did lead to the appearance of the v component of the velocity. The theory also shows that frictional effects in particular reduced the lag between maximum temperature difference and maximum intensity of the sea breeze from 4.7 h to 1.4 h. The upper level return current was also modelled, being four to five times deeper than the sea breeze layer but only barely a quarter of its intensity, a result expected from the continuity equation. The vertical velocities of about 2 cm s^{-1} per unit temperature difference are perhaps what would be expected due to the omission of sea breeze frontal characteristics in the model.

It is fitting that one of the last major contributions to linear theory of the sea breeze was written by one of its pioneers, B. Haurwitz (1959), at the end of the decade which witnessed the first substantial use of computers (and thus non-linear models) in meteorology. In turn, Haurwitz recognized the insight of an earlier pioneer with the words: "It is remarkable that its [Haurwitz's model] results resemble in many aspects quite closely those of Jeffreys (1922), even though Jeffreys considered a steady model only" (Haurwitz, 1959, p. 2).

Haurwitz's 1959 model accorded with earlier ones quoted in being constructed in a framework with a coastline parallel to the y-axis and allowing no variations in the y direction. The equations of motion in standard notation were as follows:

$$\frac{\partial u}{\partial t} + u\frac{\partial u}{\partial x} + w\frac{\partial u}{\partial z} - fv = -\frac{1}{\rho}\frac{\partial p}{\partial x} + \frac{1}{\rho}\frac{\partial}{\partial z}\left(K_z^M \frac{\partial u}{\partial z}\right), \tag{85}$$

$$\frac{\partial v}{\partial t} + u\frac{\partial v}{\partial x} + w\frac{\partial v}{\partial z} + fu = \frac{1}{\rho}\frac{\partial}{\partial z}\left(K_z^M \frac{\partial v}{\partial z}\right), \tag{86}$$

$$\frac{\partial w}{\partial t} + u\frac{\partial w}{\partial x} + w\frac{\partial w}{\partial z} = -\frac{1}{\rho}\frac{\partial p}{\partial z} - g. \tag{87}$$

In equation (86) there is no pressure gradient force because of the assumption of independence from y. Consequently the sea breeze component v parallel to the coast is entirely due to the action of the coriolis acceleration.

The airflow was considered to be incompressible and thus the continuity equation took on the form

$$\frac{\partial u}{\partial x} + \frac{\partial w}{\partial z} = 0. \tag{88}$$

Consequently a stream function ψ exists such that

$$u = -\frac{\partial \psi}{\partial z}, \qquad w = \frac{\partial \psi}{\partial x}.$$

The system of equations was completed by an equation governing the temperature variation:

$$\frac{\partial \theta}{\partial t} + u\frac{\partial \theta}{\partial x} + w\frac{\partial \theta}{\partial z} = \frac{\partial}{\partial z}\left(\frac{K_z^H}{\rho}\frac{\partial \theta}{\partial z}\right). \tag{89}$$

The above equations were then linearized by considering the sea breeze to cause small deviations in u, w and v from an equilibrium state—i.e. the atmosphere at rest. The deviations were supposed to be small enough so that terms made up of them could be neglected if they were of higher order than the first. The introduction of the linearization assumption and of the assumption that $K_z^M/\rho = K_z^H/\rho = k = $ const resulted in the following equations:

$$\frac{\partial v}{\partial t} - f\frac{\partial \psi}{\partial z} = k\frac{\partial^2 v}{\partial z^2}, \tag{90}$$

$$-\frac{\partial}{\partial t}(\nabla^2 \psi) - f\frac{\partial v}{\partial z} = -\frac{g}{\bar{\theta}}\frac{\partial \theta}{\partial x} - k\frac{\partial^3 \psi}{\partial z^3}, \tag{91}$$

$$\frac{\partial \theta}{\partial t} + \bar{J}\frac{\partial \psi}{\partial x} = k\frac{\partial^2 \theta}{\partial z^2}, \qquad (92)$$

where $\bar{\theta}$ is the mean daily potential temperature in the absence of the sea breeze and of the diurnal temperature variation, θ represents deviations from the mean and $\bar{J}$ is $\partial \bar{\theta}/\partial z = \text{const.}$

The boundary conditions were $u = v = w = 0$ at $z = 0$, $\theta = F(x, t)$ at $z = 0$ and u, v, w, θ finite as $z \to \infty$. Haurwitz noted that solutions to equations (90)–(92) exist in the following form:

$$\psi = A\,e^{ivt} \cos mx\,e^{\alpha z},$$
$$v = B\,e^{ivt} \cos mx\,e^{\alpha z},$$
$$\theta = C\,e^{ivt} \sin mx\,e^{\alpha z}.$$

Here v is the circular frequency of the sea breeze which Haurwitz put equal to the Earth's rotation rate Ω. The quantity m depended on the horizontal extent, π/m, of the sea breeze. The value of α had to be found from the differential equations of the problem and the values of A, B and C from the equations and the boundary conditions. Despite several physically sensible simplifications, Haurwitz's subsequent analysis results in expressions which are, to use his own words, "very complicated" and which require much "tedious labour" to produce the elements of the sea breeze circulation. Even then, to quote Haurwitz (1959, p. 18), "the expressions which describe the sea breeze circulation according to the simplified model assumed here are still too involved to permit a simple discussion". However, with further simplification, Haurwitz showed that, with a surface potential temperature amplitude of 5 °C, the sea breeze was about 400 m deep with its maximum velocity (about $2\,\text{m s}^{-1}$) at just under 200 m. The return current lay between 400 and 1200 m. The maximum vertical velocities (about $6\,\text{cm s}^{-1}$) occurred at about 400 m, the height of the boundary between sea breeze and return current.

Several lessons were learnt from this difficult exercise. The assumption that the surface temperature variation is described as follows,

$$\theta = \theta_0 \cos vt \sin mx,$$

was found to be inappropriate. More complicated functions could be introduced with an associated increase in the labour for solutions. If the coriolis force was included in the model the horizontal dimensions of the breeze could not be arbitrarily chosen. This result was physically plausible since the deflecting effect of the coriolis force depends on the horizontal distance travelled. The most important simplification made in the model was the omission of the advective effects on temperature. Thus no allowance was made for the fact that the sea breeze itself modifies the temperature field which provides its driving force. Despite this, the linear model does not give as bad an approximation as might be expected. The effects of eddy viscosity, static stability and land–sea temperature difference on the intensity of the sea breeze circulation and on the height of the return current aloft were all included in this model.

In concluding this section on analytical investigations of both frictional and coriolis effects on sea breezes it is appropriate to mention Neumann's (1977) results which relate principally to a modification of the latter effects. Neumann noted that the rate of turning of the sea and land breezes is not uniform over the diurnal cycle. He showed that the rate of local turning is given by

$$\frac{\partial \alpha}{\partial t} = -f + \frac{1}{V^2}\left[\frac{v}{\rho}\frac{\partial p_m}{\partial x} + f(uu_g + vv_g)\right], \qquad (93)$$

where α is the angle between the positive direction of the x-axis (perpendicular to the shore) and the direction of the wind, $V^2 = u^2 + v^2$ and p_m is the meso-scale pressure excess due to the diurnal heating/cooling of air over and relative to that over the sea.

Clearly the rate of turning is due to three effects. The first one is the coriolis effect elucidated by Haurwitz and others. The second is the cross-product of the horizontal meso-scale pressure gradient and the velocity of the breeze. The third involves the cross-product of the horizontal, large scale pressure gradient, assumed not to be affected by the diurnal heating, and the velocity of the breeze. All three terms on the right-hand side of equation (93) represent rotation about the vertical but, while the first term is constant, the other two are variable in both magnitude and sign. Neumann stressed that these two variable terms change the rate of turning in an important manner, and presented a few of the many possible combinations of effects.

B. Numerical results

In common with other meteorological theoreticians, those interested in the land/sea breeze were constrained prior to the 1950s by the necessity to linearize the equations of motion in order to ease their solutions. All the authors mentioned above were very aware of the limitations of their work, the primary one being the omission of the advective terms in the equations of motion. Any satisfactory theory of the sea breeze must consider the feedback effects between the velocity and temperature distributions. This became possible with the advent of the computer which allowed the calculation of solutions to the non-linear equations of motion by numerical methods. Probably the first attempt at a sea breeze calculation using the non-linear equations was that by Pearce (1955).

In his model Pearce calculated the velocity, temperature and pressure distributions of the sea breeze produced when an initially isothermal and static atmosphere is heated differentially across a long, straight coastline. Heat was assumed to be distributed vertically from the land surface by convection currents; internal friction was included but surface drag was taken as zero. In contrast to the works mentioned above, which started with assumed temperature distributions, Pearce postulated only the distribution in space and time of the heating and cooling sources and the initial and boundary conditions. The fields of wind, temperature and pressure were then all derived using the equations of heating and motion. To achieve a solution of the non-linear equations of motion Pearce used the fact that the velocity field was both rotational and divergent. It was by analysing this property that the equations were broken down into a form adaptable to numerical solution. The velocity field was found to comprise two components: one was a tidal motion of small velocity that affects a region of continental dimensions, whereas the other was a vertical circulation, centred on the coastline, with larger velocities than the tidal motion. This latter component corresponds to the observed sea breeze. In the original the whole calculation was within the Cartesian framework of Ox being the coastline, Oy being positive inland and Oz being the vertical direction; u, v and w are the velocities in these directions. For consistency and to aid comparison with other studies the coordinates have been changed to give Ox perpendicular to the coast positive inland and Oy along the coast.

In deriving the heating equations Pearce assumed that heat was distributed vertically by convection currents such that inland, well away from the sea breeze, the lapse rate was always dry adiabatic in the heated layer of depth H. The change in heat content of unit vertical column of the atmosphere not affected by the sea breeze was considered to be

$$\frac{\partial Q}{\partial t} = c_p \frac{\partial \theta_c}{\partial t} H \rho_{H/2}, \qquad (94)$$

where Q is the quantity of heat, θ_c is the potential temperature of the region of convection, H is the depth of the layer in which heat is distributed by convection currents and the subscript means the value at that height.

Because horizontal advection of heat occurs in the convection region, heat is taken up by air particles that are moving and thus $H(\partial \theta_c/\partial t)$ was replaced by

$$\int_0^H \left(\frac{\partial \theta_c}{\partial t} + u \frac{\partial \theta_c}{\partial x} \right) dz;$$

but $\partial \theta_c/\partial t$ and $\partial \theta_c/\partial x$ are independent of z, so this expression was written as

$$H \left(\frac{\partial \theta_c}{\partial t} + \bar{u} \frac{\partial \theta_c}{\partial x} \right)$$

where

$$\bar{u} = \frac{1}{H}\int_0^H u\,dz.$$

If H does not exceed 1.5 km the neglect of variation of ρ with z does not significantly affect the issue. The heat entering the base of a moving column of air was then given by

$$\frac{\partial Q}{\partial t} = c_p \rho_{H/2} H \left(\frac{\partial \theta_c}{\partial t} + \bar{u}\frac{\partial \theta_c}{\partial x}\right), \tag{95}$$

where Q is a given function of time and is taken to be independent of x (>0). Equation (95) becomes of more value later in the form

$$\frac{d\theta_c}{dt} = \frac{1}{c_p \rho_{H/2} H}\frac{\partial Q}{\partial t} + (u - \bar{u})\frac{\partial \theta_c}{\partial x}. \tag{96}$$

The equations of motion were taken to be

$$\frac{du}{dt} - fv - F_x = -\frac{1}{\rho}\frac{\partial p}{\partial x}, \tag{97}$$

$$\frac{dv}{dt} + fu - F_y = 0, \tag{98}$$

$$\frac{dw}{dt} + g = -\frac{1}{\rho}\frac{\partial p}{\partial z}, \tag{99}$$

where F_x and F_y are internal friction terms taken as $K_z^M(\partial^2 u/\partial z^2)$ and $K_z^M(\partial^2 v/\partial z^2)$. Operating on equation (97) by $\partial/\partial z$ and equation (99) by $\partial/\partial x$, substracting and resubstituting equations (97) (neglecting F_x) and (99) (neglecting dw/dt) gave the equation for the y component of vorticity:

$$\left(\frac{\partial u}{\partial x} + \frac{\partial w}{\partial z} + \frac{d}{dt}\right)\left(\frac{\partial u}{\partial z} - \frac{\partial w}{\partial x}\right)$$

$$= -\frac{g}{T}\frac{\partial T}{\partial x} + \left(\frac{\partial u}{\partial t} - fv\right)\frac{1}{T}\frac{\partial T}{\partial z} + f\frac{\partial v}{\partial z} + \frac{\partial F_x}{\partial z}, \tag{100}$$

where ρ has been eliminated using the gas equation $p = R\rho T$. Pearce showed that the second term on the right-hand side and the

$$\left(\frac{\partial u}{\partial x} + \frac{\partial w}{\partial z}\right)\left(\frac{\partial u}{\partial z} - \frac{\partial w}{\partial x}\right)$$

term on the left-hand side could justifiably be neglected to give

$$\frac{\partial \xi}{\partial t} = -u\frac{\partial \xi}{\partial x} - w\frac{\partial \xi}{\partial z} - \frac{g}{T}\frac{\partial \theta}{\partial x} + f\frac{\partial v}{\partial z} + \frac{\partial F_x}{\partial z}, \tag{101}$$

where

$$\zeta = \frac{\partial u}{\partial z} - \frac{\partial w}{\partial x} \quad (102)$$

and $\partial T/\partial x$ has been replaced by $\partial \theta/\partial x$. The continuity equation gives

$$\frac{\partial u}{\partial x} + \frac{\partial w}{\partial z} = -\frac{1}{\rho}\frac{d\rho}{dt} = -\frac{1}{\gamma p}\frac{dp}{dt} + \frac{1}{\theta}\frac{d\theta}{dt}, \quad (103)$$

where γ is the ratio of specific heats of air. Neglecting $\partial p/\partial t$ and $u(\partial p/\partial x)$ in the comparison with $w(\partial p/\partial z)$ gave

$$\frac{\partial u}{\partial x} + \frac{\partial w}{\partial z} = \frac{gw}{c^2} + G, \quad (104)$$

where

$$G = \frac{1}{\theta}\frac{d\theta}{dt}. \quad (105)$$

Equations (98) and (105) for v and θ may be written as

$$\frac{\partial v}{\partial t} = -u\frac{\partial v}{\partial x} - w\frac{\partial v}{\partial z} - fu + F_y, \quad (106)$$

$$\frac{\partial \theta}{\partial t} = -u\frac{\partial \theta}{\partial x} - w\frac{\partial \theta}{\partial z} + \theta G. \quad (107)$$

The five equations (96), (101) (with (102)), (104), (106) and (107), with suitable boundary and initial conditions, can be solved for u, v, w, θ and G. The boundary conditions were $w = 0$ at $z = 0$ for all y and t and u and $v \to 0$ as $x \to \pm\infty$ and $z \to \infty$ for all t. The initial conditions were $u, v, w = 0$; $\theta = \Gamma z$, where $\Gamma = 10\,\text{K}\,\text{km}^{-1}$; surface pressure $p = 1000\,\text{mb}$. With the aid of a theorem due to Helmholtz, Pearce manipulated equation (101) into the following form:

$$\frac{\partial}{\partial t}(\nabla^2 \psi) = -u\frac{\partial}{\partial x}(\nabla^2 \psi) - w\frac{\partial}{\partial z}(\nabla^2 \psi) - \frac{g}{T}\frac{\partial \theta}{\partial x} + f\frac{\partial v}{\partial z} + \frac{\partial F_x}{\partial z}, \quad (108)$$

where ψ is the stream function: $u = \partial\psi/\partial z$; $w = -\partial\psi/\partial x$. v and θ are given by equations (106) and (107). Regarding G as known, equations (108), (106) and (107) constitute a set of non-linear simultaneous equations to be solved for v, θ and ψ. Pearce noted that analytical solutions could not be obtained without neglecting the advection terms which are of fundamental importance and so he devised a numerical solution.

Pearce's results are summarized in Fig. 72 which shows the development of the vertical circulation between 1.4 and 12.6 h of simulated time. At 4.6 h, a typical pattern, the streamlines represent the spread inland of a tongue of cold

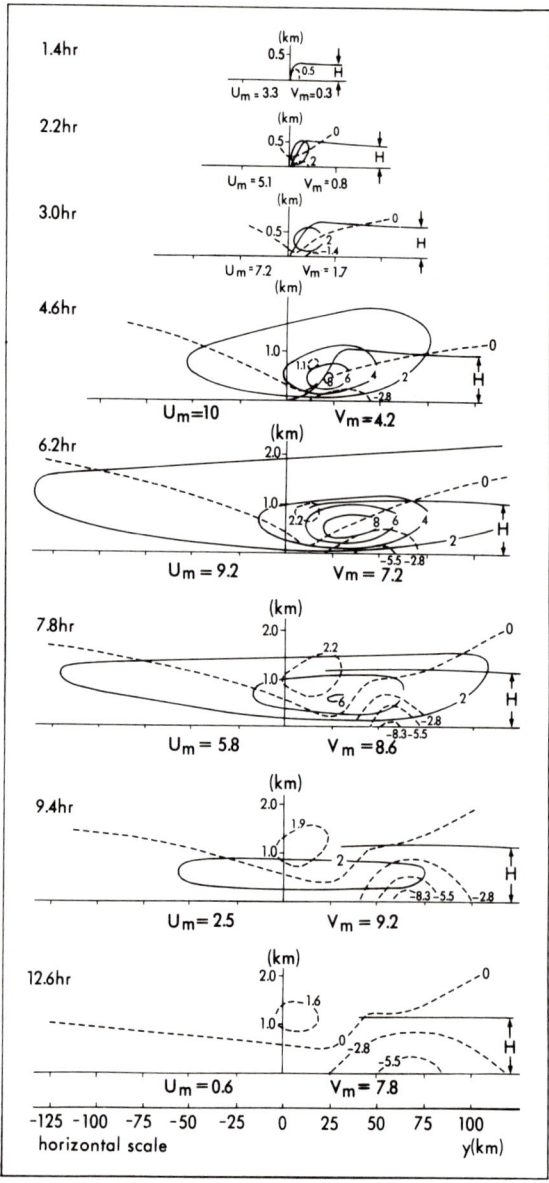

Fig. 72. Development of the sea breeze in Pearce's model. Full lines are streamlines of the motion in the plane of the diagram with values of ψ in square kilometres per hour. Dashed lines represent velocity component v in metres per second, negative out of the diagram. Maximum values of the horizontal velocity components u_m and v_m are given in metres per second for each pattern. (After Pearce, 1955.)

air. The maximum return velocity occurs at about 0.7 km, just below the level $z = H$ and just above the maximum surface velocity. The effects of the coriolis force are clearly seen in the large v component after about 6 h.

Fisher (1961a) also used a model that was governed by the equations of motion, the equation of continuity and a heating equation. Whereas his dynamical equations were virtually identical to these used by Pearce (1955), Fisher included surface drag and his heating function was different. The initial temperature field and its variation with time at the ground were specified to accord with observations. Initially there were no horizontal gradients, a desirable condition as it fitted the assumption that the velocity field was initially zero everywhere. As regards temporal variations, the sea surface temperature was held constant during the entire computation but over the land the temperature was assumed to vary near the ground as observed at an inland station unaffected by the sea breeze. Fisher used the following heat diffusion equation to describe temperature changes in his model:

$$\frac{\partial \theta}{\partial t} = -u\frac{\partial \theta}{\partial x} - v\frac{\partial \theta}{\partial y} - w\frac{\partial \theta}{\partial z} + \frac{\partial}{\partial z}\left(K_z^H \frac{\partial \theta}{\partial z}\right), \tag{109}$$

where x is the direction perpendicular to the coastline, positive inland, and y is parallel to the coastline, positive to the left of the x-axis. Fisher integrated his model within a domain in the vertical plane just over 7.5 km high and 256 km in the horizontal. He employed a grid structure in which the grid mesh increased logarithmically with height and horizontal distance from the coastline. In this way Fisher retained detail near the coastline yet allowed the model to cover a large area within the constraints of 2 min time steps for an integration period of 12–14 h. The model atmosphere was initially at rest but the initial temperature distribution was isothermal from 465 m downward and adiabatic from that point upward. This contrasted with Pearce's isothermal initial state but accorded with Fisher's (1960) earlier observations. In a further attempt to simulate reality Fisher allowed K_z^H to vary in the vertical in accord with the majority of observations available at the time. They indicated that K_z^H increased from zero to $5 \times 10^5 \text{ cm}^2 \text{ s}^{-1}$ between the ground and a height of 50–60 m then fell to about $2 \times 10^5 \text{ cm}^2 \text{ s}^{-1}$ at a height of 120 m, above which the value remained constant.

Figure 73 shows the u, v, w and θ distributions in the mature sea breeze circulation—that is after 6–8 h since its origin at 0700 hours. The computed velocity maximum occurred at 200–400 m—the same height as the observed maximum (Fisher, 1960). In addition the boundary between the inland flow at low levels and the offshore flow at upper levels occurred at about the same height as was observed (approximately 950 m). The computed maximum sea breeze velocity of 7 m s^{-1} agreed quite well with the observations in Table 18, particularly when we remember that Table 18 contains surface observations. The vertical velocities and development of velocity parallel to the coast as a

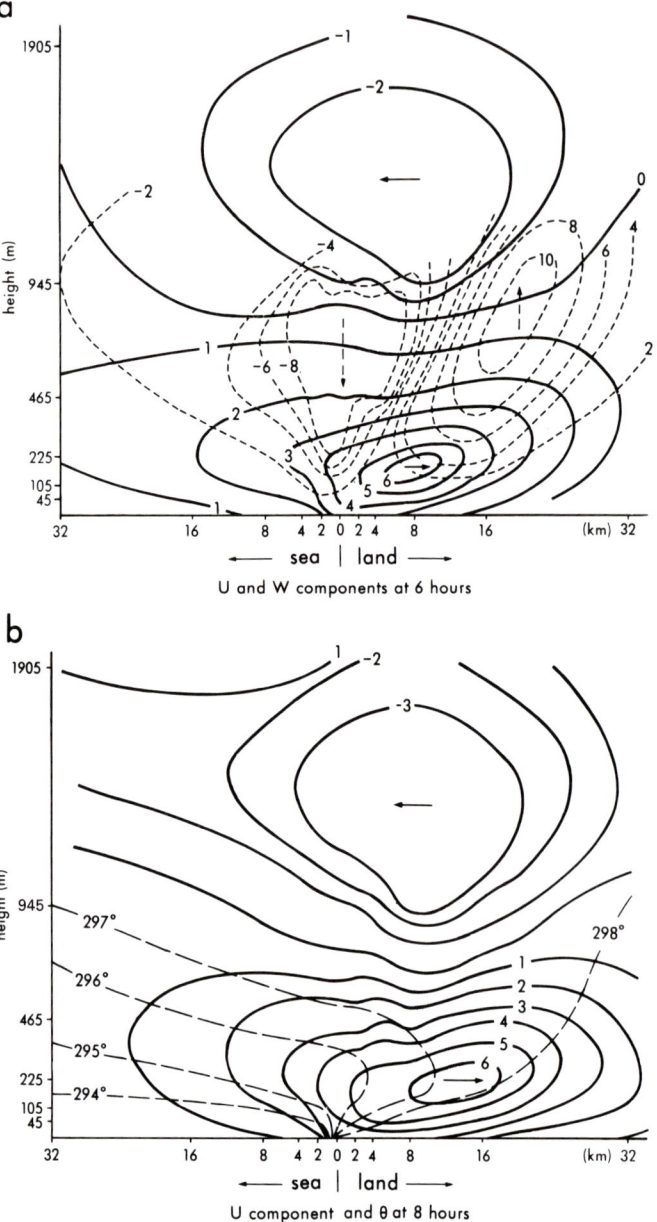

FIG. 73. Vertical cross-sections of theoretical sea breeze showing velocity components and temperatures.

(a) Wind speed component normal to the coast (u in metres per second), solid lines and vertical speed (w in centimetres per second) dashed lines, after 6 h simulated time. u is positive towards land and w positive upward.

(b) Wind speed component normal to the coast (u in metres per second), solid lines,

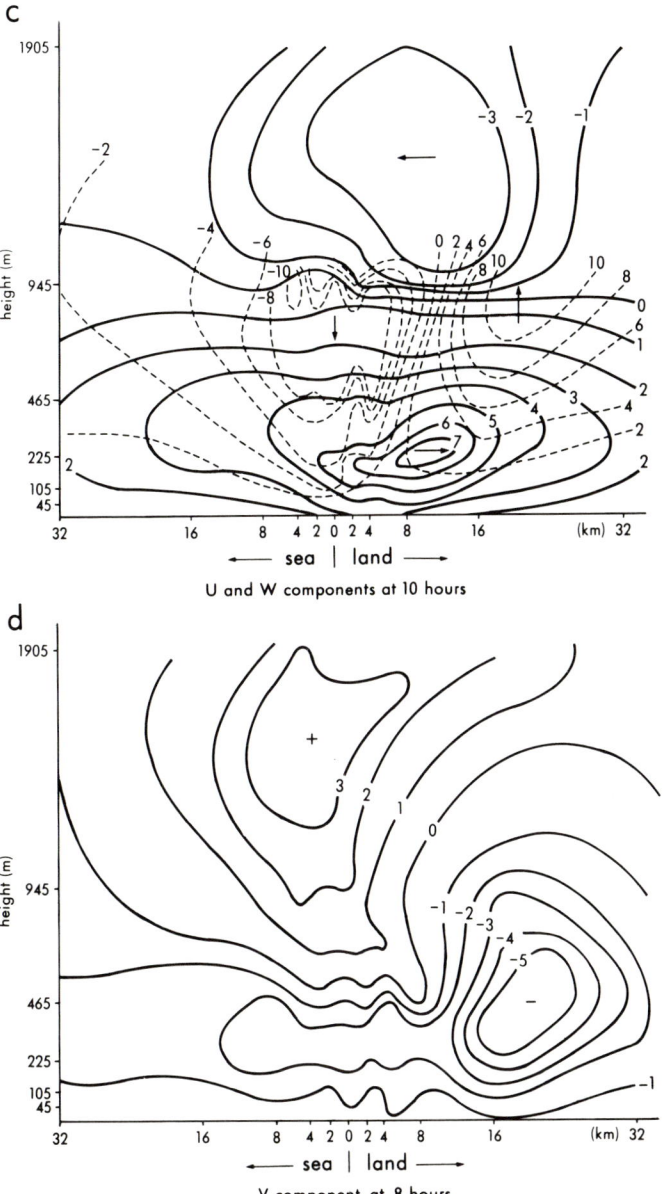

and potential temperature (θ in kelvins), dashed lines, after 8 h of simulated time, u is positive towards land.

(c) Wind speed component normal to the coast (u in metres per second), solid lines, and vertical speed (w in centimetres per second), dashed lines, after 10 h of simulated time. u is positive towards land and w positive upward.

(d) Wind speed component parallel to coast (v in metres per second) after eight hours simulated time. v is positive into the page. (After Fisher, 1961a.)

result of the coriolis force were also well simulated in location and magnitude by the model. The temperature distribution also seemed to be quite realistic.

Fisher was aware of the simplifying compromises made in the dynamics and integration procedures employed in the model, outlining them in a later report (Fisher, 1961b). Unfortunately he died before he was able to refine the model. Nevertheless, his two major papers on the observation and theory of sea breezes remain substantial contributions to our knowledge and fine examples of lucid, scientific exposition.

Contemporaneous with Fisher's study was that by Estoque (1961) who employed a different approach to that used by Pearce and Fisher and laid the foundation for nearly a dozen subsequent theoretical analyses of the sea breeze. Estoque's prime concern was to improve on the heating functions used by Pearce and Fisher as their mechanisms implied vertical heat transfers which did not depend directly on the existing temperature and velocity fields. Estoque noted that the intensity, duration and dimensions of the sea breeze must be governed largely by the amount of heat supplied by the ground to the atmosphere and also by the prevailing large scale, synoptic conditions. His model was an attempt to simulate these factors. As this model has been extended in later papers to include features omitted by Estoque, the most complete relevant form of the governing equations are given immediately below and abbreviated forms included later where necessary.

Estoque considered a vertical domain with the x-axis normal to the shoreline (positive inland) and the z-axis being in the vertical. He assumed no variation in the sea breeze along the shore line (y-axis). The vertical cross-section was bounded at the bottom by the Earth's surface and at the top by the level $z = H = 2 \text{km}$. The lateral boundaries were at $x = \pm D = \pm 200 \text{km}$. Estoque subdivided this area into two horizontal layers: a thin layer, $0 \leqslant z \leqslant h \approx 50 \text{m}$, characterized by constancy with height of the vertical eddy fluxes of heat and momentum; and an overlying layer, called the "transition" layer, $h \leqslant z \leqslant H$, where the effect of eddy fluxes decreased with elevation. In the lower layer the equations were solved analytically whereas in the upper layer they were solved numerically. The solutions of both layers were matched by imposing certain conditions on the velocity and temperature fields at the internal boundary $z = h$.

In the lower layer the wind and temperature profiles were expressed by

$$\frac{\partial}{\partial z}\left(K_z^M \frac{\partial U}{\partial z}\right) = 0, \tag{110}$$

$$\frac{\partial}{\partial z}\left(K_z^H \frac{\partial \theta}{\partial z}\right) = 0, \tag{111}$$

where $U = (u^2 + v^2)^{1/2}$, u and v being the horizontal velocity components along the x- and y-axes respectively. Estoque allowed the eddy diffusivities of heat and momentum, considered to be equal, to vary with the stability of the layer as follows:

$$K_z = [k_v(z+z_0)(1+\alpha \mathrm{Ri})]^2 \frac{\partial U}{\partial z}, \qquad \mathrm{Ri} \geqslant (\mathrm{Ri})_c \simeq -0.03, \qquad (112)$$

$$K_z = \lambda z^2 \left(\frac{g}{T}\left|\frac{\partial \theta}{\partial z}\right|\right)^{1/2}, \qquad \mathrm{Ri} < (\mathrm{Ri})_c, \qquad (113)$$

where k_v is von Kármán's constant (0.4), z_0 is the roughness length and $\alpha \simeq -3$ and $\lambda \simeq 0.9$ are empirical constants. Equations (110) and (111) parameterize the physical processes in the boundary layer for the purpose of obtaining reasonable and consistent values of u, v and θ at $z = h$ to serve as lower boundary conditions for the numerical integration necessary in the upper layer. Without this parameterization a very high resolution of the grid lattice in the vertical would have been necessary to represent the large vertical gradients which develop in the boundary layer.

In the upper "transition" layer, the three-dimensional equations of motion were as follows:

$$\frac{\partial u}{\partial t} + u\frac{\partial u}{\partial x} + v\frac{\partial u}{\partial y} + w\frac{\partial u}{\partial z} - fv = -\frac{RT}{p}\frac{\partial p}{\partial x} + \frac{\partial}{\partial z}\left(K_z^M \frac{\partial u}{\partial z}\right)$$

$$= -c_p\theta\frac{\partial \pi}{\partial x} + \frac{\partial}{\partial z}\left(K_z^M \frac{\partial u}{\partial z}\right); \qquad (114)$$

$$\frac{\partial v}{\partial t} + u\frac{\partial v}{\partial x} + v\frac{\partial v}{\partial y} + w\frac{\partial v}{\partial z} + fu = -\frac{RT}{p}\frac{\partial p}{\partial y} + \frac{\partial}{\partial z}\left(K_z^M \frac{\partial v}{\partial z}\right)$$

$$= -c_p\theta\frac{\partial \pi}{\partial y} + \frac{\partial}{\partial z}\left(K_z^M \frac{\partial v}{\partial z}\right) \qquad (115)$$

(π is the Exner function, i.e. $\pi \equiv (p/P)^{R/c_p} = T/\theta$, where P is 1000 mb);

$$\frac{\partial w}{\partial t} + u\frac{\partial w}{\partial x} + v\frac{\partial w}{\partial y} + w\frac{\partial w}{\partial z} = -\frac{1}{\rho}\frac{\partial p}{\partial z} - g. \qquad (116)$$

The thermodynamic equation was

$$\frac{\partial \theta}{\partial t} + u\frac{\partial \theta}{\partial x} + v\frac{\partial \theta}{\partial y} + w\frac{\partial \theta}{\partial z} = \frac{\partial}{\partial z}\left(K_z^H \frac{\partial \theta}{\partial z}\right). \qquad (117)$$

The equation of continuity was

$$-\frac{1}{\rho}\frac{dp}{dt} = \frac{\partial u}{\partial x} + \frac{\partial v}{\partial y} + \frac{\partial w}{\partial z}. \qquad (118)$$

Equations (114)–(118) describe the three-dimensional structure of the velocity and temperature fields. Estoque restricted himself to two dimensions;

consequently all the $\partial/\partial y$ terms are missing from his equations. Further, Estoque considered that the vertical accelerations in equation (116) were negligible and reduced the equation to the hydrostatic relationship

$$\frac{\partial p}{\partial z} = -\frac{pg}{RT}. \qquad (119)$$

Later authors (e.g. McPherson, 1970) used the form

$$\frac{\partial \pi}{\partial z} = -\frac{g}{c_p \theta}. \qquad (120)$$

Estoque then modified the equations in the interests of computational stability by assuming an incompressible atmosphere and differentiating the two-dimensional continuity equation to give

$$\frac{\partial^2 w}{\partial z^2} = -\frac{\partial}{\partial z}\left(\frac{\partial u}{\partial x}\right). \qquad (121)$$

Finally

$$\theta = T(P/p)^{R/c_p}. \qquad (122)$$

Within the upper, transition layer Estoque assumed the value of K_z to decrease linearly from its value $K_z(h)$ at $z = h$ to zero at $z = H$, the top of the integration area, as follows:

$$K_z = K_z(h)\frac{H-z}{H-h}. \qquad (123)$$

The effects of different vertical distributions of K_z were evaluated for four cases by Yoshikado and Asai (1972). They concluded that the simple distribution of equation (123) was preferable to any of the three assumptions that the vertical variations of K_z be zero, be described by a cubic polynomial or be a complicated function of θ, u, v, z, K_{z0} (the surface value) and f.

Estoque's model was run on a grid system with equal spacing of 100 m along the vertical and horizontal spacing that started with a value of 2 km at the shoreline and increased with distance from this point to a maximum spacing of 34 km at the lateral boundaries. Starting with an atmosphere at rest and temperatures assumed to be a linear function of height, i.e. $T = 283 - 8z \times 10^{-5}$, the differential heating of the Earth's surface was prescribed as through the surface temperature field as

$$T = 283 + 10 \sin(15t - 120), \quad x > 0 \text{ (land)},$$
$$T = 283, \quad x < 0 \text{ (sea)},$$
$$T = \tfrac{1}{2}[T(x > 0) + T(x < 0)], \quad x = 0 \text{ (coastline)},$$

where t is the number of hours after midnight.

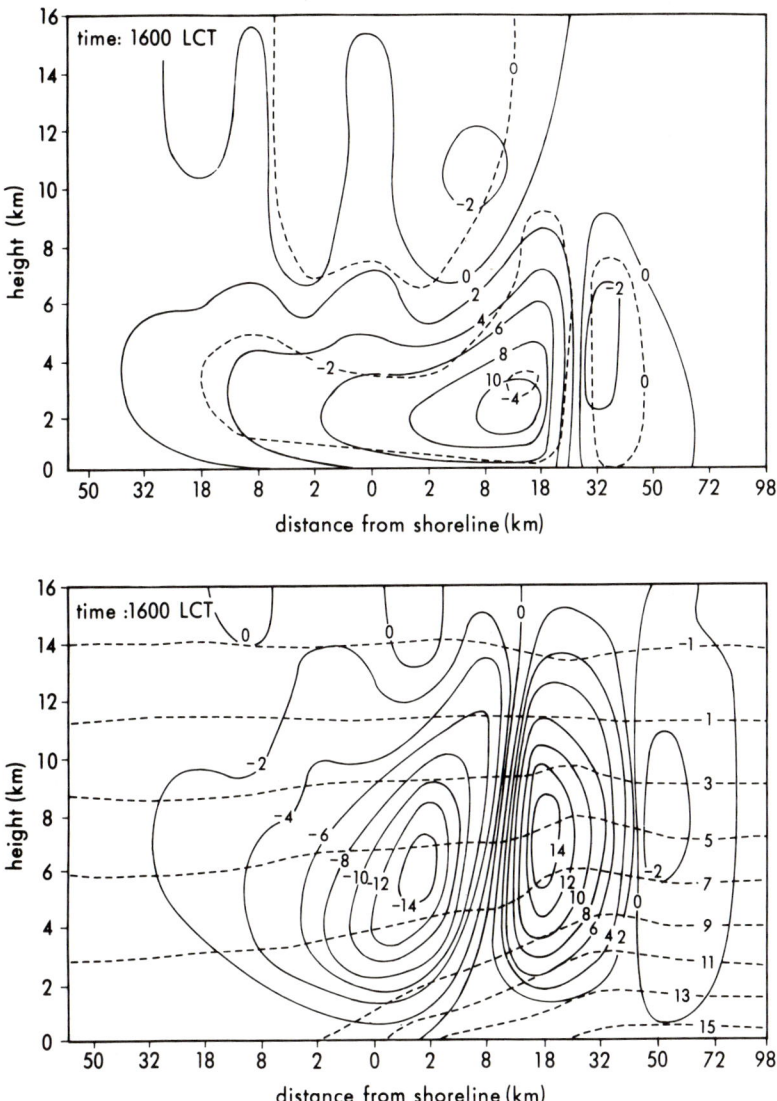

FIG. 74. Vertical cross-sections of theoretical sea breeze after 8 h simulated time as derived by Estoque. *Top*: Computed distributions of velocity components (in metres per second) normal (solid lines) and parallel (dashed lines) to the shoreline. Normal components positive toward land (right) and parallel components positive into the page. *Bottom*: Computed distributions of vertical velocity (in centimetres per second, solid lines) and temperatures (in degrees Celsius, dashed lines). Vertical velocities positive upward (After Estoque, 1961.)

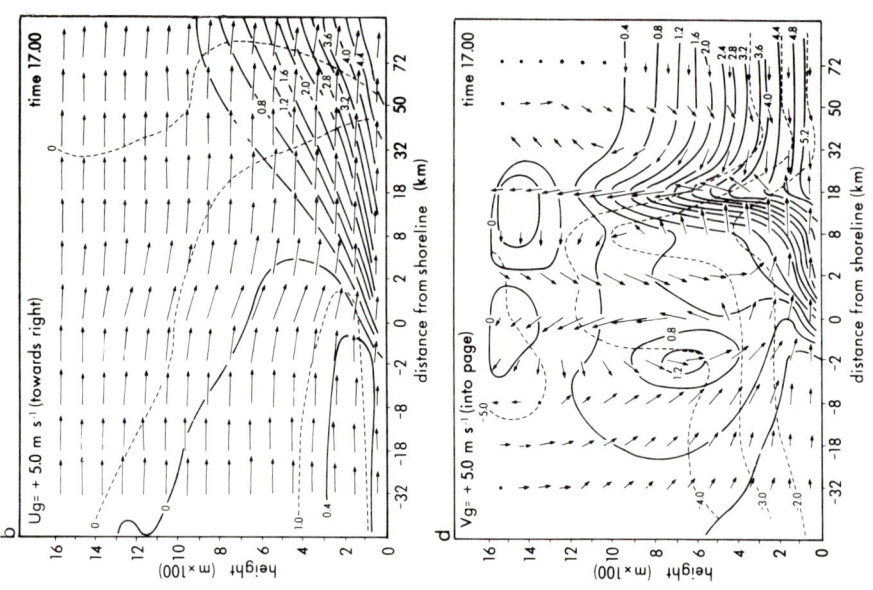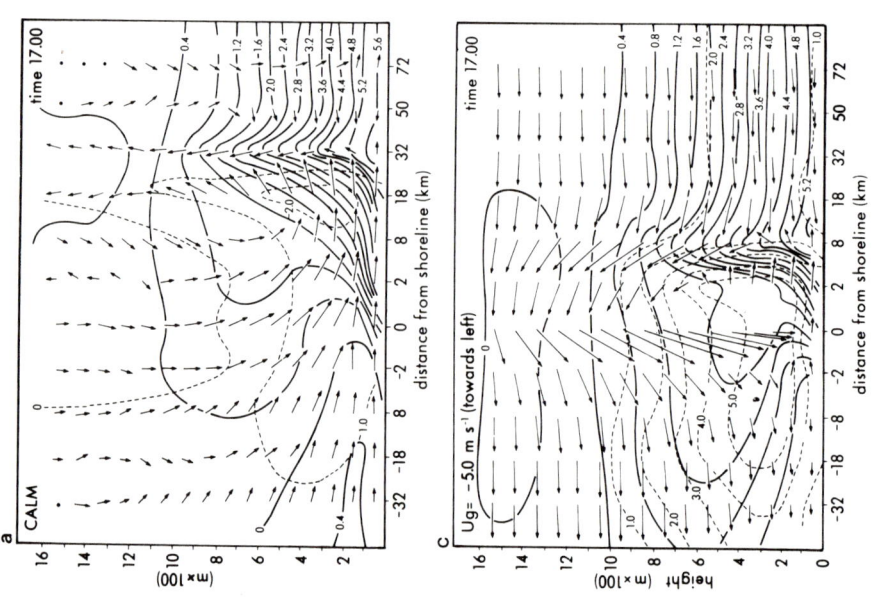

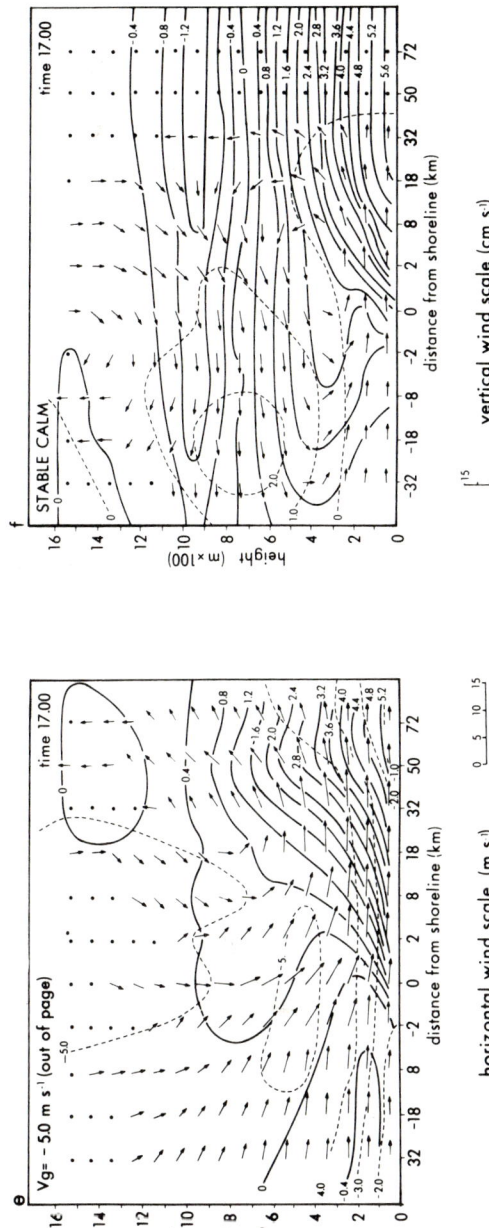

FIG. 75. Vertical cross-sections of theoretical sea breeze, after 9 h simulated time, showing effect of geostrophic wind. Arrows show direction and magnitude of flow at each point. Solid lines show the temperature change (degrees Celsius) since the beginning of the simulation. Dashed lines show the velocity component (metres per second) parallel to the coast and positive into the page. Land is on the right. (a) with zero wind; (b) with an onshore geostrophic wind of 5 m s^{-1}; (c) with an offshore geostrophic wind of 5 m s^{-1}; (d) with a geostrophic wind of 5 m s^{-1} into the page; (e) with a geostrophic wind of 5 m s^{-1} out of page; (f) with zero wind and an isothermal surface layer. (After Estoque, 1962.)

The resultant velocity and temperature fields near the time of maximum development are shown in Fig. 74, in which the horizontal scale is compressed. The sea breeze attained a maximum speed of about $10\,\text{m\,s}^{-1}$ near an altitude of 250 m. A weak return flow existed aloft and negative values of v in regions of positive u and vice versa, revealed the effect of the coriolis force. The model also simulated a low level convergence zone at the leading edge—between 18 and 32 km at 1600 hours—of the landward current in the afternoon, but this is exaggerated on the figure by the compressed horizontal scale. This was probably the nearest that the model could approach to a description of the sea breeze front because of the size of the horizontal grid spacing. But it was the first time that this feature had appeared at all in sea breeze simulations.

Estoque's model provided the basis for several later experiments on the character of the sea breeze. The effects of synoptic scale winds and their vertical shear, vertical variations in K_z, condensation, and variations in coastline on the sea breeze, together with a closer look at the sea breeze front, were all investigated with models closely based on Estoque's original.

Estoque himself used his model to assess the effects of the geostrophic wind on sea breeze behaviour (Estoque, 1962). By imposing a pressure distribution that determined a geostrophic wind at the upper boundary of the model Estoque simulated the effects on the breeze of onshore and offshore winds and winds parallel to the shore, with low pressure over both land and water. He also investigated the effects of stability by prescribing an isothermal layer in the lowest kilometre. The results of the integrations are shown in Fig. 75, where Fig. 75(a) is the case with zero geostrophic wind for comparative purposes.

The difference between the on- (Fig. 75(b)) and offshore (Fig. 75(c)) cases is clearly seen and is primarily due to the intensification of the horizontal temperature (and thus pressure) gradient by the large scale wind moving warmer, land air towards the sea. Offshore winds of $5\,\text{m\,s}^{-1}$ also reduce inland penetration to less than 18 km in contrast to about 32 km when the geostrophic wind is zero. When the geostrophic wind blows along the shoreline with low pressure over the ocean (Fig. 75(d)) the seaward frictional inflow in the lowest layers leads to a strengthening of the horizontal pressure gradient in the same way as that resulting from an offshore prevailing geostrophic wind. When the parallel geostrophic flow is in the opposite direction (Fig. 75(e)), the low level frictional inflow is onshore and a weak circulation is induced. The component of the sea breeze circulation parallel to the coast shows the coriolis effect in all cases, but to different degrees.

The evolution of the temperature field on Fig. 75 is indicated by isolines of temperature changes which have occurred since the initial time (0800 hours). As the horizontal temperature gradient everywhere was initially zero, the distribution of the isolines gives a picture of the actual temperature gradients.

The temperature field shows a clear response to the velocity field and clear differences are revealed between the "calm" case and those cases representing the effects of the geostrophic wind. The general effect of increased stability (Fig. 75(f)) is to decrease both the depth and speed of the sea breeze and the convergence zone is not apparent.

Estoque realized the several deficiencies of his model, particularly the lack of mass conservation attributed to the use of the continuity equation in the form of equation (121) and the omission of the non-adiabatic processes of radiation and the release of latent heat by condensation. These non-adiabatic processes were included in Magata's (1965) model by dispensing with Estoque's constant flux layer. The equations describing the velocities and temperatures of the sea breeze were virtually identical to those used by Estoque except for the heating equation where Magata added a term (H_θ) to represent the changes of latent heat by condensation and evaporation. Thus equation (117) (without the $\partial/\partial y$ terms as Magata's model was also two-dimensional) became

$$\frac{\partial \theta}{\partial t} + u \frac{\partial \theta}{\partial x} + w \frac{\partial \theta}{\partial z} = H_\theta + \frac{\partial}{\partial z}\left(K_z^H \frac{\partial \theta}{\partial z}\right). \tag{124}$$

In addition Magata included equations describing the changes in the specific humidity, the liquid water content, the sea and land temperatures as follows:

$$\frac{\partial q}{\partial t} + u \frac{\partial q}{\partial x} + w \frac{\partial q}{\partial z} = \frac{\partial}{\partial z}\left(K_z^W \frac{\partial q}{\partial z}\right) + C_q, \tag{125}$$

$$\frac{\partial q_w}{\partial t} + u \frac{\partial q_w}{\partial x} + w \frac{\partial q_w}{\partial z} = \frac{\partial}{\partial z}\left(K_z^W \frac{\partial q_w}{\partial z}\right) - C_q, \tag{126}$$

where q is the specific humidity of the air, q_w is the liquid water content of the air and C_q is the variation of q and q_w by condensation and evaporation:

$$\frac{\partial T_1}{\partial t} = K_1 \frac{\partial^2 T_1}{\partial z^2}, \tag{127}$$

$$\frac{\partial T_2}{\partial t} = K_2 \frac{\partial^2 T_2}{\partial z^2}, \tag{128}$$

where T_1 and T_2 are the temperatures of sea and land respectively and $T_1 = T_1(z)$, $T_2 = T_2(z)$ for $z < 0$. K_1 is the diffusion coefficient for sea and K_2 is the coefficient of heat conduction of land. With the aid of suitable boundary conditions, equations (125)–(128) allowed the incorporation of transfers of water and heat from the Earth's surface to the overlying atmosphere without resort to Estoque's constant flux layer. The primary determinant of the boundary conditions was the necessity for heat balance at both the land and sea surface. Thus, over the sea surface,

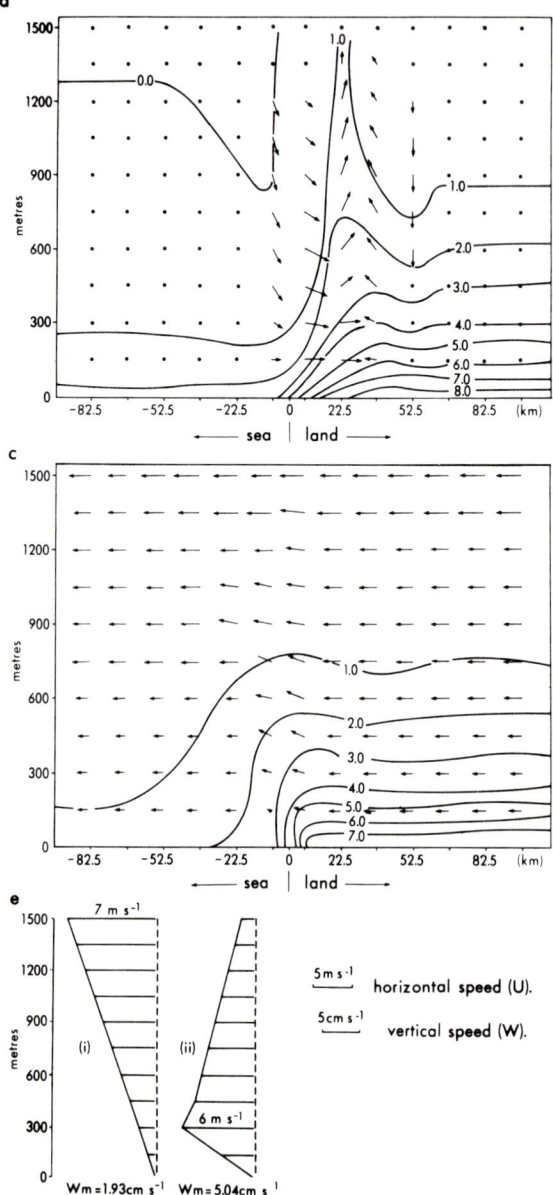

Fig. 76. Vertical cross-sections of theoretical sea breeze after 9 h simulated time as derived by Magata. Arrows show direction and magnitude of flow at each point. Solid lines show the temperature change (in degrees Celsius) since the beginning of the simulation: (a) with zero wind and no condensation; (b) with zero wind but

5 SEA/LAND BREEZE CIRCULATION

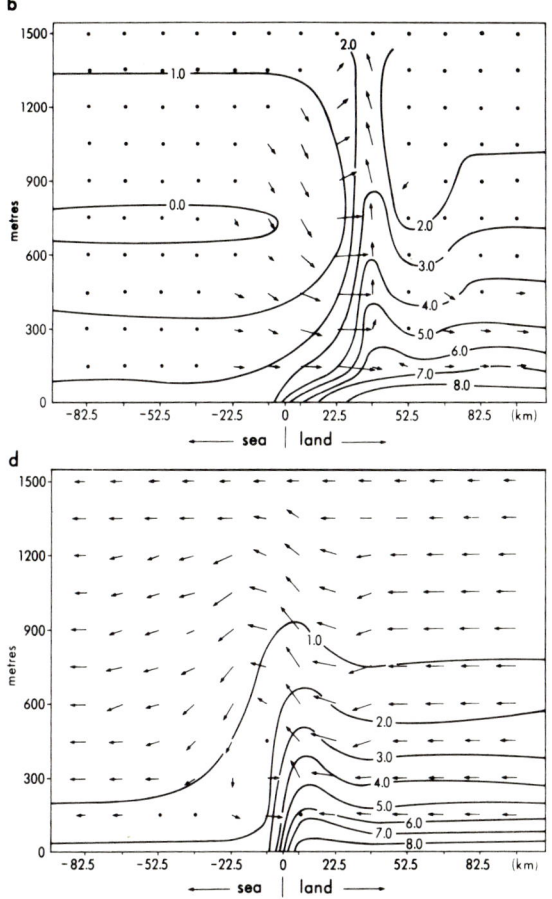

a: after 9 hours simulation, general current. U = 0 effect of condensation neglected.
b: as 'a' but conditionally unstable and including effects of condensation.
c: as 'a' but $\partial U / \partial Z$ 0.5 m s^{-1} / 100m (approx ● (i)).
d: as 'a' but general current corresponding to ● (ii).

conditionally unstable and including the effects of condensation; (c) as (a) but with vertical profile of wind as in (e)(i); (d) as (a) but vertical profile of wind as in (e)(ii); (e) two types of initial vertical profiles of wind. Maximum vertical velocities (w_{max}) appearing after 9 h simulated time. (After Magata, 1965.)

$$\theta = T_1, \qquad (129)$$

$$S - h(T_1 - C_h) - c_1\rho_1 K_1 \frac{\partial T_1}{\partial z} + c_p\rho K_z^H \frac{\partial \theta}{\partial z} = 0, \qquad (130)$$

$$q = \alpha_1 q_s, \qquad (131)$$

and over the land surface,

$$\theta = T_2, \qquad (132)$$

$$S - h(T_2 - C_h) - c_2\rho_2 K_2 \frac{\partial T_2}{\partial z} + c_p\rho K_z^H \frac{\partial \theta}{\partial z} = 0, \qquad (133)$$

$$q = \alpha_2 q_s, \qquad (134)$$

where S represents insolation given as a function of time, h is a cooling coefficient, C_h is the specific temperature at which there is neither cooling nor heating, q_s is the saturated specific humidity and α_1 and α_2 are constants: $0 \leq \alpha_1, \alpha_2 \leq 1$.

The third and fourth terms of equations (130) and (133) represent the heat transfer downward in sea or land and the eddy transfer of heat upward in the air respectively. These boundary conditions formed the necessary complement to those of the equations of motion. Magata derived his value for K_z^H (10^4 cm^2 s^{-1}) from observational studies of inversions. By assuming values of K_1, K_2, $c_1\rho_1$ and $c_2\rho_2$ in equations (127), (128), (130) and (133) he was able to integrate all the necessary equations to produce the two-dimensional sea breeze circulation for three different situations, viz. (1) with no general wind and no condensation effect; (2) with no general wind but including condensation; (3) with no condensation but incorporating vertical shear in the general wind. Figure 76 shows the sea breeze at 1500 hours under these different conditions. In similar fashion to Estoque's diagrams, temperature is shown by the deviation from the initial value. The release of latent heat had a clear effect on the intensity of the breeze, strengthening the frontal properties (Fig. 76(a) and (b)). An offshore wind with positive shear had the effect of reducing vertical velocities (maximum vertical velocity of 1.93 cm s^{-1}) and indeed preventing the sea breeze as such (Fig. 76(c)). Yet a strong temperature gradient formed over the coast. An offshore wind with a maximum velocity of 6 m s^{-1} at a height of 300 m deepened the layer through which a strong horizontal temperature gradient existed (Fig. 76(d)) and increased vertical velocity values. The maximum vertical velocity was 5.04 cm s^{-1}. Clearly, not only does the presence of a prevailing wind affect the structure of the sea breeze, as shown by Estoque (1962), but also the characteristics of the vertical shear may induce significant changes in velocity and temperature distribution.

Nearly a decade after Estoque's first two-dimensional model appeared, it was extended and applied to three dimensions by McPherson (1970). The incorporation of variations along the coastline meant that the $\partial/\partial y$ terms were no longer omitted and thus McPherson's model comprised equations (114), (115), (117), (120), (121) and (122). The other major change made by McPherson was to assume an exponential as opposed to a linear decrease of K_z in the "transition" layer as used by Estoque. Thus instead of equation (123), McPherson used the following:

$$K_z(z) = K_z(h) \exp\left[-m\left(\frac{z-h}{H}\right)^2\right], \quad (135)$$

where $K_z(z)$ is the value of the exchange coefficient within the transition zone at level z, $K_z(h)$ is the maximum value of the exchange coefficient occurring at the internal boundary $z = h$, H is the top of the volume of integration and m is a parameter that governs the rate of decrease of $K_z(z)$, found by observation to be $m = 4.75$.

At the onshore and offshore boundaries, u and v and the normal gradients of w, π and θ were required to vanish. At the lateral boundaries perpendicular to the coastline conditions of symmetry were imposed and at the upper boundary temporal invariance of all quantities was imposed.

The model was run on a grid of a constant horizontal spacing of 4 km with 20 levels, each 200 m apart. McPherson simulated the effects of a square bay (roughly equal in size to Galveston Bay, Texas) and found interesting deformations in the three-dimensional flow due to the thermal effects of the bay. Figure 77 shows both the horizontal (at a height of 250 m) and vertical (at a height of 850 m) components of the flow 4 and 10 h respectively after the start of the integration. In the early stages the flow was symmetric around the bay with two centres of ascent immediately east and west of the bay and a single centre of subsidence over the bay itself. After 10 hours considerable asymmetry had developed with large vertical velocities to the north-west of the bay. These values were nearly twice as large as those calculated by Estoque (1961). The asymmetry of the flow was considered to be due to the coriolis effect. On the west side of the bay the pressure gradient force that drove the sea breeze acted from east to west in an opposite sense to the coriolis force which acted toward the east, at right angles to *overall* onshore flow. The result was a relative enhancement of convergence west of the bay which was absent to the east because the local pressure gradient force and coriolis force acted in the same direction there. McPherson suggested that this convergence zone would be the most likely source of cloud and possibly shower development.

Further modifications to the Estoque model allowed Neumann and Mahrer (1971, 1974a, b, 1975) to investigate the diurnal variation of the sea/land breeze circulation, the sea breeze front in more detail and the characteristics of the breeze around both circular islands and circular lakes. Neumann and Mahrer retained the acceleration terms in the equation for

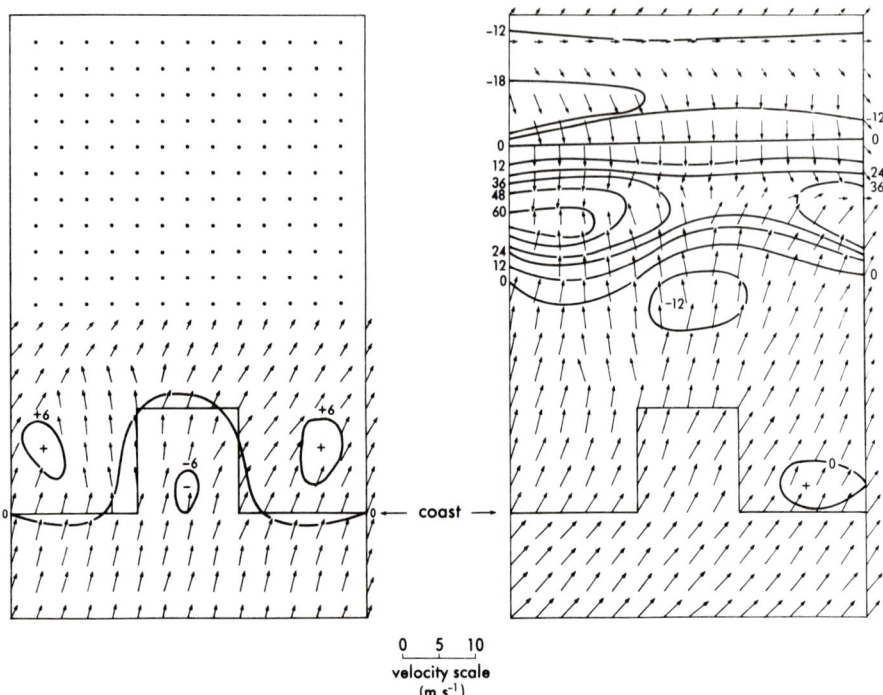

FIG. 77. Theoretical flow over a regular shaped bay, sea at foot of diagram. The horizontal motion field at a height of 250 m is described by the arrows which give magnitude (in metres per second) and direction. The vertical motion field (in centimetres per second) at a height of 850 m is shown by the solid lines. *Left*: flow after 4 h simulated time. *Right*: flow after 10 h simulated time. (After McPherson, 1970.)

vertical motion (equation (116)), abandoning the hydrostatic approximation. In addition they retained the equation of continuity in its original form (equation (118)) "to prevent violation of the mass conservation law" (Neumann and Mahrer, 1971, p. 532).

Whilst their model clearly simulated the diurnal reversal of the sea/land breeze circulation, Neumann and Mahrer (1971) concentrated on the land breeze, justifiably claiming that it had been hitherto much neglected. With the aid of a horizontal grid interval of 4 km, a vertical interval of 100 m and a time step of 3 min, the land breeze was well simulated, as shown in Fig. 78. The speed, depth and turning with height and time of the land breeze all accorded well with the available observations. Neumann and Mahrer claimed that the cells of vertical velocity in Fig. 78(b) were in good agreement with observed conditions in the coastal area of Israel for early morning hours of the summer months.

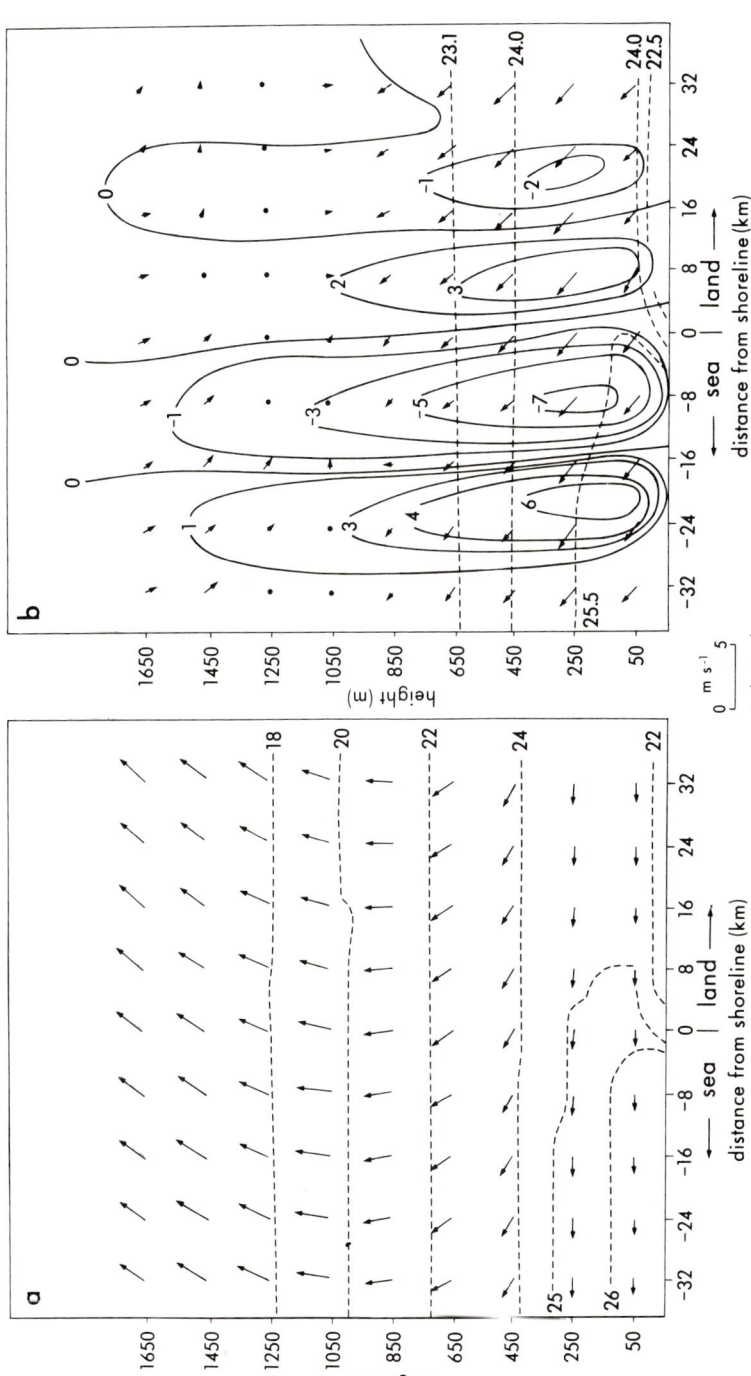

Fig. 78. Theoretical flow in a land breeze. Horizontal winds are shown by the arrows at the height being placed at the appropriate height, the base points of the arrows being placed at the height in question. Their direction is appropriate to a shoreline directed north–south. The lengths of the arrows are proportional to the speed (in metres per second) as shown in scale. Vertical speeds (in centimetres per second) are shown by the solid lines, positive upward. Temperatures (in degrees Celsius) are shown by dashed lines. (a) Situation at 0200 hours; (b) situation at 0600 hours. (After Neumann and Mahrer, 1971.)

In their analyses of breezes around circular islands and lakes, Neumann and Mahrer (1974b, 1975) found further variations in structure. Over an island about 100 km in diameter the sea breeze front was much better developed than in the case of a straight coast. A strong horizontal convergence in advance of, and divergence behind, the front occurred up to an altitude of 400–500 m. The maximum computed upward velocity occurred at the front and reached 50 cm s^{-1}. The land breezes were horizontally divergent as one would expect. Outward flow from a similar sized lake was only weakly frontal, probably due to the divergent nature of the flow.

Despite the most encouraging results from the above-quoted studies, the use of relatively coarse grids for the numerical calculation meant that some details of the sea breeze circulation were lost. In particular the sea breeze front was not always clearly simulated. Neumann and Mahrer (1974a) integrated their model

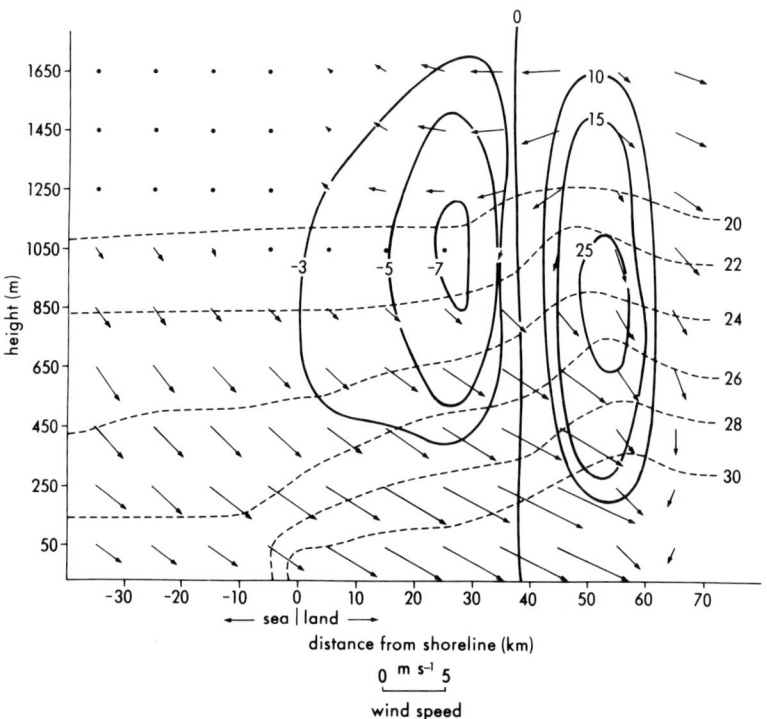

FIG. 79. Sea breeze front as theoretically analysed by Neumann and Mahrer. Diagram shows the sea breeze circulation at 1530 hours. Arrows denote the horizontal wind speed (in metres per second) in similar fashion to Fig. 78. Vertical speeds (in centimetres per second) are shown by solid lines; temperatures (in degrees Celsius) by dashed lines. (After Neumann and Mahrer, 1974.)

over a straight coastline with grid intervals of 5 km in the horizontal and 100 m along the vertical and based their study on print-outs at 15 min intervals (the time step of integration was 3 min) for the period between 1000 and 1600 hours. Figure 79 shows the flow at 1530 hours and Neumann and Mahrer claimed to see the front up to a height of 700 m about 50–60 km inland. They did not clarify what criteria they used to identify the front but most probably it was the area where horizontal convergence of the wind and vertical velocities were greatest. The rate of advance of the front was about $2.2 \, \text{m s}^{-1}$, a speed which corresponded with the horizontal wind velocity in advance of it near the surface. Behind the front, from the surface up to an altitude of a few hundred metres, the winds were nearly twice as strong so that there was considerable convergence of horizontal winds around the front. In association with this velocity divergence distribution, upward motion of up to $25 \, \text{cm s}^{-1}$ and downward motion of about $7 \, \text{cm s}^{-1}$ occurred ahead of and behind the front respectively.

Neumann and Mahrer suggested that their horizontal grid spacing of 5 km might have been too large to specify the sea breeze front in any more detail. This is probably true in the light of Lambert's (1974) findings. Using Estoque's model on a grid with horizontal spacing of 1 km, vertical spacing of 75 m and a time step of 40 s, Lambert simulated the sea breeze front in calm conditions and with an offshore geostrophic wind of $3 \, \text{m s}^{-1}$. In the case of offshore wind the sea breeze front penetrated 17 km inland, having stagnated at 7 km inland in the early afternoon. Figure 80(a) shows the structure of the front at the time of maximum activity (1500 hours). The breeze was about 400 m deep with maximum velocities of $3 \, \text{m s}^{-1}$ over the coastline. Maximum vertical velocities were $30 \, \text{cm s}^{-1}$ above the front. In the calm situation (Fig. 80(b)) the breeze was about 750 m deep and the front penetrated 30 km inland. Vertical velocities reached $55 \, \text{cm s}^{-1}$. Despite being the highest values yet reached by modelling, they are still well below the $7 \, \text{m s}^{-1}$ estimated from a glider by Wallington (1961). Lambert was aware of the underestimate of the vertical velocities and his results suggest that even greater horizontal resolution is necessary to achieve more accurate predictions. This is not surprising in the light of Wallington's (1961) observations that the front is only about 200 m wide.

Further insight into the behaviour of the sea breeze front and the overall intensity of the land/sea breeze circulation has in fact been forthcoming from both simpler numerical non-linear models and even linear models. The effects on the speed of the sea breeze front of the heat transfer from the land were found to be comparatively simple (Pearson, 1973). For a fixed total heat input to the atmosphere, the speed of the front was independent of the initial vertical potential temperature profile, the vertical distribution of heat and the way in which the final potential temperature profile was obtained. The speed of the front was found to increase with the square root of the total heat input from the land. Pearson also tested the effects of surface drag and diffusion and

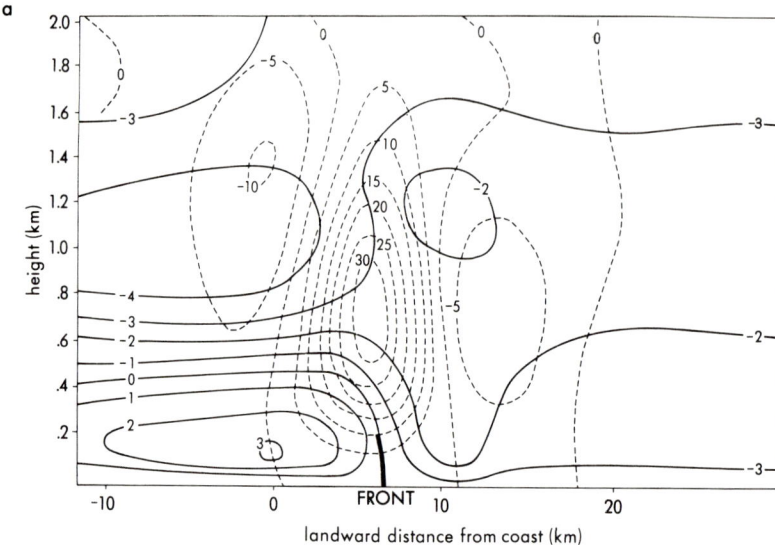
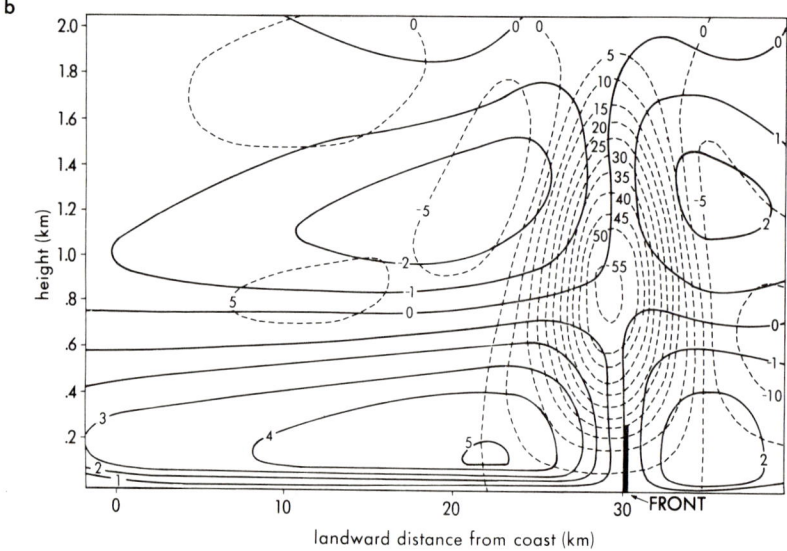

FIG. 80. Sea breeze front as theoretically analysed by Lambert. Solid lines are components of wind normal to coast (in metres per second, positive toward land). Dashed lines are vertical velocities (in centimetres per second, positive upward). (a) Situation at 1500 hours in an offshore geostrophic wind of $3 \, \text{m s}^{-1}$; (b) situation at 1600 hours in zero geostrophic wind. (After Lambert, 1974.)

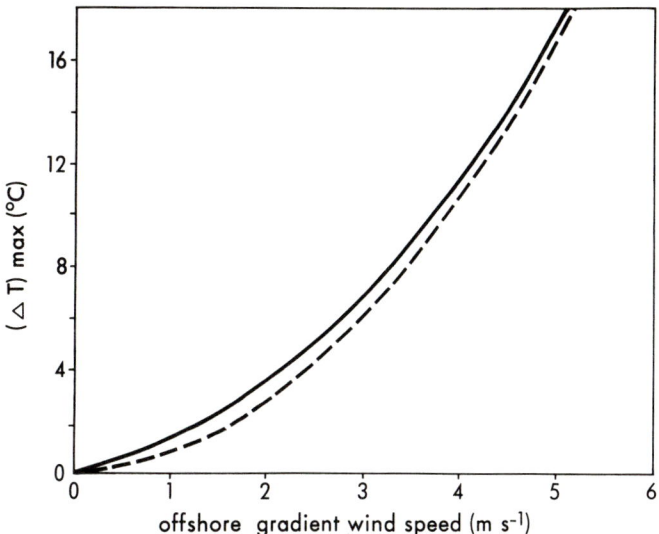

FIG. 81. The critical land–sea temperature contrast (ΔT_{max}) required for a sea breeze to occur against a given offshore gradient wind. Dashed line based on observational data from Biggs and Graves (1962). Solid line theoretically derived. (After Walsh, 1974.)

found that whereas they did significantly alter the velocity and temperature structure of the sea breeze they only slightly decreased the speed of the front. This clear re-affirmation of the dependence of the sea breeze upon temperature differences was further supported by Walsh's (1974) linear model. Among other characteristics which had previously been demonstrated (for example the effect of the coriolis force) Walsh established a theoretical relationship between the critical land/sea temperature difference required for a sea breeze to exist in opposition to a given synoptic scale wind. From Fig. 81 it is clear that for wind speeds greater than $6 \, \mathrm{m \, s^{-1}}$ very large temperature differences are required: such differences seldom occur.

Apart from Neumann and Mahrer's (1971) analyses, by far the greatest number of studies of the land/sea breeze circulation have concentrated on the evolution of the sea breeze, together with its front and associated upper level return current. The land breeze has been analyzed by Feit (1969) but the different intensities of the land and sea breezes have received scant attention. Three factors have been suggested to explain what Pearson (1975, p. 529) calls "the asymmetry of the land-breeze sea-breeze circulation", which occurs even when the land–sea temperature difference is equally strong during both sea breeze and land breeze phases: they are static stability, eddy diffusivity and available potential energy. In the opinion of Mak and Walsh (1976) the major differences between land and sea breeze circulations can be attributed to the diurnal variations in the static stability and eddy diffusivities of the air. Of the two effects, that due to stability fluctuations seems more important than that due to diffusivity variations. These results accord with both observations and intuitive qualitative reasoning. Diurnal variations of stability and diffusivity affect the energetics of the circulation by changing the depth of the layer of horizontal temperature contrast which is the circulation's driving mechanism. Pearson (1975) has evaluated this effect in terms of the available potential energies of the land and sea breezes. He found that the air which has been heated has three times the available energy of the same air which is cooled, where the heating and cooling are of equal but opposite magnitude. As the available energy is less, the strength and intensity of the land breeze is expected to be less than that of the sea breeze.

All the theoretical analyses reviewed up to this point, both analytical and numerical, were concerned with improving our understanding of the sea breeze circulation *per se*, having no regard to application to either realistic or real coastlines, with the possible exception of McPherson's (1970) pioneer three-dimensional study. Important steps in the application of a numerical model of the sea breeze to realistic coastlines were made by Mahrer and Pielke (1977), who used a two-dimensional hydrostatic model to study the circulations that develop over a mountain barrier, a flat coastline and a mountainous coastline in the absence of a larger scale flow. Their results showed that the combined sea breeze and mountain circulations produced a more intense circulation during both day and night than when they acted separately. In the case of real coastlines significant contributions were made by Estoque *et al.* (1976) who simulated the lake breeze over southern Lake Ontario, by Physick (1976) who concentrated on lakes and gulfs and by Pielke (1974a,b) who simulated the sea breeze over Florida with the aid of an eight-level, three-dimensional primitive equation model. In the last-mentioned study (Pielke, 1974b) the minimum horizontal grid mesh was 11 km and the eight levels were the surface, 50 m, 100 m, 1.22 km, 1.82 km, 2.42 km, 3.62 km and 4.82 km. Although different in several respects to Estoque's (1961) original model, the broad strategy of Pielke's model was in accord with Estoque's thinking. Two layers were used in the model, as in Estoque's, but their depth was allowed to vary, in sharp contrast to Estoque's scheme. Within each layer the values of the exchange coefficients for momentum, heat and water vapour were as follows:

$$K_z^\eta = K_z^\eta|_H + \left[\frac{(H-z)^2}{(H-h)^2}\right]\left[K_z^\eta|_h - K_z^\eta|_H + (z-h)\left\{\frac{\partial}{\partial z} K_z^\eta|_h + \frac{2(K_z^\eta|_h - K_z^\eta|_H)}{(H-h)}\right\}\right], \qquad (136)$$

where $H \geq z \geq h$;

$$K_z^\eta = K_z^\eta|_H, \qquad (137)$$

where $z > H$;

$$K_z^\eta = (z/h)K_z^\eta|_h, \qquad (138)$$

where $z < h$; where the superscript η refers to either the momentum, heat or moisture exchange coefficient; $|_H$ or $|_h$ means the value to the left of the vertical line applicable to level H and h respectively.

The value of the exchange coefficient at the top of the planetary boundary layer (height H), $K_z^\eta|_H$, was set equal to the value $1\,\text{cm}^2\,\text{s}^{-1}$. At height h,

$$K_z^M|_h = k_v u_* h/\phi_i, \qquad (139)$$

$$K_z^W|_h = K_z^H|_h = \begin{cases} K_z^M|_h & \text{in stable air,} \\ \dfrac{K_z^M|_h}{\phi_i} & \text{in unstable air,} \end{cases} \qquad (140)$$

where u_* is the friction velocity and ϕ_i is a non-dimensional windshear in unstable ($i = 1$) and stable ($i = 2$) air.

These specifications of the separate exchange coefficients were clearly far more elaborate parameterizations of sub-grid scale processes than appeared in either Estoque's or McPherson's studies. Pielke claimed that the resulting vertical distributions of the K_z values were more realistic than previously attained and thus improved the solutions of the equations of motion and heating in the upper layers of his model. In addition to velocity and temperature structure he included the specific humidity equation in similar fashion to Magata (1965) to allow a prediction of cloud base height. The basic structure of the two equations of horizontal motion, the thermodynamic equation and the specific humidity equation was that of equations (114), (115), (117) and (125). The major differences were the inclusion of horizontal diffusion terms (solely to help in reducing computational noise) and the resolution of u, v, θ and q into a synoptic, or background component and "grid-volume averaged perturbations" from the overall synoptic state. It was these latter components that comprised the sea breeze characteristics within any given synoptic situation. Once more Pielke followed the path pioneered by Estoque (1961) and followed so fruitfully by Magata (1965) and others. In contrast to McPherson (1970), Pielke included both the differences of roughness and in heating between land and water. In fact, he found that the differential roughness did not by itself play an important role in the formation of the velocity divergence fields produced by the model. Indirectly, however, the surface roughness influenced the magnitudes of convergence through the increased turbulent transfer of heat and this effect should be included in any numerical model of the sea breeze.

Pielke analysed the sea breeze development within both south-east and south-west synoptic winds of $4.2\,\mathrm{m\,s^{-1}}$. He represented his results in the form of maps of both horizontal and vertical velocities at $0.05\,\mathrm{km}$ and $1.22\,\mathrm{km}$ levels respectively for the times 3, 5, 8 and 10 h after the start of integration (simulated sunrise). In the south-east wind a strong convergence zone developed about 20 km inland of the west coast after 5 h and moved 30–40 km inland after 10 h. Associated with this convergence was a zone of uplift, reaching $40\,\mathrm{cm\,s^{-1}}$ in places. In the south-west wind similar features developed, as would be expected with identical dynamic treatment, but the west coast convergence zone moved far further inland whereas that on the east coast remained essentially stationary. The vertical velocities at maximum development of the breeze were 20–$30\,\mathrm{cm\,s^{-1}}$.

Pielke compared his model results with radar and ATS pictures of cumulus development over Florida (Pielke, 1973). A dramatically close correspondence led Pielke to conclude that on days without significant, organized, synoptic scale disturbances overlying south Florida, the sea breeze convergence patterns were the primary control of the general locations of the cloud and shower complexes at least for several hours after significant precipitation had begun. These conclusions were substantiated by the use of the model in conjunction with a one-dimensional, time-dependent cumulus model (Cotton et al., 1976). The sea breeze model altered the synoptic environment by increasing temperature and humidity below 800 mb, increasing the depth of the planetary boundary layer, inducing larger surface fluxes of momentum, heat and moisture, changing the vertical shear of the horizontal winds in lower levels of the atmosphere and developing intense, horizontal convergence regions of heat, moisture and momentum. The cumulus-scale model responded by developing a significantly deeper cloud which lived longer than in an undisturbed synoptic environment and precipitated. Compared with reality, the cloud model underpredicted cloud-top height, overpredicted precipitation, but generally it simulated reality more closely when operated within the meso-scale sea breeze circulation than within an unmodified atmosphere. Such fruitful experiments as these go a long way towards the establishment of the role of meso-scale circulations within the atmosphere and have profound implications for both theoretical understanding and practical forecasting.

IV. Hardware models

Hardware experimental studies of sea/land breezes, as opposed to density currents in general, are comparatively rare. This is probably due to the remarkable success achieved by theoretical investigations. Yet interesting results are forthcoming from comparatively simple experiments. In particular

Simpson (1969) has clearly revealed the lobe structure of the leading edge of a density current and vertical motion at this edge in a simulation of the sea breeze front. He supported his result with observations of lobes on a real sea breeze front over Hampshire. Dynamic similarity between model and atmosphere was assessed with the aid of Keulegan's (1957, 1958) law of saline fronts, viz.

$$U = Fr\,[(\Delta\rho/\rho)gh]^{1/2},$$

where U is the speed of the head of the current, ρ is the density of the denser fluid, $\Delta\rho$ is the density difference of the fluids, h is the height of the head of the current and Fr is a constant value 0.78. The constant Fr is an internal Froude number relating inertial force to bouyancy force. Using data gathered on sea breezes at Lasham in Hampshire, Simpson showed that Fr = 0.62 and he considered such a value to support the view that the breeze behaved as a density current. As in lee-wave simulation, the Froude number (with values in this case of about 0.75) is the important measure of similitude.

V. The role of sea/land breezes

Apart from their intrinsic interest, sea/land breezes attract attention for their various roles in other realms of meteorology. In the last two decades it has become clear that the breezes have profound effects on the distribution of pollution in coastal areas and, at the other extreme, of cumulus convection inland of the coast. More recently their energetic contribution to the general circulation has been estimated.

The classic area for investigation of the effects of sea/land breezes on pollution is the Los Angeles basin. As a result of long-standing monitoring of airflow and pollution over Los Angeles, the former by both surface stations and tetroons, we now know that, in general, the sea breeze tends to bring welcome relief from high pollution levels over the city. Figure 82 shows the striking decrease in ozone concentration in moving from "city" air to marine air (Angell et al., 1972). A further beneficial effect lies in the comparative stability of the sea breeze air. This may constrain the effluent from a chimney on the coast into a steady cone as it moves downstream, thus preventing pollution at the ground. Eventually the modification of the sea air by overland heating causes a deepening of the near-surface mixing layer until the effluent cone is reached and "fumigation" occurs. The unfortunate ground area under the fumigation point then experiences high concentrations of effluent (Collins, 1971). Clearly the effect of the sea breeze is not always beneficial to all areas. The reversal of the sea/land breeze circulation in very weak gradient winds may lead to the oscillation of an increasingly dirty volume of air over the city (Stephens, 1968). Trajectory analysis (Kauper,

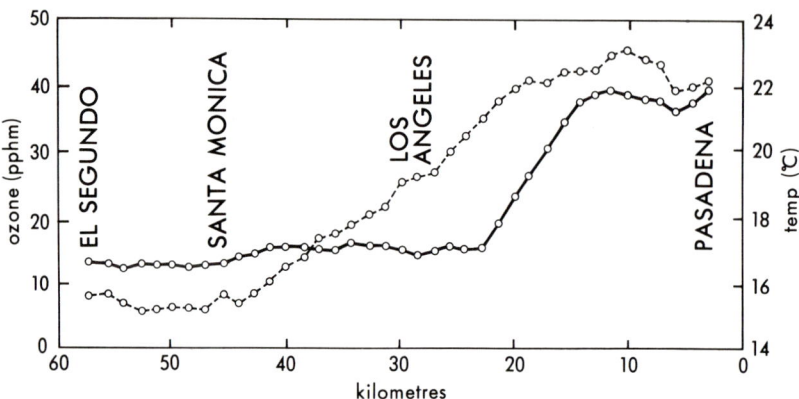

FIG. 82. Variation of air temperature (dashed line) and ozone concentration in parts per hundred million (solid line) as a helicopter moved westward through a sea breeze front lying between Pasadena and Santa Monica, California, on 25 September 1969. (After Angell et al., 1972.)

1960) and tetroon observations (Angell et al., 1972) both bear witness to this. Further, because of the distinct orientation of the coastline of the Los Angeles basin, sea breezes tend to meet in quite strong convergence zones which favour the concentration of pollution. The edge of the pollution pall is frequently well defined by convergence zones of similar dimensions (c. 25 km long and 5 km wide) and intensities. Near Los Angeles the San Fernando convergence zone leads to uplift of pollution which occasionally then returns over the basin as an extensive upper sheet of smog distinct from the polluted marine layer at the ground. Between the two layers, the subtropical inversion layer remains unpolluted, resisting contamination from either above or below (Edinger and Helvey, 1961).

The effect of the convergence zone associated with the sea breeze front was mentioned briefly earlier in this chapter. In the immediate post-war period Leopold (1949) showed how the Hawaiian sea breezes frequently interacted with the trade winds to produce streets of cumulus, some of which gave showers. A similar mechanism operates in Florida (Day, 1953; Lhermitte and Gilet, 1975), frequently triggering thundery outbreaks (Byers and Rodebush, 1948). Pielke's (1973) more thorough examination of the location of convective activity within the Florida peninsula confirms a close dependence on the movements of the east and west coast sea breezes, particularly when gradient wind speeds are low. Severe convective activity at the sea breeze front is rarer in mid-latitudes but has been recorded by Saunders (1958) who noted cumulonimbi 12 km deep some 15–25 km inland from the southern England coast on the day of a well-developed sea breeze from the English Channel.

Although the land breeze is weaker than the sea breeze, in the sub-tropics a favourable configuration of coastline (i.e. concave to the sea) may cause convergence sufficient to encourage the development of cumulus offshore at the leading edge(s) of the land breeze(s). Indeed Neumann (1951) was of the opinion that nocturnal thunderstorms off the coasts of Israel and Egypt are due to this mechanism. Preston-Whyte (1970) found a similar situation off the Natal coast of South Africa.

A less obvious effect of the sea breeze is the upward transfer of heat. Walsh (1974) has calculated that the vertical heat flux due to sea breezes throughout the globe is 2.1×10^{14} J s^{-1} which is 1–3 per cent of the total vertical flux of heat into the atmosphere. Put in another way, in terms of the vertical transport of sensible heat, the globally integrated effect of the sea breeze at a height of about 250 m is roughly equal to that at 500 mb in one extratropical cyclone (2.3×10^{14} J s^{-1}). Clearly from an energetics point of view sea breezes play a minor role in the suite of atmospheric circulations. Yet, as this chapter has attempted to show, they have significant effects on the actual weather and climate of coastal areas.

References

Air Ministry (1943). Sea breezes and land breezes. *Synoptic Division Technical Memorandum* no. 58, Meteorological Office, London.

Angell, J. K. and Pack, D. H. (1965). Study of the sea breeze at Atlantic City N.J., using tetroons as Langrangian tracers. *Mon. Weath. Rev.*, **93**, 475–493.

Angell, J. K., Pack, D. H., Machta, L., Dickson, C. R. and Hoecker, W. H. (1972). Three dimensional air trajectories determined from tetroon flights in the planetary boundary layer of the Los Angeles Basin. *J. appl. Met.*, **11**, 451–471.

Arakawa, H. and Utsugi, M. (1937). Theoretical investigation on land and sea breezes. *Geophys. Mag., Tokyo*, **11**, 97–104.

Baralt, G. L. and Brown, R. A. (1965). *The Land and Sea Breeze: An Annotated Bibliography*, Contract No. AF19 (604)-7259, Final Report (1960–65), Article F, Department of Geophysical Sciences, University of Chicago, Chicago.

Biggs, W. G. and Graves, M. E. (1962). A lake breeze index. *J. appl. Met.*, **1**, 474–480.

Blanford, H. F. (1889). *A Practical Guide to the Climates and Weather of India, Ceylon and Burmah and the Storms in the Indian Seas*, Macmillan and Co., London.

Brittain, O. W. (1978). Forecasting sea breezes at Eskmeals. *Met. Mag.*, **107**, 88–96.

Byers, H. R. and Rodebush, H. R. (1948). Causes of thunderstorms of the Florida peninsula. *J. Met.*, **5**, 275–280.

Capper, J. (1801). *Observations on the Winds and Monsoons*, C. Whittingham, London.

Clarke, R. H. (1955). Some observations and comments on the sea breeze. *Aust. Met. Mag.*, **11**, 47–68.

Collins, G. F. (1971). Predicting "sea-breeze fumigation" from tall stacks at coastal locations. *Nuclear Safety*, **12**, 110–114.

Cotton, W. R., Pielke, R. A. and Gannon, P. T. (1976). Numerical experiments on the influence of the meso-scale circulation on the cumulus scale. *J. atmos. Sci.*, **33**, 252–261.

Craig, R. A., Katz, I. and Harney, P. J. (1945). Sea-breeze cross-sections from psychometric measurements. *Bull. Am. met. Soc.*, **26**, 405–410.

Davis, W. M., Schultz, L. G. and Ward, R. De C. (1890). An investigation of the sea-breeze. *Ann. astronom. Obs. Harvard College*, **21**, 215–263.

Day, S. (1953). Horizontal convergence and the occurrence of summer precipitation at Miami, Florida. *Mon. Weath. Rev.*, **81**, 155–161.

Defant, F. (1950). Theorie der Land und Seewinde. *Arch. Met. Geophys. Bioklim. A*, **2**, 404–425.

de Felice, P. and Gasne-Tabbagh, J. (1971). A possible sea breeze effect in Brittany. *Weather*, **26**, 217–221.

Dekate, M. V. (1968). Climatological study of sea and land breezes over Bombay. *Indian J. Met. Geophys.*, **19**, 421–442.

Dixit, C. M. and Nicholson, J. R, (1964). The sea breeze at and near Bombay. *Indian J. Met. Geophys.*, **15**, 603–608.

Eddy, A. (1966). The Texas coast sea-breeze: a pilot study. *Weather*, **21**, 162–170.

Edinger, J. G. and Helvey, R. A. (1961). The San Fernando convergence zone. *Bull. Am. met. Soc.*, **42**, 626–635.

Elliott, A. (1964). Sea breezes at Porton Down. *Weather*, **19**, 147–149.

Estoque, M. A. (1961). A theoretical investigation of the sea breeze. *Q. Jl R. met. Soc.*, **87**, 136–146.

Estoque, M. A. (1962). The sea-breeze as a function of the prevailing situation. *J. atmos. Sci.*, **19**, 244–250.

Estoque, M. A., Gross, J. and Lai, H. W. (1976). A lake breeze over southern Lake Ontario. *Mon. Weath. Rev.*, **104**, 386–396.

Feit, D. M. (1969). Analysis of the Texas coast land breeze. Report no. 18, Atmospheric Science Group, University of Texas at Austin.

Fergusson, P. (1971). A sea breeze at Harrogate. *Weather*, **26**, 125–127.

Findlater, J. (1963). Some aerial explorations of coastal airflow. *Met. Mag.*, **92**, 231–243.

Findlater, J. (1964). The sea breeze and inland convection—an example of their interrelation. *Met. Mag.*, **93**, 82–89.

Fisher, E. L. (1960). An observational study of the sea breeze. *J. Met.* **17**, 645–660.

Fisher, E. L. (1961*a*). A theoretical study of the sea breeze. *J. Met.* **18**, 216–233.

Fisher, E. L. (1961*b*). Local wind circulations, final report. Vol. 1. Further studies of theoretical sea breeze models. Contract No. DA-36-039-sc-84939, Department of Meteorology and Oceanography, College of Engineering, Research Division, New York University, New York.

Fosberg, M. A. and Schroeder, M. J. (1966). Marine air penetration in Central California. *J. appl. Met.*, **5**, 573–589.

Frizzola, J. A. and Fisher, E. L. (1963). A series of sea-breeze observations in the New York City area. *J. appl. Met.*, **2**, 722–739.

Gill, D. S. (1968). The diurnal variation of the sea-breeze at three stations in north-east Scotland. *Met. Mag.*, **97**, 19–24.

Hatcher, R. W. and Sawyer, J. S. (1947). Sea breeze structure with particular reference to temperature and water vapour gradients and associated radio-ducts. *Q. Jl R. met. Soc.*, **73**, 391–406.

Haurwitz, B. (1947). Comments on the sea-breeze circulation. *J. Met.*, **4**, 1–8.

Haurwitz, B. (1959). A linear sea breeze model. Quarterly Progress Report no. 3, Project no. 3-36-05-401, College of Engineering, Research Division, New York University, New York.

Hess, S. L. (1959). *Introduction to Theoretical Meteorology*, H. Holt and Co., New York.

Hsu, S.-A. (1967). Mesoscale surface temperature characteristics of the Texas coast sea breeze. Report no. 6, Atmospheric Science Group, College of Engineering, University of Texas at Austin.
Hsu, S.-A. (1970). Coastal air-circulation system: observations and empirical model. *Mon. Weath. Rev.*, **98**, 487–509.
Jeffreys, H. (1922). On the dynamics of wind. *Q. Jl R. met. Soc.*, **48**, 29–47:
Jehn, K. H. (1973). A sea breeze bibliography 1664–1972. Report no. 37, Atmospheric Science Group, College of Engineering, University of Texas at Austin.
Jewell, C. J. (1964). Land and sea breeze circulation experienced over Durban. Dissertation, Department of Geography, University of Natal, Durban. Quoted in Tyson (1966).
Johnson, A., Jr and O'Brien, J. J. (1973). A study of an Oregon sea breeze event. *J. appl. Met.*, **12**, 1267–1283.
Kauper, E. K. (1960). The zone of discontinuity between the land and sea breezes and its importance to Southern California air pollution. *Bull. Am. met. Soc.*, **41**, 410–422.
Keen, C. S. and Lyons, W. A. (1978). Lake/land breeze circulations on the western shore of Lake Michigan. *J. appl. Met.*, **17**, 1843–1855.
Keulegan, G. H. (1957). An experimental study of the motion of saline water from locks into fresh water channels. 12th Progress Report on Model Laws for Density Currents, US National Bureau of Standards no. 5168.
Keulegan, G. H. (1958). The motion of saline fronts in still water. 13th Progress Report on Model Laws for Density Currents, US National Bureau of Standards no. 5831.
Kimble, G. H. T. and collaborators (1946). Tropical land and sea breezes. *Bull. Am. met. Soc.*, **27**, 99–113.
Kobayasi, T. and Sasaki, T. (1932). Über Land- und Seewinde. *Beitr. Phys. freien Atmos.*, **19**, 17–21.
Kopfmüller, A. (1922). Der Land und Seewinde am Bodensee. *Das Wetter*, **39**, 97–107; **40**, 33–41, 65–78, 108–115 (1923); **41**, 1–8, 33–42 (1924).
Koschmieder, H. (1936). Danziger Seewinstudien I, Nachweis und Beschreibung, sowie Beiträge zur Kinematik und Dynamik des Seewindes. *Danziger met. ForschArbeit.*, **8**.
Koschmeider, H. (1941). Danziger Seewindstudien II, Ergebnisse Gehäufter Höhenwindmessungen. *Danziger met. ForschArbeit.*, **10**, 1–39.
Lambert, S. (1974). High resolution numerical study of the sea-breeze front. *Atmosphere*, **12**, 97–105.
Leopold, L. B. (1949). The interaction of trade wind and sea breeze, Hawaii. *J. Met.*, **6**, 312–320.
Lhermitte, J. S. and Gilet, M. (1975). Dual-doppler radar observation and study of sea-breeze convective storm development. *J. appl. Met.*, **14**, 1346–1361.
Loomis, F. E. (1871). A sea breeze bibliography 1664–1972. *Conn. Acad. Trans.*, ii, 209–269.
Lyons, W. A. (1972). The climatology and prediction of the Chicago lake breeze. *J. appl. Met.*, **11**, 1259–1270.
Magata, M. (1965). A study of the sea breeze by the numerical experiment. *Pap. Met. Geophys.*, **16**, 23–36.
Mahrer, Y. and Pielke, R. A. (1977). The effects of topography on sea and land breezes in a two-dimensional numerical model. *Mon. Weath. Rev.*, **105**, 1151–1162.
Mak, M. K. and Walsh, J. E. (1976). On the relative intensities of sea and land breezes. *J. atmos. Sci.*, **33**, 242–251.

Marshall, W. A. L. (1950). Sea breeze across London. *Met. Mag.*, **79**, 165–168.
McCaffery, W. D. S. (1966). On sea breeze forecasting techniques. Memorandum no. 12, Forecasting Techniques Branch, Meteorological Office, London.
McPherson, R. D. (1970). A numerical study of the effect of a coastal irregularity on the sea breeze. *J. appl. Met.*, **9**, 767–777.
Meyer, J. H. (1971). Radar observations of land breeze fronts. *J. appl. Met.*, **10**, 1224–1232.
Mizuma, M. and Kakuta, M. (1974). Observational study on land and sea breezes in the Tokai village area. *J. met. Soc. Jap.*, **52**, 417–427.
Moroz, W. J. (1967). A lake breeze on the eastern shore of Lake Michigan: observation and model. *J. atmos. Sci.*, **24**, 337–355.
Narayanan, V. (1967). An observational study of the sea breeze at an equatorial coastal station. *Indian J. Met. Geophys.*, **18**, 497–504.
Neumann, C. and Partsch, J. (1885). *Physikalische Geographie von Griechenland*, Breslau.
Neumann, J. (1951). Land breezes and nocturnal thunderstorms. *J. Met.*, **8**, 60–67.
Neumann, J. (1973). Sea and land breezes in the classical Greek literature. *Bull. Am. met. Soc.*, **54**, 5–8.
Neumann, J. (1977). On the rotation rate of the direction of sea and land breezes. *J. atmos. Sci.*, **34**, 1913–1917.
Neumann, J. and Mahrer, Y. (1971). A theoretical study of the land and sea breeze circulation. *J. atmos. Sci.*, **28**, 532–542.
Neumann, J. and Mahrer, Y. (1974a). Evolution of a sea breeze front; a numerical study. *Bonn met. Abh.*, **17**, 481–492.
Neumann, J. and Mahrer, Y. (1974b). A theoretical study of the sea and land breezes of circular islands. *J. atmos. Sci.*, **31**, 2027–2039.
Neumann, J. and Mahrer, Y. (1975). A theoretical study of the lake and land breezes of circular lakes. *Mon. Weath. Rev.*, **103**, 474–485.
Olsson, L. E., Elliott, W. P. and Hsu, S.-I. (1973). Marine air penetration in western Oregon: an observational study. *Mon. Weath. Rev.*, **101**, 356–362.
Pack, D. H. and Angell, J. K. (1963). A preliminary study of air trajectories in the Los Angeles basin as derived from tetroon flights. *Mon. Weath. Rev.*, **91**, 583–604.
Patrinos, A. A. N. and Kistler, A. L. (1977). A numerical study of the Chicago lake breeze. *Boundary Layer Met.*, **12**, 93–123.
Pearce, R. P. (1955). The calculation of a sea breeze circulation in terms of the differential heating across the coast line. *Q. Jl R. met. Soc.*, **81**, 351–381.
Pearce, R. P. (1968). The generation of sea-breezes. *Schweiz. Aero-Rev.*, **43**, 195–200.
Pearson, R. A. (1973). Properties of the sea breeze front as shown by a numerical model. *J. atmos. Sci.*, **30**, 1050–1060.
Pearson, R. A. (1975). On the asymmetry of the land breeze/sea-breeze circulation. *Q. Jl R. met. Soc.*, **101**, 529–536.
Pedgley, D. E. (1958). The summer sea breeze at Ismailia. *Met. Rep., Lond.*, **3** (19).
Peters, S. P. (1938). Sea Breezes at Worthy Down, Winchester. *Professional Notes no. 86*, **6**, Meteorological Office.
Physik, W. (1976). A numerical model of the sea-breeze phenomenon over a lake or gulf. *J. atmos. Sci.*, **33**, 2107–2135.
Pielke, R. A. (1973). Observational study of cumulus convection patterns in relation to the sea-breeze over south Florida. Technical Memorandum, NOAA Environmental Research Laboratories, Boulder, Colorado.
Pielke, R. A. (1974a). A comparison of three-dimensional and two-dimensional numerical predictions of sea breezes. *J. atmos. Sci.*, **31**, 1577–1585.

Pielke, R. A. (1974b). A three-dimensional numerical model of the sea breezes over south Florida. *Mon. Weath. Rev.*, **102**, 115–139.

Pierson, W. J. (1950). The effects of eddy viscosity, coriolis deflection and temperature fluctuation on the sea breeze as a function of time and height. *Met. Pap.*, **1** (2), 7–29.

Prestel, M. A. F. (1864). Die jährliche and tägliche Periode in der Aenderung der Windesrichtungen über der deutschen Nordseeküste sowie der Winde an den Küsten des Rigaischen und Finnischen Meerbusens und des Weissen Meeres. *Abh. Leopold Akad.*, **30**.

Preston-Whyte, R. A. (1969). Sea breeze studies in Natal. *S. Afr. Geogr. J.*, **51**, 38–49.

Preston-Whyte, R. A. (1970). Land breezes and rainfall on the Natal coast. *S. Afr. Geogr. J.*, **52**, 38–43.

Ramakrishnan, K. P. and Jambunathan, R. (1958). Sea breeze and maximum temperatures in Madras. *Indian J. Met. Geophys.*, **9**, 349–358.

Ramanadham, R. and Subbaramayya, I. (1965). The sea breeze at Visakhapatnam. *Indian J. Met. Geophys.*, **16**, 241–248.

Ramanathan, K. R. (1931). The structure of the sea breeze at Poona. *Sci. Notes Met. Dept India*, **3**, 131–134.

Ramdas, L. A. (1931). The sea breeze at Karachi. *Sci. Notes Met. Dept India*, **4**, 115–124.

Rao, D. V. (1955). The speed and some other features of the sea breeze front at Madras. *Indian J. Met. Geophys.*, **6**, 233–242.

Ringe, C. (1882). Über San Diego an der Küste von Kalifornien. *Ann. Hydrogr. maritimen Met.*, **10**.

Roy, A. K. (1940). The sea-breeze at Madras. *Sci. Notes Met. Dept India*, **8**, 138–146.

Saunders, P. M. (1958). Sea-breeze convergence zone. *Sailplane Gliding*, **9**, 276–279.

Schmidt, F. H. (1946). On the causes of pressure variations at the ground. *K. ned. meteor. Inst. Meded.*, **2**.

Schmidt, F. H. (1947). An elementary theory of the land- and sea-breeze circulation. *J. Met.*, **4**, 9–15.

Schroeder, M. J., Fosberg, M. A., Cramer, O. P. and O'Dell, C. A. (1967). Marine air invasion of the Pacific coast: a problem analysis. *Bull. Am. met. Soc.*, **48**, 802–808.

Sen Gupta, P. K. and Chakravortty, K. C. (1947). Land breeze at Calcutta (Alipore). *Sci. Notes Met. Dept India*, **9**, 73–80.

Sherman, O. T. (1880). Observations on the height of land and sea breezes taken at Coney Island. *Z. Met.*, **15**, 446–449.

Simpson, J. E. (1964). Sea-breeze fronts in Hampshire. *Weather*, **19**, 208–215.

Simpson, J. E. (1967). Aerial and radar observations of some sea-breeze fronts. *Weather*, **22**, 306–317.

Simpson, J. E. (1969). A comparison between laboratory and atmospheric density currents. *Q. Jl R. met. Soc.* **95**, 758–765.

Simpson, J. E., Mansfield, D. A. and Milford, J. R. (1977). Inland penetration of sea-breeze fronts. *Q. Jl R. met. Soc.* **103**, 47–76.

Smith, M. F. (1974). A short note on a sea-breeze crossing East Anglia. *Met. Mag.*, **103**, 115–118.

South African Weather Bureau (1960). *Climate of South Africa*, Part 6, *Surface Winds*, Government Printer, Pretoria. Quoted in Tyson (1966).

Staley, D. O. (1957). The low-level sea breeze of north-west Washington. *J. Met.*, **14**, 458–470.

Stephens, E. R. (1968). The marine layer and its relation to a smog episode in Riverside, California. *Atmos. Environ.*, **2**, 393–396.

Sutcliffe, R. C. (1937). The sea breeze at Felixstowe. A statistical investigation of pilot balloon ascents up to 5,500 feet. *Q. Jl R. met. Soc.*, **63**, 137–146.

Taylor, A. D. (1877). Beschreibung des Mergui-Archipels, bengalischer Meerbusen. *Ann. Hydrogr. maritimen Met.*, **5**, 166.

Tyson, P. D. (1966). Examples of local air circulations over Cato Ridge during July 1965. *S. Afr. Geogr. J.*, **48**, 13–31.

van Bemmelen, W. (1922). Land und Seebrisen in Batavia. *Beitr. Phys. freien Atmos.*, **10**, 169–177.

Wallington, C. E. (1959). Structure of sea breeze front as revealed by gliding flights. *Weather*, **14**, 263–269.

Wallington, C. E. (1961). An introduction to the sea breeze front. *Schweitz. Aero-Rev.*, **7**, 393–397.

Wallington, C. E. (1965). Gliding through a sea breeze front. *Weather*, **20**, 140–143.

Walsh, J. E. (1974). Sea breeze theory and applications. *J. atmos. Sci.*, **31**, 2012–2026.

Watts, A. J. (1955). Sea breeze at Thorney Island. *Met. Mag.*, **84**, 42–48.

Wexler, R. (1946). Theory and observations of land and sea breezes. *Bull. Am. met. Soc.*, **27**, 272–287.

Woeikof, A. (1875). *Discussion and Analysis of Professor Coffin's Tables and Charts of the Winds of the Globe*, Smithsonian Institution, Washington, D.C., pp. 688, 704, 706, 742.

Yamashita, R. (1953). On land and sea breezes. *J. met. Soc. Jap.*, **31**, 157–172.

Yoshikado, H. and Asai, T. (1972). Numerical experiment of effects of turbulent transfer processes on the land and sea breeze. Contribution no. 12, University Geophysical Institute, Kyoto, Japan, pp. 33–48.

Zambakas, J. D. (1973). The diurnal variation and duration of the sea-breeze at the National Observatory of Athens, Greece. *Met. Mag.*, **102**, 224–228.

6
Slope and Valley Wind Circulation

I. Introduction

For over a century we have been aware of winds that appear to be peculiar to sloping topography. In particular the mountains and valleys of the European Alps appeared to cause airflows that aroused the curiosity of nineteenth century meteorologists. Simple observation soon revealed downslope (katabatic) flow at night and, less easily observed, upslope (anabatic) flow in the day. Despite numerous observational studies in Europe, many simply of general airflow in mountains rather than with the clear aim of analysing the valley wind system *per se*, the detailed nature of slope and valley winds remained obscure for several decades. Similarly the search for a theory of their formation led to a spate of literature in German meteorological periodicals in the 1920s and 1930s.

Most slopes (begging the question of their definition!) are at most of the order of a few square kilometres in area and consequently could generate winds covering similar sized areas. Such "local" winds may be thought to be beyond the scope of this book. Yet, of course, some slopes are truly massive (e.g. the Antarctic ice surface) and others, the majority, are but components of a mountain–valley system. Many such systems have horizontal dimensions of 10–150 km or perhaps a little larger and consequently generate meso-scale circulations. Indeed systems of this size reveal themselves in analyses of both smaller and larger scale circulations. Thus, Tyson's analyses (Tyson, 1966, 1967, 1968a; Tyson and Preston-Whyte, 1972) of airflow in the valleys to the north-west of Durban eventually revealed a larger, "regional" (at the mesoscale) circulation over the uplands of Natal. Conversely, in their studies of the apparently massive Antarctic katabatic winds Mather and Miller (1966, p. 284) indicated that

> we should not regard the high plateau as a breeding ground from which gravity winds drain radially outwards across the rest of the continent. Rather, the wind is generated locally, as in the case of glacier and valley winds, according to local conditions of slope and inversion, and its continued flow downstream depends wholly on the existence of similar favourable generating conditions downstream and not on the inertia of the air mass.

Thus, within the huge variety of land form and size, distinct meso-scale circulations and associated airflows are found world wide. This chapter outlines their structure and mechanism.

In essence winds on slopes form part of a thermally direct, reversible circulation in a vertical plane. In this respect they are identical to sea/land breezes. In what follows, the structure and life cycle of the circulation is outlined. The analysis is facilitated by a clarification of terms which abound in the literature. In order of increasing size we may recognize the following: First, the slope wind *per se*, which may occur both on individual slopes or slopes which form a mountain or a valley. In the latter case the winds are frequently known as "cross-valley" winds. Second, the aggregation of slopes into hill and valley generates the generally larger "along-valley" winds, which may blow either "up-" or "down-" valley, the latter frequently being called the "mountain wind". Third, Tyson and Preston-Whyte (1972) recognized a

Table 27 Nomenclature of slope and valley wind circulation

Term	Meaning
Slope/valley wind	Thermally induced day-time upslope/-valley wind in the lower layers
Downslope/down-valley (mountain) wind	Thermally induced night-time downslope/-valley wind in the lower layers
Anti-slope/-valley wind	Compensation wind to slope/valley wind of opposite direction, flowing just above top of the slope/valley wind
Anti-downslope/-down-valley (mountain) wind	Compensation wind to the downslope/down-valley (mountain) wind of roughly opposite direction, flowing just above the downslope/down-valley (mountain) wind
Valley wind system	The system of the above-named four winds in *valleys* as opposed to *slopes*
Anti-wind	Anti-slope/valley wind or anti-downslope/-down-valley (mountain) wind

Source: Buettner and Thyer (1959).

larger system comprising what they call "mountain–plain" and "plain–mountain" winds. In this case the horizontal dimensions are of the order of 100–200 km and the valley wind circulations operate within the mountain-plain system.

All the above terms apply to the near-surface limb of the relevant vertical circulations. They are associated with an airflow in the opposite direction at a higher level which may be classified in similar fashion. Thus up- and downslope winds at the surface are accompanied by down- and upslope (or "anti-slope"; Buettner and Thyer, 1966) flows at a higher level. Similarly, surface up- and down-valley (or mountain) winds have down- and up-valley flows aloft. Tyson (1968a) and Buettner and Thyer (1959) used the terms "anti-valley" and "anti-mountain" winds to describe these upper level compensatory flows. Although Tyson (1968a) introduced the notion of a reversible mountain–plain circulation, he did not suggest a term for the necessary upper-level return flows. Following his earlier terminology, they should, of course, be "anti-mountain–plain" and "anti-plain–mountain" winds. Clearly this above classification is somewhat subjective, but it provides a sensible and convenient framework for the analysis that follows. Table 27 summarizes these terms and their meanings as they are used in this chapter. Note that slopes and valleys have been amalgamated purely to save space.

A. Mechanism of the slope and valley wind

In common with the sea/land breeze circulation, much misunderstanding of the mechanism of the slope/valley wind circulation abounded in the literature of the late nineteenth and early twentieth centuries. Despite a painstaking and very perceptive study by Fournet (1840), Hann's (1879) theories, which in fact predicted a circulation on a sense contrary to observation even in those times, appeared to be most persuasive. But by the inter-war period, Bjerknes' circulation theorem was known to investigators of slope and valley winds and Wenger (1923) used it to very good effect in his demonstration of the inadequacies of Hann's theories.

We noted in Chapter 5 that the circulation theorem provided the basic explanation for the sea/land breeze. An identical thermally induced mechanism also explains the slope/valley wind system. For simplicity, only the slope is considered below and the initial conditions are assumed to be calm and cloudless. In the day-time, absorption of radiation by sloping ground leads to warming of air near the surface—as in any other part of the world. But the air near the sloping ground thus becomes warmer than air in the free atmosphere at the same height above sea level. Consequently, if we invoke the hydrostatic principle and consider the isobars to be initially horizontal, the vertical pressure gradient will be greater in the cooler "free" air than in the warmer air near the slope. In turn, this means that at a given height within the influence of slope heating, pressure will become higher than at a point further away from the slope. This horizontal pressure gradient causes air to move away from the slope and thus generates higher pressures than had previously been experienced at a lower level—say over an adjacent plain. Consequently a low level, horizontal pressure gradient from plain to slope is created and an upslope wind results. The circulation in the vertical plane is equivalent to that in the sea breeze. At night, the mechanism and the circulation are reversed. Surface cooling causes an upper level pressure gradient towards the slope and a cold air drainage down and away from the slope.

Temperature differences between the "slope" and "free" air of only fractions of a degree are sufficient to initiate the mechanism. Wenger (1923) calculated that the velocity around a closed vertical circulation near a slope would be $9\Delta T$ m s^{-1} (where ΔT is the temperature difference between near-slope and free air) only 3 h after initiation. Despite the approximations made in his analysis, his results clearly illustrated the remarkable effectiveness of the baroclinic overturning required by the Bjerknes circulation theorem.

Table 28 provides a useful comparison of the characteristics of the sea breezes and valley winds. Of the two circulations, the valley system appears

Table 28 Comparison of data on the sea/land breeze and valley wind systems

	Sea	Valley
Heat supply (cal cm^{-2} min^{-1})	0.5	0.2
ΔT (°C)†	10	10
Δp (mb)†	1	1
u (day) (m s^{-1})‡	3	3
u (night) (m s^{-1})‡	0.3	2
v (day) (m s^{-1})‡	3	<1
v (night) (m s^{-1})‡	0.3	<1
w (day) (m s^{-1})‡	5.7	3
Height of lower system (day) (km)	1	Ridge height
Depth of anti-system (day) (km)	1	Equal to ridge height
Begins at (summer)	1000	0800
Ends at	2000	2400
Coriolis involved	Yes	No
Velocity at which front progresses (km h^{-1})	10	—
Cumuli caused	Inland near front	Over ridges

† ΔT, Δp are differences between system and environment.
‡ u, v, w are wind components, where u is perpendicular to the coast and parallel to the long axis of the valley.
Source: Buettner (1968).

to be the most efficient, requiring much less heat input to generate a circulation of comparable size to but greater intensity than that of the sea breeze.

II. Observation

A. Climatology of slope and valley winds

The comparatively small and transient nature of slope and valley winds hinders a climatological analysis. Despite this, sufficient observations exist, particularly in areas of extreme warmth and cold, to draw a rough climatological sketch of both phenomena. In contrast to the sea/land breeze, it is the downslope and down-valley winds (equivalent to the land breeze) which are best documented: in many studies, the upslope and up-valley winds (equivalent to the sea breeze) proved to be quite elusive.

Table 29 summarizes the monthly variation in occurrence of downslope and down-valley winds in different parts of the World. At Poona almost every night in the months December to March was characterized by katabatic flows. During these four months, strong surface inversions occurred almost every night and hence conditions were altogether favourable for pronounced downslope winds. Also in these months the gradient winds were light and had a prominent up-valley component, so that the onset of the katabatic winds was shown by a well-marked reversal of the prevailing wind direction. On the other hand, the months May to August had comparatively high values of temperature minima, vapour pressure and cloud, so that conditions were not very favourable for pronounced nocturnal cooling by radiation, and downslope flow, if it occurred at all, would merge with the prevailing gradient winds which were strong and had prominent westerly (i.e. down-valley) components. At Mauna Loa and Cato Ridge, the winter figures showed good mutual agreement, but not with those at Poona. The reduction in summer frequency at Mauna Loa does, however, agree in a qualitative way with the observations at Poona.

High frequencies were also observed in high latitudes, particularly in the winter months (Table 29(c)). Indeed, in those months, conditions were favourable for downslope flow throughout the day. This is peculiar to Antarctica and is, of course, due to the predominantly ice surface. In the four warmest months (November to February), not only did downslope flows occur less frequently *in toto*, they were quite rare at 1700 hours local time. That these winds were indeed katabatic has been elegantly demonstrated by Lettau (1966) with the aid of theory developed by Prandtl (1952). In fact, as shown later in the chapter, Prandtl analysed upslope winds, but the theory is just as applicable to downslope flows.

In contrast to those in both low and high latitudes, frequencies of katabatic

Table 29 Monthly mean frequencies of downslope winds (slopes *per se*, S; valley, V)

	Jan	Feb	Mar	Apr	May	Jun	Jul	Aug	Sep	Oct	Nov	Dec
(a) *Sub-tropics*												
Poona (V): †percentage frequency of nights with downslope flow												
Per cent	93	97	90	50	0	3	0	0	20	31	77	92
Total no. of anemograms examined	62	32	62	60	62	60	62	62	60	62	60	62
Mauna Loa (S): ‡percentage frequency of days with downslope winds	60.9	—	—	—	—	—	40.8	—	—	—	—	—
Cato Ridge, South Africa (V): §percentage frequency of days with downslope winds	—	—	—	—	—	—	50	65	57	29	—	—
(b) *Mid-latitudes*												
Driffield, UK (V); ‖number of days with downslope winds	—	—	6	3	6	7	8	5	—	—	—	—
Oxfordshire, UK (V): ¶number of days with downslope winds	4	3	4	1	3	3	1	4	4	4	3	1
(c) *Polar*												
Mawson (S): ††number of occasions												
0000	11	7	5	9	5	9	2	7	11	14	15	8
0600	6	6	12	10	5	0	2	9	10	10	10	6
1200	0	2	9	9	3	11	7	6	4	7	5	0
1800	—	—	6	7	8	8	—	—	—	—	—	—
Total	17	15	32	35	21	28	11	22	25	31	30	14

† *Source*: Atmanathan (1931). Period of observations: May 1929 to April 1931.
‡ *Source*: Mendonca (1969). Period of observations: an unspecified 8 year period.
§ *Source*: Tyson (1966): Period of observations: July to October 1965.
‖ *Source*: Eldridge (1951). Period of observations: March to August 1943.
¶ *Source*: Heywood 1933. Period of observations: January to December 1930.
†† *So.urce*: Streten (1963). Period of observations: February 1960 to January 1961.
—, No information available.

Table 30   Monthly mean frequencies of upslope winds (slopes, S; valley, V)

	Jan	Feb	Mar	Apr	May	Jun	Jul	Aug	Sep	Oct	Nov	Dec
Mauna Loa (S): †percentage frequency	21	—	—	—	—	—	41.5	—	—	—	—	—
Cato Ridge (V): ‡percentage frequency	—	—	—	—	—	—	61	71	74	42	—	—

† *Source*: Mendonca (1969). Period of observations: an unspecified 8 year period.
‡ *Source*: Tyson (1966). Period of observations: July to October 1965.
—, No information available.

flow in mid-latitudes appear to be small. This may account for the sparsity of data. Table 29(b), incomplete as it is, suggests that downslope flows occur on a fifth to a quarter of all possible nights.

Climatological data on upslope winds are also rare. Table 30 presents some

Table 31(a) Time of onset of downslope/mountain winds: Poona (valley)

Time	Frequency	Percentage frequency
1800	—	—
1900	4	0.3
2000	28	2.0
2100	52	3.6
2200	96	6.7
2300	111	7.8
2400	120	8.4
0100	125	8.7
0200	143	10.0
0300	129	9.0
0400	134	9.4
0500	130	9.1
0600	125	8.8
0700	120	8.4
0800	86	6.0
0900	29	2.0

Source: Atmanathan (1931). Period of observations: May 1929 to April 1931.

Table 31(b) Time of onset of downslope/mountain winds: Cato Ridge, South Africa (valley)

	Jul	Aug	Sep	Oct
Mean onset time to nearest quarter-hour	2100	2145	2145	2215

Source: Tyson (1966). Period of observations: July to October 1965.

Table 31(c) Time of onset of downslope/mountain winds: Meadow, Kananaskis Valley, Alberta, Canada (valley)

	Jul	Aug
Mean onset time to nearest hour	2000	2100

Source: MacHattie (1968). Period of observations: 1 July to 9 September 1960.

quite contrary results. The winter figure for Mauna Loa is 21 per cent and for Cato Ridge 61 per cent. No obvious explanation presents itself and here is a clear case for further observation.

The times of onset, durations and times of cessation of the down- and upslope winds are shown in Tables 31–33. At Poona, Atmanathan (1931) claimed that katabatic flows may start at any time between 2200 and 0700 hours, with a maximum frequency at 0200 hours. In contrast, Tyson (1966) identified one clear onset time which varied between 2100 and 2215 hours from July to October. These times are in good agreement with those found by MacHattie (1968) in Alberta, Buettner and Thyer (1962) near Mt Rainier and Eldridge (1951) in Yorkshire, UK. Noting that the time of

Table 31(d) Time of onset of downslope/mountain winds: Driffield (valley)

	Time from sunset to start of wind (h)						Mean time
	1	2	3	4	5	6	
Number of days with downslope wind	—	2	5	2	1	1	3.5

Source: Eldridge (1951). Period of observations: March to August 1943.

Table 31(e) Time of onset of downslope/mountain winds: Mount Rainier (valley)

	Jul	Aug
Average time of onset	1900	1900

Source: Buettner and Thyer (1962). Period of observations: July and August 1958.

Table 31(f) Time of onset of downslope/mountain winds: Mawson (slope)

Month	Percentage frequency of times of onset—local time			
	1600–1800	1800–2000	2000–2200	2200–2400
Sep. 1960	17	61	10	0
Oct. 1960	4	46	21	14
Nov. 1960	5	11	33	33
Dec. 1960	0	4	54	25
Jan. 1961	0	8	25	50
Feb. 1961	17	17	25	17

Source: Streten (1963). Period of observations: February 1960 to January 1961.

sunset in the UK varies from about 1800 hours in March to about 2100 hours in July, we should expect the downslope flows to start between 2100 and 2400 hours. In the polar areas over 80 per cent of all onsets between September and January (summer in Antarctica) occurred between 1600 and 2400 hours. Table 31(f) suggests higher frequencies of onset at a later hour as the season advances from September to January. By February the pattern had become fairly evenly distributed within the time limits of 1600 to 2400 hours—a situation which continued into March. By April the times of

Table 32(a) Durations and times of cessation of downslope winds: Poona (valley)

Duration (h)	Frequency	Percentage frequency
<1	301	31
1.1–2.0	321	33
2.1–3.0	169	16
3.1–4.0	100	10
4.1–5.0	32	3
5.1–6.0	23	3
6.1–7.0	14	2
7.1–8.0	9	1
>8	13	1

Source: Atmanathan (1931). Period of observations: May 1929 to October 1975.

Table 32(b) Durations and times of cessation of downslope winds: Cato Ridge, South Africa (valley)

	Jul	Aug	Sep	Oct
Duration (h) (to nearest quarter-hour)	14.25	11.5	11.5	10.5

Source: Tyson (1966). Period of observations: July to October 1965.

Table 32(c) Durations and times of cessation of downslope winds: Meadow, Kananaskis Valley, Alberta, Canada (valley)

	Jul	Aug
Duration (h) (to nearest hour)	11	11

Source: MacHattie (1968). Period of observations: 1 July to 9 September 1960.

6 SLOPE AND VALLEY WIND CIRCULATION

Table 32(d) Durations and times of cessation of downslope winds: Mount Rainier (valley)

	Jul	Aug
Duration (h)	12.5	12.5

Score: Buettner and Thyer (1962). Period of observations: July and August 1958.

Table 32(e) Durations and times of cessation of downslope winds: Mawson (slope)

Local time	Percentage frequency of times of katabatic cessation†
0800–1000	13
1000–1200	19
1200–1400	23
1400–1600	17
1600–1800	16

Source: Streten (1963). Period of observations: February 1960 to January 1961.
† The data presented here cover only the period September 1960 to January 1961 (cf. Table 29(c)).

onset were no longer largely confined to the period between 1600 and 2400 hours and this situation was common to the rest of the winter months.

The rather anomalous figures for Poona in Table 31(a) lead to the similarly anomalous durations of Table 32(a), which indicates that 80 per cent of the katabatic flows lasted for less than 3 h. It becomes clear that Atmanathan considered not just "the katabatic wind" but every "katabatic movement" on each night. In his own words (Atmanathan, 1931, p. 103):

> It is further noticed that the katabatic movement consists of what may be termed a series of "floods and ebbs", the "ebb" representing an interval of calm between two "floods". The "flood" has a peculiar structure in which the wind-force rapidly increases from zero to its maximum value and then very gradually falls off. This is probably due to the drainage down the slopes of limited supplies of cold air, the absence of a constant motive force such as a pressure gradient being responsible for the peculiar structure of the "flood".

A climatology of such detailed structures is rare indeed.

In contrast, the data for Cato Ridge, Meadow and Mt Rainier (Tables 32(b)–(d)) give the duration of the general downslope flow, not its constituent parts. In all the areas the winds lasted for approximately half a day. Further poleward in the Antarctic summer the katabatic flows were most likely to

Table 33 Times of onset and duration of upslope/valley winds (combining slopes (S) and valleys (V))

	Jul	Aug	Sep	Oct
(a) Cato Ridge, South Africa (V)†				
Mean time of onset (to nearest quarter-hour)	1130	1200	1100	1045
Mean time of cessation (to nearest quarter-hour)	1930	1845	1815	1800
(b) Mount Rainier (V)‡				
Time of onset	0730	0730		
Time of cessation	1930	1930		
(c) Meadow, Kananaskis Valley, Alberta, Canada (V)§				
Time of onset (to nearest hour)	0700	0700		
Time of cessation (to nearest hour)	2000	2000		

† *Source:* Tyson (1966). Period of observations: July to October 1965.
‡ *Source:* Buettner and Thyer (1962). Period of observations: July and August 1958.
§ *Source:* MacHattie (1968). Period of observations: 1 July to 9 September 1960.

cease just after mid-day (Table 32(e)), whereas in winter they frequently blew continuously for days.

The upslope wind is less well observed. Such evidence as could be gathered (Table 33) suggests early initiation and late demise at Mt Rainier and Meadow and a much later development and slightly earlier demise on Cato Ridge. Durations consequently varied from about 7 to 13 h. Such differences strongly suggest powerful influences by the location, size, shape, orientation and vegetational cover of the landform in question.

In the papers reviewed above it is not always clear whether the data refer solely to slopes *per se*, to mountains and valleys or to both. In fact this lack of distinction is not a significant impediment to our understanding of the processes at work. Yet, some differences do exist between slope and valley winds and they are well illustrated by Fig. 83, which shows seasonal and diurnal variations in the speed of both slope and valley winds occurring at Fruitvale Flat in British Columbia. Over a period of 8 months, starting in December 1938, hourly winds were classified into one of the following categories: up- or downslope (the slopes being the valley sides) and up- or down-valley. Hourly mean values of wind speed in each of these four directions are plotted for three seasons (Fig. 83).

In winter (Fig. 83(a)) the resultant slope wind was downward throughout the day with a maximum value of $0.6 \, \text{m s}^{-1}$. The valley wind was much better developed than the slope wind. The up-valley wind occurred during daylight hours, reaching a maximum of $0.9 \, \text{m s}^{-1}$ at mid-day. A down-valley wind of similar magnitude occurred during the night. Both the slope and valley wind

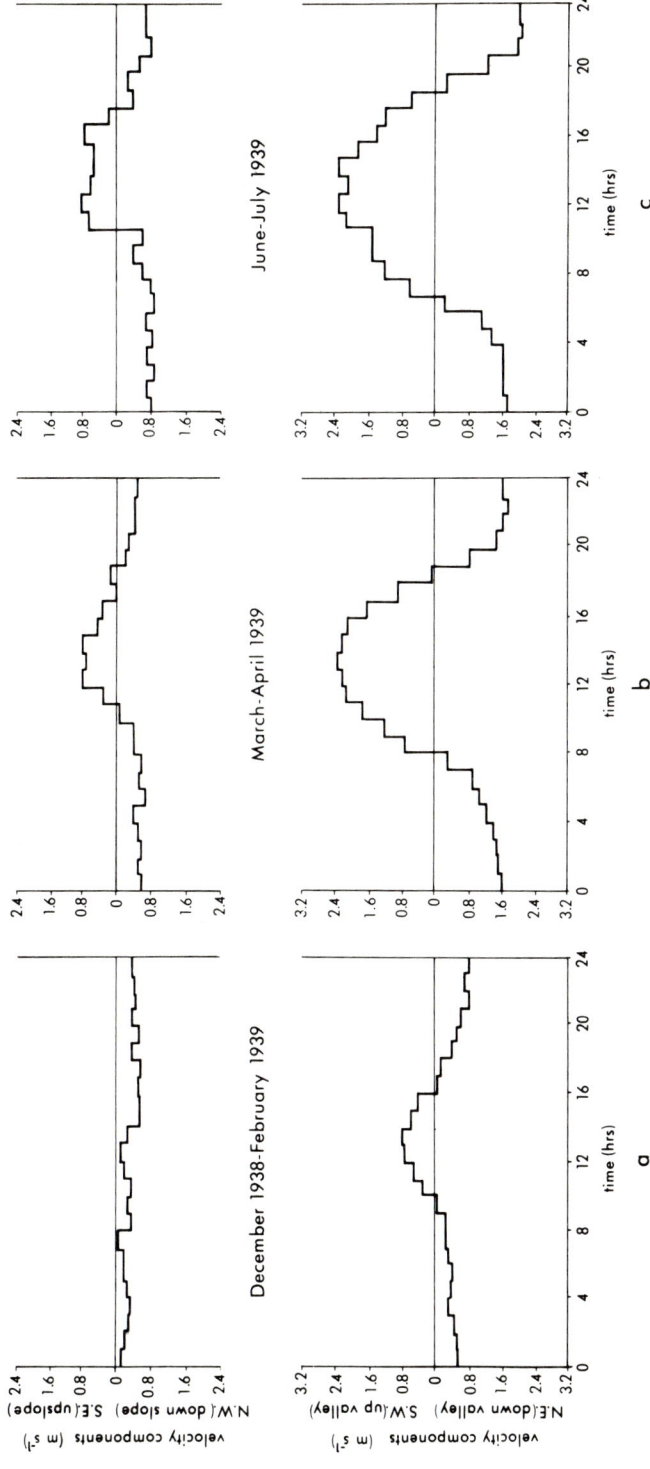

FIG. 83. Diurnal variation of slope and valley wind speed near Fruitvale Flat, British Columbia. (After Cross, 1950.)

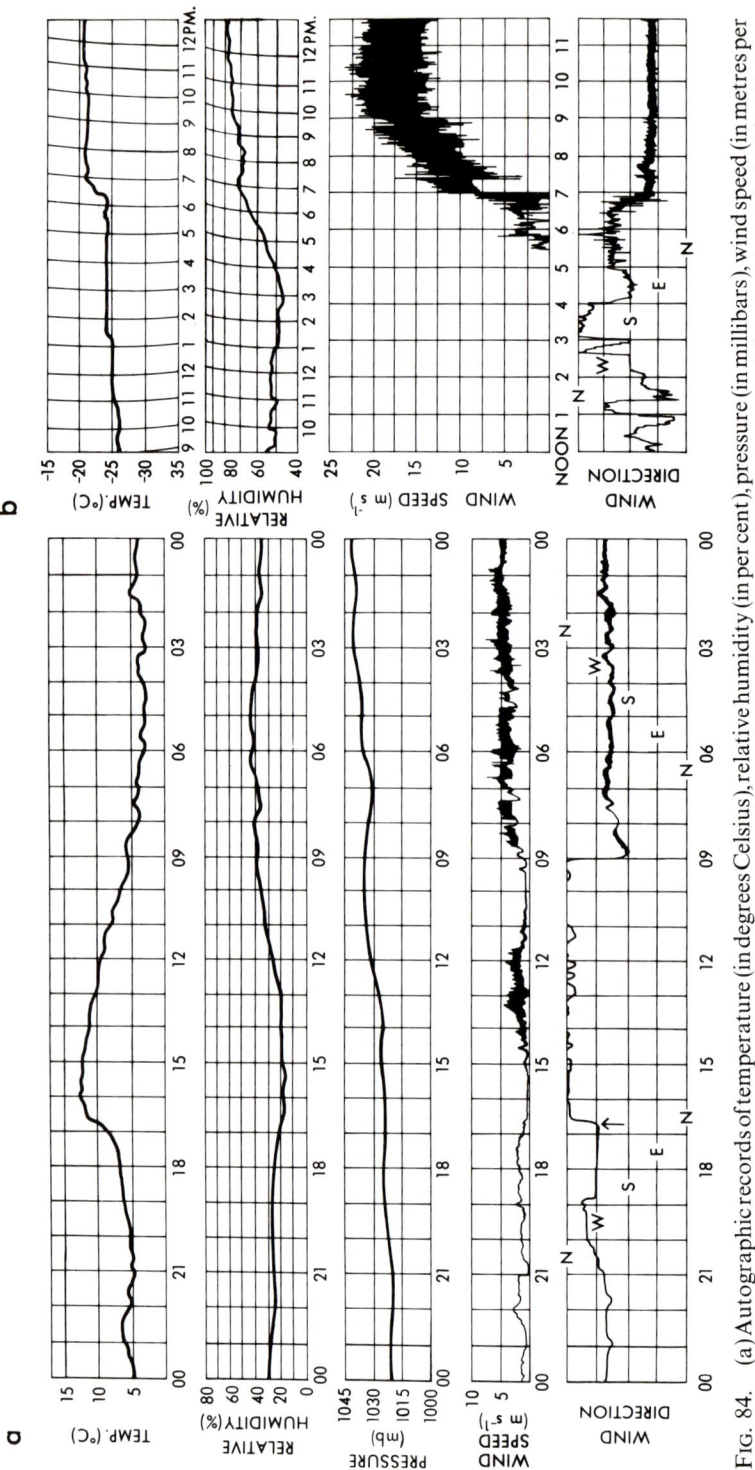

FIG. 84. (a) Autographic records of temperature (in degrees Celsius), relative humidity (in per cent), pressure (in millibars), wind speed (in metres per second) and direction at the onset of a mountain wind on 21 December 1960 in Arizona. Arrow denotes passage of the mountain wind. (After Brown, 1961.)
(b) Autographic records of temperature (in degrees Celsius), relative humidity (in per cent), wind speed (in metres per second) and direction at the onset of a typical winter katabatic wind on 21 August 1960 at Mawson, Antarctica. (After Streten, 1963.)

6 SLOPE AND VALLEY WIND CIRCULATION

systems became more pronounced in spring (Fig. 83(b)) and summer (Fig. 83(c)). A marked upslope wind of about 1 m s^{-1} occurred for about 7–8 h. Maximum speed occurred about noon. The summer graph shows a double maximum, a result similar to those of Jelinek (1937a). The valley wind system in spring and summer was regular and well developed with maximum speeds of nearly 3 m s^{-1}. The transitions from down-valley to up-valley winds and vice versa took place shortly after sunrise and before sunset. There was a closer correspondence between the up- and the down-valley winds and the periods of daylight and darkness than there was in the case of the slope wind system.

B. Surface characteristics of slope and valley winds

Early experience of slope and valley winds was restricted to a recognition of changes in the familiar elements such as wind speed and direction, temperature and humidity. Of the two types of wind, the downslope wind is the easier to identify on the records of autographic instruments, as typified by those maintained in the mountains near Fort Huachuca, Arizona (Fig. 84(a)).

Clearly the onset of a downslope wind is likely to reveal itself initially in the anemograph trace. Tables 34(a)–(d) show the speeds typical of the onset of a katabatic flow. The data for Poona (Table 34(a)) suggest that most such winds have speeds less than 3 m s^{-1}. On Mauna Loa (Table 34(b)) the speeds may be twice that value, possibly due to the huge length of the slope on which the winds develop. In the middle latitudes, as typified by a small valley in

Table 34(a) Speeds of down- and upslope flows: downslope flow at Poona (valley)

Speeds (m s^{-1})	Actual frequencies	Percentage frequencies
<0.4	123	8.6
0.49–0.89	151	10.5
0.94–1.34	159	11.1
1.39–1.79	253	17.7
1.83–2.24	216	15.1
2.28–2.69	213	14.9
2.73–3.13	156	10.9
3.17–3.58	91	6.4
3.62–4.02	37	2.6
4.07–4.47	24	1.7
4.52–4.92	9	0.6

Source: Atmanathan (1931). Period of observations: May 1929 to April 1931.

Table 34(b) Speeds of down- and upslope flows: downslope flow at Mauna Loa (slope)

Month	Average speed (m s^{-1})
Jan.	8.5
Jul.	5.3

Source: Mendonca (1969). Period of observations: an unspecified 8 year period

Table 34(c) Speeds of down- and upslope flows: downslope flows at Mawson (slope) showing frequency of wind speed increases at onset of katabatic flow

	Speed (m s^{-1})				
	<5.1	5.1–7.2	7.7–9.8	10.3–12.4	>12.4
Frequency					
Sep.–Feb.	17	18	9	7	2
Mar.–Aug.	11	26	15	9	4
Year	28	44	24	16	6

Source: Streten (1963). Period of observations: February 1960 to January 1961.

Table 34(d) Speeds of down- and upslope flows (combining slopes (S) and valleys (V)): upslope flow

	Average speed (m s^{-1})
Mauna Loa (S)†	
Jan.	3.0
Jul.	2.9
Vermont (V)‡	
Jun.–Aug.	5–10

† *Source:* Mendonca (1969). Period of observations: an unspecified 8 year period.
‡ *Source:* Davidson and Rao (1958). Period of observations: June to August 1957.

Oxfordshire, UK (Heywood, 1933), velocities of 1–2 m s^{-1} have been observed. At Mawson, the katabatics are very strong, occasionally reaching 12 m s^{-1} (Table 34(c)). Figure 84(b) illustrates strikingly the onset of a typical katabatic wind at Mawson and its effect on temperature and humidity. The onset was spectacular. Although the flow was gusty, its direction was remarkably steady, from just south to east. Frequently the Antarctic katabatics are associated with a hydraulic jump, one such example being vividly described by Lied (1964). This stationary jump resulted in a vertical wall of drift snow 30–100 m high, through which observers could easily walk in both directions. Over a horizontal distance as small as 5 m, pressure, temperature and wind characteristics changed dramatically. The theory for such hydraulic dumps in katabatic flows has been developed by Ball (1956, 1957, 1960).

In Antarctica the effects of the katabatic winds on temperatures, humidity and pressure vary with the seasons. Figure 84(b) illustrates the changes typical of a winter katabatic. In this season a strong inversion persists at and near the ground. Katabatic flow mixes the air in the inversion layer bringing down warmer air to the surface and causing quite abrupt increases in temperature (Fig. 84(b)). In summer, when the surface inversion is weak or totally absent, a fall of temperature frequently accompanies the onset of the downslope flow at Mawson as the colder air from the plateau displaces the warmer coastal air. Table 35 lists the temperature changes due to katabatic flows over a 12 month period. For the 6 months from May to October, 80 per cent of the onsets produced a temperature increase, while for the mid-summer period of November to January, 74 per cent of the onsets were accompanied by a fall in temperature. The remaining transitional months were characterized by approximately equal numbers of onsets with temperature falls on the one hand and those with rises on the other. Temperature increases at the surface in katabatic flow have also been observed in mid-latitudes and the sub-

Table 35 Temperature changes (in degrees Celsius) on katabatic onset at Mawson

Month	Total no.	Temperature rise				No change	Temperature fall			
		>1.6	1.2–1.6	0.6–1.1	0.05–0.5		0.05–0.5	0.6–1.1	1.2–1.6	>1.6
Jan.	12			2	2	1	1	2	3	1
Feb.	7					3	2	2		
Mar.	7				2	1	1	2	1	
Apr.	7			1		4		1	1	
May	14	2	2	6	1					
Jun.	15	1	4	3	1				1	1
Jul.	10	4	3		1	1		1		
Aug.	9	2	4		1			1		
Sep.	8		1	3	1	1		2	1	
Oct.	10		3	3	1			2	1	
Nov.	8						2	5	1	
Dec.	11				3		1	6	1	
Year	118	9	17	18	13	11	7	22	12	2

Source: Streten (1963). Period of observations: February 1960 to January 1961.

Table 36 Changes in relative humidity (in per cent) on katabatic onset at Mawson

Month	Total no.	Increase >10	6–10	1–5	No change	Decrease 1–5	6–10	>10
Jan.	12			3		7	2	
Feb.	7			2	3		1	1
Mar.	7		1	3	1	1	1	
Apr.	7	1	1	1	3		1	
May	14	1		3	1	4	1	1
Jun.	15		2	3	2	4	1	
Jul.	10		1	4		3	2	
Aug.	9			3	1	3	1	
Sep.	8			2		6	1	
Oct.	10			2	3	4	1	
Nov.	8			4		2		1
Dec.	11			5	1	3	2	
Year	118	2	5	35	15	37	14	3

Source: Streten (1963). Period of observations: February 1960 to January 1961.

tropics. Both Newnham (1918) and Eldridge (1951) suggested that such increases were due to stirring of the near surface air which was previously stably stratified. Atmanathan (1931) invoked the same mechanism to explain increases of up to 5 °C.

Table 37 Changes in surface pressure (in millibars) on katabatic onset at Mawson

Month	Total no.	Pressure fall ⩾0.5	0.1–0.4	No change	Apparent rise in pressure < 0.2 mb
Jan.	12	1	5	5	1
Feb.	7	4	2	1	0
Mar.	7	2	0	5	0
Apr.	7	2	2	3	0
May	14	4	7	3	0
Jun.	15	2	8	5	0
Jul.	10	4	4	1	1
Aug.	9	2	2	3	2
Sep.	8	1	4	2	1
Oct.	10	0	6	3	1
Nov.	8	0	4	2	2
Dec.	11	0	1	7	3
Year	118	22	45	40	11

Source: Streten (1963). Period of observations: February 1960 to January 1961.

6 SLOPE AND VALLEY WIND CIRCULATION

Variations in relative humidity on katabatic onset in Antarctica (Table 36), which may take place with or without temperature changes, are largely due to the presence, or otherwise of drifting snow. In the 5 months from September to January, 67 per cent of onsets were accompanied by a decrease in relative humidity or in no detectable change, as cooler and drier air from the plateau replaced coastal air. During the winter months, 35–40 per cent of the onsets occurred with drifts of snow and accordingly caused increases in relative humidity. Decreases of relative humidity on onset during winter seem to occur in the absence of drift and to be due to the mixing of drier air from above the ground inversion.

Surface pressure changes at the onset of katabatic flow are summarized in Table 37. The larger pressure falls at onset are chiefly in the winter months. Only small fluctuations of surface pressure are associated with summer katabatic flows.

C. Slope wind circulation

Despite the fundamental nature of slope winds *per se* to the understanding of all mountain–valley wind phenomena, surprisingly few good observations of slope winds exist in the literature. Certainly their existence had been confirmed by visual observations of fragmentary cloud and smoke in Alpine

Table 38 Slope wind characteristics

	Direction slope faces			
	S	E	N	W
Sunrise	0700	0518	0532	0722
Time of start of upslope wind	0722	0605	0618	0750
Difference (min)	22	47	46	23
Sunset	1833	1625	1912	1948
Time of start of downslope wind	1845	1650	—	—
Difference	12	25	—	—
Maximum speed of upslope wind (m s^{-1})	3.7	1.7	1.6	3.0
Maximum speed of downslope wind (m s^{-1})	2.1	1.8	1.6	2.4
Maximum depth of slope wind (m)	280	190	170?	—
Duration of slope wind (h)	11.38	10.75	—	—
Mean upslope wind speed (m s^{-1})	2.2	1.0	0.8	1.6
No. cases	19	14	10	19
Mean downslope wind speed (m s^{-1})	0.9	1.2	0.9	0.8
No. cases	12	8	10	6

Source: Jelinek (1937*a*).

areas, but it was not until the 1920s and 1930s that instrumental observations were made in earnest. Within the Alps the glaciers provided excellent sources for downslope winds that did not experience the normal daily reversal because of the persistent cooling, even in day-time, of the lower atmospheric layers by the ice surface. Such "glacier winds" were often strong enough to undercut the day-time upslope winds (Tollner, 1931, 1935). The German observational programmes in the inter-war period set a pattern to be followed by the significant American projects of the 1950s and 1960s. The basic method involved the frequent release and tracking of known-lift balloons from as many stations as possible in the research area. Such a procedure was followed by Jelinek (1937a) in his analysis of winds on slopes that faced, respectively, north, east, west and south. Table 38 summarizes Jelinek's results, clearly demonstrating the shallowness and weakness of the slope winds (on slopes 2–5 km long over a vertical distance of 1500–2000 m) and their close relationship to times of sunrise and sunset. Later observations by Jelinek and Riedel (1937) confirmed these results, which were in accord

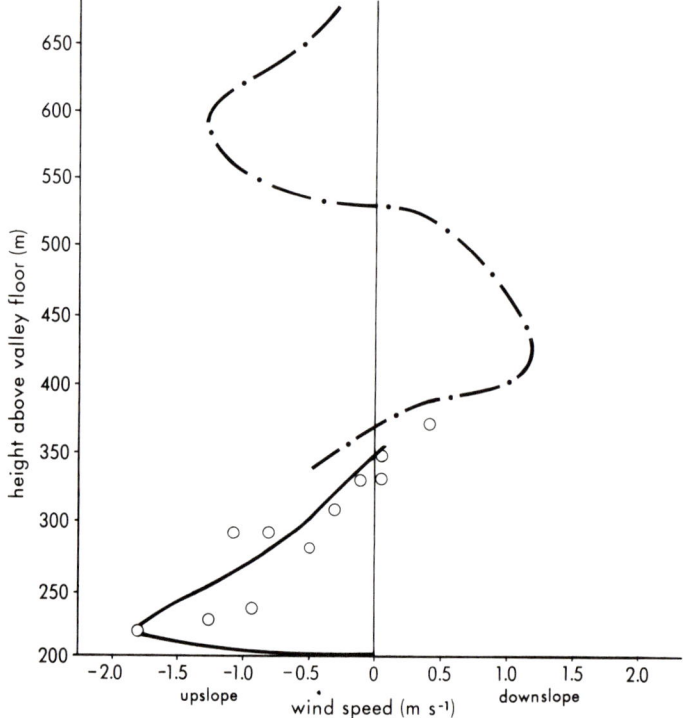

FIG. 85. Vertical distribution of wind speed (in metres per second) in an almost pure upslope wind system at 0840–1010 hours on 14 August 1959 in a Vermont valley. (After Davidson, 1961a.)

with those found by Moll (1935). The "anti-slope" wind was difficult to discern in Jelinek's ascents but Buettner and Thyer (1966) claimed to have observed it in the valleys around Mt Rainier.

Both the slope and anti-slope winds were clearly observed by Davidson (1961a) in the Vermont valleys. On a clear day with an overhead wind of $2\,\mathrm{m\,s^{-1}}$, the vertical profile of velocity (Fig. 85) revealed an upslope wind 150 m deep with a sharp maximum of nearly $2\,\mathrm{m\,s^{-1}}$ at a height of 20 m above the ground. The anti-slope wind lay between 150 and 300 m above the slope with maximum velocities of just over $1\,\mathrm{m\,s^{-1}}$. Below 150 m the form of the upslope profile agrees well with Prandtl's (1942) equation—a subject more fully covered in the section on valley winds below. Jelinek (1937b) found similar shaped profiles for the downslope winds in Alpine valleys. On a larger scale, the eastern slopes of Hawaii, 30–100 km long over a vertical distance of 3400 m, generate gusty upslope winds 600 m deep and stable downslope winds 50 m deep. Figure 86 clearly illustrates the unchanging depth of the downslope wind, contrasting with the rapid increase in depth of the upslope wind. During the day the warm ground promotes convective motion while at

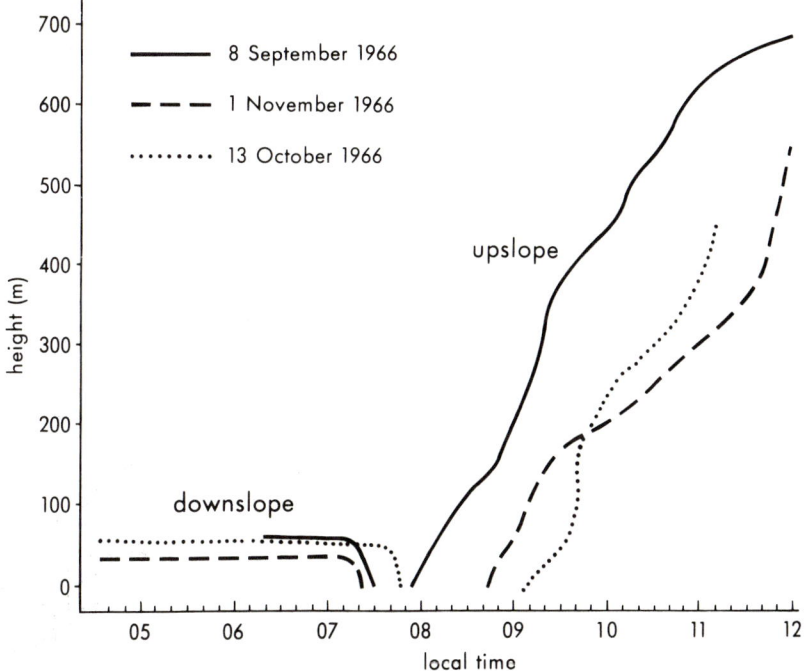

FIG. 86. Depths of downslope and upslope winds as measured at the Mauna Loa Observatory. (After Mendonca, 1969.)

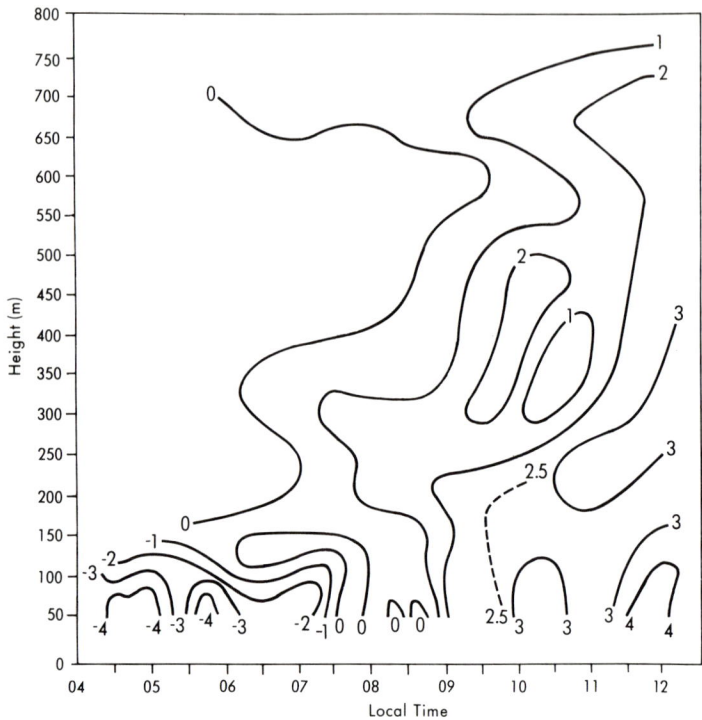

FIG. 87. Variation with height and time of wind speeds (in metres per second) of slope winds on 8 September 1966 on Mauna Loa. Upslope winds positive; downslope winds negative. (After Mendonca, 1969.)

night the cool ground promotes thermally stable layers and a more shallow laminar flow. We should note, however, that these observations were made at the Mauna Loa observatory which is virtually at the top of a huge slope. Consequently, it lies at the "end" of the upslope wind, where it should be well developed, and at the beginning of the downslope wind, where it is perhaps not so well developed. In association with these well-developed surface winds there was a suggestion of a reverse flow above them in the movement of orographic cumulus, but Mendonca (1969) warned that this reverse flow was heavily influenced by the free air flow.

The evolution of the slope wind circulation throughout the day is illustrated in Fig. 87. A complete reversal of the wind at the ground occurred on all observational days. The cessation of the downslope wind occurred almost exactly at the time the ground temperature first equalled the free air temperature. The brief interval of calm and variable winds corresponded to the period when the ground and air temperatures were approximately equal. Upslope winds started as the ground temperature increased and exceeded the

6 SLOPE AND VALLEY WIND CIRCULATION

air temperature by several degrees. The calm and variable winds that followed the cessation of the downslope flow, extended to the previous height of the downslope flow. The upslope flow, initially contained within this layer, increased in depth in proportion to the heating of the ground and the relative velocity of the overlying free air flow.

The observation of "pure" slope winds within valleys has usually been a secondary aim of an investigation of the whole-valley wind circulation. In such a project Davidson and Rao (1958) took 10 min mean winds for hourly intervals over 3 days on eastern and western slopes of a valley in Vermont. The slope observations were taken 150 m above the valley floor. Their results (Fig. 88) showed well-developed downslope winds despite a westerly gradient wind speed of up to $10 \, \text{m s}^{-1}$. In the night hours of 6–7 August, simultaneous downslope motion occurred on both slopes but the air was calm in the valley bottom. Wind directions reversed on the eastern slope in day-time, but not on the western slope—no doubt due to the gradient wind. Yet, whilst Davidson and Rao saw the downslope winds as confirmation of theory, they

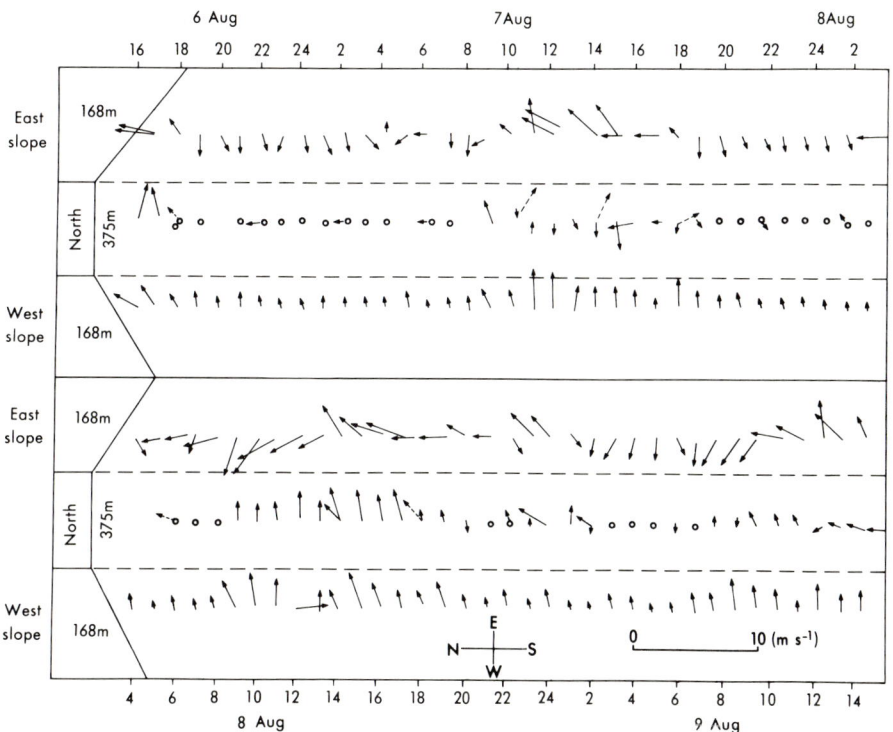

FIG. 88. Hourly wind directions on east slope, west slope and valley bottom of a Vermont valley in August 1957. (After Davidson and Rao, 1958.)

remained sceptical about the existence of the upslope winds, particularly with gradient winds of the order of $5\,\mathrm{m\,s^{-1}}$.

D. The valley wind circulation

The change of scale involved in transferring our attention from slopes *per se* to valleys and their surrounding hills has no effect on the mechanism of the ultimately thermally induced air flows. Cross (1950, p. 79) clearly explains as follows:

> On a sunny day the same amount of insolation is received on similar horizontal areas at any level over the valley and over the plain. Consider now the relative volumes of air in the columns below these given horizontal areas. It will be seen that, owing to the space occupied by the land mass which makes up the valley side and floor, the volume of air in the column over the valley will be considerably smaller than the corresponding value over the plain. The insolation first heats the ground which in turn warms the air above by turbulent conduction and convection. Since the same amount of heat is added to the small volume of air over the valley and the larger volume over the plain, the temperature of the valley air becomes progressively greater than that of the air at the same level over the plain, resulting in the development of a local pressure gradient. In response to the resulting pressure gradient force there is a flow of air from the plain toward and up the valley. This flow of air is known as the up-valley wind. Similarly, nocturnal radiation leads to a loss of heat which reduces the temperature of the smaller body of air over the plain. This leads to a reversal of the pressure gradient and a movement of air down the valley, called the down-valley wind. The above discussion assumes, of course, that no pressure gradient other than the locally induced one is present.

1. *Along-valley section*

Implicit in Cross's explanation is the concept of a thermally direct vertical circulation oriented along the axis of the valley. The vertical structure of such a valley wind circulation was first fully observed by Ekhart (1932*a, b,* 1934); in particular he confirmed the existence of return currents, or "anti-winds" above the surface flows. Yet, surprisingly, very few "synoptic" observations exist of the vertical structure of the valley wind when viewed as a profile along the valley axis. Buettner and Thyer (1966) provided such observations (Fig. 89) in their detailed study of airflow in the Carbon River valley near Mt Rainier. In virtually all the cases illustrated, three layers clearly exist: the near-surface up- or down-valley flow, the "anti-wind" and the gradient wind above. The valley and "anti-wind" layers were approximately the same depth and maximum velocities range from 3 to $5\,\mathrm{m\,s^{-1}}$.

FIG. 89. Longitudinal sections of down- and up-valley winds and their associated anti-winds in Carbon River valley near Mt Rainier, USA. Top four diagrams show the depths of the different types of wind at different times; bottom two show their typical speeds. (After Buettner and Thyer, 1966.)

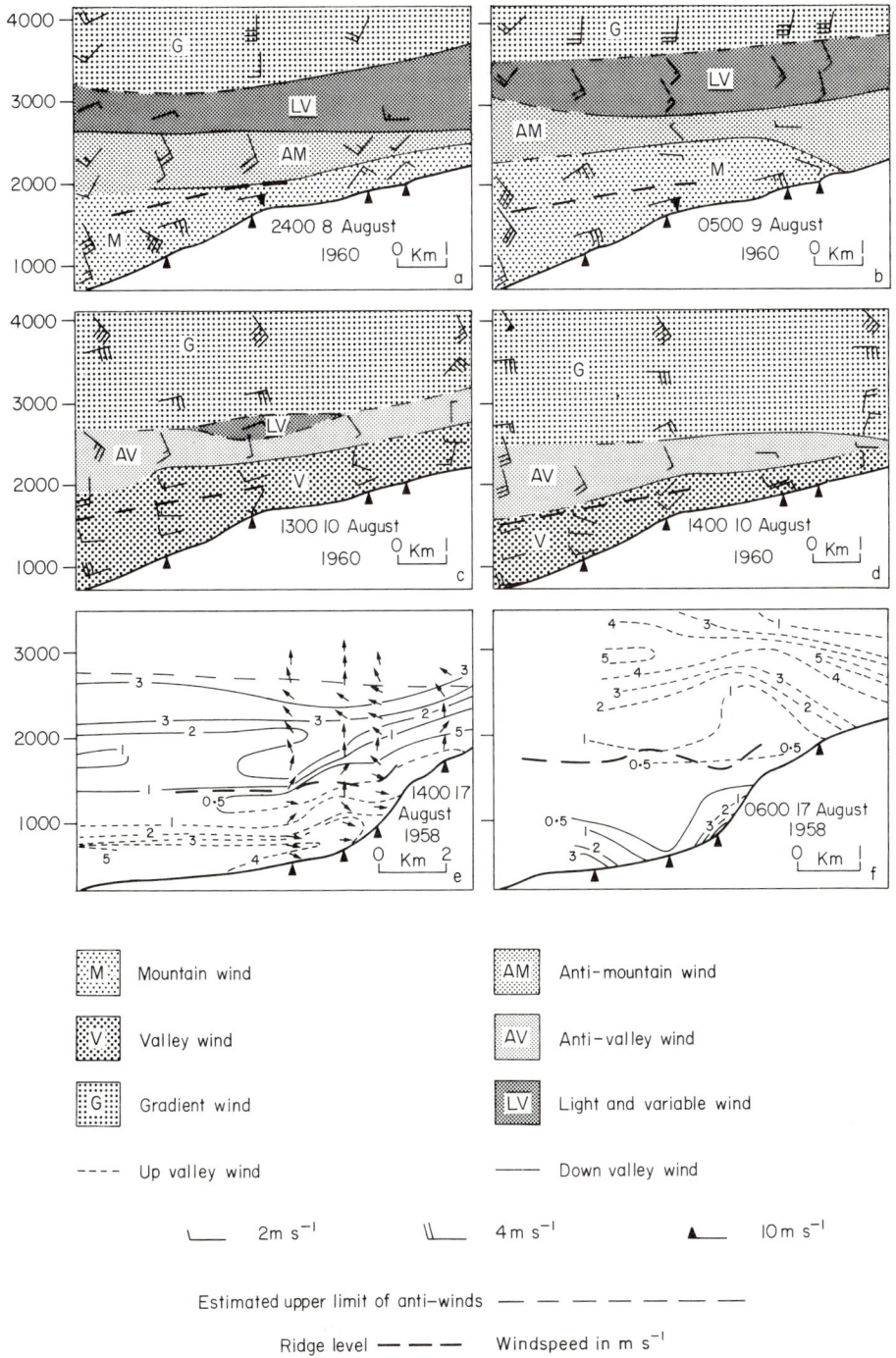

2. Cross-valley section

Buettner and Thyer (1966) also provided most useful cross-valley sections of both up- and down-valley winds. Four pilot-balloon stations were employed—one in the valley bottom, one on each of the valley sides and one

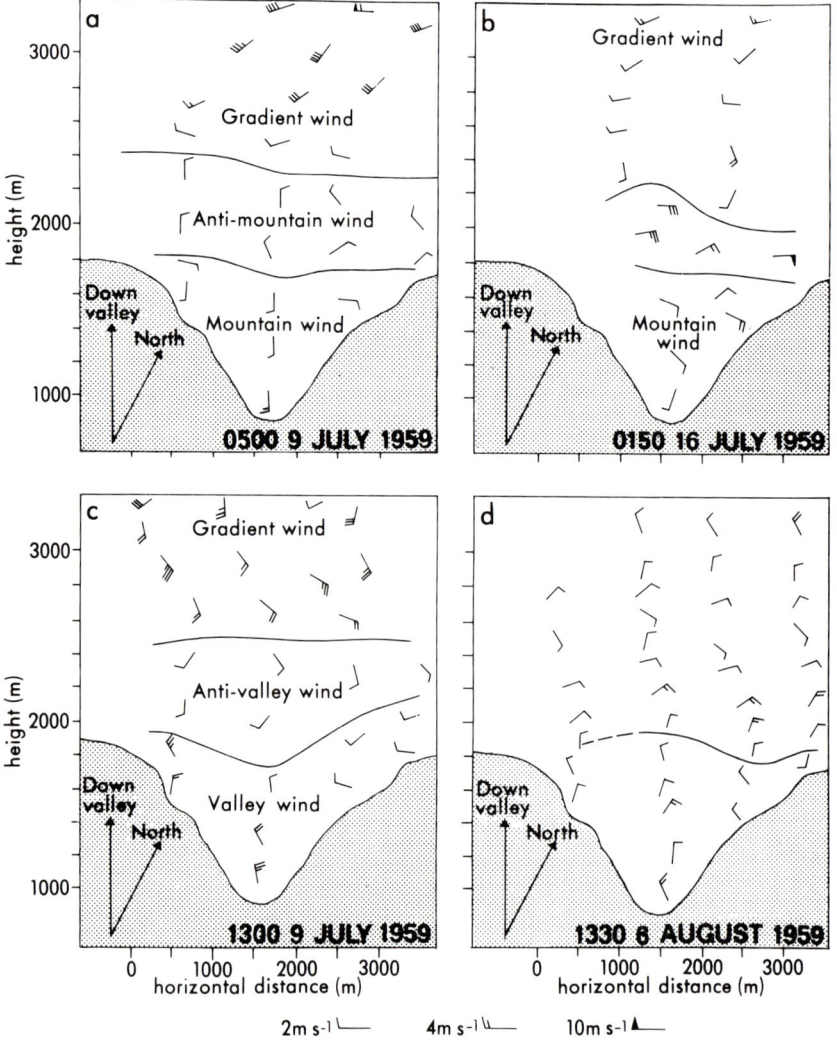

FIG. 90. Cross-sections of down- and up-valley winds and their associated anti-winds in Carbon River valley near Mt Rainier, USA. The four diagrams show typical changes in the winds with time. (After Buettner and Thyer, 1966.)

6 SLOPE AND VALLEY WIND CIRCULATION   241

on top of one of the bounding ridges. Figure 90 clearly illustrates the relationships of the up- and down-valley winds to their "anti-winds", with the gradient winds overlying the valley wind circulation. Tyson's (1966) Cato Ridge observations provided similar results.

3. *Depth*

Figures 89 and 90 strongly suggest that the depth of the valley wind system is closely related to the relative relief of the valley, i.e. the height difference between valley bottom and ridge top. In both the up- and down-valley wind systems the bulk of the evidence (Davidson and Rao, 1958; Buettner and Thyer, 1959) suggests that the surface winds are as deep as the valley and that the overlying "anti"-winds are of a similar depth. Figure 91 shows the height of the down-valley wind at 0200 hours on 8 August 1957 as a function of the height above the valley floor of the ridge line. The prevailing wind speed was 5 m s^{-1}. However, in Vermont Davidson and Rao (1958) found little evidence of either up-valley winds or anti-down-valley winds. This was probably because they studied the air-flow in a rather open-ended valley which had neither obvious upper end nor slope. In the well-defined Alpine valleys of

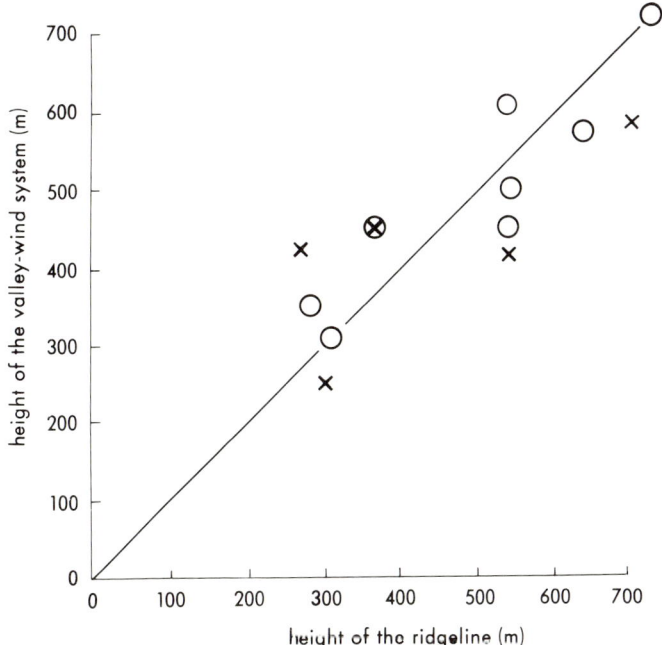

FIG. 91. Height of the valley wind as a function of height of the valley-side range. (After Davidson and Rao, 1958.)

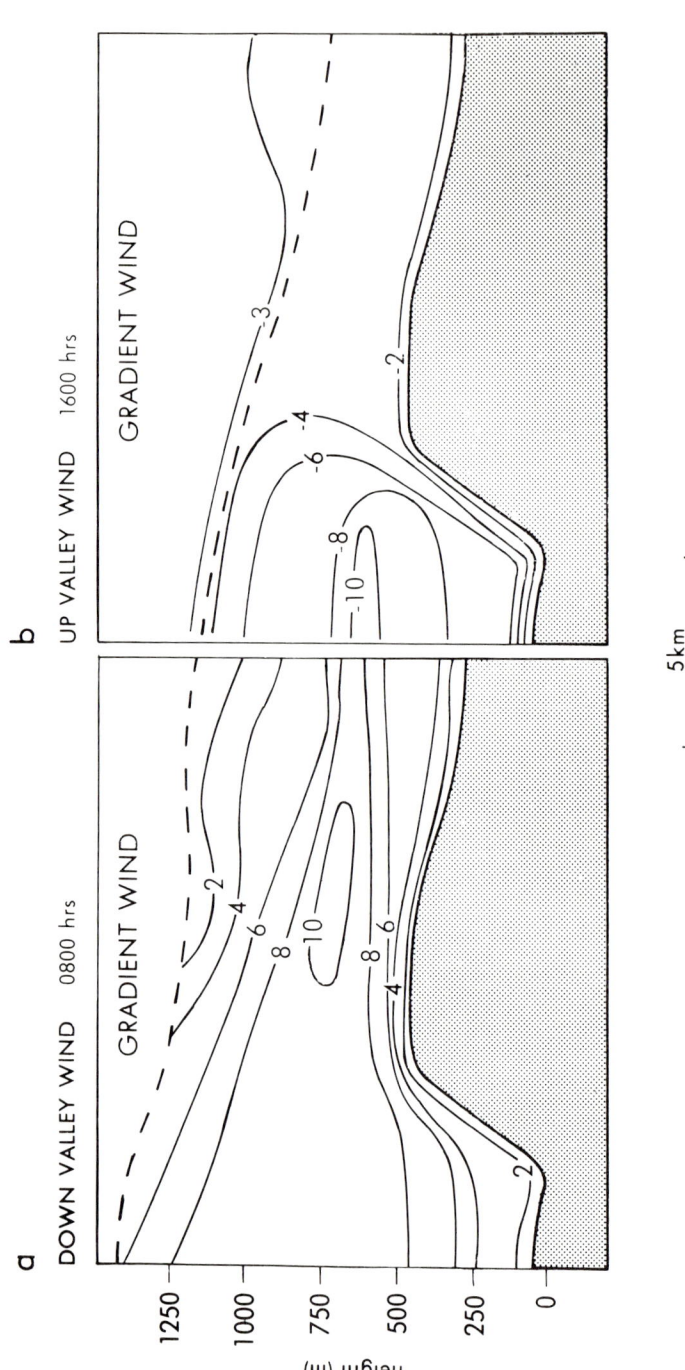

FIG. 92. Down- and up-valley winds near the Cato Ridge plateau, South Africa. Isotachs in metres per second, positive values indicating down-valley motion. Pecked lines indicate transitional boundary layers. (a) Down-valley wind in morning of 16 July 1965; (b) up-valley wind in evening of 27 July 1965. (After Tyson, 1966.)

Europe, such observations as were made (e.g. Ekhart, 1932a) hinted at the picture painted nearly 30 years later in American valleys.

In contrast to the above studies some of Tyson's (1966) observations suggested that both the up- and down-valley winds may "overspill" the valley. Figure 92 illustrates such situations. The lack of agreement with other studies suggested that perhaps winds other than "pure" valley winds were being observed and led Tyson to conceive the larger scale "mountain–plain" wind, to be considered later.

4. *Velocity variations*

Within the valley wind circulation, the near surface down-valley flow is generally more readily observed than the day-time up-valley flow. This difference has meant that analysis of the velocity variations within the valley wind system has been concerned largely with the down-valley flow. Close observation has revealed interesting variations in the vertical profile of velocities, down-valley accelerations and a rather complicated gust structure within the down-valley wind.

For the earliest good observations we must rely once more on the Austrian studies in the inter-war period. Figure 93(a) illustrates Ekhart's (1932a,b) findings in both up- and down-valley winds, suggesting that the up-valley wind was stronger than the down-valley flow and that the maximum of the former usually occurred at a lower level than in the latter. Yet two years later Ekhart (1934) (Fig. 93(b)) found the maximum up-valley flow of 3–$4\,\text{m}\,\text{s}^{-1}$ to occur 600–700 m above the ground in a "U"-shaped valley whereas the maximum down-valley speed of $4\,\text{m}\,\text{s}^{-1}$ occurred at a height of 250 m. In a "V"-shaped valley the maximum down-valley velocity lay 400 m above the ground. In both valleys the relative relief was 1200–1300 m. Clearly no consistent picture emerged from these observations, but valley shape may have played some role in causing the differences.

Later work on velocity profiles has concentrated upon the down-valley winds. Prandtl (1942) suggested that the typical profile could be described as follows:

$$U = \frac{\exp(\pi/4)}{\sin(\pi/4)} U_m \sin\left(\frac{z\pi}{H}\right) \exp\left(-\frac{z\pi}{H}\right),$$

where U is the velocity of down-valley flow at height z above the ground. U_m is the maximum velocity—found at a height h above the ground. H is the depth of the down-valley wind.

In such a profile the ratio h/H should equal 0.25. Observed profiles in very close accord with the Prandtl formula were found by Tyson (1966) (Fig. 93(c)). The fit was particularly good below the maximum velocity at 0600 hours and above the maximum velocity at 1000 hours. We should note in passing that this "down-valley" wind in mid-morning is considered by Tyson to be part of a larger mountain–plain wind which flowed over the Cato Ridge (the site of these observations) and was dissipating at this time.

An alternative description of vertical profiles of horizontal velocity is provided by the equation for a parabola. In this case the ratio h/H should equal 0.50. Both Tyson (1968a) and Davidson and Rao (1963) found fairly good fits for such a profile, the former less so than the latter. Tyson fitted an approximately parabolic profile to both the mean of 75 hourly (Fig. 93(d)) observations and to certain 5 h means on particular occasions (Fig. 93(d)). In the case of the mean picture the parabolic profile underestimates the wind speed, but for particular occasions the fit is far better. Also of interest is that the "anti-wind" profile is quite well described by a quasi-parabolic curve. Tyson attributed this to "strong shear and thermal discontinuity between the mountain wind and return current act[ing] as the upper boundary to enforce the symmetry for parabolic flow" (Tyson, 1968a, p. 26). In the light of the above results, it is somewhat surprising to find that in the same analysis, Tyson (1968a) found a mean value of h/H of 0.36 with a standard deviation of 0.20. Such a ratio implies a profile nearer in shape to Prandtl's than to a parabola. In a later paper Tyson (1968b) showed that 39 per cent of down-valley winds near Pietermaritzburg in 1964–65 had an h/H value between 0.20 and 0.29, i.e. effectively are in accord with the Prandtl profile. This was by far the largest percentage.

The parabolic profile was found most frequently by Davidson and Rao (1963) (Fig. 93(e)). Using 214 vertical profiles, they found the mean value of h/H to be 0.50 with a standard deviation of 0.11. Further analysis suggested

---

FIG. 93. Vertical distribution of valley wind speed (in metres per second) (a) Dashed line shows up-valley wind. Solid curves show two measurements of down-valley wind. Observed in European Alps. (Up-valley and down-valley (1) after Ekhart, 1932a; down-valley (2) after Ekhart, 1932b.) (b) Mean up-valley wind on 6 August 1933 in Langenfeld, European Alps. (After Ekhart, 1934.) (c) Two profiles (solid lines) of down-valley wind on 16 July 1965 near Cato Ridge, South Africa. Dashed line is theoretical Prandtl profile. (After Tyson, 1966.) (d) *Left-hand diagram*: mean observed down-valley and anti-winds based on more than 75 hourly ascents. *Right-hand diagrams*: profiles on two representative nights. In each case the curve gives the parabolic profile of best fit. Observed near Cato Ridge, South Africa. (After Tyson, 1968a.) (e) Typical speed profiles in three valleys in Vermont, USA. (After Davidson and Rao, 1963.) (f) Profiles of both up- and down-valley winds and their anti-winds in Carbon River valley near Mt Rainier, USA. Measurements taken on 17–18 August 1959. Squares and crosses represent winds at 1400 hours showing day-time up-valley wind in the layer 0–700 m and down-valley anti-wind between 1000 and 1800 m. Triangles and dashed line represent winds at 2130 hours showing night-time down-valley wind in the layer 0–700 m and up-valley anti-wind between 1000 and 1800 m. Open and closed circles represent winds at 0200 hours showing night-time down-valley wind in the layer 0–700 m and up-valley anti-wind between 1000 and 1800 m.
(After Buettner and Thyer, 1959.)

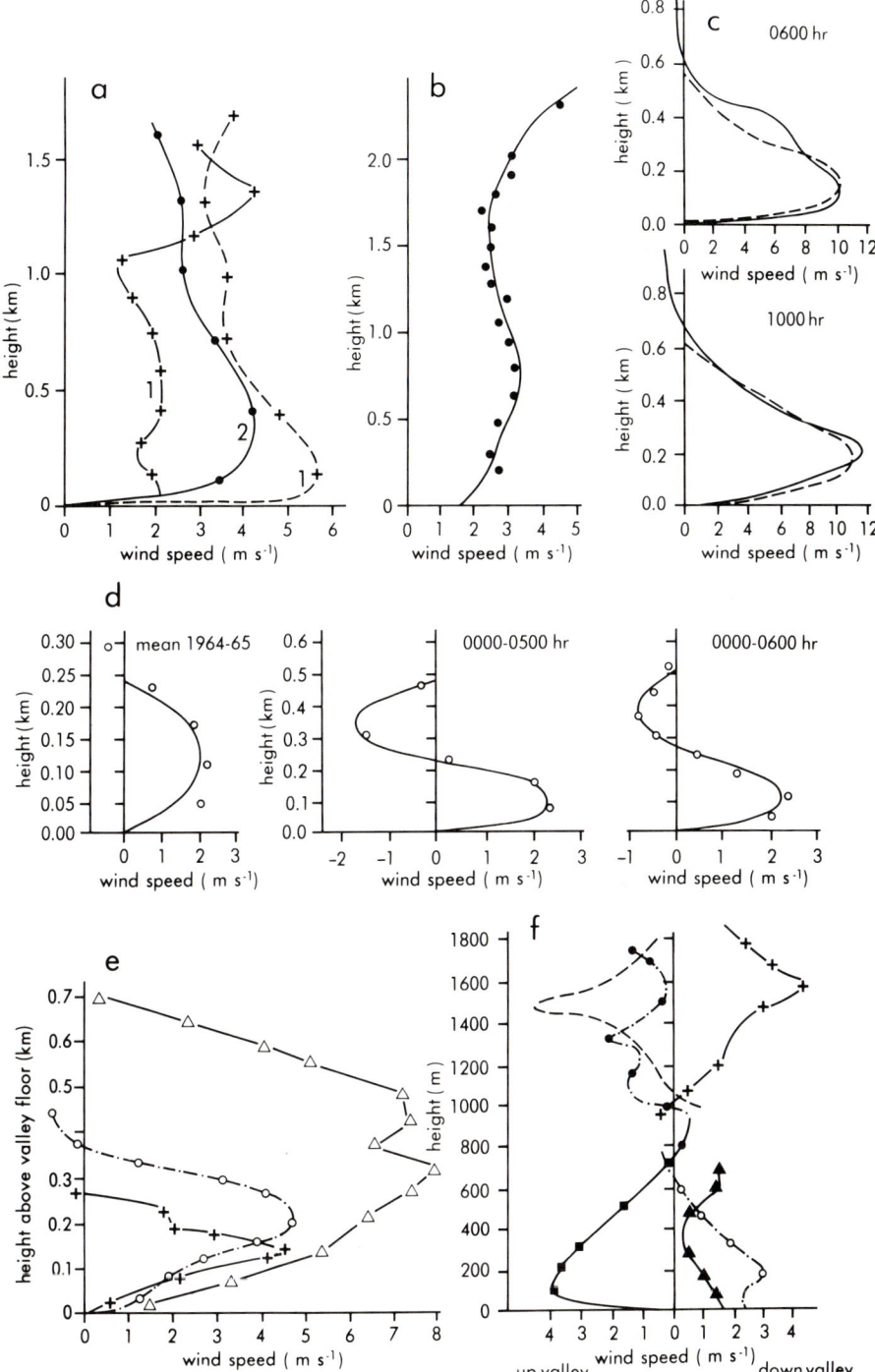

that the Prandtl profile cannot hold for slopes of less than 8° but that it may be approached as the slope of the valley floor becomes very large.

In contrast to the above studies, Buettner and Thyer (1959) found neither parabolic nor Prandtl profile near the ground or aloft (Fig. 93(f)). In fact the gentle slope (2–5°) of the valley bottom would preclude the Prandtl profile according to Tang (1960), but with geostrophic winds of about $1 \, \text{m s}^{-1}$ there appears no obvious reason for the large discrepancy between these and other observations of the parabolic profile.

As noted earlier, by far the greatest number of profile analyses have been for the down-valley wind. In one of the very few studies of the profile of the up-valley wind, Tyson's (1968a) observations showed very good agreement with the Prandtl equation.

5. Down-valley acceleration

In their preliminary investigation of airflows in the Vermont valleys, Davidson and Rao (1958) discovered a marked down-valley acceleration of the wind in the direction from the head of the valley to its mouth. In the case of a narrow gorge with a floor sloping at some 8° they found that the maximum wind speed increased from virtually zero at the valley head to $3 \, \text{m s}^{-1}$ at about 5 km down-valley and to $5.4 \, \text{m s}^{-1}$ at about 7 km down-valley. In less steeply sloping valleys, the down-valley velocity (in metres per second) meaned over the depth of the flow increased approximately with the square root of the distance (in kilometres) from the head of the valley (Davidson and Rao, 1963).

The detailed observation of the valley wind system soon revealed a fine structure of air flow, particularly within the near-surface limb of the circulation. In the Carbon River Valley Buettner and Thyer (1966) found pulsations in the down-valley flow from 1.5 to $6.5 \, \text{m s}^{-1}$ over periods of about 25 min. The fluctuations were most marked in amplitude and regularity at about 100 m above the valley bottom (i.e. deep down in the trough of the valley but clear of the treetops). In his studies in South Africa Tyson (1968a,b) also noted surges in the down-valley wind. Spectral analysis revealed speed changes of from $1 \, \text{m s}^{-1}$ to about $4 \, \text{m s}^{-1}$ with a period of 75 min. In general the wavelength of such surges increased from about 5 km at a height of 0.6 m to 8 km at 6 m. Around Pietermaritzburg, the deeper and more steep-sided the valley, the narrower was the spectrum of velocity fluctuations on the mountain wind. In the Bushman's Valley the spectra were much broader and densities were lower than in airflows near Pietermaritzburg. No obvious general pattern emerged from Bushman's Valley observations, but surges with a period of 2–3 h were quite frequent (Tyson, 1968a). The "surges" found in both the North American and South African studies have periods roughly equal to those found by Atmanathan (1931) (Table 31(a)). In all cases the surging is probably ultimately due to adiabatic warming of the downward flowing air. Such warming would reduce the pressure gradient which "drives" the airflow and consequently the wind would decelerate. This would continue until radiational cooling had built up a sufficient temperature gradient to generate a pressure gradient which would cause further acceleration of air—the surge. Such a mechanism could recur many times in one night, as suggested by the observations outlined above.

6. *Vertical velocities*

The thermally driven vertical circulation that comprises the valley wind system must contain areas of vertical air motion. To quote Buettner and Thyer (1966, p. 104): "Somehow the lower valley or mountain wind has to feed the anti-wind system". In the up-valley flow that is observed to decelerate near the valley head, the "lost" air can only be moved upwards to feed the antiwind. In the Carbon River Valley, Buettner and Thyer calculated on a mass continuity basis that a vertical wind of $0.11 \, m \, s^{-1}$ must result at the interface between valley wind and the anti-valley wind. Observations with no-lift balloons from the bounding ridges revealed upward motion of $2.5 \, m \, s^{-1}$. Following their argument, "should this wind extend to only the fraction 0.11/2.5 or 4.4 per cent of the valley width or a 150 m wide chimney, the slope wind alone could explain the exchange" (Buettner and Thyer, 1966, p. 141). Despite the difficulty in measuring these vertical motions, whether by balloons or aircraft, the feeling remains that the currents are strong, very localized and easily achieve the necessary vertical mass transport.

7. *Evolution of the valley wind circulation*

The above outline of the nature of the valley winds is purposely restricted to the well-developed circulation. This section describes the evolution of the circulation, concentrating on the phases of wind reversal.

Despite the fairly large number of observations on valley winds, very few cover the whole diurnal cycle of the reversible circulation: by far the majority concentrate their attention on either the up-valley (e.g. Ekhart, 1932a) or the down-valley (e.g. Tyson, 1968a) wind. The most usual way of illustrating the evolution of the winds is a time–height section at one point. The more desirable method of showing several long and cross profiles over a period of time requires a number of observations beyond the capacity of virtually all experimental projects.

Figure 94 clearly illustrates the diurnal changes in the up- and down-valley winds and their associated "anti-winds", in a valley about 1000 m deep. Both types of wind showed very rapid reversals in direction from down- to up-valley and vice versa. In the up-valley wind maximum velocities of over $4 \, m \, s^{-1}$ lay at a height of 200 m above the valley floor. In the anti-up-valley

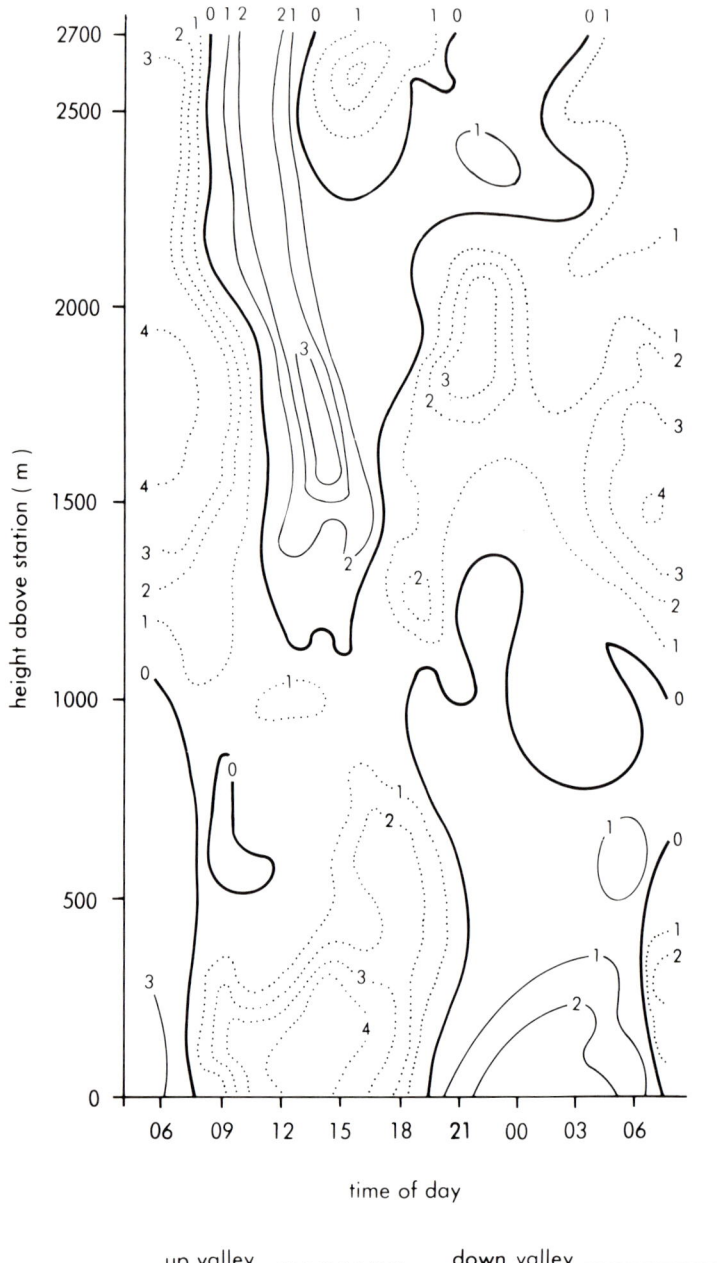

FIG. 94. Variation with height and time of valley wind and anti-wind speeds (in metres per second) in Carbon River Valley, near Mt Rainier, USA. (After Buettner and Thyer, 1959.)

flow, velocities of $3\,\mathrm{m\,s^{-1}}$ existed between 1600 and 1900 m above the valley floor. The nocturnal flows near the surface had speeds of about $2.5\,\mathrm{m\,s^{-1}}$ and were overlain by anti-winds of about $3\,\mathrm{m\,s^{-1}}$. The sharp reversal between the flows illustrated in Fig. 94 for one station occurred in fact simultaneously everywhere in the valley, about an hour after sunset and sunrise.

Both Davidson and Rao (1963) and Tyson (1968a) were of the opinion that the nocturnal down-valley wind developed rapidly. In Vermont a down-valley wind of $1.5\,\mathrm{m\,s^{-1}}$ at a height of 200 m above ground developed within 1 hour of a clearly observed up-valley wind. Within a further 90 min the maximum speed increased to about $3\,\mathrm{m\,s^{-1}}$ at a height of 280 m. In South Africa Tyson observed the onset of the down-valley wind in Bushman's Valley to occur from the ground upwards and to be associated with a local front of cool air that advanced down-valley at a rate dependent partly on the size and gradient of the valley. Temperature contrasts of 3–6 °C were found across these fronts. The first cool surge in the Bushman's Valley usually had a duration of 2–4 h, deepened at an average rate of $320\,\mathrm{m\,h^{-1}}$ ($0.09\,\mathrm{m\,s^{-1}}$) and had a depth about 200 m greater than that of the "all night" average down-valley wind. The subsequent surges in the flow have already been considered in the previous section.

Studies of the dissipation of the down-valley wind have led to rather surprising results. *A priori* one would perhaps expect that the wind would be reversed from the surface upwards as insolation warms the ground. In fact, observations by Pollak (1924), Davidson and Rao (1959, 1963), Ayer (1961) and Tyson (1968a) all showed that the down-valley wind was eroded from above, at rates varying from 40 to $300\,\mathrm{m\,h^{-1}}$. Most authors agreed that the most likely reason for this dissipation was the instability caused at ridge level by early morning insolation. Ayer (1961) amplified the point by noting that mixing at the top of the down-valley drainage current takes place with air of different characteristics and hence erodes the top of the drainage current, while the mixing in the valley bottom occurs within the drainage current itself at first and consequently has far less effect on the wind. Some support for this suggested mechanism was forthcoming in the theoretical study by Shieh (1971). Ayer also offered an alternative idea which has not yet been tested. In any katabatic layer where active drainage is taking place, destruction of the source of fresh, cold air by the onset of morning insolation would result in a drainage-away of the cold layer and an automatic lowering of the top. It is of course possible that both mechanisms operate.

E. Mountain–plain circulation

It was suggested earlier that a hierarchy of scale provided a useful framework for analysis of winds in sloping topography. Just as slope winds may exist within valleys, so the typical valley wind circulation may itself lie within a larger scale, topographically forced system. These larger circulations usually have horizontal dimensions ranging from 150 to 900 km (Burger and Ekhart, 1937; Hawkes, 1945) and depths of 1200–1800 m. A particularly good example emerged from Tyson's analyses of the valley winds near Cato Ridge and Grants Castle which lie between the main mass of the Drakensberg mountains and the Indian Ocean in South Africa. Although the primary concern of the investigations was the valley wind, it became clear that an organized larger scale flow, other than the gradient wind, existed above the local valley circulation. This larger flow was recognized by Tyson and Preston-Whyte (1972) as a mountain–plain and plain–mountain wind directed toward the Drakensberg by day and away by night and occurring above ridge level and the constraining influence of smaller scale topography. The nocturnal mountain–plain wind ranged from 400 to 800 m deep with velocities about 4 m s^{-1}: the day-time plain–mountain wind was 500–750 m deep with speeds of about 2 m s^{-1}. The plain–mountain winds were better developed in summer, whereas in winter the mountain–plain winds were more frequent.

As Tyson and Preston-Whyte (1972, p. 649) point out, such systems are not very well documented in the literature and yet "they frequently have a regional significance as important as that of the general circulation".

F. Effects of stability and gradient wind

None of the investigators mentioned in this chapter analysed the effects of stability on slope and valley winds in any real way. All admitted that observation of the required parameters would be very difficult, as would the isolation of effects solely due to instability. The problem remains. The situation is little better for gradient winds. Tyson (1968a) found that strong cross-valley gradient winds restricted the depth of the down-valley wind and MacHattie (1968) concluded that day-time instability allowed a downward transfer of gradient wind momentum which disrupted the upslope winds. Davidson and Rao (1958) found that with clear skies, and reasonably steady wind direction, slope and valley winds could develop in a gradient wind of 5 m s^{-1}. Speeds above 10 m s^{-1} prohibited the development of local winds. In a later study (Davidson and Rao, 1963) they related the depth of the down-valley wind to the free atmosphere wind speed at 1000 m—about ridge height in their research area. They found the following:

$$H = H_0(1 - 0.05 U),$$

where H is the depth of the down-valley wind, H_0 is the height of the ridge line and U is the wind speed in metres per second at about ridge height but in the free atmosphere. This equation predicts the absence of a down-valley wind with a 1000 m wind of 20 m s^{-1}, just twice their earlier estimate.

Davidson and Rao were careful to point out that their calculations were rather crude and wrote that: "It is our feeling or intuition that...a 1000 m wind speed of 10–12 m s^{-1} represents the upper limit for which a detectable and independent valley wind exists" (Davidson and Rao, 1963, p. 916). In large measure this feeling is supported by Jaffe's (1960) observations of the relations between valley and geostrophic winds. Certainly both down- and up-valley flows were most frequent with opposing geostrophic winds of less than 6 m s^{-1}. Geostrophic winds greater than 6 m s^{-1} tended to "swamp" the valley winds, regardless of the direction of the former relative to the valley axis.

III. Theory

Earlier in this chapter a brief qualitative outline of the slope/valley wind mechanism provided a context for the observational studies. We are now in a position to review theoretical developments in more detail.

Somewhat surprisingly the development of a satisfactory theory of slope and mountain/valley winds is still in the future. Despite the close attention of many eminent meteorologists over the last century, no fully three-dimensional, time-dependent model which realistically simulates the structure and diurnal changes of the slope/valley wind circulation has been produced. Indeed, for several decades around the turn of the twentieth century much of the theoretical literature (most of it in German) was essentially qualitative, leaving ample scope for lengthy criticisms of the extant theories. It was really only after World War II that theoretical analysis became quantitative and this in turn highlighted the difficulties of satisfactorily modelling the slope/valley wind system.

The literature on qualitative theories is well exemplified by Kleinschmidt (1921), Schmauss (1931), Ekhart (1932a,b) and Wagner (1932a), the bulk of their contribution usually comprising criticisms of previous theories or comparisons of them with observations. Other articles (e.g. Hann, 1919) combined such criticism with an over-simplified and inadequate attempt at a new theoretical viewpoint on the subject. In the period prior to 1940 the main contributions were made by Jeffreys, Wenger and Wagner. Although the titles of their articles, in common with many others at that time, suggested that their theories explained mountain/valley winds, in reality they explained slope winds. It is of course true that valleys comprise slopes, but it is precisely the complicated combination of individual slopes within valleys that hinders the development of a comprehensive model of the valley wind system. These authors were in fact analysing the "building blocks" of a theory for the mountain/valley wind.

In his classic, and already oft-quoted paper, Jeffreys (1922) considered the "mountain and valley" (in reality "slope") winds to be "antitriptic", and assumed an equilibrium between the forces of pressure gradient, gravity and friction. With the further assumption that the air is incompressible, Jeffreys showed that the wind speed varied directly with horizontal temperature gradient and slope of the ground. The essentially baroclinic nature of the slope wind system is clear in Jeffreys' (1922, pp. 42–43) statement that

> ...equilibrium is possible only if the surfaces of equal temperature are horizontal. For if they are inclined, even though the pressure may be uniform over one level surface, the fact that the temperature is not uniform over the same surface will cause the density to be variable, and therefore the pressure will be variable over another surface.

One year later Wenger (1923), apparently unaware of Jeffreys' paper, refuted Hann's theory and demonstrated the value of Bjerknes' circulation theorem to the mountain/valley wind problem. Despite later criticism, particularly by Ekhart (1932a,b) who disbelieved the theoretical requirement of anti-winds, Wenger's ideas remain essentially sound, particularly for the slope wind *per se*. Subsequent observation and theory have, in essence, provided the details of the baroclinic overturning.

In an attempt to find some order in the array of theories, Wagner (1932b) introduced the classification of slope/mountain/valley winds used earlier in this chapter, viz. the slope wind, the mountain/valley wind, and the "mountain–plain" wind. Wagner accepted Wenger's suggestions for the slope wind and agreed that, in general, they provided the essential mechanism for all three circulations. Wagner went on to show that the tilt of the valley floor would cause the up-valley flow to cool adiabatically and thus create a pressure gradient force down-valley, i.e. in opposition to the valley wind. He argued that such a pressure gradient force would rapidly stop the up-valley wind. Clearly, if such a pressure gradient is to be avoided, air must escape from the valley and Wagner suggested that the slope wind was the vital mechanism for this removal. Here for the first time was a simple, but important link between the slope and valley winds. We have seen earlier in the chapter that this theoretical idea agrees well with the observations. Wagner's further suggestion that the slope wind air eventually moved towards the valley axis and subsided has been more difficult to observe in nature. A corollary of Wagner's theory is that the steeper the valley floor the stronger will be the counteracting pressure gradient force and the less likely will be the valley wind. Conversely, in wide, flat bottomed valleys with virtually no upward gradient, no counteracting pressure gradient force is generated—yet the slope winds are well developed and consequently so is the valley wind. Such a contrast between valleys with different gradients was in fact observed in the Vermont experiments (Davidson and Rao, 1958).

Before reviewing the quantitative theories of slope-induced motion it is profitable to examine the nature of the forces at work by consideration of a

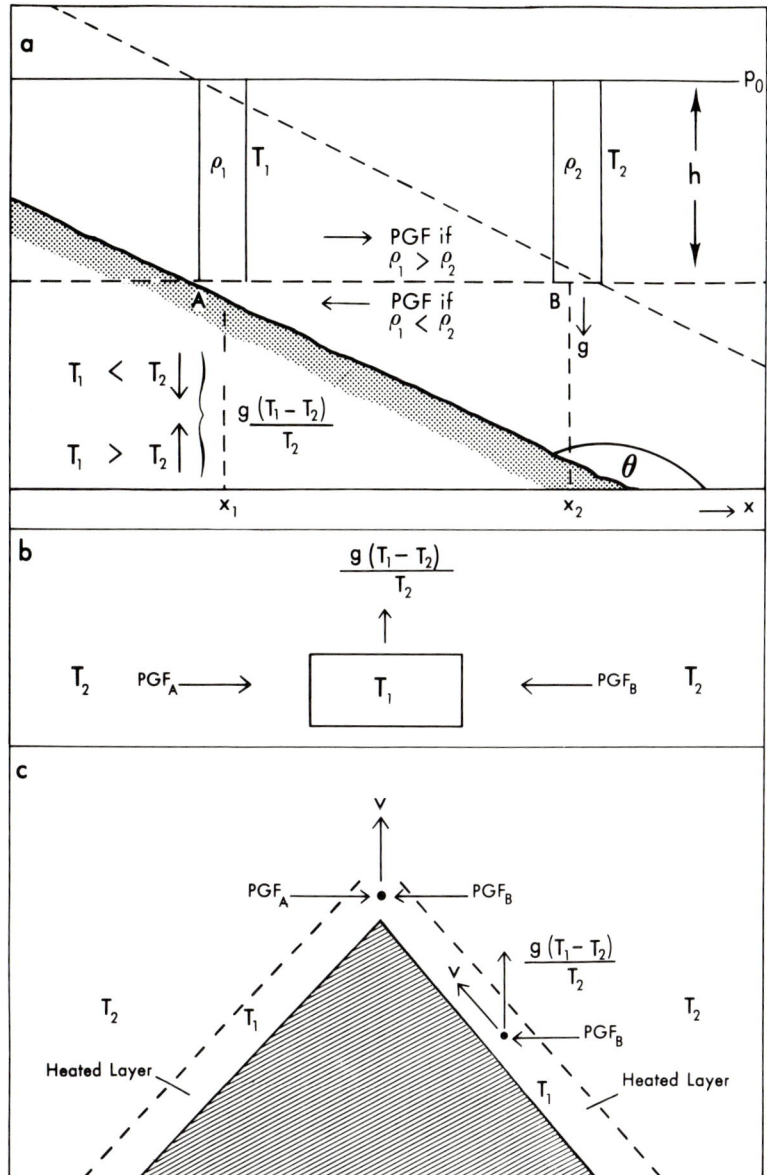

FIG. 95. Mechanism of slope winds. See text for explanation.

small volume of air near a typical slope (Fig. 95(a)) which may be either heated or cooled. We take the heated/cooled layer to be h in vertical extent. We also assume the existence of a level where pressure is constant at (say) p_0. As a result of heating/cooling, ρ_1 will be less/greater than ρ_2 and consequently the pressure at A will be $p_0 + \rho_1 gh$ and at B will be $p_0 + \rho_2 gh$. The relative magnitudes of these pressures clearly depend upon the relative magnitudes of the densities. On the reasonable assumption that the unheated/uncooled air is in hydrostatic equilibrium, then the air at A will not be in hydrostatic equilibrium to a degree given by the difference in pressure of $gh(\rho_1 - \rho_2)$. For a parcel of unit height and unit mass, the resultant force per unit mass or acceleration on the parcel is $g(\rho_1 - \rho_2)/\rho_2$ and this force acts in all directions. If temperatures are substituted for densities the force becomes the well-known buoyancy expression $g(T_1 - T_2)/T_2$, where T_1 is the temperature of the air with density ρ_1 and T_2 is the temperature of the air with density ρ_2. When the slope air is cooled, $T_1 < T_2$ and the force is negative (downward); conversely when the slope is heated $T_1 > T_2$ and the force is positive (upward). The isotropic nature of this density effect is important to the understanding of the papers reviewed below and indeed of the whole mechanism of slope winds. For the heated or cooled parcel is not only out of hydrostatic equilibrium in the vertical, thus having either an upward or downward component of acceleration but also, due to the density differences, it simultaneously finds itself in a horizontal pressure gradient which accelerates it either away from the slope (if cooled) or toward the slope (if warmed). The nature of the horizontal force is explained below in the section on drainage winds.

In summary, the heating and cooling of near-slope air has the effect of creating a force which moves the air both horizontally and vertically. Depending upon the relative sizes of these horizontal and vertical "components", air particles may move parallel to the slope or deviate from parallel somewhat. In the upslope case the vertical component of acceleration may carry thermals upward slightly faster than the horizontal acceleration pushes the air parcel toward the slope, with the effect that the parcel does indeed drift upslope, but along a trajectory not perfectly parallel to it. The effects of the buoyant force may be further clarified if we assume a volume of air to be heated by some mechanism other than by slope. In this case, two equal and opposite horizontal pressure gradient forces are created (Fig. 95(b)) and consequently there is no horizontal motion. The parcel is thus free to rise vertically as a "thermal". When the slope provides the heating (or cooling, the argument works in reverse) as in Fig. 95(b), then PGF_A in Fig. 95(b) cannot exist because the ground occupies this space. Clearly, the air must then move upslope. When the air reaches the top of the slope, it immediately experiences the additional PGF_A force which stops horizontal motion and allows the buoyant parcel to rise vertically (Fig. 95(c)). In this way, cumulus development over hills is explained. In the downslope wind, the situation is different due to the lower boundary of the ground. Rapidly sinking air can sink only so far and then has to "await" removal by the "outward" horizontal force.

A. Downslope winds

1. *Analytical results*

This interplay of the vertical and horizontal components of acceleration is far from clear in most of the literature. We shall see below that several authors concentrated on the purely horizontal pressure gradient caused by density changes and yet call the resultant flow a "drainage wind". Foremost in this respect is Fleagle (1950) who presented one of the first post-war quantitative theories of air drainage, with the aim of analysing the effects on drainage velocity of compressibility and unbalanced forces. No previous theory had accommodated both these characteristics, but Prandtl (1942) and Defant (1949) did consider compressibility. In common with all previous studies, Fleagle ignored the coriolis force and the effects of limited extent of the slope.

Starting once more from the equations of motion, Fleagle wrote

$$\frac{du}{dt} + F_x = -\frac{1}{\rho}\frac{\partial p}{\partial x}, \tag{141}$$

where u is the horizontal velocity in the x direction (i.e. downslope—see Fig. 95). Fleagle referred to u as the "drainage velocity". To make the problem easier Fleagle differentiated equation (141) with respect to time to give

$$\frac{d^2 u}{dt^2} + \frac{d}{dt}(F_x) = -\frac{d}{dt}\left(\frac{1}{\rho}\frac{\partial p}{\partial x}\right). \tag{142}$$

The value of this operation becomes evident below.

Reference to Fig. 95 together with the above general discussion shows that

$$p_1 - p_2 = gh(\rho_1 - \rho_2), \tag{143}$$

where h is the vertical thickness of the layer that undergoes the temperature change. Consequently

$$\frac{1}{\rho}\frac{p_1 - p_2}{x_2 - x_1} = g\frac{h}{x_2 - x_1}\left(\frac{\rho_1 - \rho_2}{\bar{\rho}}\right), \tag{144}$$

where $\bar{\rho}$ is the mean of ρ_1 and ρ_2. If the proportional difference in pressure is neglected in comparison with proportional difference in temperature, equation (144) may be written as

$$-\frac{1}{\bar{\rho}}\frac{\Delta p}{\Delta x} = -g\left(\frac{T_1 - T_2}{\bar{T}}\right)\tan(\pi - \theta), \tag{145}$$

where the mean densities have been replaced by the corresponding mean temperatures. We note here that the horizontal pressure gradient force is due to the "buoyancy" effect multiplied by the tangent of the slope angle $(\pi - \theta)$, emphasizing the isotropic nature of the effect.

With an error of less than 5 per cent, equation (145) leads to

$$-\frac{d}{dt}\left(\frac{1}{\rho}\frac{\Delta p}{\Delta x}\right) = -g \tan(\pi - \theta)\frac{1}{T_1}\frac{dT_1}{dt}. \tag{146}$$

The individual temperature change may be written as

$$\frac{1}{T}\frac{dT}{dt} = \left(\frac{1}{T}\frac{dT}{dt}\right)_{ad} + \left(\frac{1}{T}\frac{dT}{dt}\right)_{n} \tag{147}$$

where the subscripts "ad" and "n" refer to adiabatic and non-adiabatic components of temperature change. But,

$$\left(\frac{1}{T}\frac{dT}{dt}\right)_{ad} = \frac{1}{T}\left(\frac{dT}{dz}\right)_{ad}\left(\frac{dz}{dt}\right) \simeq \frac{1}{T}\left(\frac{g}{c_p}\right)w = \frac{1}{T}\left(\frac{g}{c_p}\right)u \tan(\pi - \theta). \tag{148}$$

Fleagle showed that the non-adiabatic temperature change is given by

$$\left(\frac{1}{T}\frac{dT}{dt}\right)_{n} = -\frac{\sigma}{\bar{\rho}hc_pT}((1-r)(T_s^4 - 0.65T_a^4)), \tag{149}$$

where σ is the Stefan–Boltzmann constant, r is the proportion of sky that does not receive net outgoing radiation due to the obstruction of the sloping ground, T_s is the temperature of the ground surface and T_a is the temperature above the cooled layer. Introducing equations (148) and (149) into equation (142) gave

$$\frac{d^2u}{dt^2} + \frac{d}{dt}(F_x) + \left(\frac{ug^2}{c_pT}\right)\tan^2(\pi - \theta) = \left(\frac{Cg}{hT}\right)\tan(\pi - \theta), \tag{150}$$

where

$$C = \frac{\sigma}{\rho c_p}(1-r)(T_s^4 - 0.65T_a^4).$$

Equation (150) expresses the mean "drainage velocity" (u) as a function of friction, slope of ground and mean rate of non-adiabatic cooling in the layer of height h. The solution is dependent upon the form of the frictional term and whether or not the term d^2u/dt^2 is neglected. Analysis that included the latter term was referred to as the dynamic theory; analysis that neglected that inertia term was referred to as the equilibrium theory.

In the dynamic theory, use of a frictional force proportional to velocity changed equation (150) into

$$\frac{d^2u}{dt^2} + k_f\frac{du}{dt} + \frac{g^2\tan^2\theta}{c_pT}u = -\frac{C}{hT}g\tan\theta. \tag{151}$$

This equation is analogous to the equation of a damped, linear oscillator with constant force and its solution is of the form

$$u = -\frac{c_p C}{hg \tan \theta} + A \exp\left(-\tfrac{1}{2}k_f t\right)\cos(\omega t - B), \tag{152}$$

where A and B are constants to be determined from the initial conditions and

$$\omega = \left(-\frac{k_f^2}{4} + \frac{g^2 \tan^2 \theta}{c_p T}\right)^{1/2}.$$

Fleagle showed that

$$A = \tfrac{1}{2}c_p C(k_f^2 + 4\omega^2)^{1/2}(\omega g h \tan \theta)^{-1},$$
$$B = \tan^{-1}(\tfrac{1}{2}k_f \omega^{-1}),$$

and therefore

$$u = -\frac{c_p C}{gh \tan \theta}\left\{1 - \frac{(k_f^2 + 4\omega^2)^{1/2}}{2\omega} e^{-k_f t/2}\cos(\omega t - B)\right\}. \tag{153}$$

Equation (153) predicts that the mean horizontal velocity in a cooled layer of air near a slope is proportional to the net outgoing radiation, inversely proportional to the thickness of the layer, begins by varying periodically and gradually becomes constant and is inversely proportional to the slope of the ground. Fleagle explained the periodicity of, and slope effects on flow as follows. Compressibility results in adiabatic warming as air sinks downslope. Associated with this, the accelerated downslope flow increases in velocity until the driving pressure gradient vanishes. To quote Fleagle (1950, p. 230):

> Continued downslope flow and consequent adiabatic heating result in an oppositely directed pressure gradient which slows the particle until radiational cooling again reverses the pressure gradient. In this way, the periodically varying velocity is produced. The final constant velocity is that which compensates the constant rate of cooling; thus the vertical component of velocity is independent of slope and the horizontal component is inversely proportional to the slope.

We noted in the first part of this chapter that downslope flows occur in surges, and Fleagle's dynamic theory goes some way towards explaining these features.

However, it is not clear from Fleagle's arguments exactly why, in physical terms, the periodicity of the drainage flow should cease. Atmanathan's (1931) observations in fact revealed surges throughout the night with no obvious damping with time. Similarly, no clear physical reasoning is given for the inverse relationship between velocity and slope angle. Marvin's (1914) qualitative argument is in clear disagreement with Fleagle's results. Marvin (1914, p. 584) wrote:

> The hillside may be replaced by a tremendous cliff with a vertical face, but the effect on the flow of air will be immaterial in so far as the change in the angle of the slope is concerned. The function of the hill-side in connection with the phenomenon of air drainage is simply that of a cooling agent.

Fleagle's equilibrium theory, neglecting d^2u/dt^2, gave the following solution to equation (151):

$$u = -\frac{c_p C}{gh\tan\theta}\left\{1 - \exp\left(-\frac{g^2\tan^2\theta}{c_p T k_f}t\right)\right\}. \tag{154}$$

Equation (154) shows that the drainage velocity is directly proportional to the net outgoing radiation, is inversely proportional to the thickness of the layer and the angle of slope and that it approaches its final value more rapidly with steep slope and large coefficient of friction than with a relatively flat slope and small coefficient of friction. Fleagle's result that the drainage velocity is proportional to the cotangent of the slope contrasts with Prandtl's (1942), who found that the cosine of slope angle was critical. This difference could be due to the fact that net radiation (used in Fleagle's theory) is nearly independent of slope at small values of slope angle, whereas Prandtl's temperature disturbance probably depends to a considerable degree on the slope and on the lapse rate in the undisturbed state.

With the aid of a slightly simpler, yet conceptually similar model to Fleagle's, Petkovšek and Hočevar (1971) showed that the drainage velocity u in an antitriptic model where friction balanced the pressure gradient force was given by

$$u = \frac{C}{(\Gamma-\gamma)\sin\theta}\left\{1 - \exp\left[-\frac{g}{k_f T}(\Gamma-\gamma)\sin^2\theta t\right]\right\}, \tag{155}$$

where $C = -(1/c_p)(dQ/dt)$, γ is the lapse rate in the air outside the cooled layer, θ is the slope angle and T is the temperature of the air outside the cooled layer. This solution is formally similar to Fleagle's, differing in that the wind velocity is given for a fixed point and the solution involves the factor $(\Gamma-\gamma)$. Calculated results from equation (155) revealed that the time period from the beginning of cooling until the time when the maximum wind velocity was attained is proportional to the stability of the atmosphere outside the cooled layer and inversely proportional to the friction coefficient and steepness of slope. The maximum velocity itself was proportional to the net radiation loss, independent of the friction coefficient and inversely proportional to the slope angle.

The theoretical ideas which above have been applied to slopes can be slightly modified for application to the flow down the slope that comprises a valley floor. In this way Rao (1970) analysed the change in wind speed along the axis of a valley. Using essentially the same arguments as Fleagle, Rao averaged the equation of motion over the cross-section of a typical valley to give

$$\frac{\partial \bar{u}^2}{\partial x} = \frac{P_x}{a} - \frac{c_d}{aH}\bar{u}^2, \tag{156}$$

where $\bar{u}$ is the cross-section average down-valley velocity, x is the horizontal axis positive along the down-valley direction and P_x is the pressure gradient force, given by

$$P_x = g \sin \theta \frac{(\rho - \rho_0)}{\rho},$$

where θ is the slope angle, ρ is air density within the valley and ρ_0 is the air density above the bounding ridge tops. a is the weighting function employed to eliminate deviations of u from $\bar{u}$ in the derivation of the equation, c_d is the drag coefficient and H is the height of the ridges above the valley floor. Equation (156) was derived on the assumption of stationary conditions and that the frictional force is proportional to the square of the mean velocity: thus this force is represented by the last term on the right-hand side of equation (156). The solution of equation (156) is

$$\bar{u}^2 = \frac{P_x H}{c_d} \left[1 - \exp\left(-\frac{c_d}{a} \frac{x}{H} \right) \right]. \tag{157}$$

Clearly, for large values of x, the distance along the down-valley direction,

$$\bar{u}^2 \simeq P_x H / c_d, \tag{158}$$

i.e. the mean velocity reaches a constant value and is no longer a function of distance down-valley. Equation (158) also shows that the mean velocity is larger in deep valleys than in shallow valleys. For small values of x, the expansion of the exponential term in equation (157) leads to

$$\bar{u}^2 = P_x x / a \tag{159}$$

when terms of higher order are neglected. Equation (159) shows that for short distances along the axis of the valley the mean velocity over the cross-section increases linearly with the square root of the distance down-valley. This acceleration of down-valley flow leads to a divergence in the volume transport and, to maintain continuity of mass, there must be a net air flow into the valley at the top. If such inflow occurs at the valley top and close to its sides over a 50 m wide strip, as suggested by Rao, there is a good agreement between observed and theoretical values of speed of about 2 m s^{-1}. On the other hand, if it is assumed that the wind at the top comes over the entire width of the valley, then its theoretical magnitude is of the order of centimetres per second and it is not easily derived from balloon observations.

B. Upslope winds

1. Analytical results

"So far as the author knows, no quantitative investigation has been made of winds at a slope" (Prandtl, 1952, p. 425). Whilst not quite true even when written, the opinion accurately reflects the sparsity of theoretical work on upslope winds. Prandtl wrote the above words in conclusion to one of the first attempts to rectify the situation.

We have noted above that the upward force generated by heating a volume of air relative to its surroundings is $(g/\rho^2)(\rho_1 - \rho_2)$, where ρ_1 is the density of the heated air and ρ_2 is the density of the ambient air. If the absolute temperature of the ambient air is T_2 and of the heated parcel $T_2 + T'$, then

$$\rho_1 = \rho_2 T_2/(T_2 + T').$$

This may be approximated as

$$\rho_1 \simeq \rho_2(T_2 - T')/T_2.$$

Substituting this into the buoyancy force gives gT'/T_2, which acts vertically upward. From the equation of state, the coefficient of thermal expansion ($\alpha = (1/V)(\partial V/\partial T)_p$, where V is volume and T is temperature) reduces to $1/T$. Consequently, the upward force upon a heated parcel of air is given by the expression $g\alpha T'$ when T' is positive. If T' were negative, the expression would describe the downward force acting on the air.

Prandtl (1952) examined several flows resulting from localized heating, one of the cases being the upslope wind. Figure 96 illustrates the situation, in

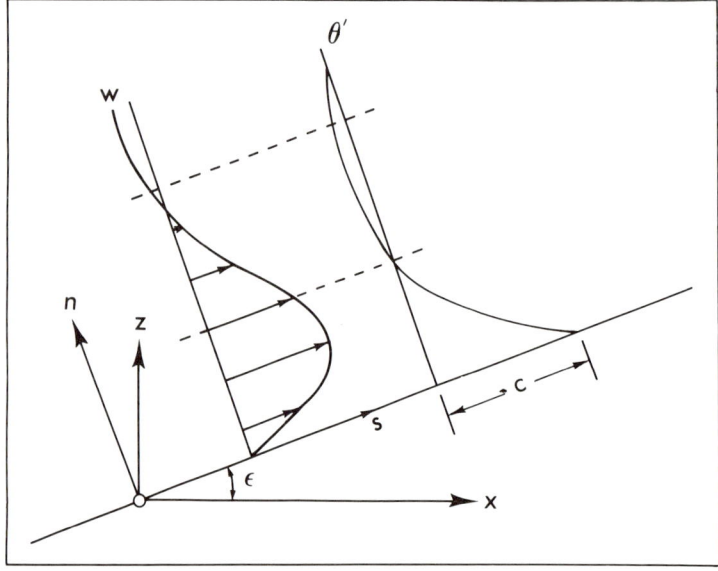

FIG. 96. Vertical profiles of vertical velocity (w) and temperature (θ') resulting from Prandtl's theory of upslope winds. (After Prandtl, 1952.)

6 SLOPE AND VALLEY WIND CIRCULATION

which the heating by the slope creates a "temperature disturbance" θ which is assumed to be a function of the distance from the slope. Thus the potential temperature θ at any point above the slope is given by

$$\theta = A + Jz + \theta'(n), \tag{160}$$

where A is a constant at $z = 0$, J is the lapse rate of potential temperature, n is the normal to the slope and θ' is the disturbance in potential temperature caused by heat transfer in the direction n. Prandtl assumed that the buoyancy would be balanced by the frictional/viscosity effects and thus wrote

$$0 = g \sin \chi \alpha \theta' - v \frac{\partial^2 w}{\partial n^2}, \tag{161}$$

where χ is the angle of slope, v is the kinematic viscosity and w is the air velocity parallel to the slope. Changes in temperature were described as follows:

$$w \frac{\partial \theta}{\partial s} = K^H \left(\frac{\partial^2 \theta}{\partial n^2} + \frac{\partial^2 \theta}{\partial s^2} \right), \tag{162}$$

where s is the direction parallel to the slope. As $z = s \sin \chi + n \cos \chi$, use of equation (160) led equation (162) to give

$$wJ \sin \chi = K^H \frac{\partial^2 \theta'}{\partial n^2}. \tag{163}$$

Differentiating equation (163) twice with respect to n and substituting for $\partial^2 w / \partial n^2$ in equation (161) gives

$$0 = g \sin \chi \alpha \theta' + \frac{vK^H}{J \sin \chi} \frac{\partial^4 \theta'}{\partial n^4}. \tag{164}$$

Such an equation has a solution of the form

$$\theta' = C e^{-n/l} \cos (n/l), \tag{165}$$

where

$$l = \left(\frac{4vK^H}{g \alpha J \sin^2 \chi} \right)^{1/4}$$

and C is a constant equivalent to the temperature disturbance at the surface. From equation (161) we see that

$$w = C \left(\frac{g \alpha K^H}{vJ} \right)^{1/2} e^{-n/l} \sin (n/l). \tag{166}$$

Figure 96 illustrates these solutions for the temperature disturbance and resultant vertical distribution of upslope velocity. Equation (166) reveals that the wind speed is proportional to the amplitude of the potential temperature disturbance at the ground and the rather unexpected result that the velocity is independent of slope angle. Prandtl considered this to be related to the fact that although for small values of χ the upward thrust resulting from the density differences due to heating is less, the opposition to motion in the direction s arising from the stratification is likewise diminished. This result agrees with Marvin's qualitative arguments.

2. Numerical results

In common with most linear, analytical models derived prior to 1955, the slope wind models were largely confined to investigation of the wind field at one point. Occasionally the vertical structure was elucidated, as by Prandtl (1952), but the inclusion of the other space dimensions and time made the equations intractable. Once more, the application of numerical methods of solution allowed the construction of non-linear models. One of the first models of this type was constructed as but part of an overall investigation of the origins of orographic cumulus. In his study Orville (1964) developed a two-dimensional model of airflow over a steep mountain slope in an attempt to simulate cumulus growth over the mountain crest. In so doing, he developed a most useful model of the upslope wind.

Basing his equations on those developed by Ogura (1962), Orville analysed the upslope wind in terms of the lateral component of vorticity, that is, the components of the vorticity involving the horizontal and vertical components of velocity. His equations were as follows:

$$\frac{\partial \eta}{\partial t} = -u\frac{\partial \eta}{\partial x} - w\frac{\partial \eta}{\partial z} + \frac{g}{\theta_0}\frac{\partial \theta'}{\partial x} + K_z^M\left(\frac{\partial^2 \eta}{\partial x^2} + \frac{\partial^2 \eta}{\partial z^2}\right), \quad (167)$$

$$\frac{\partial \theta'}{\partial t} = -u\frac{\partial \theta'}{\partial x} - w\frac{\partial \theta'}{\partial z} + K_z^H\left(\frac{\partial^2 \theta'}{\partial x^2} + \frac{\partial^2 \theta'}{\partial z^2}\right), \quad (168)$$

$$\frac{\partial u}{\partial x} + \frac{\partial w}{\partial z} = 0, \quad (169)$$

$$\eta = \frac{\partial w}{\partial x} - \frac{\partial u}{\partial z}. \quad (170)$$

In these equations η is the vorticity, θ' is the potential temperature deviation from a reference state θ_0, and the eddy coefficients for heat and momentum are equal. Orville considered the mountain ridge (one side of which comprised the slope in question) to be infinite in the y direction.

The vorticity equation (equation (167)) was derived from the "u" and "w" equations of motion and coriolis terms were neglected. Changes in vorticity are due to horizontal gradients of potential temperature deviation from the reference state. This is equivalent to the density gradient because of the relationship

$$d\rho/\rho_0 = -d\theta/\theta_0$$

from the equation of state, and so, in turn, the third term on the right of equation (167) represents the pressure gradient force due to the heating effect. The fourth term represents diffusion of momentum. Equation (168) describes the changes in the potential temperature deviation due to eddy heat diffusion and advection. The continuity equation (equation (169)) is valid in two dimensions, assuming an incompressible fluid.

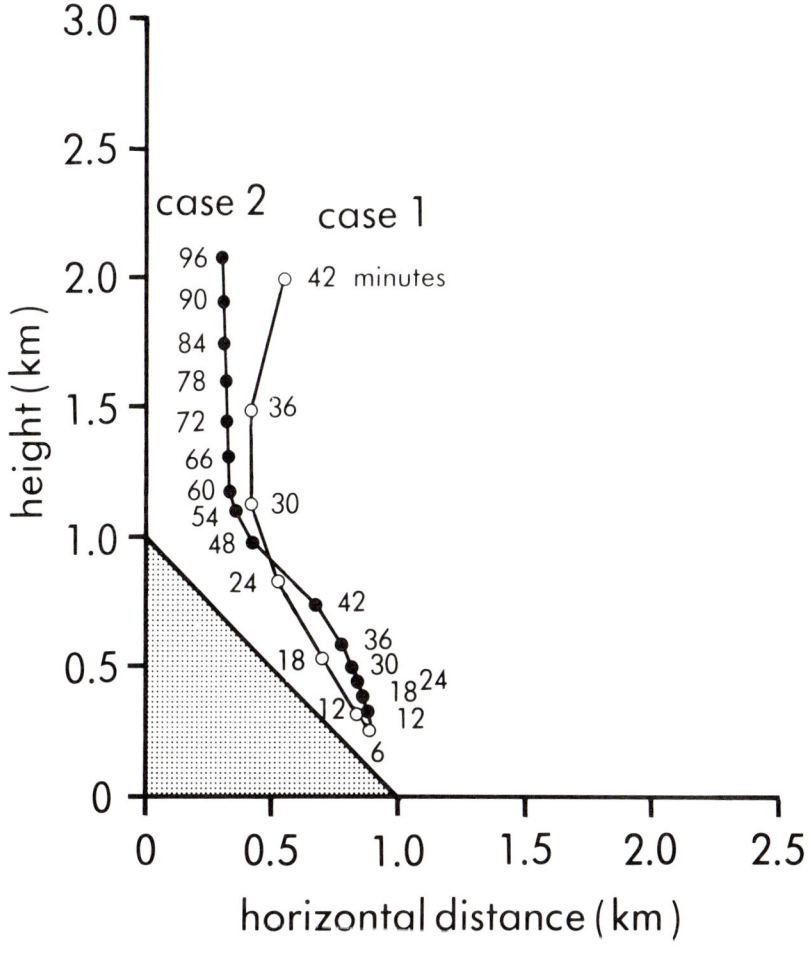

FIG. 97. Trajectories of centres of theoretical horizontal vortices for two simulated upslope winds. (After Orville, 1964.)

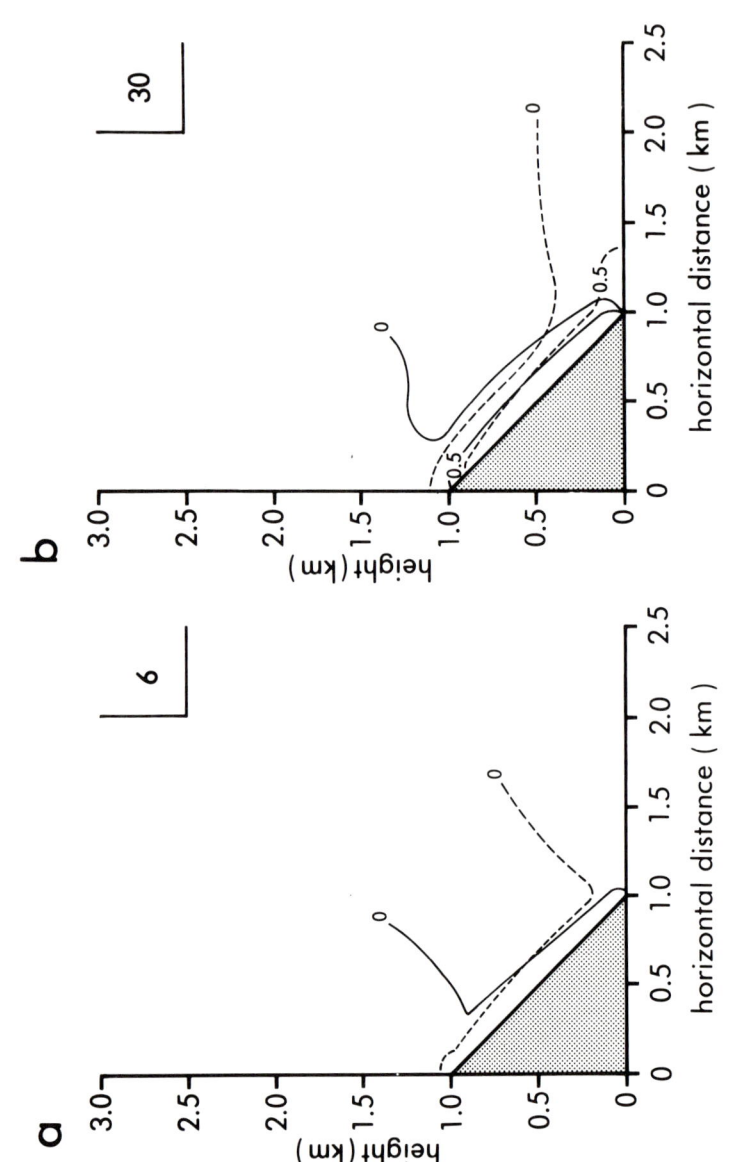

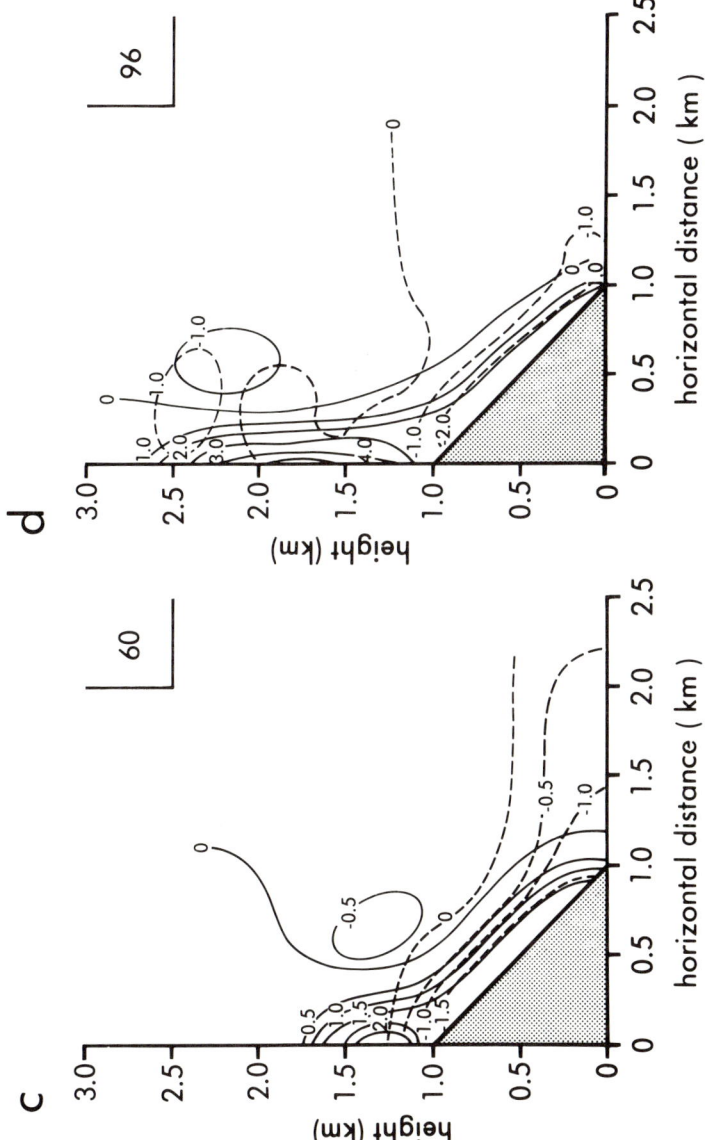

FIG. 98. Four stages of evolution of a theoretical upslope wind over a period of 96 min. Horizontal component shown by dashed lines (in metres per second), positive towards the right; vertical component shown by solid lines (in metres per second), positive upwards. (After Orville, 1964.)

Orville integrated his non-linear slope wind model using the following structure. His mountain was 1 km high and abutted a plain 2 km long. The mountain slope was 45°, admittedly large, but this made programming much easier. The height of the volume of integration was 3 km. In both the x and z directions the grid length was 100 m. Orville employed methods quite similar to those of Fisher (1961) and Estoque (1961) as reviewed in Chapter 5. He considered the potential temperature change to occur within a metre or so of the Earth's surface and to be propagated upward by the Fickian diffusion term to the lowest grid points of the model, which were situated 10 m above the surface. The potential temperature change itself was based upon observations. A sinusoidal potential temperature change was postulated at the surface with a period of 24 h and an amplitude of 7 °C along the level plain, decreasing linearly to 3 °C at the summit.

Initially the atmosphere was at rest and no horizontal potential temperature gradients existed. Motion began when the potential temperature at the Earth's surface increased and this change was transmitted to the lower boundary grid points by the diffusion term. Subsequently both diffusion and advection affected the potential temperature at these lowest grid points.

Orville ran the model with time increments of 15 and 30 s for two cases. Case 1 had a neutral environment and Case 2 a potential temperature increase with height of 1 K km^{-1}. Figure 97 shows how the centre of vorticity (in essence the centre of the thermal, or bubble of air that is rising) moved away from the slope in Case 1, but remained roughly parallel to the slope in Case 2, in close correspondence with the schematic illustration of Fig. 95(c). Clearly the stability of the atmosphere effects the nature of the upslope wind; the less the stability, the greater the likelihood of thermals "breaking away" from the slope, giving an "erratic", gusty, upslope flow. The stability also determines the rate of development of the wind. In Case 2, the wind developed one-half to one-third as rapidly as Case 1.

Figure 98 illustrates the development of the u and w components of the upslope wind for Case 2 over a period of $1\frac{1}{2}$ h. Initially the upslope winds extended only 100–150 m above the surface. As the motion developed the slope wind thickened. Also at all times the depth of the upslope winds increased with distance up the slope. In the later stages the upslope winds occupied a layer 100 m thick near the bottom of the slope increasing to 350 m at the top of the slope. Final velocities reached 3.8 m s^{-1}. In association with this quite realistic upslope wind, the model produced a flow from the plain which feeds the upslope flow, together with (after 96 min) a suggestion of an upper level return flow at a height of 2.5 km.

Despite the two-dimensional nature of the model and the difficulties of the transfer properties at the lower surface, Orville's model produced most realistic results and remains one of the few numerical models of slope winds.

C. Valley and mountain winds

Quantitative theoretical studies of the whole valley wind system are rare. As hinted earlier, the reasons for this probability lie in the difficulty in satisfactorily modelling winds on one slope let alone the many slopes that together comprise a valley. Nevertheless both analytical and numerical attempts have been made.

1. Analytical results

The analytical approach is exemplified by Gleeson's (1951, 1953) studies of valley winds. In truth, Gleeson analysed airflow over a slope that could be just as justifiably a valley-side slope, but he chose to call it the valley floor and thus considered that he was modelling the true valley wind. An expression was derived for the horizontal components of the valley wind as functions of the slope of the valley floor, turbulent heat conduction and turbulent friction. The diurnal variation of the valley wind was given and the influence of the Earth's rotation on this wind was considered. Gleeson closely followed Pierson's (1950) methods which we noted in Chapter 5 on the sea breeze.

Gleeson's model was a linear one. He represented temperature as a function of time, elevation and horizontal distance. From the postulated temperature function, a horizontal pressure gradient force was derived which served to accelerate the air along the valley in one direction at low levels and in the opposite direction at upper levels. The equations of motion containing the factors mentioned previously, were combined with the derived pressure gradient force as follows:

$$\frac{\partial u}{\partial t} - fv + qv = K_z^M \frac{\partial^2 u}{\partial z^2}, \tag{171}$$

$$\frac{\partial v}{\partial t} + fu - qu = K_z^M \frac{\partial^2 v}{\partial z^2} - \frac{1}{\rho}\frac{\partial p}{\partial y}, \tag{172}$$

where x and y are the axes across and along the valley respectively. The pressure gradient term was a rather complicated function of temperature, height and time, derived in detail by Gleeson (1953, p. 263). In equations (171) and (172) Gleeson introduced the terms qv and qu as

> components of a postulated horizontal pressure gradient force which opposes the coriolis force. It may be reasoned that such a force would develop in the following manner. As the valley wind is deviated away from the axis of the valley by the coriolis force, there will be air flow toward one side of the valley away from the other side. This cross valley flow (analysed in detail by Gleeson (1951)) should result in an accumulation of air at the first side and a depletion at the other side,

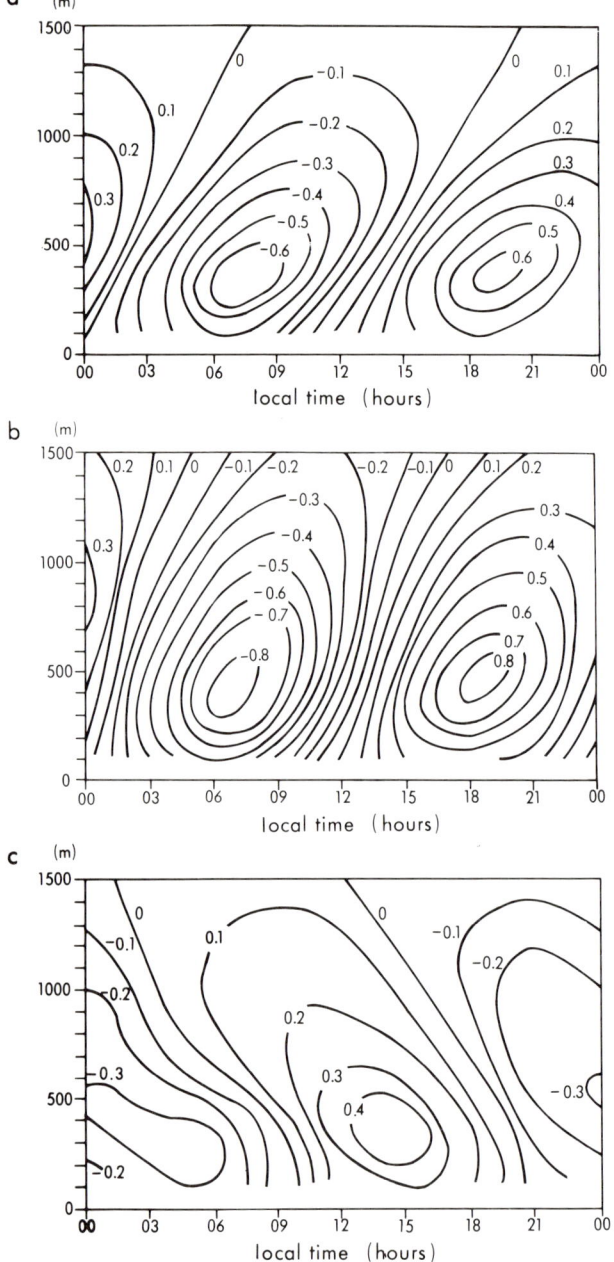

FIG. 99. (a) Variation with time and height of a theoretical valley wind speed (in metres per second). Positive values indicate up-valley flow. The case for $q = f$. (b) As (a), but $q = 0.5f$; (c) As (a) but $q = 0$. (d) Hodograph of computed wind for the case of $q = 0.5f$. Each curve is labelled in hours of local time. Dots on each curve refer to levels

6 SLOPE AND VALLEY WIND CIRCULATION

d

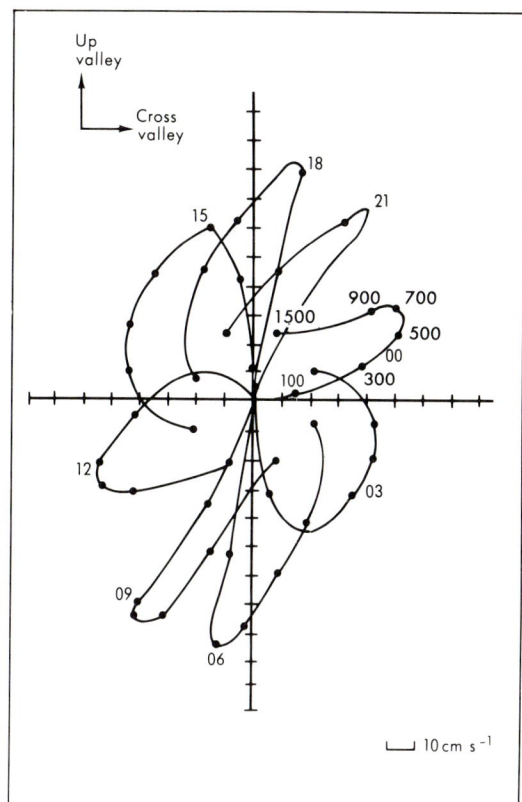

e

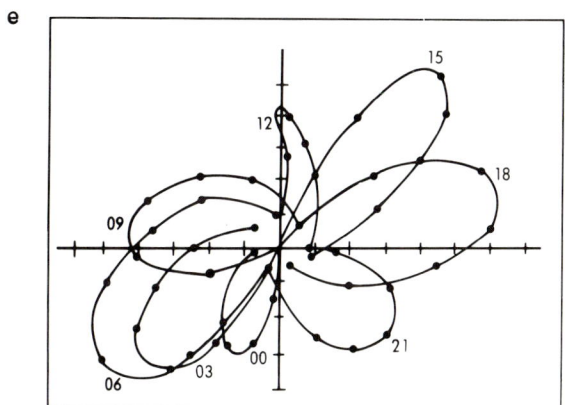

100, 300, 500, 700 and 1500 m above the surface. Direction of wind is from the origin to the appropriate dot. Up- and cross-valley directions are indicated. (e) as (d) but $q = 0$. (After Gleeson, 1953.)

since the valley walls and the air beyond these walls would act as barriers. The horizontal pressure gradient force thus created would oppose the Coriolis force, tending to restore the valley wind parallel to the valley (Gleeson, 1953, p. 264).

This restoring force was assumed to be proportional to the wind speed and the value of q was limited to $0 \leqslant q \leqslant f$.

Combination of equations (171) and (172) into one equation describing a complex wind allowed a solution for the components u and v. These solutions were very lengthy expressions included in an appendix to the original paper, where interested readers may readily consult them. Examination of the solutions gave few obvious insights: principally, they indicated that the theoretical valley wind was proportional to the amplitude of the diurnal temperature variation, to the horizontal temperature gradient and to the slope of the valley, this latter point being in agreement with Jeffreys but disagreement with Prandtl. Gleeson's other results are illustrated on Fig. 99, which includes time–height sections of the along-valley wind (positive values indicating up-valley flow) and hodographs of the "total" wind on which the dots refer to the levels 100, 300, 500, 700, 900 and 1500 m above the surface.

The time sections with q varying from $q = f$ to $q = 0.5f$ to $q = 0$ (Fig. 99(a)–(c)) reveal that the lag with height decreases to zero and reverses as q tends to zero. In the case of $q = f$, the coriolis force is balanced by the restoring force so that the wind is always parallel to the valley. The lag in maximum speed with elevation is identical with the lag of maximum temperature with elevation. The variation in lag is probably explained by the varying differences between the coriolis force and restoring force. When the difference is small, the wind backs with elevation (Fig. 99(d)); whereas when the difference is large, the wind veers with elevation (Fig. 99(e)).

In comparison with reality (in this case, the Inn Valley) Gleeson found that his velocities were an order of magnitude too small. He attributed this to his use of too small a valley slope and temperature amplitude. Perhaps more important, Gleeson found his best agreement with observations in the case where $q = 0.5f$, showing some influence of the coriolis force. This was the first demonstration of such an effect but was not followed up in any detail. Later writers have tended to be somewhat sceptical about this conclusion. In particular, Thyer (1966) has reservations about Gleeson's "arbitrary force to counteract [the coriolis force]".

2. Numerical results

Gleeson was not alone, however, in being criticized by Thyer. All the earlier qualitative and quantitative theories have obvious defects that are difficult to remedy. The principle deficiencies were that the models were two-dimensional and, in all cases except Orville's, linear. To quote Thyer (1966, p. 321):

> An ideal theory of valley and mountain winds would give, as functions of all three space co-ordinates and of time, all three velocity components, the temperature,

pressure and density, in the valley, above it and over the adjacent plain. Slope winds, valley/mountain winds, their respective anti-winds and vertical currents would all be explained simultaneously, and thus would all be deduced from the minimum initial conditions of geographical location, the topography of the area, the nature of the surface of the ground (especially its albedo, specific heat and thermal conductivity), the insolation on the surface, or radiation from it, as a function of position and time and the prevailing atmospheric conditions, such as gradient wind, stability, humidity and cloud. Further, it would not ignore the fact that actual valleys are finite in size.

Perhaps this rather daunting list of requirements provides the explanation for the lack of such models!

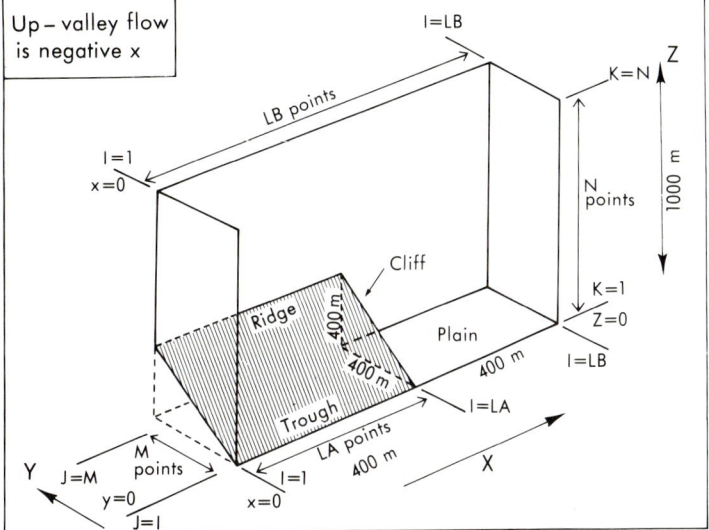

FIG. 100. Domain of Thyer's numerical simulation of slope and valley winds. (After Thyer, 1966.)

In an attempt to satisfy the above requirements Thyer developed a three-dimensional non-linear model of the up- and down-valley winds for the valley illustrated in Fig. 100. In contrast to Gleeson's model the coriolis terms were omitted but viscosity and heat diffusion were included in the form of an eddy viscosity and an eddy diffusivity which were assumed to be both equal and constant everywhere. Thyer's model was derived from the following equations: the equations of motion,

$$\frac{\partial u}{\partial t} + u\frac{\partial u}{\partial x} + v\frac{\partial u}{\partial y} + w\frac{\partial u}{\partial z} = -\frac{1}{\rho}\frac{\partial p}{\partial x} + K^M\left(\frac{\partial^2 u}{\partial x^2} + \frac{\partial^2 u}{\partial y^2} + \frac{\partial^2 u}{\partial z^2}\right), \quad (173)$$

$$\frac{\partial v}{\partial t} + u\frac{\partial v}{\partial x} + v\frac{\partial v}{\partial y} + w\frac{\partial v}{\partial z} = -\frac{1}{\rho}\frac{\partial p}{\partial y} + K^M\left(\frac{\partial^2 v}{\partial x^2} + \frac{\partial^2 v}{\partial y^2} + \frac{\partial^2 v}{\partial z^2}\right), \quad (174)$$

$$\frac{\partial w}{\partial t} + u\frac{\partial w}{\partial x} + v\frac{\partial w}{\partial y} + w\frac{\partial w}{\partial z} = -\frac{1}{\rho}\frac{\partial p}{\partial z} - g + K^M\left(\frac{\partial^2 w}{\partial x^2} + \frac{\partial^2 w}{dy^2} + \frac{\partial^2 w}{\partial z^2}\right); \tag{175}$$

the heat conduction equation,

$$\frac{\partial \theta}{\partial t} + u\frac{\partial \theta}{\partial x} + v\frac{\partial \theta}{\partial y} + w\frac{\partial \theta}{\partial z} = K^H\left(\frac{\partial^2 \theta}{\partial x^2} + \frac{\partial^2 \theta}{\partial y^2} + \frac{\partial^2 \theta}{\partial z^2}\right); \tag{176}$$

the continuity equation,

$$\frac{\partial \rho}{\partial t} + u\frac{\partial \rho}{\partial x} + v\frac{\partial \rho}{\partial y} + w\frac{\partial \rho}{\partial z} + \rho\left(\frac{\partial u}{\partial x} + \frac{\partial v}{\partial y} + \frac{\partial w}{\partial z}\right) = 0; \tag{177}$$

the equation of state,

$$p = \rho RT; \tag{178}$$

and Poisson's equation,

$$T/\theta = (p/1000)^{R/c_p}; \tag{179}$$

where x is along the axis of the valley.

In developing his three-dimensional model, Thyer devised three versions of a two-dimensional slope model. The third and most successful version, which was subsequently extended to three dimensions, was similar in conception to Orville's (1964) in its use of vorticities. Equations (174) and (175) were cross-differentiated to give an equation in

$$\xi = \frac{\partial w}{\partial y} - \frac{\partial v}{\partial z},$$

the x-component of vorticity:

$$\frac{\partial \xi}{\partial t} = -\frac{\partial u}{\partial y}\frac{\partial w}{\partial x} + \frac{\partial u}{\partial z}\frac{\partial v}{\partial x} + \xi\frac{\partial u}{\partial x} - u\frac{\partial \xi}{\partial x} - v\frac{\partial \xi}{\partial y} - w\frac{\partial \xi}{\partial z} + \frac{g}{\theta}\frac{\partial \theta}{\partial y} + K_z^M \nabla^2 \xi. \tag{180}$$

Similarly equations (173) and (175) were cross-differentiated to give an equation in

$$\eta = \frac{\partial u}{\partial z} - \frac{\partial w}{\partial x},$$

the y-component of vorticity:

$$\frac{\partial \eta}{\partial t} = -\frac{\partial v}{\partial z}\frac{\partial u}{\partial y} + \frac{\partial v}{\partial x}\frac{\partial w}{\partial y} + \eta\frac{\partial v}{\partial y} - u\frac{\partial \eta}{\partial x} - v\frac{\partial \eta}{\partial y} - w\frac{\partial \eta}{\partial z} - \frac{g}{\theta}\frac{\partial \theta}{\partial x} + K_z^M \nabla^2 \eta. \tag{181}$$

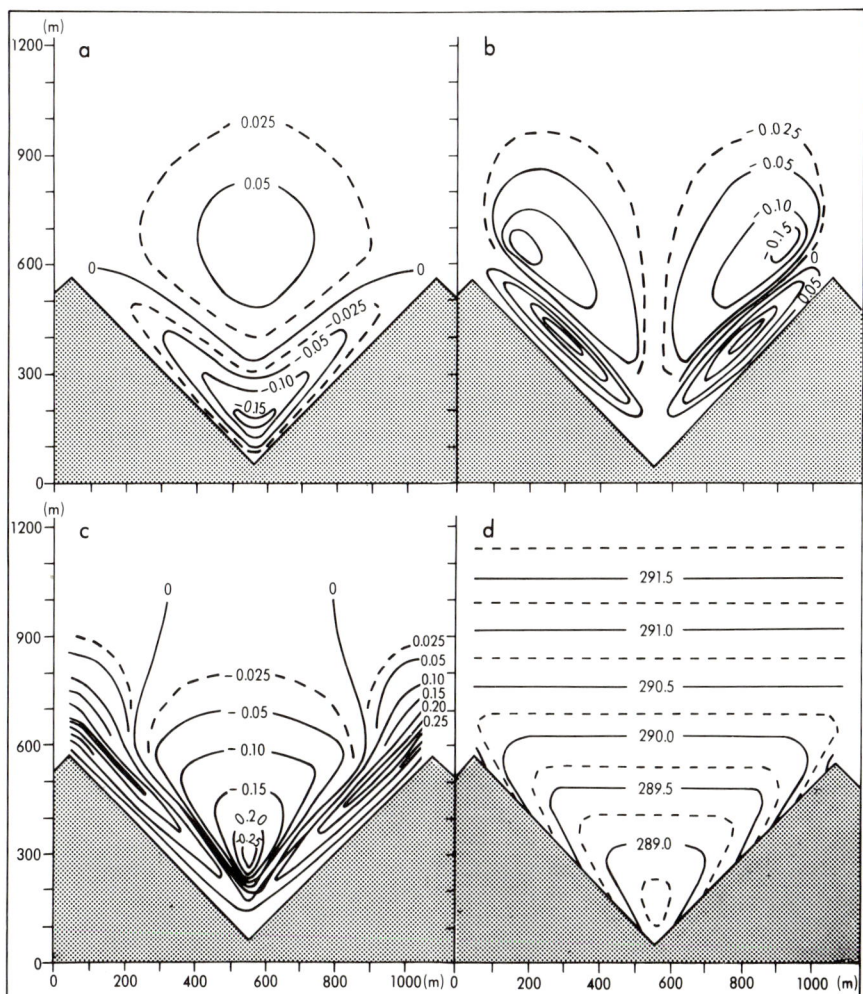

FIG. 101. Cross-section of computed winds and temperatures at the valley entrance. (a) Longitudinal wind component (in metres per second), positive down valley. The negative values indicate a maximum up-valley flow of $0.15\,\mathrm{m\,s^{-1}}$. (b) Cross-valley wind component (in metres per second), on right-hand side of diagram, positive from valley bottom to right-hand ridge. Left-hand side of the diagram is included by symmetry. (c) Vertical wind speed (in metres per second), positive upward. (d) Potential temperature (in kelvins). (After Thyer, 1966.)

If the three velocities u, v and w are expressed in terms of two stream functions ϕ and ψ such that

$$u = -\frac{\partial \psi}{\partial z}, \qquad v = \frac{\partial \phi}{\partial z}, \qquad w = \frac{\partial \psi}{\partial x} - \frac{\partial \phi}{\partial y},$$

then

$$\frac{\partial u}{\partial x} + \frac{\partial v}{\partial y} + \frac{\partial w}{\partial z} = 0,$$

the condition for incompressibility.

Equations (176), (180), (181) and the assumption of incompressibility allow solutions for the four unknowns u, v, w and θ, basically through the stream functions ϕ and ψ. The model was run in a valley with dimensions shown in Fig. 100, with original trough temperature of 288 K (equal to the constant plain temperature), an original lapse rate of 6.5 K km^{-1}, an eddy viscosity of 10 m^2 s^{-1} and a heat input of 90.6 W m^{-2}. Results of the runs, which simulated only 2 min of real time, are shown in Figs 101 and 102. The up-valley wind and its associated "anti-" wind were well simulated (Fig. 101(a)) and the important associated slope wind circulations also clearly appeared (Fig. 101(b)). The v-component pattern is very similar to Orville's results.

The vertical velocity field (Fig. 101(c)) showed the uplift over the slopes which, according to the model, is compensated by sinking over the valley axis. Thyer admitted that this motion has been difficult to measure in the field if it actually exists, but his model results were in agreement with Wagner's (1932b) qualitative reasoning. Along the valley (Fig. 102) both the valley wind and anti-wind decreased with distance from the valley. The vertical velocity showed a very strong gradient at a height of about 200 m above the valley floor: below that level air was rising at the foot of the slopes of the valley side; above that level, compensatory mid-valley sinking occurred.

Night-time conditions were simulated with a lapse rate of 10 K km^{-1} and heat flow reversed with a magnitude of 181 W m^{-2}. The most striking feature of the results was their similarity to the day-time distributions, apart from a reversal of sign in every case.

Thyer's model must be one of the very few to simulate quite realistically several characteristics of the valley wind system, e.g. the thin layer of slope wind, updraughts over the ridges, valley wind below ridge level having a maximum speed near the trough, an anti-wind layer of about the same thickness, situated around and above ridge height, both valley wind and anti-wind having their maxima near the valley entrance. Thyer recognized the major deficiencies as the neglect of gradient wind and coriolis force and, in particular, the short length of time (2 min) represented by each run, due to shortage of computer time. Nevertheless, his model is probably the most advanced to date.

IV. Hardware models

In common with sea breezes, few hardware models of the valley wind system exist. A good example is provided by Thyer and Buettner (1961), who

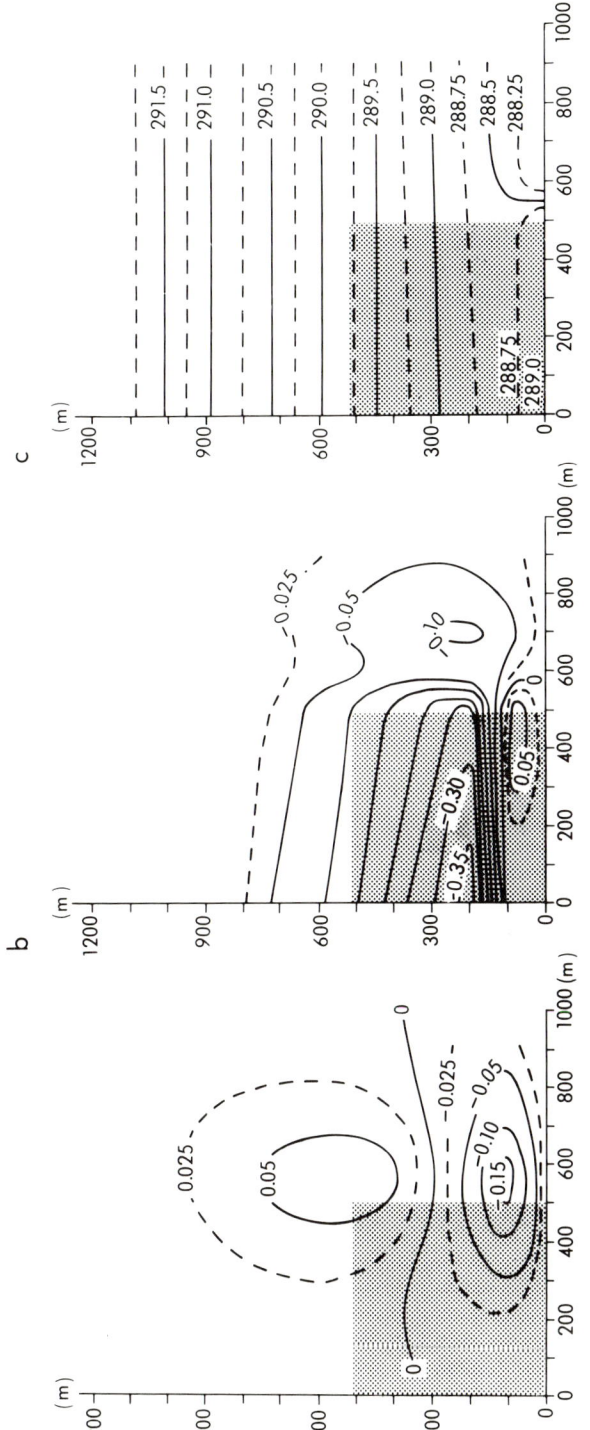

FIG. 102. Longitudinal section of computed winds and temperatures along valley trough. Shaded area represents valley. (a) Longitudinal wind component (in metres per second), positive down valley. (b) Vertical speeds (in metres per second), positive upwards. Note the marked subsidence over the valley axis as shown in Fig. 101(c). (c) Potential temperature (in kelvins). (After Thyer, 1966.)

constructed a V-shaped valley 76 cm long, having 45° slopes of length 15 cm. The valley sides could be heated at a rate of about 49 W m^{-2} or cooled with solid carbon dioxide. The authors could also close off one end of the "valley" and impose a lid, 15 cm above the ridge tops, in a crude attempt to simulate the effect of general atmospheric stability on the vertical extent of the valley wind system. Air movements were traced with smoke.

When cooling was applied, downslope winds were observed and in the lower part of the valley, there was a well-defined down-valley wind (of about 10 cm s^{-1}) filling the valley to a depth of 6 cm and continuing beyond the valley as a jet protruding into the adjacent "plains". When heating was applied to simulate a sunny day, upslope winds were general. With the "lid" off, no anti-wind was detected. With the "lid" on, valley and slope winds still occurred, accompanied by an anti-wind. Valley and upslope winds were of the order of 20 cm s^{-1}. The artificiality of the "lid" was one of the major drawbacks of this experiment, but it did further indicate that a sloping valley floor is not a necessary condition for a valley wind system yet that slope winds are important.

V. The role of slope and valley winds

Perhaps more than any other type of meso-scale circulation, slope and valley winds have both "pure" and "applied" roles. We may conceive their "pure" role as being but one of a multitude of meso-scale circulations within the atmosphere, each necessarily interacting with larger and smaller circulations. In contrast, their "applied" role may profitably be viewed as one of creating meteorological conditions that have particular effects upon man's activities. Both roles are exemplified below.

We noted earlier that Orville's theoretical study of the upslope wind was but part of an investigation of orographic cumulus. In the observational part of the overall project Braham and Draginis (1960) showed clearly for the first time that the convective clouds which frequently develop over mountains have their origin in the valley wind circulation. The upslope winds continued as free convection above the mountain ridges (as illustrated in Fig. 95) and produced small cumulus clouds. Braham and Draginis were of the strong opinion that the lifting due to this upslope wind effect was far greater than that due to the "obstacle" effect of the mountains. A second, less obvious effect of slope winds—in this case downslope winds—was pointed out by Ball (1957). He estimated that the total loss to Antarctica due to the outward flowing katabatic winds was 1.1×10^{14} J s^{-1} in comparison to a total rate of heat loss of 1.7×10^{15} J s^{-1}: that is, despite their extreme shallowness, katabatic winds account for about 6 per cent of the energy balance of Antarctica.

The effects of slope winds on man's activities probably represent some of the most familiar aspects of applied meteorology, in particular frost hollows and the accumulation of fog and pollutants (Davidson, 1961b), the two most notable possible effects of downslope flow. Countless studies of these effects exist in the literature and no attempt to summarize them is made here. Yet early investigations by Cornford (1938) on the prevention of frost damage and by Hewson and Gill (1944) on the effects of valley winds on pollution from the Trail smelter in British Columbia still merit recognition.

References

Atmanathan, S. (1931). The katabatic winds of Poona. *Indian Met. Dept Sci. Notes*, **4**, 101–115.

Ayer, H. S. (1961). On the dissipation of drainage wind systems in valleys in morning hours. *J. Met.*, **18**, 560–563.

Ball, F. K. (1956). The theory of strong katabatic winds. *Aust. J. Phys.*, **9**, 373–386.

Ball, F. K. (1957). The katabatic winds of Adelie land and King George V land. *Tellus*, **9**, 201–208.

Ball, F. K. (1960). Winds on the ice slopes of Antarctica. *Proc. Symp. Antarct. Met.*, Melbourne, 9–16.

Braham, R. R. and Draginis, M. (1960). Roots of orographic cumuli. *J. Met.*, **17**, 214–226.

Brown, H. A. (1961). Horizontal structure of mountain winds. Mesometeorology Project Research Paper No. 4, Contract DA-36-039-SC-78901, Department of Geophysical Sciences, University of Chicago, Chicago.

Buettner, K. J. K. (1968). Valley winds, sea-breeze and mass fire: three cases of quasi-stationary airflow. Colorado State University Department of Atmospheric Science Paper no. 122, pp. 103–129.

Buettner, K. J. K. and Thyer, N. (1959). Valley and mountain winds I. AF contract 19 (604)-2289, Department of Atmospheric Science, University of Washington.

Buettner, K. J. K. and Thyer, N. (1962). Valley and mountain winds III. AF contract 19 (604)-7201, Department of Atmospheric Science, University of Washington.

Buettner, K. J. K. and Thyer, N. (1966). Valley winds in the Mount Rainier area. *Arch. Met. Geophys. Bioklim. B*, **14**, 125–147.

Burger, A. and Ekhart, E. (1937). Über die tägliche Zirkulation der Atmosphäre im Bereiche der Alpen. *Beitr. Geophys.*, **49**, 341–367.

Cornford, C. E. (1938). Katabatic wind and prevention of frost damage. *Q. Jl R. met. Soc.*, **64**, 553–587.

Cross, C. M. (1950). Slope and valley winds in the Columbia river valley. *Bull Am. met. Soc.*, **31**, 79–84.

Davidson, B. (1961a). Local wind circulations: final report. Vol. II. Studies of the field of turbulence in the lee of mountain ridges and tree lines. Contract no. DA-36-039-sc-84939, College of Engineering, Research Division, New York University.

Davidson, B. (1961b). Valley wind phenomena and air pollution problems. *Air Pollut. Control Assoc. J.*, **11**, 364–386.

Davidson, B. and Rao, P. K. (1958). Preliminary report on valley wind studies in Vermont: final report. Contract no. AF 19 (604)-1971 AECRC-TR-58-29, College of Engineering, Research Division, New York University.

Davidson, B. and Rao, P. K. (1963). Experimental studies of the valley–plain wind. *Air Water Pollut.*, **7**, 907–921.

Defant, F. (1949). Zur Theorie der Hangwinde nebst Bemerkungen zur Theorie der Bergund Talwinde. *Arch. Met. Geophys. Bioklim. A*, **1**, 421–450.

Ekhart, E. (1932a). Zur Aerologie des Berg und Talwindes, Ergebnisse von Pilotballon-Aufstiegen in Innsbruck. *Beitr. Phys. freien Atmos.*, **18**, 1–26.

Ekhart, E. (1932b). Weitere Beiträge zum Problem des Berg und Talwindes. *Beitr. Phys. freien Atmos.*, **18**, 242–252.

Ekhart, E. (1934). Neuere Untersuchungen zur Aerologie der Talwinde: Die periodischen Tageswinde in einem Quertale der Alpen. *Beitr. Phys. freien Atmos.*, **21**, 245–268.

Eldridge, R. H. (1951). Katabatic wind at Driffeld. *Met. Mag.*, **80**, 288–293.

Estoque, M. A. (1961). A theoretical investigation of the sea breeze. *Q. Jl R. met. Soc.*, **87**, 136–146.

Fisher, E. L. (1961). A theoretical study of the sea breeze. *J. Met.*, **18**, 216–233.

Fleagle, R. E. (1950). A theory of air drainage. *J. Met.*, **17**, 227–232.

Fournet, M. J. (1840). Des brises de jour et de nuit autour des montagnes. *Annls Chim. Phys.*, **74**, 337–401.

Gleeson, T. A. (1951). On the theory of cross-valley winds arising from differential heating of the slopes. *J. Met.*, **8**, 398–405.

Gleeson, T. A. (1953). Effects of various factors on valley winds. *J. Met.*, **10**, 262–269.

Hann, J. von (1879). Zur Theorie der Berg- und Talwinde. *Z. öst ges. Met.*, **14**, 444–448.

Hann, J. von (1919). Über die Theorie der Berg- und Talwinde. *Met. Z.*, **36**, 287–289.

Hawkes, H. B. (1945). Mountain and valley winds. Pt. 1 local winds. US Army Signal Corps Report no. 982, Weather Division.

Hewson, E. W. and Gill, G. C. (1944). Meteorological investigations in Columbia River Valley near Trail, BC. US Department of the Interior, Bureau of Mines Bulletin no. 453, pp. 23–228.

Heywood, G. S. P. (1933). Katabatic winds in a valley. *Q. Jl R. met. Soc.*, **59**, 47–57.

Jaffe, S. (1960). Effect of prevailing wind on the valley wind regime at anemometer level. Observation and theory of local wind systems: final report, 1 Jan. 1959–30 Sept. 1960. College of Engineering, Research Division, New York University.

Jeffreys, H. (1922). On the dynamics of the wind. *Q. Jl R. met. Soc.* **48**, 29–47.

Jelinek, A. (1937a). Beiträge zur Mechanik der periodischen Hangwinde. *Beitr. Phys. freien Atmos.*, **24**, 60–84.

Jelinek, A. (1937b). Über den themischen Aufbau der periodischen Hangwinde. *Beitr. Phys. freien Atmos.*, **24**, 85–97.

Jelinek, A. and Riedel, A. (1937). Über die Schichtdicke der periodischen Lokalwinde un Inntal. *Beitr. Phys. freien Atmos.*, **24**, 205–215.

Kleinschmidt, E. (1921). Zur theorie der Talwinde. *Met. Z.*, **38**, 43–46.

Lettau, H. H. (1966). A case study of katabatic flow on the south Polar plateau. *Antarctic Res. Ser. Am. geophys. Union*, **9**, 1–11.

Lied, N. T. (1964). Stationary hydraulic jumps in a katabatic flow near Davis Antarctica 1961. *Aust. Met. Mag.*, **47**, 40–51.

MacHattie, L. B. (1968). Kananaskis valley winds in summer. *J. appl. Met.*, **7**, 348–352.

Marvin, C. F. (1914). Air drainage explained. *Mon. Weath. Rev.*, **42**, 583–585.

Mather, K. B. and Miller, E. S. (1966). Wind drainage off the high plateau of eastern Antarctica. *Nature, Lond.*, **209**, 281–284.

Mendonca, B. G. (1969). Local wind circulation on the slopes of Mauna Loa. *J. appl. Met.*, **8**, 533–541.

Moll, E. (1935). Aerologische Untersuchungen periodischer Gebirgswinde in V-förmingen Alpentälern. *Beitr. Phys. freien Atmos.*, **22**, 177–199.
Newnham, E. V. (1918). Notes on examples of katabatic wind in the valley of the upper Thames at the aerological observatory of the meteorological office at Benson, Oxon. *Profess. Notes Met. Office, Lond.*, **1** (2).
Ogura, Y. (1962). Convection of isolated masses of a buoyant fluid: a numerical calculation. *J. Atmos. Sci.*, **19**, 492–502.
Orville, H. D. (1964). On mountain upslope winds. *J. atmos. Sci.*, **21**, 622–633.
Petkovšek, Z. and Hočevar, A. (1971). Night drainage winds. *Arch. Met. Geophys. Bioklim. A*, **20**, 353–360.
Pierson, W. J. (1950). The effects of eddy viscosity, coriolis deflection and temperature fluctuation on the sea breeze as a function of time and height. *Met. Pap. New York Univ.*, **1** (2), 7–29.
Pollak, L. W. (1924). Berg- und Talwinde un Becken von Trient. *Met. Z.*, **41**, 18–21.
Prandtl, L. (1942). *Führer durch die Strömungslehre*, Braunschweig Vieweg und Sohn.
Prandtl, L. (1952). *Essentials of Fluid Dynamics*, Blackie and Son Ltd, London (translation of Prandtl (1942)).
Rao, P. K. (1970). Theoretical investigation of the change of wind speed along the axis of a valley. *Arch. Met. Geoph. Bioklim. A*, **19**, 59–70.
Schmauss, A. (1931). Zur Enstehung der Tal- und Bergwinde. *Met. Z.*, **48**, 511–512.
Shieh, L. J. (1971). An investigation of the morning dissipation of the valley wind by means of a one-dimensional model. *Boundary Layer Met.*, **2**, 38–51.
Streten, N. A. (1963). Some observations of Antarctic katabatic winds. *Aust. Met. Mag.*, **42**, 1–23.
Tang, W. (1960). Temperature and wind distribution over sloping terrain. Part 1. Interim Final Report, 1 July–31 Dec. 1959. Analysis of wind structure. Phase I. Valley wind theory. Contract no. DA-36-039-SC-78091, College of Engineering, Research Division, New York University.
Thyer, N. H. (1966). A theoretical explanation of mountain and valley winds by a numerical method. *Arch. Met. Geophys. Bioklim. A*, **15**, 318–347.
Thyer, N. and Buettner, K. (1961). Valley and mountain winds II. AF Contract 19 (604)-5578, Department of Atmospheric Science, University of Washington.
Tollner, H. (1931). Gletscherwinde in den Ostalpen. *Met. Z.*, **48**, 414–421.
Tollner, H. (1935). Gletscherwinde auf der Pasteze. *XLIV Jahresbericht des Sonnblick-Vereines*, 38–54.
Tyson, P. D. (1966). Examples of local air circulation over Cato Ridge during July 1965. *S. Afr. Geogr. J.*, **48**, 13–31.
Tyson, P. D. (1967). Some characteristics of the mountain wind over Pietermaritzburg. *Proc. S. Afr. Geogr. Soc. Jubilee Conf.*, 103–128.
Tyson, P. D. (1968a). Nocturnal local winds in a Drakensberg valley. *S. Afr. Geogr. J.*, **50**, 15–32.
Tyson, P. D. (1968b). Velocity fluctuations in the mountain wind. *J. atmos. Sci.*, **25**, 381–384.
Tyson, P. D. and Preston-Whyte, R. A. (1972). Observations of regional topographically-induced wind systems in Natal. *J. appl. Met.*, **11**, 643–650.
Wagner, A. (1932a). Hangwind-Ausgleichsströmung—Berg- und Talwinde. *Met. Z.*, **49**, 209–217.
Wagner, A. (1932b). Neue Theorie des Berg- und Talwindes. *Met. Z.*, **49**, 329–341.
Wenger, R. (1923). Zur Theorie der Berg- und Talwinde. *Met. Z.*, **40**, 193–204.

Part III

Free atmosphere circulations

A. Non-convective circulations

7

Moving Gravity Waves

I. Introduction

In this chapter waves of a type called "vertical transverse" by Thompson (1961) are considered. In such waves air particles move up and down as the wave propagates horizontally (Fig. 103(a)). In contrast, longitudinal or compression waves are those in which the particle trajectories lie in lines parallel to the direction of wave propagation (Fig. 103(b)); and horizontal-transverse waves are those in which the particles move meridionally while the waves are propagated zonally (Fig. 103(c)). The force that causes the upward and downward movement in the vertical-transverse waves is buoyancy (Brunt, 1927) and, in turn, this force is ultimately due to the action of gravity. For this reason vertical-transverse waves are widely known as gravity waves, despite Gossard and Hooke's (1975) preference for the term "buoyancy waves".

> Atmospheric gravity waves first appeared in the literature in the nineteenth century, but with conceptual and mathematical limitations that hampered their elementary identification, characterization and exploitation. They emerged explicitly as a mathematically distinct type of wave some forty years ago, but made little immediate impact on the study of atmospheric dynamics (Hines, 1972, p. 73).

The lack of impact remains today, particularly when gravity waves are compared with the other meso-scale circulations considered in this book. Yet the obscurity of gravity waves is rapidly being swept away, with some enthusiasts going so far as to say that "...the atmosphere from its lowest levels to its highest must be suspected of being permeated by gravity waves" (Hines, 1972, p. 74). This faith in the ubiquity of gravity waves was nurtured by the discovery that many variations in the high (75–100 km) atmosphere

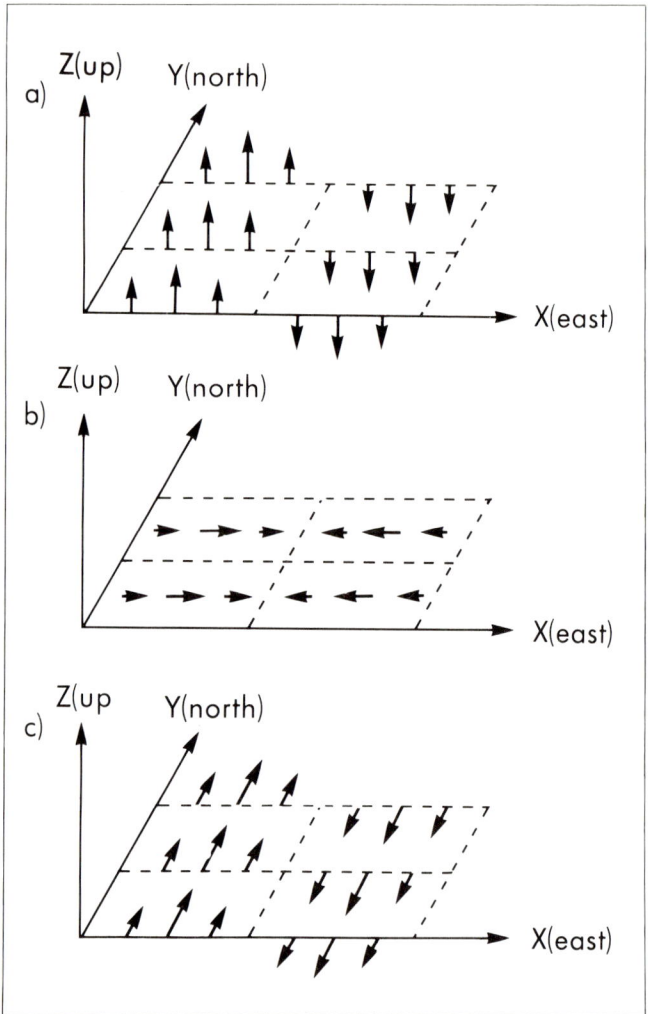

FIG. 103. Schematic distribution of velocity perturbations associated with (a) eastward-moving vertical transverse waves; (b) eastward-moving compression waves; (c) eastward-moving horizontal-transverse waves. (After Thompson, 1961.)

could be explained quantitatively by the single mechanism of a gravity wave. The most obvious source of these very high level waves was thought to be the lower atmosphere, since most internal waves grow exponentially with height. The last two decades have witnessed remarkable growth in the study of gravity waves at these lower levels and, as a result, although much remains to be understood, it does appear that the waves are far more frequent than hitherto believed, that they may have an important role in vertical transfer of momentum and also that they may occasionally directly affect weather. For these reasons our comparatively scant knowledge is reviewed below. Only travelling waves are covered as the stationary (relative to the ground) lee waves are considered in Chapter 2.

Observations have now been made of waves in the three parts of the atmosphere defined as follows: at the very high levels (above 20 km); at the very low levels (below 500 m); and in the layer in between. Waves in the very high atmosphere are beyond the scope of this book and, in any case, are thoroughly covered by Gossard and Hooke (1975). Small waves at very low levels have been the object of many studies, too numerous to mention here: they are well exemplified by the works of Hicks and Angell (1968) and Gossard and Richter (1970). In these near-surface layers, wavelengths are usually less than 5 km, occasionally as short as 100 m or less, and periods are of the order of 2–3 min. Frequently these waves are of the Kelvin–Helmholtz type (considered in more detail later) and are seen by many as a useful link between the meso- and micro-scale air motions. The third part of the atmosphere—essentially the bulk of the troposphere—has, rather ironically, been less well observed than the other two, from the point of view of gravity waves. Such information as is available suggests that gravity wave lengths in this part of the atmosphere may vary from less than 50 to about 500 km, but observations of the latter very large wavelengths are rather sparse (Wagner, 1962; Eom, 1975; Uccellini, 1975). Consequently most of the observations have revealed meso-scale gravity waves, and it is these which form the object of this chapter.

II. Observation

The recent emergence of gravity waves in the literature is no doubt partly due to improvements in observational techniques. Yet, half a century ago Johnson (1929) revealed such waves by analysis of conventional surface microbarograph records and today surface pressure measurements, albeit with more sensitive recorders, still provide us with the bulk of our observations on meso-scale tropospheric gravity waves. Figure 104 illustrates two typical examples, Fig. 104(a) showing large amplitude, irregular patterns whereas Fig. 104(b) reveals smaller amplitude, more regular waves. Striking as these records are, they are often but a very indirect indication of wave activity some distance, both horizontal and vertical, away from the recording site. Attenuation of the waves as they travel from source to recorder, together with the fact that pressure measurements themselves are but a surrogate for air motion, necessitates careful interpretation of surface barograph recordings.

In contrast to the microbarograph, radar and satellite provide direct, visual

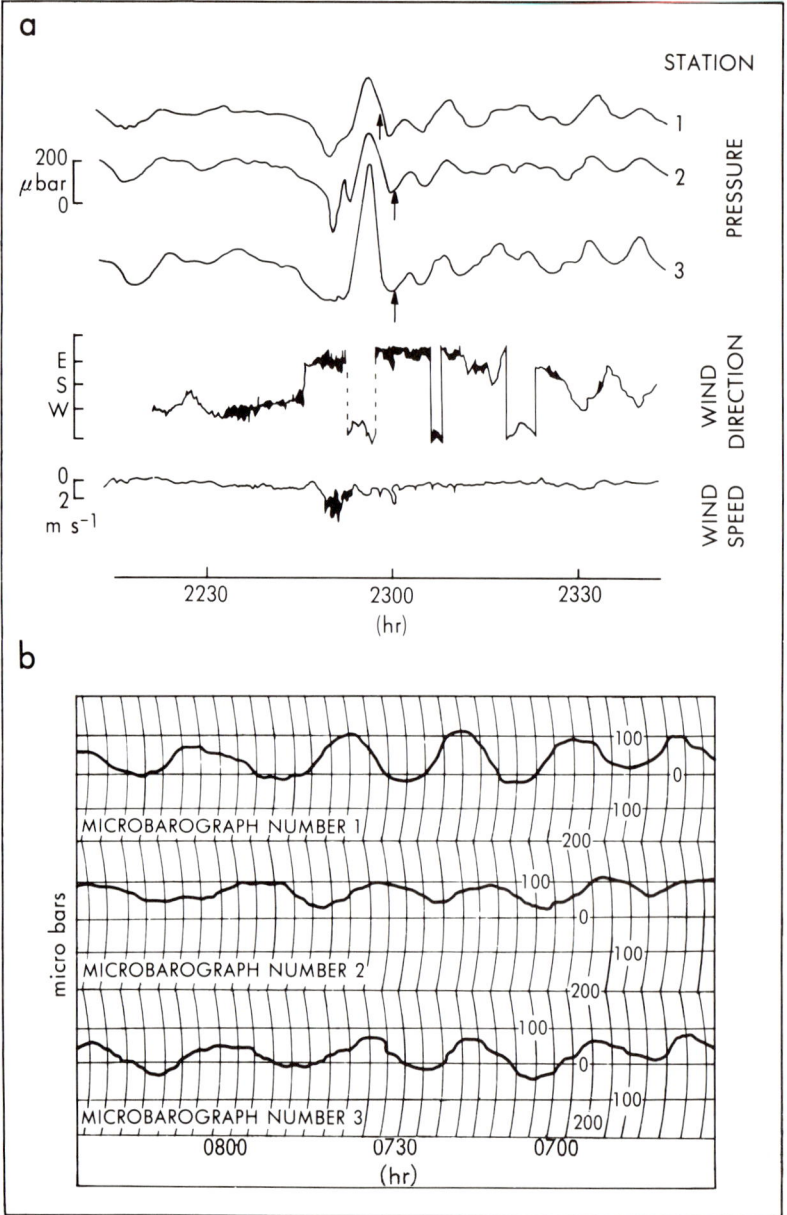

FIG. 104. Surface recordings of gravity waves. (a) Large amplitude, irregular patterns shown in pressure (in microbars) and wind traces. Pressure is shown for three stations and increases downward on the chart. The oscillations to the right of the arrows result from gravity waves generated by a thunderstorm. (After Jordan, 1972.) (b) Small amplitude, regular patterns shown on three microbarographs of one gravity-wave occasion. Units: microbars. (After Curry and Murty, 1974.)

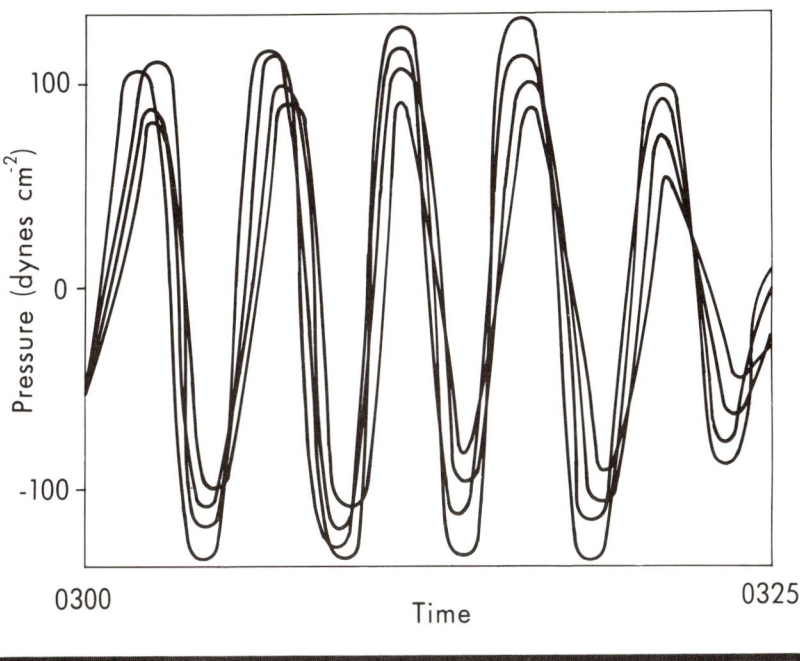

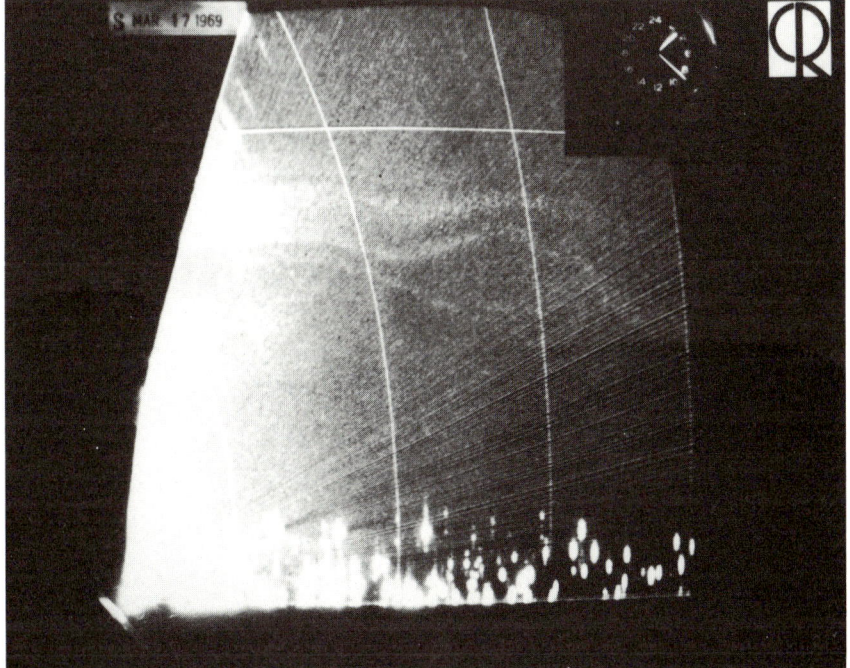

FIG. 105. Comparison of microbarograph and radar data of gravity waves at a height of about 10 km. (After Hooke and Hardy, 1975.)

Table 39 Wave characteristics

Source	Wavelength (km)	Crest to trough amplitude (m)	Crest to trough pressure amplitude (mb)	Period (min)	Phase speed (ms^{-1})	Height of occurrence† (km)	How identified	Inversion height (km)	Temperature difference, $\Delta\theta$ (°C)
Johnson (1929)				7–28		Surface	Pressure at a point	0.3–2	
Ali (1931)	6.0		0.37	17.0	5.9	Surface	Anemograph at a point	0.7	
Gossard and Munk (1954)	4.4–9.6		0.14–0.86	5.5–14.0	11–12	Surface	Pressure at a point	0.2–1	10–20
Pothecary (1954)	12.0–13.3	324–402	1.0–1.4	12.5–20.0	11–13	Surface	Pressure at a point	1.6	
Yamamoto (1957)	12.0		0.27	9.4	21.4	Surface	Pressure at three points		
	16.9		0.53	17.4	16.4	Surface	Pressure at three points	1.5	
	8.7		0.27	9.8	14.8	Surface	Pressure at three points	2–3	

Reference								
Balachandran and Donn (1964)	12.0–18.0		0.7	4–5	50–60	Surface	Pressure at a point	2	Virtually isothermal
Herron and Tolstoy (1969)	20–300			30–90	10–50	Surface	Pressure at a point		
Reed and Hardy (1972)	15–30	2000			44	10	Radar	8–10	
Bosart and Cussen (1973)	55–60		2–5	160	10–13	Surface	Detailed surface pressure maps		
Erikson and Whitney (1973)	12					c. 5	Satellite		
Cunning (1974)	30–90		0.4	30–90	30	Surface	Pressure at a point		
Curry and Murty (1974)	20		0.1	10	33.2	Surface	Pressure	c. 10	
Merceret and Black (1975)	20			12	28	c. 1	Satellite	0.3	
Hooke and Hardy (1975)	12–14	680		4–5	40–50	Surface and 8–12	Pressure and radar		
Gedzelman and Rilling (1978)	18		0.1	10	30	Surface	Pressure		

† Although waves certainly did occur at the surface in those cases where so indicated, as shown in the text, these waves, usually identified by pressure variations, are probably but "shadows" of their former selves at their origins.

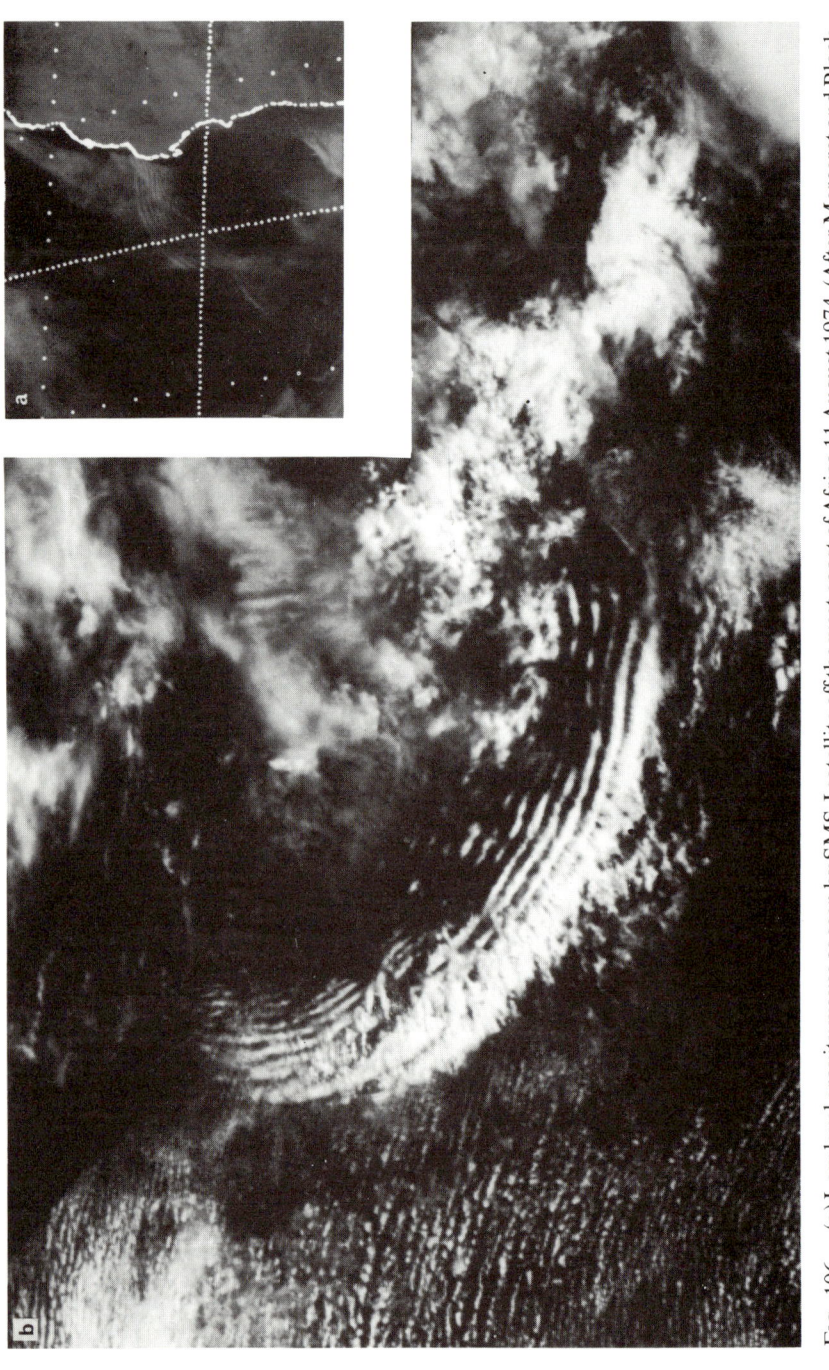

FIG. 106. (a) Low level gravity waves as seen by SMS-I satellite off the west coast of Africa 11 August 1974. (After Merceret and Black, 1975.) (b) High level gravity waves as seen by the Very High Resolution Radiometer aboard NOAA No. 2 satellite over the eastern United States, 22 May 1973. (After Erickson and Whitney, 1973.)

evidence of the waves, particularly those occurring in the mid- and upper troposphere. Figure 105 shows how radar information of waves at heights of 8–12 km compares with surface microbarograph measurements: the agreement is most satisfactory. In this case the radar was registering waves in a cirrus layer. In similar fashion, provided cloud is present, the satellite may

Table 40 Waves that occurred when the atmosphere contained distinct shear layers and low level inversions. A question mark next to the wavelength indicates a broad spectrum

Wavelength (km)	Pressure amplitude (μb)	Minimum Richardson no.	Height of shear layer (km)	Height of stable layer (km)
22.6	52.5	0.24	9.0	2.8
56.0?	192.5	0.08	8.2	3.2
43.3	116.6	0.03	8.8	2.2
14.7?	106.1	0.11	8.4	1.6
35.7	99.1	0.18	6.2	2.6
33.6	87.5	0.17	8.0	2.7
31.2?	142.8	0.11	2.8	1.0
21.6?	72.6	0.21	12.5	2.6
23.0	81.6	0.19	4.4	1.2
26.1	81.6	0.38	6.0	1.3
18.5?	69.1	0.23	7.0	1.5
10.9	76.0	0.47	2.0	0.7
18.3	64.1	0.27	5.6	2.2
13.9	154.7	0.04	8.0	1.6
16.8	96.3	0.14	10.0	2.5
12.2?	84.7	0.31	9.2	1.6
38.1	107.8	0.24	9.3	2.3
30.8	52.5	0.36	5.3	1.8
20.0	67.2	0.44	7.4	2.3
6.0	41.0	2.03	7.0	2.0
10.8?	35.0	0.41	7.0	2.6
23.3	96.3	0.37	6.4	3.2
29.8	111.0	0.13	4.6	2.9
20.0	35.0	0.50	3.6	3.6
110.8?	35.0	0.27	8.0	2.5
31.7?	32.2	0.16	7.5	3.3
14.6?	70.0	0.24	7.7	2.2
12.0?	111.0	0.22	1.3	0.4
18.9	99.1	0.23	3.4	2.3
24.2	64.1	0.18	4.6	3.1
39.0?	140.0	0.18	4.7	1.2
34.9	67.2	0.14	7.6	1.4
31.4?	128.5	0.08	3.0	2.6
30.6?	122.5	0.07	4.6	3.0

Source: Gedzelman and Rilling (1978).

reveal a whole train of gravity waves as shown in Fig. 106. In Fig. 106(a) the waves are at a height of about 300 m and in Fig. 106(b) they are at a height of about 5 km. This difference in height reflects their different origins, on which more is said later. Before leaving this brief look at observational methods, we should note that specially instrumented aircraft have provided measurements on high level, short gravity waves (Axford, 1970) and, at levels below 500 m, acoustic sounding is providing a mass of most useful information on similar sized features (e.g. Hooke *et al.*, 1972; Emmanuel *et al.*, 1972).

The main characteristics of meso-scale travelling gravity waves as observed by the methods outlined above are shown in Table 39. Wavelengths vary from the very short (*c*. 5 km) to the very long (300 km) but the majority of values lie in the 10–30 km range. Characteristic periods are 5–30 min, but the very long waves have periods of up to 90 min. Phase speeds vary from 10 to 50 m s^{-1}.

Observational experience shows that wave occurrence is frequently associated with vertical stability and wind shear, a point noted in Chapter 2. Table 40 shows some 34 cases in which distinct inversions or stable layers were present in the lower troposphere and in which distinct shear layers were also present. Pressure variations and wavelengths are given together with the minimum value of the Richardson number. The significance of this parameter

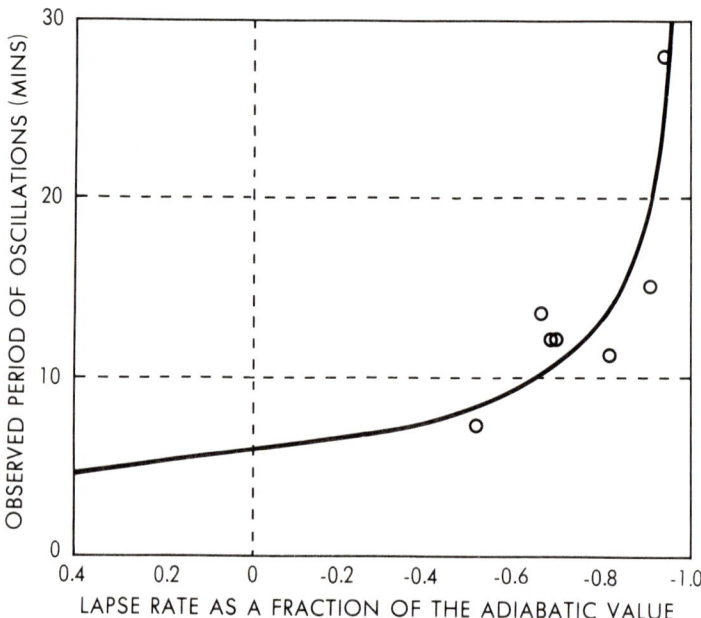

FIG. 107. Relationship between period of vertical oscillations and lapse rate of temperature, negative value being decrease of temperature with height. Line results from Brunt's theory. Circles result from observations. (After Johnson, 1929.)

is noted below. Gedzelman and Rilling (1978) were of the opinion that high level shear and low level stable layers are together particularly favourable to constrained rather than freely propagating waves.

In his early analysis Johnson (1929) compared the observed wave period in given lapse rate conditions to the Brunt–Väisälä periods (Brunt, 1927) for the same lapse rates (Fig. 107) and found a reasonable agreement. Noting an inversion at a height of 1 km he wrote:

> We have seen that the periods which have been found in microbarograms lie in the same region as the natural period of oscillation of the atmosphere. Indeed, it seems clear that we are here dealing with a case of true resonance. If these oscillations originate at a height of about a kilometre at the interface of two currents, it is evident that pressure variations of appreciable amplitude will be transmitted to ground level most frequently when the period of the wave motion agrees with the natural period of oscillation of the kilometre depth of atmosphere at the time. If the period of wave motion differs considerably from the period of oscillation of the atmosphere at the time, then the original disturbance will not be recorded at the ground unless it is of exceptionally large amplitude (Johnson, 1929, p. 26).

Subsequent observations have shown that waves do occur when their frequency is less than or equal to the Brunt–Väisälä frequency. Merceret and Black (1975) and Hooke and Hardy (1975) both found wave frequencies less than the Brunt–Väisälä frequency, the former in the lowest 300 m of the atmosphere, the latter at heights of 8–12 km. Curry and Murty (1974) found that the wave frequency was comparable with the Brunt–Väisälä frequency at heights of about 10 km. In these three cases, the Brunt–Väisälä period varied from 4 to 25 min.

The second major feature associated with the waves is wind shear. Typical values in the wave layer are $16-30 \, \text{m s}^{-1} \, \text{km}^{-1}$, as measured by Reed and Hardy (1972), Hooke and Hardy (1975) and Balachandran and Donn (1964). The relationship between vertical shear and wave occurrence was well illustrated by Keliher's (1975) analyses of some 280 meso-scale gravity waves observed in Colorado. Figure 108 shows a close correlation between the directions of wind shear and observed wave motion; wave speeds, while showing some correlation with the upper tropospheric wind speeds, tended to be lower by $20-40 \, \text{m s}^{-1}$.

Figure 109(a), (b), (c) provides a neat summary of the relationships between meso-scale gravity waves on the one hand and lapse rate, Brunt–Väisälä frequency (N) and wind shear on the other. Both the lapse rate and the Brunt–Väisälä frequency histograms show a peak near the dry adiabatic lapse rate. The wind shear histogram is rather more flat than the histograms for the dry adiabatic lapse rate or N, a feature that Keliher (1975) interpreted as showing that the production of gravity waves by wind shear layers is not strongly controlled by the value of the wind shear but is more

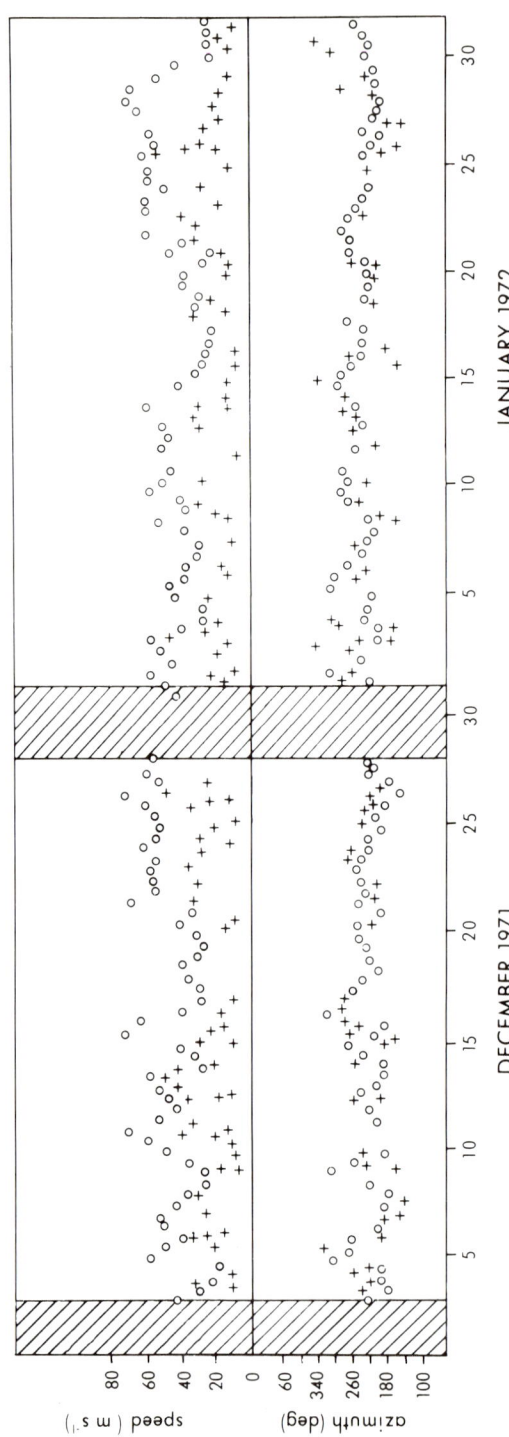

Fig. 108. Comparison between speed and direction of gravity waves (detected by microbarographs), shown by plus signs, and upper tropospheric wind maxima, shown by open circles, for winter months 1971–72 near Boulder, Colorado, USA. (After Keliher, 1975.)

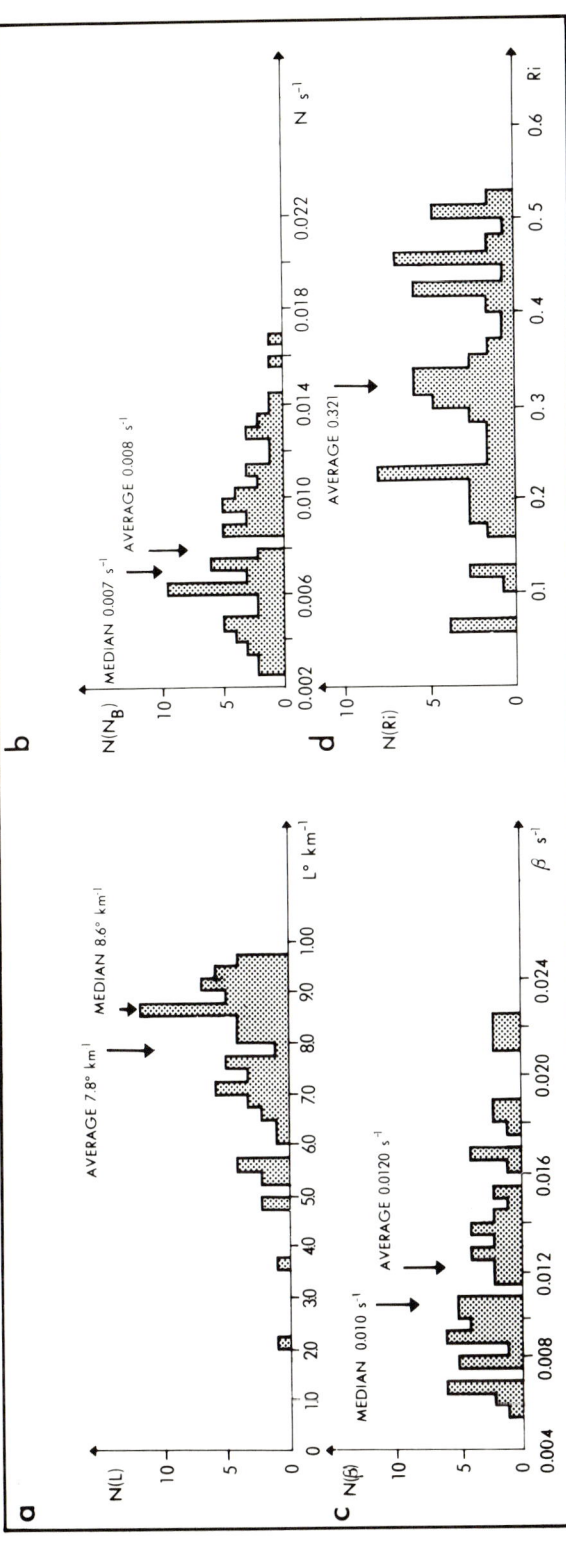

FIG. 109. Histograms of four parameters associated with gravity waves generated by wind shear, near Boulder, Colorado, USA, winter 1971–72: (a) lapse rate; (b) Brunt–Väisälä frequency; (c) wind shear; (d) Richardson number. (After Keliher, 1975.)

strongly influenced by the static stability of the atmosphere, the production being more likely for an almost neutrally stable stratification.

The combined effects of stability and shear are conveniently expressed in the form of the Richardson number (Ri) as follows:

$$\text{Ri} = \frac{g}{\theta}\frac{\partial \theta}{\partial z} \bigg/ \left(\frac{\partial U}{\partial z}\right)^2 = \left(\frac{N}{S}\right)^2,$$

where N is the Brunt–Väisälä frequency, i.e.

$$N = \left(\frac{g}{\theta}\frac{\partial \theta}{\partial z}\right)^{1/2}$$

and S is the vertical wind shear $\partial U/\partial z$. Figure 110 shows that, in the case analysed by Reed and Hardy (1972), wave activity between 8 and 10 km was clearly associated with an upper level frontal zone, a local maximum of vertical shear and a minimum of Richardson number. These and other observations suggest that waves form in layers with Richardson numbers of less than 0.50 (Table 40 and Fig. 109(d)), probably less than 0.25. In addition wave amplitude is related to minimum values of the Richardson number. The

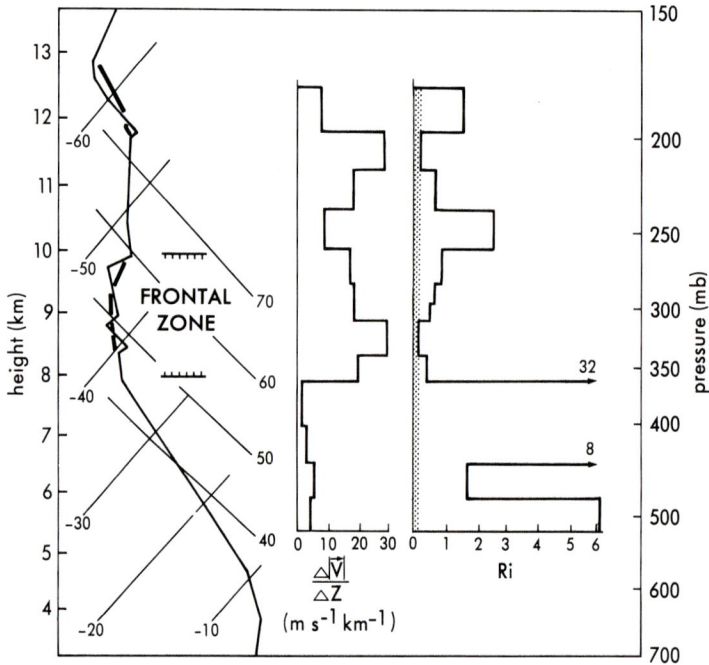

FIG. 110. Temperatures, wind shears and Richardson numbers associated with gravity waves over the eastern USA 18 March 1969. Heavy segments of temperature sounding denote smoothed values used in computing Ri. (After Reed and Hardy, 1972.)

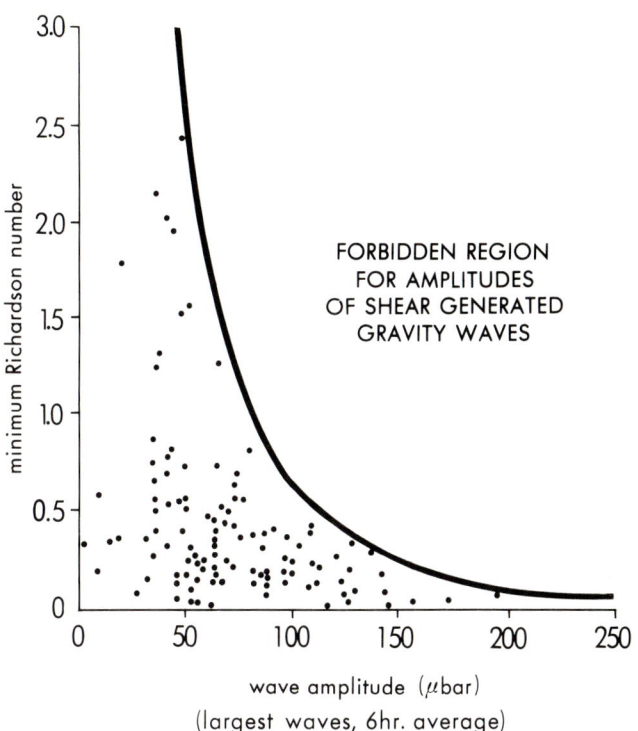

FIG. 111. Wave amplitudes as a function of minimum Richardson number. The envelope indicates that large amplitude waves are not expected from shear instability when $Ri_{min} > 0.25$. (After Gedzelman and Rilling, 1978.)

wave amplitudes used in Fig. 111 were calculated by averaging the largest wave each hour over 6 h periods. Gedzelman and Rilling (1978) drew an envelope for possible wave amplitudes as a function of minimum Richardson number. The envelope was drawn in such a manner that large amplitude shear waves were not permitted when the Richardson number was large or were possible although not necessary when the Richardson number was small. The increasing belief that vertical shear is the critical factor in generating these waves is further considered in Section III of this chapter. Gedzelman and Rilling (1978, p. 201) provided a fitting conclusion of this section of the chapter:

> It is apparent that although the existence of the waves ultimately can often be related to shearing instability, we still remain largely ignorant of the finer details of the wave generation process... recent theoretical papers... have addressed the perplexing question of why the observed wavelengths are so long. At present the question remains unanswered. The problem of verification is closely tied to obtaining detailed measurements of the vertical structure of the waves. Needless to say, such data are still sorely lacking.

III. Theory

We noted in the first part of this chapter that meso-scale gravity waves frequently occurred in association with a stable stratification and a marked vertical wind shear. This association is not peculiar to meso-scale waves—indeed both long-standing theory and observation show it to be particularly evident in the case of waves with much smaller wavelengths. These smaller waves result from the interplay between the stability of the fluid and the shear, frequently occurring on a surface inside the fluid across which shear is strong. If the Richardson number should fall to a value of about 0.25 the surface becomes unstable and is deformed into waves, which grow spontaneously. These waves are known as Kelvin–Helmholtz (KH) (or billow) waves, and should they eventually "roll up" and break, the phenomenon is known as Kelvin–Helmholtz instability (KHI). In recent years, investigation of both KH waves and KHI has proceeded apace, refining the original theories suggested by the eminent scientists who gave their names to the phenomena. But the very nature of KH waves means that such theoretical developments have not yet solved the main problems posed by the waves that are the central concern of this chapter. The problem is twofold:

(1) Why are the waves so long?
(2) Are they really gravity as opposed to shear waves?

In the context of the first question we should note that KH wavelengths are of the order of hundreds of metres, whereas Table 39 reveals wavelengths of 30–300 km. The second question arises from Thorpe's (1978) careful experiments into waves in sheared flows. Some of his results are presented below.

Thorpe (1978) generated beautiful KH waves on the interface between layers of dense brine and fresh water by tilting and then quickly restoring to the horizontal the tube containing the fluids. Figure 112 shows the situation when the tube has been returned to the horizontal position after being tilted down to the left, shortly after the value of the Richardson number reached 0.25 but before the waves have had time to grow appreciably. A second experiment yielded interesting differences to this classic picture. The same apparatus was used but a train of waves was generated artificially by a wavemaker at the right-hand end of the tube before it was tilted. The internal motion in these waves was clearly in accord with that in the vertical-transverse waves and as such they were "true" gravity waves. When the tube was tilted down to the right and shear flow generated between the downward flowing brine and the upward flowing water, the waves continued to propagate, although they changed their shape and speed much as do waves at sea as they approach a beach. Unlike the KH waves, however, they

7 MOVING GRAVITY WAVES

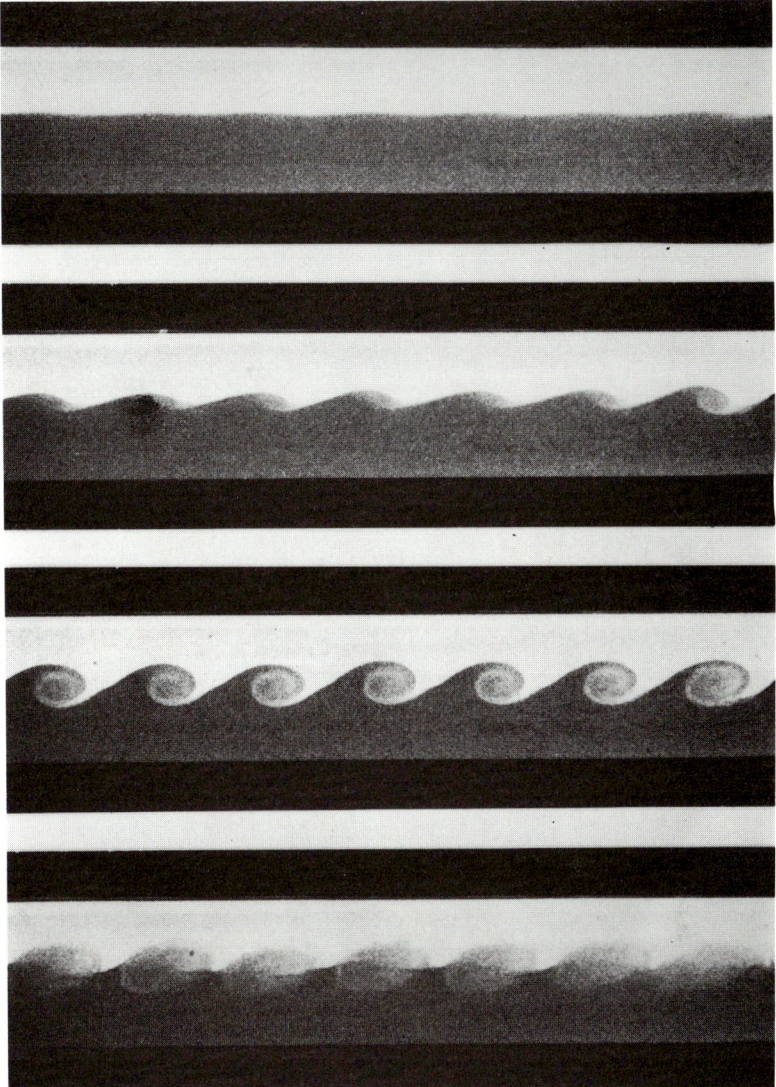

FIG. 112. Kelvin–Helmholtz waves in a laboratory tube. (After Thorpe, 1979.)

propagated with a velocity which lay outside and not between the velocities of the layers. When the tube was returned to the horizontal at an early stage in the experiment, the waves continued to propagate without further change in form. If the tube remained tilted, so that the shear continued to increase, the waves "broke" at their crests (Fig. 113) releasing jets of brine that moved

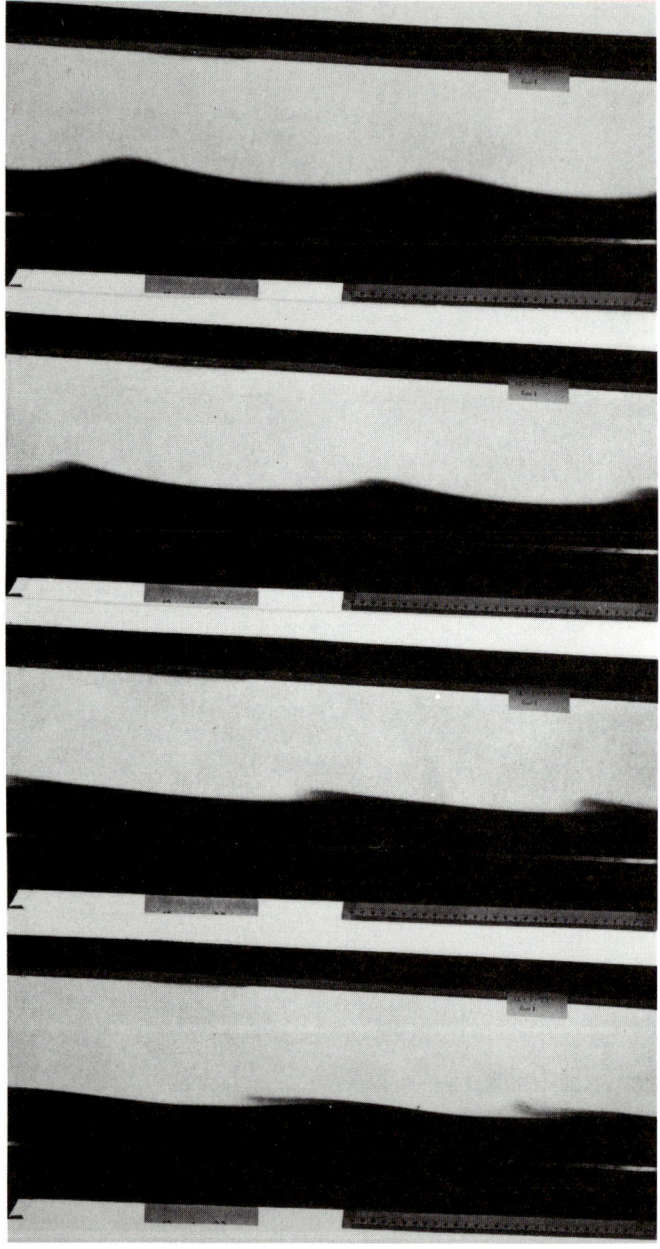

Fig. 113. Gravity waves affected by shear in a laboratory experiment. (After Thorpe, 1979.)

forward more rapidly than the waves were advancing. The fluid in the jets was heavier than that below and was thus unstable. Thorpe noted that this breaking could occur before the Richardson number of the mean flow fell below the critical value for KHI and that it was distinct both in appearance and form from KHI. There was little or no increase in wave amplitude as breaking occurred and there was no "roll up". The difference between the two results is summed up by Thorpe as follows: in the case of KHI, "the basic 'ingredient' is a sufficiently strong shear"; in the second case "the ingredients... are waves and an increasing shear" (Thorpe, 1979). If the second type is of any relevance to real situations we must discover natural atmospheric "wave-makers".

Notwithstanding Thorpe's results there is a wide belief in the literature that shear itself is indeed a "wave-maker" and that KH waves are a type of gravity wave. But the literature is vague about the precise way in which a shear comprising purely horizontal flow may generate wavy motion in the absence of any other factor. Until this is solved we are left with the undoubted observation that shear and waves frequently co-exist, together with a mass of

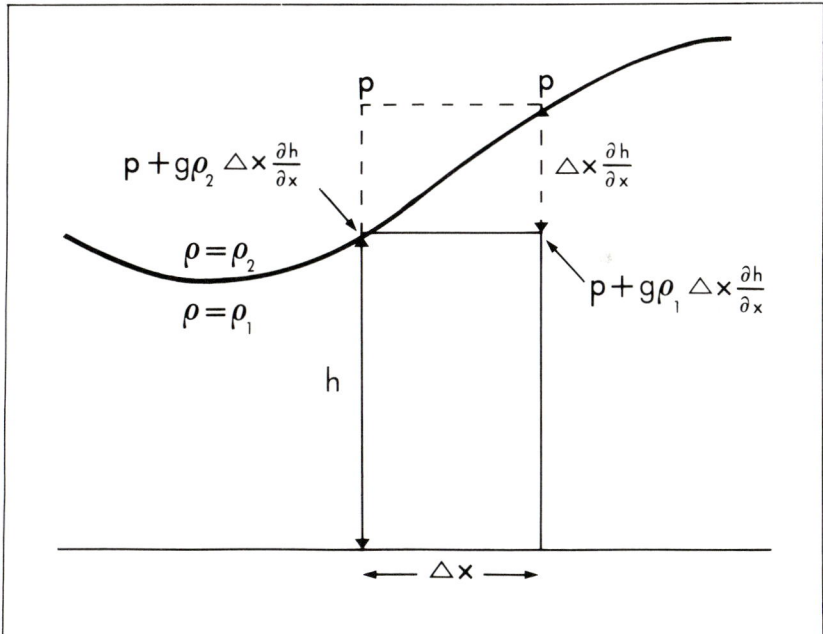

FIG. 114. A vertical cross-section in the (x, z) plane showing the gradient of pressure in a gravity wave due to variations in the height of a density discontinuity. (After Thompson, 1961.)

theory about waves on interfaces and within sheared flow that appears to "explain" some of these observations.

Putting aside for the moment these problems of nomenclature and mechanism, the character of gravity waves is now investigated, drawing heavily on Thompson (1961). By considering only the vertical (x, z) plane, horizontal-transverse waves are excluded; and similarly, by regarding the atmosphere as incompressible, compression or sound waves are excluded. If the atmosphere is considered to comprise two homogeneous layers of fluid with uniform densities ρ_1 and ρ_2 separated by a surface of discontinuous density, a portion of a gravity wave may be schematically shown as in Fig. 114. Using the hydrostatic assumption Thompson inferred that the horizontal pressure gradient in the lower layer is

$$\frac{\partial p}{\partial x} = g(\rho_1 - \rho_2)\frac{\partial h}{\partial x}, \tag{182}$$

where h is the height of the surface of density discontinuity. Recalling the equation of motion for the u-component,

$$\frac{du}{dt} - fv + \frac{1}{\rho}\frac{\partial p}{\partial x} = 0, \tag{183}$$

setting both v and $\partial u/\partial z$ equal to zero and substituting from equation (182) for $\partial p/\partial x$, the following equation for the lower layer results:

$$\frac{\partial u}{\partial t} + u\frac{\partial u}{\partial x} + g\left(1 - \frac{\rho_2}{\rho_1}\right)\frac{\partial h}{\partial x} = 0. \tag{184}$$

By integrating the continuity equation,

$$\frac{\partial u}{\partial x} + \frac{\partial w}{\partial z} = 0, \tag{185}$$

with respect to height from $z = 0$ to $z = h$ and making some reasonable assumptions, Thompson showed that

$$\frac{\partial h}{\partial t} + u\frac{\partial h}{\partial x} + h\frac{\partial u}{\partial x} = 0. \tag{186}$$

Equations (184) and (186) constitute a complete system in which the variables are u and h. These equations were solved by the perturbation method, with an equilibrium state in which $u = \bar{u}$, $h = H$, so that $\partial \bar{u}/\partial t = \partial \bar{u}/\partial x = \partial H/\partial t = \partial H/\partial x = 0$. The resultant wave solution required that the phase speed c is given by

$$c = \bar{u} \pm \left[gH\left(1 - \frac{\rho_2}{\rho_1}\right)\right]^{1/2}. \tag{187}$$

Thompson noted that, for values of H ranging from 1 to 10 km and for values of ρ_2/ρ_1 ranging from 0.90 to 0.99, the speed of the internal gravity waves

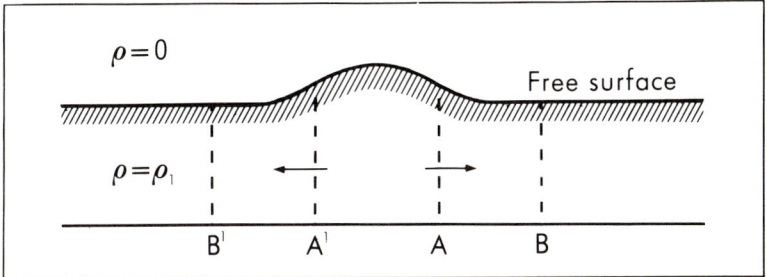

FIG. 115. The set-up for an experiment illustrating the propagation of gravity waves. Fluid of uniform density is originally held at rest by placing a rigid form over its surface, but is later left free to move by taking away the form. (After Thompson, 1961.)

relative to the medium varies from 9 to 90 m s^{-1}, depending on the vertical variation in density. When the wind is very strong, the speed of the gravity waves relative to the ground may be as great as 150 m s^{-1}. In common with sound waves, gravity waves may travel in both directions at once.

As regards the physical mechanism responsible for the existence and propagation of gravity waves we can do little better than quote Thompson (1961, pp. 58–59):

> ...let us imagine a long deep channel containing a layer of fluid of finite density ρ_1, surmounted by a fluid of zero density, as shown by the cross-section in [Fig. 115]. The fluid is initially at rest, but the interface between the fluids of density ρ_1 and zero is deformed as shown in the figure, by placing a rigid form over the surface. Now suppose that the form is suddenly removed. What motions are induced in the lower layer of fluid?
>
> Since a greater depth of fluid lies to the left of the line A than to the right, the mass above a point to the left of A and lying on any given horizontal surface is greater than the mass above a point to the right of A and lying on the same horizontal surface. Thus, according to the hydrostatic principle, the horizontal pressure gradient along A is directed toward the right and accelerates the fluid in that direction. At a very short time Δt later, therefore, the fluid along A is moving toward the right, while the fluid along the line B is still at rest. This implies, however, that the total transport of fluid across A is more rapid than across B. Thus, since the fluid is incompressible, the interface between A and B must be rising at time Δt. At a slightly later time, then, the pressure gradient along B is directed toward the right and accelerates the fluid at B in that direction, and so on. Thus, the original deformation of the interface is propagated to the right of A and, by symmetry, to the left of A'. The process described here is exactly the one expressed mathematically in [equations (184) and (186)].

In both oceanography and meteorology at least two types of gravity wave are recognized. Gravity waves at the surface of the sea are constrained to propagate horizontally. In their elementary form their vertical variation consists of an exponential decay of amplitude away from the interface upon which they form. Such waves have been called "evanescent" (Hines, 1972). Similar waves are often found on a density discontinuity within the ocean.

Whenever the density of the water varies continuously with height, "internal" waves may form. Their distinction from the evanescent waves is that they are no longer constrained to horizontal propagation but may progress obliquely upwards or downwards in the fluid. As the atmosphere displays a continuous decrease of density with height it is able to support gravity waves of both types, but particularly internal ones. Hines showed that an intrinsic feature of internal gravity waves in the atmosphere is that they tend to grow stronger the further they rise above their source. This occurs because the waves carry energy upwards or downwards to regions of different density. Upward propagation carries energy to regions where fewer gas molecules are available to carry the energy onwards and these molecules achieve the task only by oscillating with greater magnitude.

The bulk of the evidence outlined so far strongly suggests that wind shear is an important aid to the generation and maintenance of meso-scale gravity waves. This argument is further developed below but it is important to recognize that other features may induce gravity waves. Notably, severe convection may force waves at upper tropospheric levels (Uccellini, 1975; Bosart and Cussen, 1973); cold downdraughts similarly may cause low level waves of the type observed by Pothecary (1954): Jordan (1972) claimed that typical synoptic weather fronts may have the same effect; and finally, topography can, of course, trigger gravity waves that move relative to the ground. But of all the factors that probably spawn meso-scale moving gravity waves, the jet stream has received most attention.

In the first part of this chapter we saw that surface pressure fluctuations as measured on the microbarograph were important for the description of gravity waves. These lower atmosphere pressure fluctuations at the ground, with periods of a few minutes to several hours and horizontal wavelengths of tens of metres to a few hundred kilometres, have been successfully correlated to gravity waves generated by the previously mentioned sources. In particular, Flaurand et al. (1954), Madden and Claerbout (1968), Claerbout and Madden (1968), Cook (1968), Tolstoy (1968), Herron and Tolstoy (1969), Tolstoy and Herron (1969), Herron et al. (1969), Keliher (1975) and others have presented evidence (Fig. 108) that pressure fluctuations monitored with an array of microbarographs at the ground can be related to the overhead tropospheric jet stream. The correlation consisted mainly of showing that the horizontal disturbance phase velocities of the order of $10–100 \text{ m s}^{-1}$ were well within the range of velocities of the jet stream aloft and that the direction of the pressure disturbance roughly coincided with that of the jet stream. Herron and Tolstoy (1969) showed that a good correlation can persist for weeks between the direction of the jet stream winds and the pressure fluctuations recorded by the microbarograph array. The order of magnitude and power spectra for microbarographic fluctuations (in the 5–60 min period range) that would be expected on the ground due to disturbances of known spectra in the

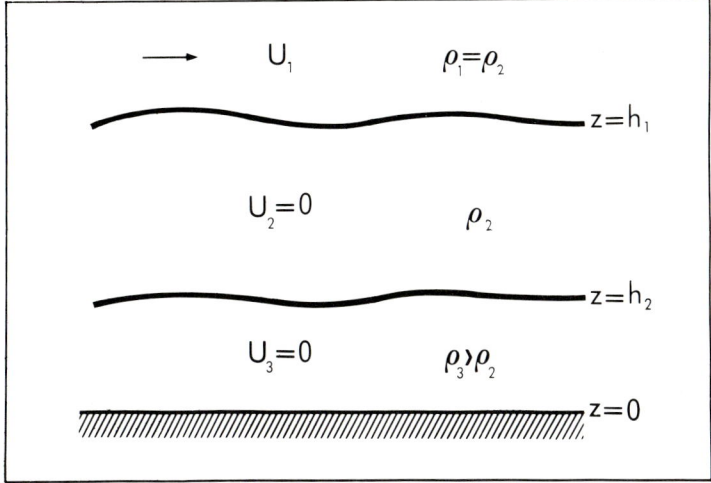

FIG. 116. A three-layer fluid model for the analysis of gravity waves. At the upper interface there is a velocity discontinuity while at the lower surface there is a density discontinuity. (After Gedzelman and Rilling, 1978.)

jet stream (Kao and Woods, 1964) have been calculated by Tolstoy and Herron (1969) by means of a linear model and were found to be in reasonable agreement with observations. In a subsequent analysis Mastrantonio *et al.* (1976) confirmed the idea that a tropospheric jet stream can in general support meso-scale gravity waves.

Theoretical modelling of this type of wave is still very much in its infancy. The few existing models are primarily concerned with deriving analytical expressions for the surface pressure perturbations that reflect the passage of gravity waves at higher levels. The model developed by Gedzelman and Rilling (1978) is typical of this approach. In essence they modelled an atmosphere that contained high level shear and a low level stable layer—a situation that they had observed to be particularly favourable for the ground level measurement of surface effects of upper level gravity waves. The model configuration (Fig. 116) comprised a three-layer fluid with a velocity difference at the upper interface and an inversion at the lower interface. At all other levels the fluid had no shear and was assumed isentropic.

The governing equation for linearized perturbations of an inviscid liquid in shearing flow was derived initially by Taylor (1931); an abbreviated derivation appears in Gossard and Hooke (1975). The equation, as used by Gedzelman and Rilling (1978) was as follows:

$$[k^2(U-c)^2 - N^2]W = (U-c)\frac{d}{dz}\left\{\left[-\frac{dU}{dz} + (U-c)\frac{d}{dz}\right]W\right\}. \quad (188)$$

In this equation wave solutions of the form $e^{ik(x-ct)}$ were assumed, where U is the background wind speed, c is the complex wave speed and W represents $\rho^{1/2}w$. In order to simplify the problem Gedzelman and Rilling (1978) assumed $N^2 = 0$ and $d^2U/dz^2 = 0$ except at the appropriate interfaces. Generally the term N^2 was of the same order of magnitude as the term $k^2(U-c)^2$ for most of the waves they observed. With these assumptions, equation (188) reduced to

$$\frac{d^2W}{dz^2} - k^2 W = 0. \tag{189}$$

This equation generates solutions of the form $W = A e^{kx} + B e^{-kx}$ and the constants and complete wave speed were determined by the boundary conditions that there sould be no flow across the interfaces, that the vertical velocity approached zero at the ground and remained finite as z tended to infinity and, finally, that the pressure was continuous across each interface. Setting $\rho_1 = \rho_2$ and $U_2 = U_3 = 0$ and retaining $\rho_2 \neq \rho_3$ for the low level inversion and $U_1 \neq U_2$ for the high level shear allowed the derivation of a fourth-order equation in c, the wave speed. By judicious approximations this equation was reduced to a quadratic that yielded the following approximate solution for the growing root:

$$c \simeq \tfrac{1}{2}U_1[(1 - e^{-2kh_1}) + (e^{-4kh_1} - 1 + G'')^{1/2}], \tag{190}$$

where

$$G'' \equiv \frac{2g(\rho_3 - \rho_2)}{U_1^2 \bar{\rho} k} \frac{(e^{kh_2} - e^{-kh_2})^2}{(e^{kh_1} - e^{-kh_1})e^{kh_1}}. \tag{191}$$

Equation (190) is a good approximation whenever $G'' \ll 1$. In fact G'' approaches unity only when the inversion is both strong and at a height near the shear layer.

Gedzelman and Rilling further developed the model to show what factors affected the magnitude of the surface-measured pressure perturbations. They expressed the pressure perturbation as follows:

$$P = \text{Real}\left[\rho_0^{1/2} \frac{(U-c)}{k} \frac{dW}{dz} e^{ik(x-ct)}\right], \tag{192}$$

where ρ_0 is the undisturbed density value. After computing the pressure perturbation at the ground $(P(0))$ for a given pressure perturbation at the shear level $(P(h_1))$ they found that

$$\frac{P(0)}{P(h_1)} = \left[\frac{\rho_0(0)}{\rho_0(h_1)}\right]^{1/2} 2e^{-kh_1} \frac{A\cos k(x-ct) + B\sin k(x-ct)}{C\cos k(x-ct) + D\sin k(x-ct)}. \tag{193}$$

Evaluation of the constants A, B, C, D and the assumption that $G'' \ll 1$ led to the following first-order approximation for the magnitude of the pressure perturbation at the surface:

$$|P(0)| \simeq \left[\frac{\rho_0(0)}{\rho_0(h_1)}\right]^{1/2} 2e^{-kh_1}\left[1 + \left\{\frac{g(\rho_3-\rho_2)}{\bar{\rho}kU_1^2}\right\}^2\right]^{1/2} |P(h_1)|. \quad (194)$$

In Equation (194) the factor e^{-kh_1} provides a so-called natural atmospheric filter. Because of this filter the surface pressure perturbations of waves less than 5 km in length which are generated at mid- and upper tropospheric heights are so reduced at ground level that they are completely masked by the background noise. As Gedzelman and Rilling noted, this explains why they had not detected any pressure perturbation from such features as short wavelength, transverse altocumulus cloud bands. Equation (194) also confirmed the view that surface-measure gravity waves tend to be larger when there is large, low level static stability. In a comparison with the case studied by Hooke and Hardy (1975), equation (194) predicted a surface pressure perturbation of 49 µb as opposed to observed values of 100–120 µb. In view of the relative simplicity of the model the discrepancy is not unreasonable. Gedzelman and Rilling recognized that the neglect of the Brunt–Väisälä frequency was a disadvantage, particularly because it required the model to give evanescent solutions. In reality, large N^2 or stratification tends to be associated with internal waves and this leads to a larger amplitude signal at ground level. To counteract this deficiency Gedzelman and Rilling (1978) noted that the low level stability acts as a "sounding board", an effect which increases the surface pressure perturbation.

The comparative brevity of this chapter reflects our ignorance of travelling meso-scale gravity waves. Whereas the smaller "true" Kelvin–Helmholtz waves have received close attention over the last two decades, and indeed continue to do so, their larger meso-scale relatives are, as yet, both inadequately observed and understood.

Consequently we remain confronted by the problems of the size and mechanisms of these waves. In addition we know very little about the dispersion of such waves, their role in momentum transfer and their possible effects on weather. Elucidation of these features will no doubt emerge in the 1980s — a decade which may see meso-scale meteorology as a whole taking a central position within atmospheric science.

References

Ali, B. (1931). A remarkable instance of waves in an anemograph trace. *Q. Jl R. met. Soc.*, **57**, 300–303.

Axford, D. N. (1970). An observation of gravity waves in shear flow in the lower stratosphere. *Q. Jl R. met. Soc.*, **96**, 273–286.

Balachandran, N. K. and Donn, W. L. (1964). Short and long period gravity waves over the north east United States. *Mon. Weath. Rev.*, **92**, 423–426.

Bosart, L. F. and Cussen, J. P. (1973). Gravity wave phenomena accompanying east coast cyclogenesis. *Mon. Weath. Rev.*, **101**, 446–454.

Brunt, D. (1927). The period of simple vertical oscillations in the atmosphere. *Q. Jl R. met. Soc.*, **53**, 30–32.

Claerbout, J. F. and Madden, T. C. (1968). Electromagnetic effects of atmospheric gravity waves. In *Proceedings of the Acoustic–Gravity Waves Symposium.* (T. M. Georges, ed.), US Government Printing Office, Washington, D.C., pp. 135–155.

Cook, R. H. (1968). Subsonic atmospheric oscillations. In *Proceedings of the Acoustic–Gravity Waves Symposium* (T. M. Georges, ed.), US Government Printing Office, Washington, D.C., pp. 209–213.

Cunning, J. B., Jr (1974). The analysis of surface pressure perturbations within the meso-scale range. *J. appl. Met.*, **13**, 325–330.

Curry, M. J. and Murty, R. C. (1974). Thunderstorm gravity waves. *J. atmos. Sci.*, **31**, 1402–1408.

Emmanuel, C. B., Bean, B. R., McAllister, L. E. and Pollard, J. R. (1972). Observations of Helmholtz waves in the lower atmosphere with an acoustic sounder. *J. atmos. Sci.*, **29**, 886–892.

Eom, J. (1975). Analysis of the internal gravity wave occurrence of 19 April 1970 in the mid-west. *Mon. Weath. Rev.*, **103**, 217–226.

Erickson, C. O. and Whitney, L. F., Jr (1973). Gravity waves following severe thunderstorms. *Mon. Weath. Rev.*, **101**, 708–711.

Flaurand, E. A., Mears, A. H., Crowley, F. A., Jr and Crary, A. P. (1954). Investigations of microbarometric oscillations in eastern Massachusetts. Technical Report no. 54-11, Geophysical Research Paper no. 27, Air Force Cambridge Research Centre.

Gedzelman, S. D. and Rilling, R. A. (1978). Short-period atmospheric gravity waves: a study of their dynamic and synoptic features. *Mon. Weath. Rev.*, **106**, 196–210.

Gossard, E. E. and Hooke, W. H. (1975). Waves in the atmosphere. Atmospheric infrasound and gravity waves: their generation and propagation. *Devlmts Atmos. Sci.*, **2**.

Gossard, E. and Munk, W. (1954). On gravity waves in the atmosphere. *J. Met.*, **11**, 257–269.

Gossard, E. E. and Richter, J. H. (1970). The shape of internal waves of finite amplitude from high-resolution radar sounding of the lower atmosphere. *J. atmos. Sci.*, **27**, 971–973.

Herron, T. J. and Tolstoy, I. (1969). Tracking jet stream winds from ground level pressure signals. *J. atmos. Sci.*, **26**, 266–269.

Herron, T. J., Tolstoy, I. and Kraft, D. W. (1969). Atmospheric pressure background fluctuations in the meso-scale range. *J. geophys. Res.*, **74**, 1321–1329.

Hicks, J. J. and Angell, J. K. (1968). Radar observations of breaking gravitational waves in the visually clear atmosphere. *J. appl. Met.*, **7**, 114–121.

Hines, C. O. (1972). Gravity waves in the atmosphere. *Nature, Lond.*, **239**, 73–78.

Hooke, W. H. and Hardy, K. R. (1975). Further study of the atmospheric gravity waves over the eastern seaboard on 18 March 1969. *J. appl. Met.*, **14**, 31–38.

Hooke, W. H., Young, J. M. and Beran, D. W. (1972). Atmospheric waves observed in the planetary boundary layer using an acoustic sounder and a microbarograph array. *Boundary Layer Met.*, **2**, 371–380.

Johnson, N. (1929). Atmospheric oscillations shown by the microbarograph. *Q. Jl R. met. Soc.*, **55**, 19–30.

Jordan, A. R. (1972). Atmospheric gravity waves from winds and storms. *J. atmos. Sci.*, **29**, 445–456.

Kao, S.-K. and Woods, H. D. (1964). Energy spectra of meso-scale turbulence along and across the jet stream. *J. atmos. Sci.*, **21**, 513–519.

Keliher, T. E. (1975). The occurrence of microbarograph-detected gravity waves compared with the existence of dynamically unstable wind shear layers. *J. geophys. Res.*, **80**, 2967–2976.

Madden, T. R. and Claerbout, J. F. (1968). Jet-stream associated gravity waves and implications concerning jet stream stability. In *Proceedings of the Acoustic–Gravity Waves Symposium* (T. M. Georges, ed.), US Government Printing Office, Washington, D.C., pp. 121–134.

Mastrantonio, G., Einaudi, F., Fua, D. and Lalas, D. P. (1976). Generation of gravity waves by jet streams in the atmosphere. *J. atmos. Sci.*, **33**, 1730–1738.

Merceret, F. J. and Black, P. G. (1975). Low altitude internal waves north of the GATE array. *Mon. Weath. Rev.*, **103**, 167–169.

Pothecary, I. J. W. (1954). Short-period variations in surface pressure and wind. *Q. Jl R. met. Soc.*, **80**, 395–401.

Reed, R. J. and Hardy, K. R. (1972). A case study of persistent, intense clear air turbulence in an upper level frontal zone. *J. appl. Met.*, **11**, 541–549.

Taylor, G. I. (1931). Effect of variation in density on the stability of superposed streams of fluid. *Proc. R. Soc. A*, **132**, 499–523.

Thompson, P. D. (1961). *Numerical Weather Analysis and Prediction*, Macmillan, London.

Thorpe, S. A. (1978). On the shape and breaking of finite amplitude internal gravity waves in a shear flow. *J. Fluid Mech.*, **85**, 7–31.

Thorpe, S. A. (1979). Instability and waves. *Weather*, **34**, 102–105.

Tolstoy, I. (1968). Meso-scale pressure fluctuations in the atmosphere. In *Proceedings of the Acoustic–Gravity Waves Symposium* (T. M. Georges, ed.), US Government Printing Office, Washington, D.C., pp. 107–120.

Tolstoy, I. and Herron, T. J. (1969). A model for atmospheric pressure fluctuations in the meso-scale range. *J. appl. Sci.*, **26**, 270–273.

Uccellini, L. W. (1975). A case study of apparent gravity wave initiation of severe convective storms. *Mon. Weath. Rev.*, **103**, 497–513.

Wagner, A. J. (1962). Gravity wave over New England April 12, 1961. *Mon. Weath. Rev.*, **90**, 431–436.

Yamamoto, R. (1957). A study of microbarographic waves (III). Part II. Verification of the various theories by observations. *J. met. Soc. Jap.*, **35**, 26–36.

Part III

Free atmosphere circulations

B. Convective circulations

8

Severe Local Storms

I. Introduction

Of all the meso-scale circulations considered in this book, severe local storms are most widely recognized as having pride of place. Not only did they provide the context for Ligda's (1951) first use of the prefix "meso-", but their violent and spectacular nature has ensured a massive interest over the last three decades, particularly by British and North American meteorologists. In addition to their intrinsic interest as atmospheric phenomena, they demand attention through their causing, for example, an annual average of 100 deaths and 2000 injuries in the United States alone (Atlas, 1976).

There is no set definition of a "severe local storm", or indeed of a thunderstorm. Yet there is wide agreement that giant cumulonimbus clouds which give lightning, thunder, heavy precipitation (often including hail) and high winds provide the essential object of study (Winston, 1956). As indicated by Ludlam (1963) and Barnes (1976) our earliest understanding of thunderstorms came as a result of visual observations, a few ground-based measurements and a large amount of deduction and speculation. Ludlam (1963) provided a most illuminating review of the different models of cumulonimbus ranging from the characteristically morphologically accurate picture by Davis (1894) to the (then) recently introduced Browning and Ludlam model. Within the 70 years spanning these two models the most important observational study of thunderstorms was the Thunderstorm Project (Byers and Braham, 1949) which gathered data on both Florida and Ohio thunderstorms. In Ludlam's words, the results "... did not so much modify previous concepts as for the first time support them with measurements of the distribution of up- and down-draughts, temperature and precipitation type" (Ludlam, 1963,

p. 5). In addition the project clarified the life cycle of thunderstorms. Barnes (1976, p. 414) summarized the project's main results as follows:

(1) A thunderstorm is composed of one of more cells each of which evolves through three stages: cumulus, in which the air currents are all upward and precipitation begins to fall; mature, with warmer updraught and colder downdraught existing side by side and precipitation reaching the surface; and dissipating, wherein the storm has abated and a weakening cold downdraught prevails throughout the cell.

(2) Surface weather conditions associated with the three stages were found to be equally distinctive. Gently converging moist warm winds accompany the cumulus stage. As the storm matures, the down rush of precipitation-cooled air forms a micro-cold front, a zone of abrupt change in temperature and wind at the leading edge of the outrushing air. As the storm decays, the cold dome of downdraught air subsides and winds decrease. Then conditions at the surface gradually return to what they were before the storm developed, except in cases when thunderstorms are associated with passages of large scale fronts commonly depicted on conventional weather maps.

(3) The lifetime of individual cells in a thunderstorm is on the order of 1 h during which time the storm moves 10–20 mi (16–32 km), on average, in the direction of the upper winds in which it is embedded.

Despite these valuable results many questions about the structure of mid-latitude severe local storms remained unanswered, leading to several massive observational programmes that have attempted to provide some answers in the following three decades. We owe most of our observational evidence to four projects, namely those run by Imperial College, London, between 1959 and 1964, the National Severe Storms Laboratory (NSSL) in Norman, Oklahoma, since the early 1960s, Alberta Research (including the Atmospheric Environment Service, the National Research Council of Canada, and the Stormy Weather Group of McGill University) since 1957 and the National Centre for Atmospheric Research, particularly the National Hail Research Experiment (NHRE), since 1972. Of these four the first two were primarily concerned with the dynamical structure of cumulonimbi whereas the last two were more interested in the physics of hail growth. Clearly, as Ludlam has indicated on several occasions, notably in his 1958 paper (Ludlam, 1958), the micro-physics of precipitation growth and the dynamics of the parent cloud are closely linked, but the different emphases of the projects are reflected in their publications, which are voluminous. In addition to this flood of literature, the American Meteorological Society sponsors conferences on both severe local storms and radar meteorology, which, over the last 20 years, have produced over 25 conference volumes.

The observation and analysis of severe local storm behaviour benefited immensely from the widespread use of weather radar which became possible in the 1950s. The fruits of radar observation of the atmosphere are reviewed by Battan (1973) and more recent developments, such as the value of doppler radar to the study of severe local storms, are covered by Lhermitte (1964) and

Battan (1975). Despite the inability of conventional radar to "see" inside a cloud, it is fair to say that the bulk of our understanding of internal flow of cumulonimbus is derived from the careful scrutiny of many PPI, CAPPI and RHI pictures. The conference volumes mentioned above abound with studies of convective radar echoes and Donaldson's (1958, 1959) papers typify their value. In these studies Donaldson showed that a high frequency of hail occurrence at the Earth's surface was associated with storms whose radar echo attained high altitudes and large reflectivities of 10^5 mm^6 m^{-3}.

Contemporaneous with the blossoming of "radar meteorology" were the beginnings of aircraft observation of severe storms (e.g. Fujita and Arnold, 1963) (limited primarily by lack of finance) and the detailed analysis of surface data from dense networks (Fujita *et al.*, 1956; Fujita, 1963). Fujita (1955) and Pedgley (1962) have produced classic examples of meso-scale analysis, in which time traces from autographic instruments are converted into spatial records by the reasonable assumption of constant meso-system velocity.

Roughly a decade of observation and analysis meant that by the early 1960s cumulus and cumulonimbus convection were appreciated to be quite different (Ludlam, 1966), the latter seeming to thrive in strong wind shear (of both speed and direction), the former being inhibited by such shear. In cumulus convection the primary element is the thermal, a volume of air which "turns itself inside out" as it rises and expands, thus thoroughly mixing its own warmer air with the ambient colder air. The familiar "cauliflower"-shaped cumulus cloud comprises many such thermals. Once the cumulus cloud produces a shower, usually when cloud top is in mid-troposphere, a marked transformation occurs, particularly in the size and vigour of the parent cloud. Within half an hour of precipitation formation successive towers penetrate into the high trophosphere and the more intense and prolonged convection leads to their persistent glaciated residues spreading out into a mushroom or anvil cloud characteristic of the mature cumulonimbus (Fig. 117). This has a shape quite different from that of the typical large cumulus, because the admixture of the cold clear air of the upper troposphere into the thermals is very inefficient in evaporating their condensed water and removing their buoyancy, so that they spread out and persist after reaching an equilibrium level.

The existence of large, long lived, severe thunderstorms in strongly vertically sheared winds became increasingly clear as a result of the many radar studies in the 1950s. This phenomenon led to the conception of several models around the turn of the decade yet remains one of the major problems of severe storm investigation. The analysis of storms in sheared environments forms a major part of this chapter.

The progress from 1949 to 1963 was thoroughly reviewed by Atlas *et al.* (1963). By that date several two-dimensional models of cumulonimbus had been suggested (e.g. Ludlam, 1961) together with some plausible dynamical analyses (e.g. Newton, 1963). In the ensuing decade the major advances included the construction of both observational and theoretical three-dimensional models (Newton, 1967) and the first steps towards direct observation of internal structure by doppler radar. Even satellites became valuable tools in meso-scale analysis (e.g. Ninomiya, 1971). Yet problems remained. Lilly (1975a) emphasized our ignorance of the relationships between individual cumulonimbus cells on the one hand and both meso-scale agglomerations of cells and other associated meso-scale circulations on the other. The limited nature of theoretical results was also clarified. In concluding his review Lilly posed several questions under the following general headings: (a) What are the most important synoptic and meso-scale factors leading to formation and guiding the propagation of a severe storm array? (b) What is the entire sequence of events leading to tornado formation and other damaging winds? Both questions cover the problems of relationships between storms and larger circulations (a) on the one hand and small circulations (b) on the other. They are supplemented by Ludlam's (1976) deceptively simple questions about the existence, growth, location, intensity and travel of cumulonimbus clouds in his perceptive comments on the planned SESAME programme (Lilly, 1975b).

This chapter reviews our present knowledge of the severe local storm circulation. For convenience we accept Winston's (1956) definition of severe local storms as thunderstorms accompanied by very strong surface winds or large hailstones. The word "local" distinguishes these storms from cyclones. With this definition we find that many "hailstorms" may be accurately called severe local storms. Despite their regular occurrence in the tropics our knowledge of such local storms is rudimentary and most of what follows relies heavily on observations of extra-tropical storms.

II. Observation

A. Climatology

The relative transience of convective systems inhibits an easy climatological compilation of data. Rather than noting each individual thunder- and hailstorm, operational observations procedures require the recording of the number of days on which any such activity occurred. Consequently there is no complete record of, for example, duration and intensity of the convective activity. Nevertheless, such necessarily simple observations of occurrence only do provide interesting insights into the climatology of severe convective activity. As this chapter is primarily concerned with severe local storms we

Table 41 Monthly mean number of days with hail

	Month												
	Jan	Feb	Mar	Apr	May	Jun	Jul	Aug	Sep	Oct	Nov	Dec	Year
(a) *Tropics*													
West Kenya†	5	6	13	9	9	13	10	15	15	10	5	4	114
(b) *Mid-latitudes*													
UK‡	1	1.1	1	0.8	0.6	0.3	0.1	0.1	0.1	0.3	0.6	0.9	—
Green Bay, Wisconsin§	0	1	8	12	20	20	8	12	12	5	3	1	102
Alberta‖	—	—	—	—	4.5	18.8	20.5	14.5	3.1	—	—	—	—
Illinois¶	1.1	1.8	5.4	5.9	7.6	5.4	3.8	2.9	1.5	1.7	1.6	0.4	39.1

—, No information.
† *Source:* Sansom (1966). Period of observations: 1960–64.
‡ *Source:* Champion (1948). Period of observations: 1937–46.
§ *Source:* Burley et al. (1964). Period of observations: 1887–1962.
‖ *Source:* Wojtiw (1975). Period of observations: 1957–73.
¶ *Source:* Huff and Changnon (1959). Period of observations: 1925–48.

restrict our climatological attention to those storms that produce hail. Using this criterion, some interesting distributions are found.

As Sansom (1966) noted, the incidence of thunderstorms over parts of Africa is as high or higher than anywhere else in the world, but reports of hail are relatively rare. Appleman (1959) suggested the existence of a natural "hail suppression mechanism" in certain areas of the world and at certain seasons. He postulated that when a thick, dense layer of cloud exists between cloud base and the freezing level, the rising large droplets will often grow to precipitation size and fall out as rain before freezing, thus inhibiting significant hail formation. Appleman's idea is particularly relevant to the tropics. Nevertheless, Table 41 shows that in western Kenya hailstorms occurred on roughly one day in three, with a slightly higher frequency between July and September. Throughout the remainder of Africa the average frequency was 0.2–1.0 hail day per year. In continental mid-latitude areas (as exemplified by Green Bay, Alberta and Illinois, see Table 41(b)) hail occurred more frequently, occasionally being of giant (i.e. golf ball to grapefruit) size. Ludlam (1963) suggested that the differences in airflow between tropical and mid-latitude storms largely explains the occurrence of very large hail in latter areas and its absence in the former areas. Maritime mid-latitude areas (as exemplified by the UK in Table 41(b)) experienced far fewer hailstorms. In all areas hailstorms tended to have maximum frequency in the mid-afternoon. Hailfall duration was usually less than 5 min (Burley *et al.*, 1964; Wojtiw, 1975) but occasionally extended to 25 min, as in Wisconsin, and 40 min, as in Alberta. Data on the duration of severe local storms are beset with problems of the definition of starting and ending times and defy comprehensive climatological analysis.

B. Extra-tropical severe local storms

1. *Synoptic background*

Atmospheric convection results from the release of vertical instability and the severity of the convection depends partially upon the magnitude of the instability and the particular method of release. Armed with this knowledge, the problems of predicting cumulonimbus development appear to have simple answers, particularly if the distribution of instability is known. Indeed, early investigations of cumulonimbus laid strong emphasis on the discovery of relationships between thundery activity and stability distribution on the synoptic scale. In practice this frequently meant establishing links between the severe weather and various familiar synoptic structures—in particular, fronts and air masses. For several decades thunderstorms were simply classified as being of the air-mass or frontal type. In the former case the instability was considered to be caused by insolation whereas in the latter case, frontal uplift was considered to release instability in an unspecified manner. Namias (1940) provided a useful summary of this classification.

Attempts to clarify further the relationships between storms and their parent synoptic scale circulations took on two basic forms: the analysis of vertical soundings of temperature and humidity which were known to produce severe convection; and the more detailed inspection of thundery outbreaks in spatial relation to the distribution of several features of cyclones. Many examples of the former approach exist: only a few are cited here. Chalker (1949) showed that shower occurrence depended upon not only instability, but also the relative humidity of the lower troposphere. Fawbush and Miller (1954) and Beebe (1958) analysed the type of air mass in which tornadoes (which of course require severe convective activity) may form. In both cases the aim was to provide information of value to the forecaster. Closely associated with this approach was that leading to the derivation of "instability indexes", the most notable being that devised by Showalter (1953). The index is defined as follows: 500 mb temperature minus the temperature of a parcel of air lifted adiabatically to the 500 mb level from 850 mb. Negative values indicate potential instability.

Most examples of the second approach employed in these early synoptic studies were embodied in forecasting research. They are well exemplified by Winston (1956), Means (1952), Sartor (1962), Porter et al. (1955). In all these studies, emphasis lay upon the distribution of convective development in relation to, for example, the location of the centre of the depression and fronts, surface dew point temperature values, direction of flow at 850, 700 and 500 mb, to name but a few of the many (Sartor (1962) lists 24) possible parameters.

Valuable as these studies are, particularly when undertaken for a specific area, their very number provides a welter of detail that may obscure the essential relationships between severe local storms and synoptic circulations. A return to the need for atmospheric instability keeps these essentials in view. Ludlam and his colleagues analysed several storms in north-western Europe and the central United States with the particular aim of elucidating the stability and configuration of the large scale flow within which the storms developed. The stability analysis was in terms of the wet bulb potential temperature (θ_w) and a saturated potential temperature (θ_s) which is defined as the wet bulb potential temperature corresponding to a state of saturation at the observed dry bulb temperature.

Carlson and Ludlam (1968) showed that the production of severe local storms depended upon the production of a substantial excess of the value of θ_w near the ground over the value of θ_s in the middle and upper atmosphere (Fig. 118). This is likely to occur due to an increase in θ_w rather than a decrease in θ_s, because the distribution of the latter is controlled mainly by the large scale circulation while the distribution of the former is influenced by

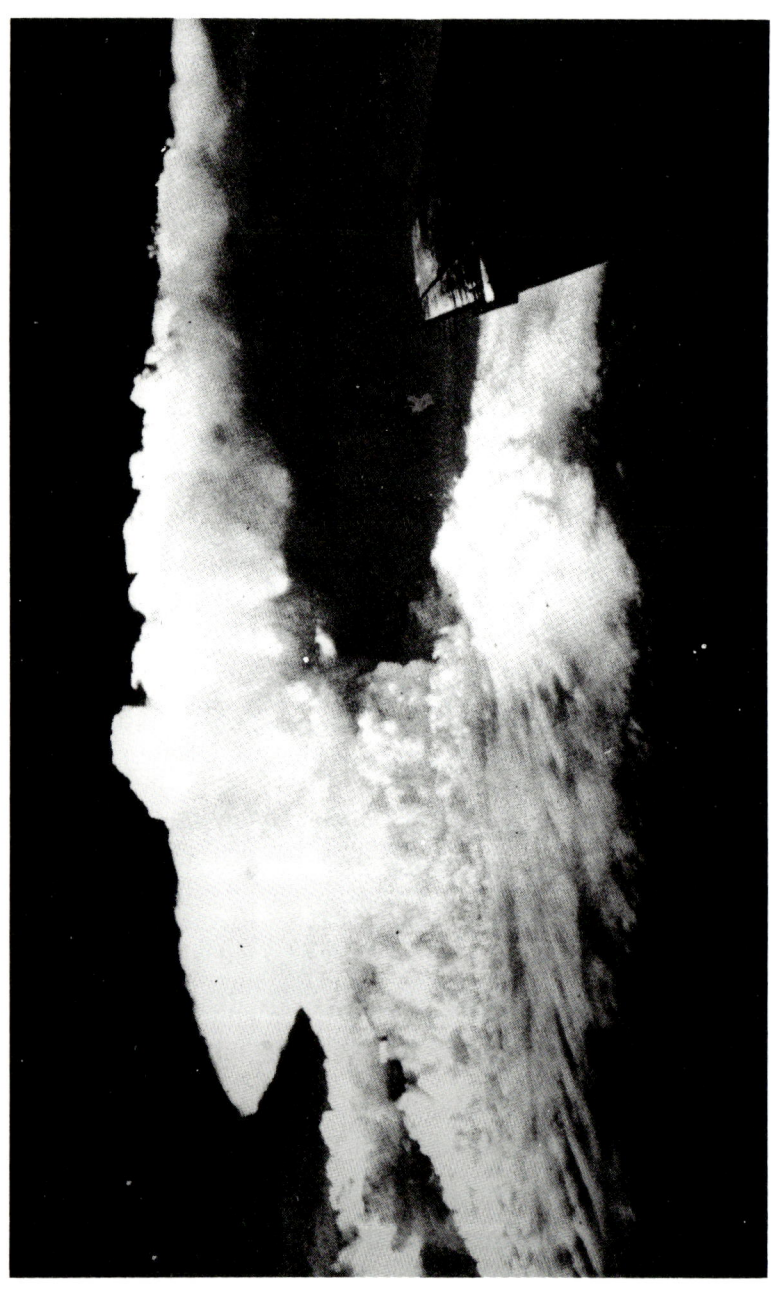

FIG. 117. An isolated thunderstorm on 21 April 1961 over the American mid-west. Photographed by T. T. Fujita. (After Fujita and Arnold, 1963.)

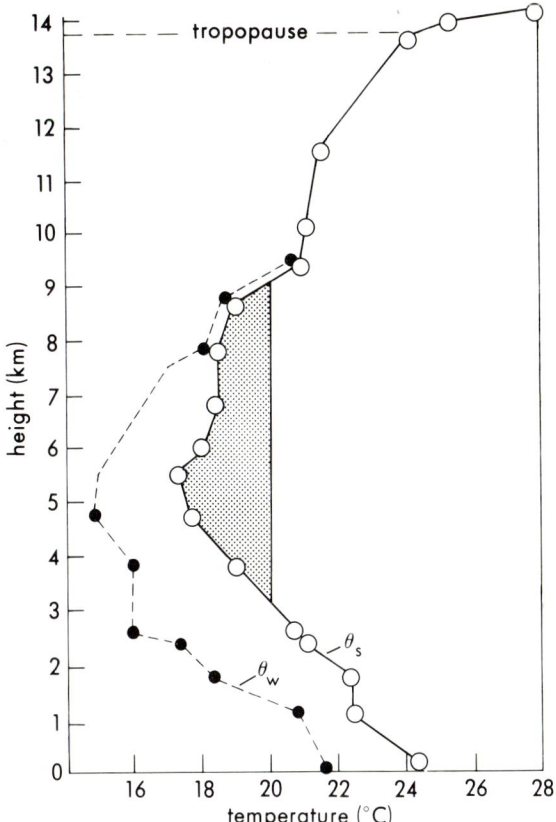

FIG. 118. Distribution with height of θ_w, the wet bulb potential temperature and θ_s, the wet bulb potential temperature if the air were saturated, on the occasion of the severe storm near Geary, USA on 4 May 1961. Air rising through an unmodified environment acquires an excess temperature where its θ_w exceeds θ_s at the new level, provided it has become saturated. Assuming the updraught was characterized by a θ_w of 20 °C, temperature excesses of 2 °C or more would occur between 4.3 and 7.9 km levels (hatched area). (After Browning and Donaldson 1963.)

the peculiarities of the terrain. In north-western Europe, high near-surface θ_w values are primarily due to heating of the bottom 1–2 km by the ground over a period of a day or two. In the central United States on the other hand, the high near-surface θ_w values are frequently due to advection of moist warm air from the Gulf of Mexico. Strong instability is allowed to build up in both areas by plumes of warm dry air from the Spanish and Mexican plateaux. These plumes act as "lids" at heights of about 1–2 km allowing heating to build up below that height which, in turn, increases θ_w values there. When the air with high values of θ_w near the surface is blown out from under the

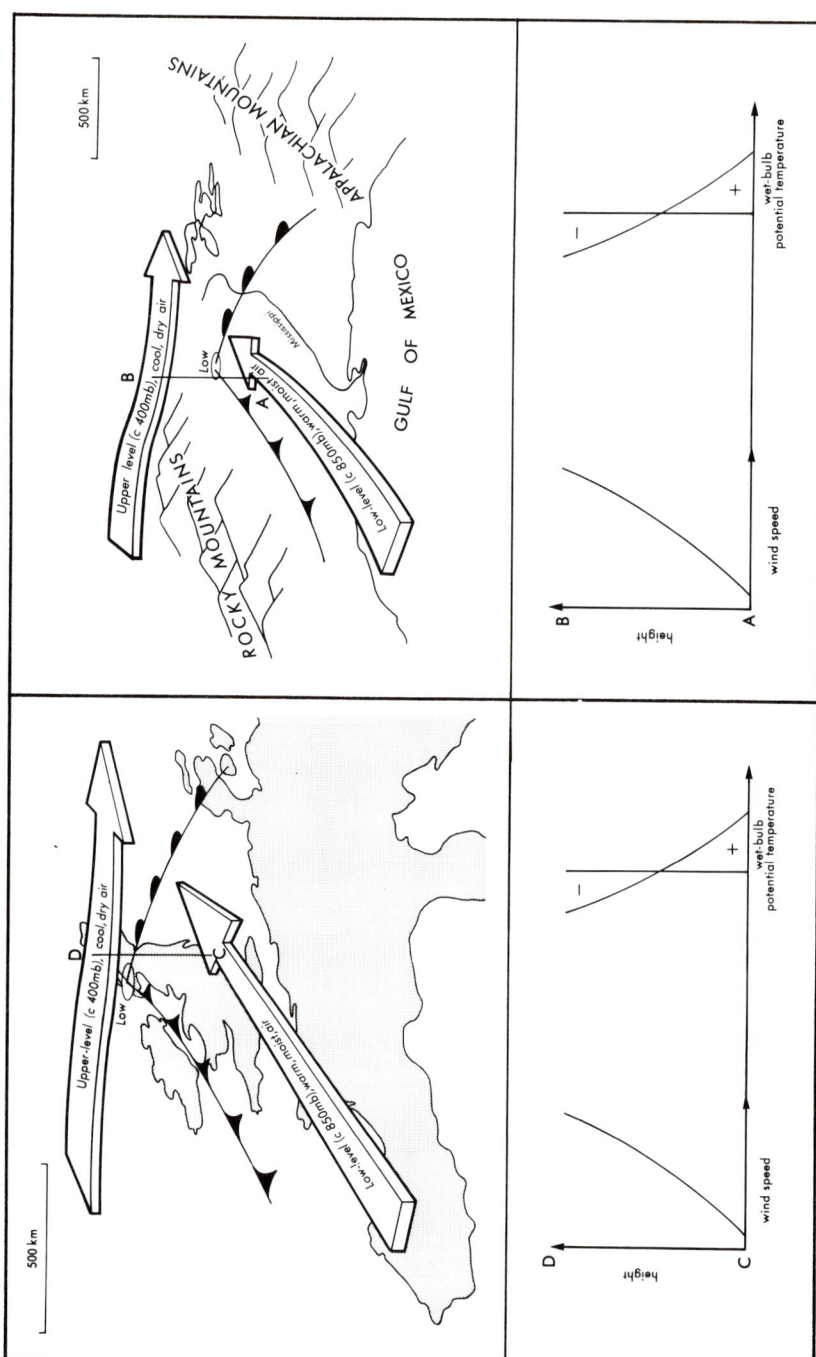

Fig. 119. Creation of vertical instability by differential advection over the American mid-west and north-western Europe.

plume, the instability is released in the form of severe convection. Alternatively the plume may be lifted by larger scale vertical motion with an associated reduction in its ability to "hold down" the warm, moist air near the surface. Figure 119 illustrates how large scale differential advection may create the "potential" instability required for severe storm development. The existence of this differential advection has been documented by Newton (1967) in North America and Carlson and Ludlam (1968) in north-western Europe. Miller (1955) has shown how such flows favour destabilization in the vertical.

The creation of instability is but one important synoptic scale antecedent of severe storm development: the instability must of course be released. The widely accepted mechanism for this release is upward motion (Beebe and Bates, 1955; Derome, 1965), but the different forms of the upward motion themselves are by no means fully documented. "Frontal uplift" was for long accepted as the main mechanism, but now low level jets (Pitchford and London, 1962; Bonner, 1966), convergence zones (Ogura and Chen, 1977) and dry lines (Rhea, 1966; Weston, 1972) appear to be important sources of uplift at the right scale. Details of the precise relationships between cumulonimbus and dry lines await further investigation.

2. *Classification and storm circulation*

Upon release of the instability, severe convection takes on a bewildering variety of shapes and sizes which produces problems not only for initial investigation of its nature but also for orderly presentation of results. One way of instilling some order into the apparent chaos is simply to recognize the different sizes of severe convective systems. Most of the systems extend at least throughout the depth of the troposphere so horizontal dimensions provide the more useful criteria. At one extreme lies the individual convective cell, about 5–10 km across, and at the other extreme is the synoptic scale squall line which may be as long as a conventional front. Between these extremes lie the meso-scale convective systems that are the main focus of interest in this chapter. Whilst having a high degree of validity and the advantage of simplicity, the framework outlined above appears to be little used in the literature. Over the last 20 years many studies have suffered from an inadequate appreciation of scale, particularly when referring to other studies produced by a different research team. Consequently comparison of results produced by different research projects is often quite difficult. Even with an appreciation of scale, the varied nature of convective systems and of the tools employed to observe them (radar, balloons, aircraft) ensure that attempts at classification are not easy. The most appropriate classification for this book was produced by Chisholm and Renick (1972) and the remainder of this section draws heavily on their article. The basis of their classification is the radar structure and behaviour of storms, rather than a certain knowledge of internal and ambient airflows, but this truly reflects the state of the art in the study of severe local storms rather than deficiencies on the part of the authors. Browning (1962, 1968) and Marwitz (1972a,b,c) have also suggested schemes of classifying severe storms along very similar lines to those of Chisholm and Renick (1972). Before outlining the classification it is pertinent to mention one general aspect of "radar meteorology".

A major problem in the analysis of storms observed by radar is that, in the absence of doppler and clearly recognizable targets, no explicit evidence on

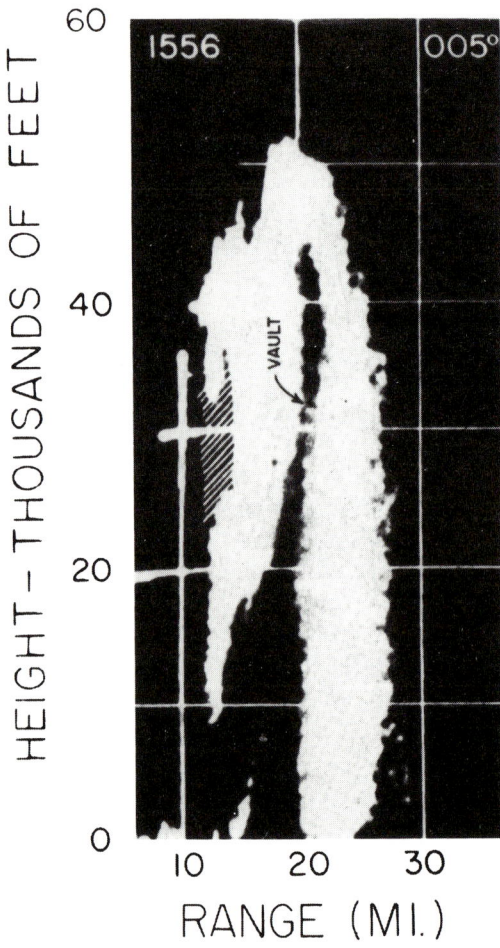

FIG. 120. Photograph of range height indicator (RHI) vertical section through a convective storm showing the vault. (After Browning, 1965.)

airflow is forthcoming. The kinematics of severe storms must be inferred from radar echoes. Indeed, much of "radar meteorology" is concerned with this inferential problem. Yet it was only in the early 1960s that the problem of elucidating internal airflow within severe local storms using conventional weather radar data was tackled. In the 1950s emphasis had been laid on the development and movement of radar echoes *per se* with little regard for the airflows within the echoes and their interactions with ambient flows.

In their important pioneering paper, Browning and Ludlam (1962) suggested that certain kinematic characteristics of severe local storms could be deduced from the shape of RHI echoes. In particular they identified a region within the storm which was devoid of radar echo and they called this region a "vault" (Fig. 120). Chisholm (1973) preferred the term "weak echo region" (WER). Whilst making very good use of the existence of this "vault" in the construction of their storm model, there was a curious lack of explanation by Browning and Ludlam as to exactly why it was so important. Only in the following year did Browning and Donaldson (1963) clearly state that the vault was probably due to "adiabatic concentrations of water in the form of rather small cloud droplets which, owing to the high velocities within the core of the updraught, would have had insufficient time to attain radar-detectable sizes" (Browning and Donaldson, 1963, p. 537). Clearly, if their suggestion is true, the importance of the "vault" is that it identifies the updraught in the storm. In fact combined radar, aircraft and photographic observations (Chisholm and Warner, 1969; Marwitz *et al.*, 1969; Chisholm, 1970a,b) have clearly shown that cloud-filled weak echo regions observed by radar are accompanied by extensive, smooth, uniform updraughts at cloud base of the order of 4–6 m s^{-1}. Observations (Marwitz, 1972a) of chaff within the weak echo region have revealed that the updraughts observed at cloud base continue upward through the full extent of the weak echo region. This information on the nature of airflow in the WER is vital to the construction of kinematical models of severe local storms and consequently to their classification.

Chisholm and Renick's (1972) classification is threefold: single cell storms; multi-cell storms; and supercell storms. The word "cell" is used primarily because it describes well the areas of high echo intensity which appear on radar screens. Meteorologically it is generally accepted to mean a cloud element comprising at least a well-established updraught and, frequently, a downdraught also. Individual cells go through a fairly well-defined lifetime, as do the storms, which are made up of cells. Of the three types of storm, the multi-cellular occurs most frequently. Supercell storms were first so-called by Browning (1962) and may occur in the mature phase of a development of a multi-cellular storm. Despite frequently representing but one stage in overall multi-cellular storm development, their structure and movement are so different to those of "ordinary" cells that they merit separate treatment.

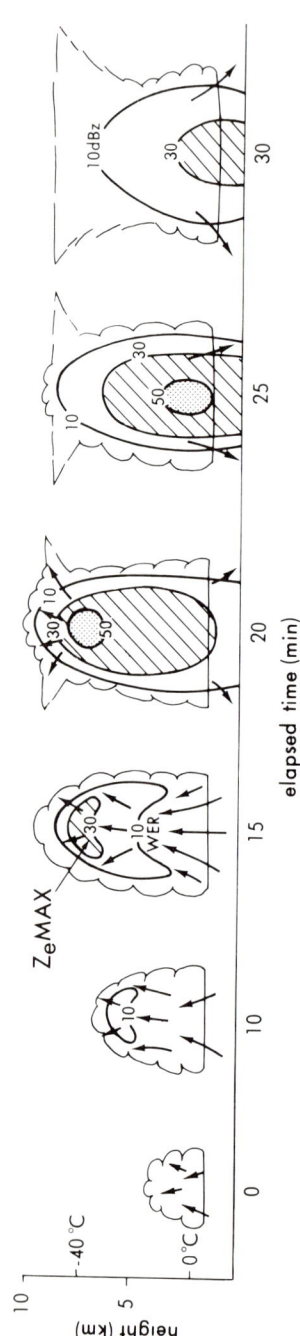

FIG. 121. Schematic vertical cross-sections through a single-cell storm at various stages of its life cycle. Contours of echo intensity labelled in dBz; arrows depict the major airflow components. (After Chisholm and Renick, 1972.)

8 SEVERE LOCAL STORMS

(a) *Single-cell storms* Whilst not being of primary interest in this chapter, these storms do provide a convenient introduction to the more complex types. Following Chisholm and Renick (1972) we note that single-cell storms are usually about 5–10 km in horizontal extent, short-lived (less than 1 h) and change markedly with time. Figure 121 illustrates schematically the life cycle of a typical cell. In the early stages of growth an inverted cup-like WER appears, indicating an updraught of less than $15\,\mathrm{m\,s^{-1}}$. This updraught exists for about 10 min, its demise being due to the fall of precipitation particles which have grown in the upper part of the cell. Without the supporting updraught, the precipitation falls to the ground with hail reaching the ground less than 20 min after the first appearance of radar echo. Throughout their lifetime single-cell storms retain their symmetry and vertical stance because the ambient winds are usually light and have little vertical shear (Fig. 122).

(b) *Multi-cell storms* An ordinary multi-cell storm consists of a sequence of evolving cells, each of which may go through a life cycle as outlined above. The multi-cellular nature was first fully recognized by Byers and Braham (1949) but structural and mechanical details continue to be unravelled (Browning *et al.*, 1976) despite massive efforts by the major projects named earlier. Our appreciation of their cellular structure rests primarily upon radar

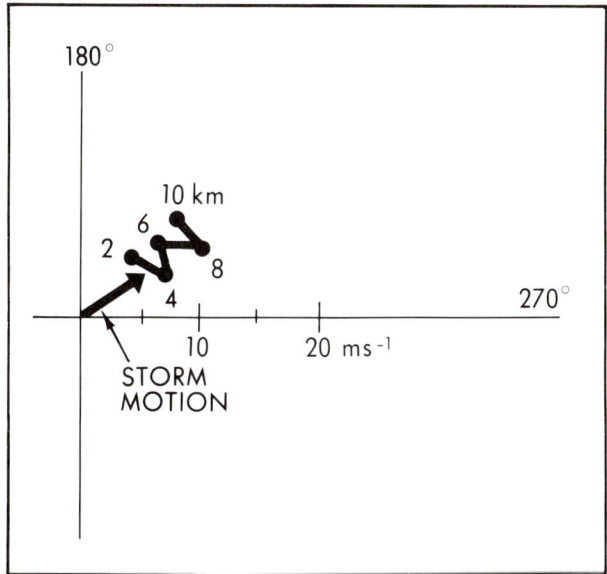

FIG. 122. Typical wind hodograph for a single-cell storm, showing light winds with little shear. (After Chisholm and Renick, 1972.)

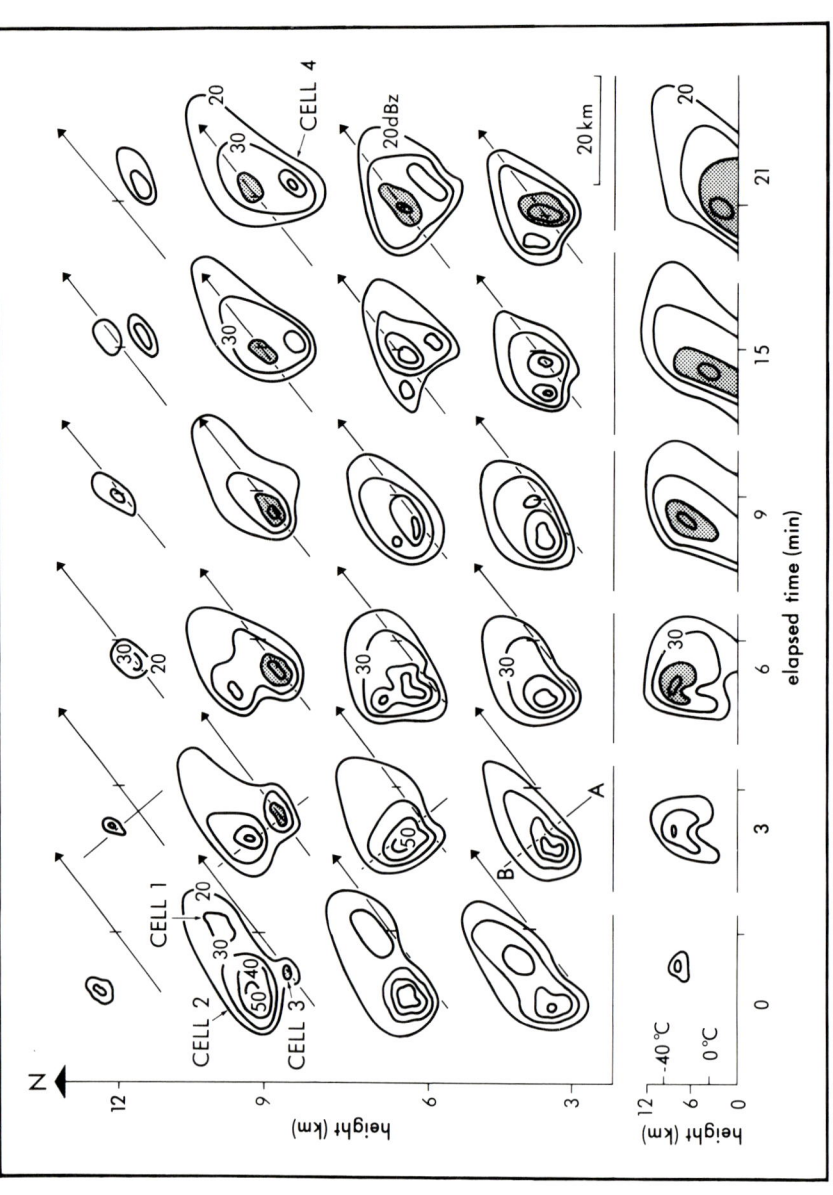

FIG. 123. Schematic plan position indicator (PPI) sections and vertical cross-sections for a mult-cell storm at various stages of its life cycle. PPI sections are illustrated for four heights (3, 6, 9 and 12 km) at six different times. The heavy arrows depict the direction of cell motion and are also geographical reference lines for the vertical cross-sections shown along the bottom of the figure. Cell 3 is shaded to emphasize its history. (After Chisholm and Renick, 1972.)

evidence. Other studies using, for example, rain areas (Newton and Newton, 1959) or the pressure field (Fujita, 1959a,b), whilst being most valuable in other ways, gave little explicit information on internal storm structure. Nevertheless, whichever view of the storms is taken there is agreement that they are frequently 30–50 km in horizontal extent. Virtually all severe storms are of tropospheric depth and Roach (1967), among others, has shown that they frequently extend for several kilometres into the stratosphere.

Figures 123 and 124 schematically illustrate the structure and evolution of a multi-cell storm as seen by radar. New cells of 3–5 km in diameter rise at speeds of 10–15 m s^{-1} in a preferred region on the right-hand side of the storm (e.g. cell 3 in Fig. 123). According to Chisholm and Renick (1972) the newly formed cell does not move into the storm complex, but rather grows rapidly and becomes the storm centre. The previous cell (cell 2) begins to decay and yet another new cell (cell 4) forms. In contrast, Dennis et al. (1970) claimed that new cells form at distances up to 30 km away from the hailstorm core and that each cloud grows rapidly as it approaches and merges with the

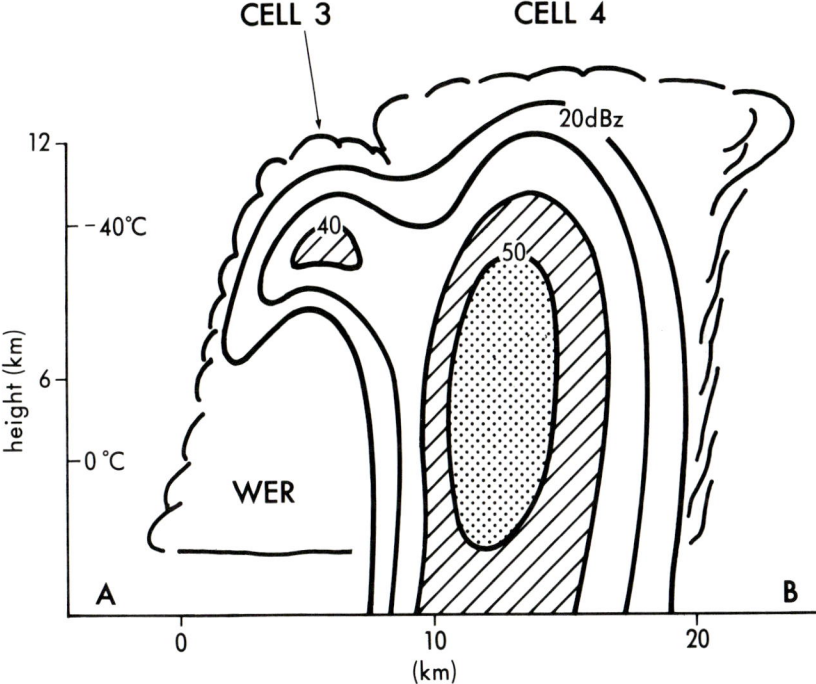

FIG. 124. Schematic vertical cross-section normal to the plane of storm motion for a multi-cell storm. The cross-section is taken along line AB in Fig. 123 at time 3 min. Echo contours are in dBz. Cell 3 is developing with a weak echo region and Cell 4 on its left hand flank is fully developed. (After Chisholm and Renick, 1972.)

Table 42 Thermodynamic stability, windshear and movement of certain well-documented multi-cell storms

Case study	(°C)	Veering in subcloud (deg)	Mean wind in subcloud (deg/m s^{-1})	Mean wind from surface to 10 km (deg/m s^{-1})	Storm motion (deg/m s^{-1})	Shear in cloud layer (s^{-1})	Propagation Individual cells	Propagation Discrete
Browning and Ludlam (1960)	+1	160	150/08	210/21	225/18	2.5×10^{-2}	No propagation	Right
Chisholm (1966), 18 July 1964	+4	40	240/07	235/26	250/12	—	No propagation	Right
Chisholm (1966), 21 July 1964	+4	−90	250/06	230/17	250/10	—	No propagation	Right
Alhambra storm, 12 July 1969	+2	30	020/30	245/11	300/09	2.0	Right	Right
Rimbey storm, 16 July 1969	+4	30	150/04	240/11	240/11	2.0	Left	Right
Benalto storm, 17 July 1968	+3	45	150/04	265/07	305/09	1.5	Right	Right
Sylvan lake storm, 25 July 1968	+6	80	010/04	275/13	315/16	2.0	Right	Right
Carstairs storm, 17 July 1969	+4	120	250/03	265/15	295/12	4.0	Right	Right
Butte storm, 11 July 1970	+7	10	140/06	235/16	310/07	4.5	Right	Right

Source: Marwitz (1972b).

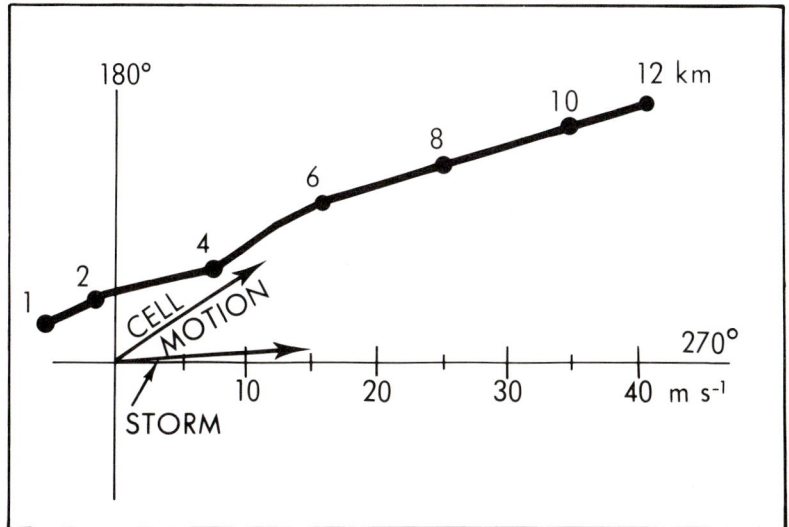

Fig. 125. Typical wind hodograph for a multi-cell storm, showing substantial shear in only one direction, that of storm motion. (After Chisholm and Renick, 1972.)

main cumulonimbus cloud mass. The cloud is first seen as a radar echo just before it merges fully with the main cloud mass. Perhaps the difference in opinion between these authors is due to one group considering clouds, whereas the other considered echoes. Cells typically form every 5–10 min and last for 30–45 min (Chandrashekhar Aiya and Sonde, 1963). Chisholm and Renick (1972) claimed that a total of 30 or more cells may develop during a typical storm's lifetime.

Table 42 and Fig. 125 show typical stabilities and wind shears before the outbreak of multi-cell storms. Marwitz (1972b) was of the opinion that light winds in the subcloud layer are an important precursor of multi-cell storms but Table 42 reminds us that instability and vertical wind shear in the cloud layer are also significant. Chisholm and Renick (1972) emphasized the existence of vertical shear (Fig. 125) but stressed its occurrence in one plane, parallel to the direction of storm motion.

The tremendous variety in shape, size and lifetime of multi-cell storms means that it is difficult to produce a valid model of the internal airflow which is a real improvement upon that suggested by Byers and Braham (1949). As a result of an observational programme which included the use of conventional and doppler radars, surface and upper air networks and instrumented aircraft, a new model (Fig. 126) of a multi-cell storm has been suggested by personnel involved in the National Hail Research Experiment (Browning et al., 1976). As noted by the authors this figure may be interpreted in two ways.

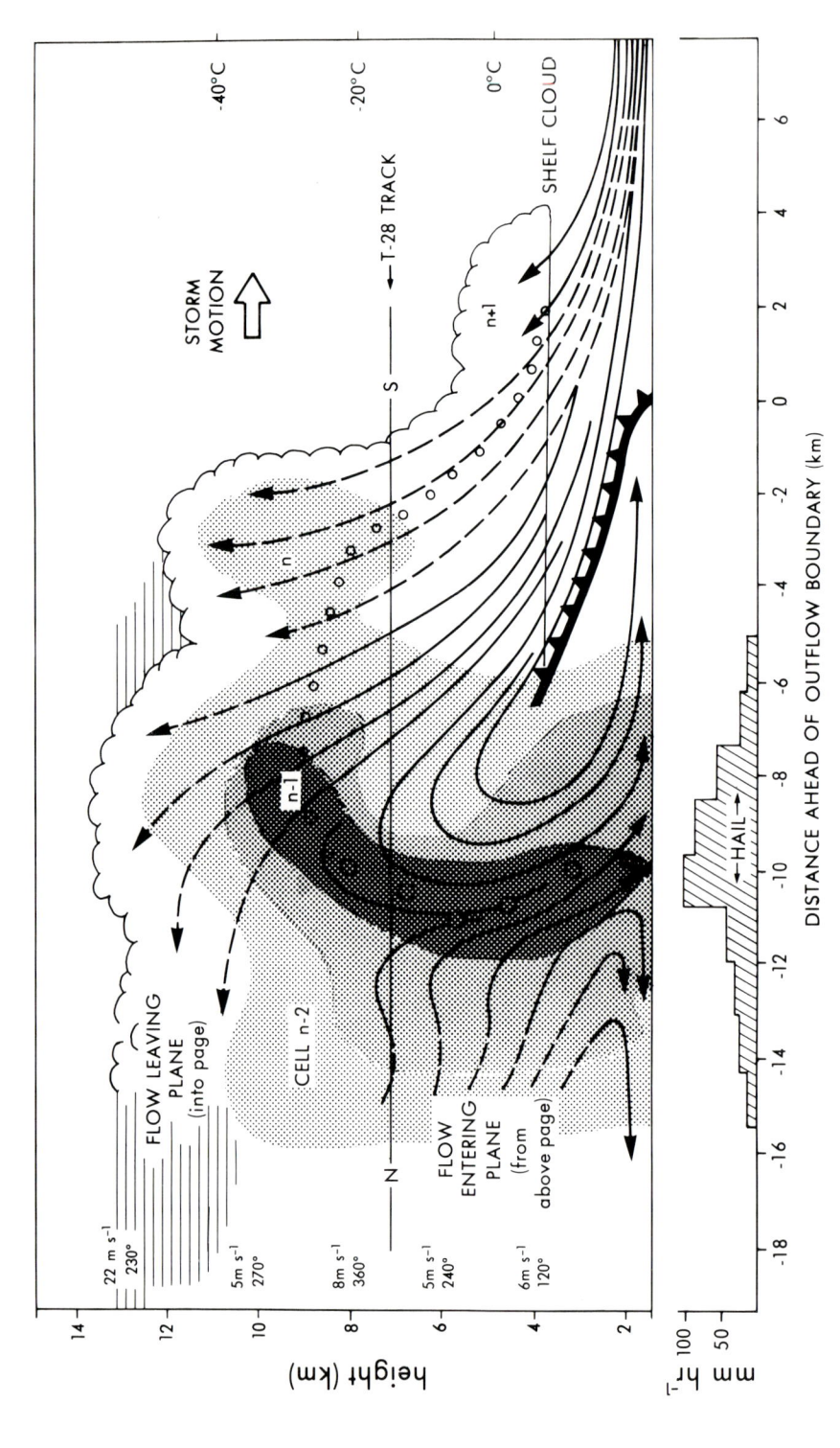

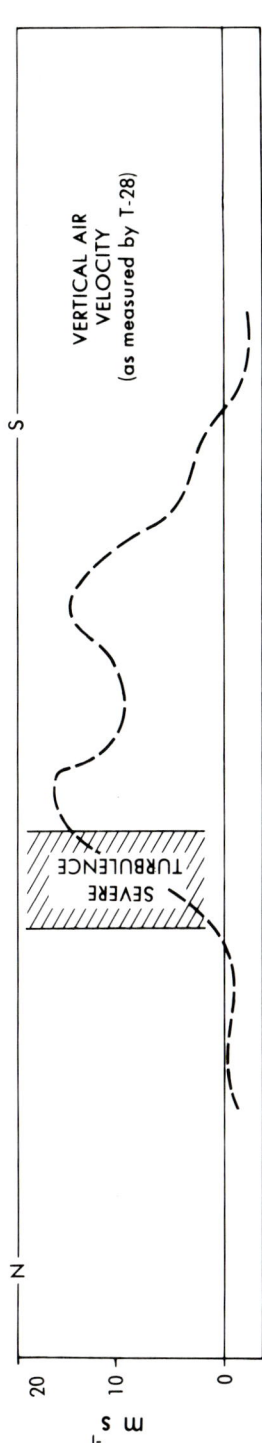

FIG. 126. Schematic model of a severe multi-cell hailstorm showing a vertical section along the storm's direction of travel through a sequence of evolving cells. Solid lines are streamlines of flow relative to the moving system; they are shown broken on the left side of the figure to represent flow into and out of the plane and on the right side of the figure to represent flow remaining within a plane a few kilometres closer to the reader. The open circles represent the trajectory of a hailstone during its growth from a small droplet at cloud base. Scalloped line represents the extent of cloud and the three shades of stipple represent radar reflectivities of 35, 45 and 50 dBz. The temperature scale on the right represents the temperature of a parcel lifted from the surface. Winds (speeds in metres per second; directions in degrees) on the left side are environmental winds relative to the storm. Surface rainfall rate averaged over 2 min intervals during the passage of the storm is plotted below the section. The horizontal line NS through the section at 7.2 km shows the track of the T-28 aircraft, smoothed data from which are plotted at the foot of the figure. (After Browning *et al.*, 1976.)

It can be regarded as either an instantaneous view of a typical structure with four different cells at different stages of evolution or as showing four stages in the evolution of an individual cell. Cell n began growing out of the so-called "shelf cloud" as a distinct "daughter" (after Browning et al., 1976) cloud ($n+1$) about 15 min earlier. Cell $n-1$, which has almost reached its maximum reflectivity, is in its mature stage with both vigorous up- and downdraughts. The decaying cell $n-2$ has weak downdraughts at most levels. The time interval between successive cells was 15 ± 2 min; it took 15 min for n to evolve to the stage of development of $n-1$ and similarly for $n-1$ to evolve to $n-2$. The total lifetime of each cell was about 45 min.

This model, derived for a hailstorm which passed near Raymer, Colorado, on 9 July 1973, illustrates most of the significant characteristics of multi-cell storms—in particular the up- and downdraughts, which, it most be noted, exist in three dimensions, the gust front and the anvil. One facet not covered by this model is the rotation of the storm or indeed of its constituent cells. Some of these characteristics are now considered in more detail.

(i) *Updraught*   The updraught in a storm originates as a largely horizontal inflow ahead of the cloud system. In this model the whole inflow occurs within 500 m of the ground and can be identified at a distance of 20 km from the storm. The inflow rises unmixed to cloud base, in accord with Auer et al.'s (1970) observations of laminar flow below cloud base. The lateral dimensions of individual updraught cells are about 8 km (Browning et al., 1976; Auer and Marwitz, 1968), decreasing to 5 km in the middle troposphere in the case of the Raymer storm. Successive updraught cells may be separated by an area of weak subsidence at cloud base level but they are contiguous at higher levels, giving rise to a fairly broad region of general updraught above. Table 43 summarizes many observations of updraught speed at different heights. Clearly a wide range of values occurs, but at or below cloud base values of about 5 m s^{-1} are typical, whereas at cloud top, speeds of 20–25 m s^{-1} have been quite frequently observed. Davies-Jones and Henderson (1974) have clearly demonstrated that the vertical variation in updraught speed is closely related to the difference in virtual temperature between updraught and environmental air. Temperature differences of up to 5 °C have been observed. The outflow from the updraught forms an anvil which, in the current model, is directed mainly toward the left rear of the storm. The entire updraught is tilted toward the rear of the storm and provides little or no opportunity for precipitation particles grown within one cell to be recycled into a younger cell.

(ii) *Downdraught*   Whereas the existence of downdraughts has been known for at least three decades (Normand, 1946), direct measurements of their

8 SEVERE LOCAL STORMS

Table 43 Updraught speeds in multi-cell storms

Source	Updraught speed (m s^{-1})	Height of measurement above ground (km)	Remarks
Auer and Marwitz (1968)	3–5	3.3, approx. cloud base	Area of updraught ranged from 10 to 172 km^2
Auer and Sand (1966)	6	Approx. cloud base	Laminar flow; smooth steady updraught for 1 h
Battan and Theiss (1966)	5 20	8.0 9.5	Tilted updraught
Browning et al. (1976)	6–8 20	7, approx. cloud base	Inflow 500 m deep; 8 km wide
Byers and Braham (1949)	25	7	Maximum observed value; updraught diameter 1.5 km
Davies-Jones and Henderson (1974)	1.3 (34) 3.6 (34) 6.6 (33) 8.3 (33) 5.1 (29) 3.8 (16) 3.0 (9) 0.4 (6)	0.9–2.1 2.1–3.3 3.3–4.5 4.5–5.7 5.7–6.9 6.9–8.1 8.1–9.3 9.3–10.5	Average values; number of cases in brackets
Dennis et al. (1970)	11	Approx. cloud base	Maximum observed value
Easterbrook (1967)	4	6	
Ellrod and Marwitz (1976)	4–10	1.2	5–10 km ahead of radar echo
Hart and Cooper (1968)	2.5–4 26	7, approx. cloud base	Maximum observed value
Ludlam (1959)	27		Radar turret ascent rate
Marwitz (1972b)	4	Approx. cloud base	
Musil et al. (1973)	20	4–6	
Roach (1967)	25	Cloud top	Stratospheric penetrations of up to 7 km
Saunders (1962)	50	Cloud top	Stratospheric penetrations of 2.5 km
Sinclair (1973)	26	9.1	Maximum observed value
Steiner and Rhyne (1962)	63	12.2	Storm 15.2 km high
Wichman (1951)	2–6 20–30	8, approx. cloud base	

characteristics are few. Byers and Braham (1949) were probably the first to demonstrate their overall characteristics in multi-cell storms but much of the subsequent literature concerns their occurrence in supercell storms. In the recent Raymer multi-cell storm model part of the downdraught originated in the mid-troposphere (about 6 km) and descended unmixed to the surface; this air entered the storm on its right rear flank. Some of the downdraught was also probably generated within former updraught air. The maximum observed downdraught velocity of $15\,\mathrm{m\,s^{-1}}$ was located in the region of highest radar reflectivity, close to cloud base level. Downdraught velocities greater than $10\,\mathrm{m\,s^{-1}}$ occurred in a region 2 km wide extending from a height of 2–6 km. Pibal, as opposed to radar, measurements allowed the calculation of vertical velocities in and near storms by integration of the continuity equation (Ragette, 1973). The results were necessarily applicable to larger areas than those from radar because, despite 27 ascents being made in 4 h, data density from this method was far smaller than that from radar. Yet, it was clear from these observations that downdraughts of up to $1\,\mathrm{m\,s^{-1}}$ could extend over 40 km horizontally mainly behind the storms and over 2.5 km vertically.

(iii) *Outflow and gust front* When the cool downdraught air reaches the Earth's surface it feeds an outflow both ahead of and to the rear of the storm. Byers and Braham (1949) termed this outflow the "cold-air dome". At the surface the onset of the outflow usually means a rise in pressure (2–4 mb), a fall in temperature (perhaps up to 5 °C) and a marked change in both wind speed and direction (the gust front). Despite the importance to aviation (several aircraft have crashed in severe downdraughts and outflows (Fujita and Byers, 1977; Goff, 1976*a*)) comparatively few detailed investigations of outflows and gust fronts (e.g. Charba, 1974; Goff, 1976*b*) appear in the literature. In his study, Goff (1976*b*) recognized that at present it is impossible to distinguish clearly between the gust front characteristics of multi-cell and supercell storms, but suggested that his observations were probably more typical of multi-cell storms—particularly those occurring in the Great Plains region of the USA.

Goff suggested a fourfold classification of gust fronts: (1) gust fronts associated with intensifying storms or accelerating outflow; (2) gust fronts associated with mature intense storms or quasi-steady outflow; (3) gust fronts associated with dissipating storms or decelerating outflow, with respect to the storm; (4) gust fronts in the final stage of their life-cycle. Table 44 characterizes each type. To a greater or lesser degree, all four tend to "run ahead" of the storm and cause a localized "gust frontal uplift" of the order of $5\,\mathrm{m\,s^{-1}}$. This uplift is not to be confused with the main updraught of the storm, although the two may combine where the outflow's leading edge is near to the storm core. In the Raymer storm (Browning *et al.*, 1976), the depth of the surface outflow exceeded 1 km ahead of the storm, but behind

8 SEVERE LOCAL STORMS

Table 44 General quantitative gust front information

Type	Gust front average speed (m s^{-1})	Smoothed maximum vertical velocity (m s^{-1}) at height of 444 m	Number of cases
1	8.1	5.0	4
2	11.3	6.2	8
3	9.0	4.3	4
4	18.6	2.3	4

Source: Goff (1976*b*).

the storm the depth of downdraught air that was directed rearward relative to the storm was less than 500 m. Surface divergence beneath the strongest downdraught was 4×10^{-3} s^{-1}. In accord with both Goff (1976*b*) and Auer *et al.* (1969) the gust front extended, on average, 5 km ahead of the leading edge of the surface precipitation.

(iv) *Movement* Within the last 30 years storm movement has been investigated primarily with the aid of PPI records or detailed rainfall observations. In the former case it was comparatively easy to monitor the motion of echoes from a sequence of PPI pictures. In the latter case, "rainstorms" were similarly tracked through a map sequence.

In an early study of the movements of convective cells seen by radar, Brooks (1946) observed that small cells (5–12 km in diameter) tended to move with the wind in the lowest part of the cloud layer, while the "steering level" for larger storms was at a higher elevation. Byers and Braham (1949) found that small radar echoes observed in the Thunderstorm Project generally moved with the mean wind in the cloud layer up to 6 km. More exhaustive studies by Newton and Fankhauser (1964*a,b*) related echo motion to a mean wind defined as the vector mean of the 850, 700, 500 and 300 mb winds. In synoptic situations where the wind veered strongly with height, storms moved as much as 60° to the right or 30° to the left of the direction of this mean wind. Figure 127 shows that there was discrimination according to size: storms with diameters greater than 15 km tended to move to the right of, and usually slower, but occasionally faster, than the mean winds. Other studies, for example those by Chalon *et al.* (1976) and Fenner (1976), suggest that motion to the right of the mean winds probably occurs more frequently than motion to the left.

Using the hourly precipitation values from more than 2000 rain gauges throughout the mid-west of the USA, Newton and Katz (1958) were able to track large convective rain areas and to correlate their motion with mean winds between 850 and 500 mb. Their results indicated that the rainstorms

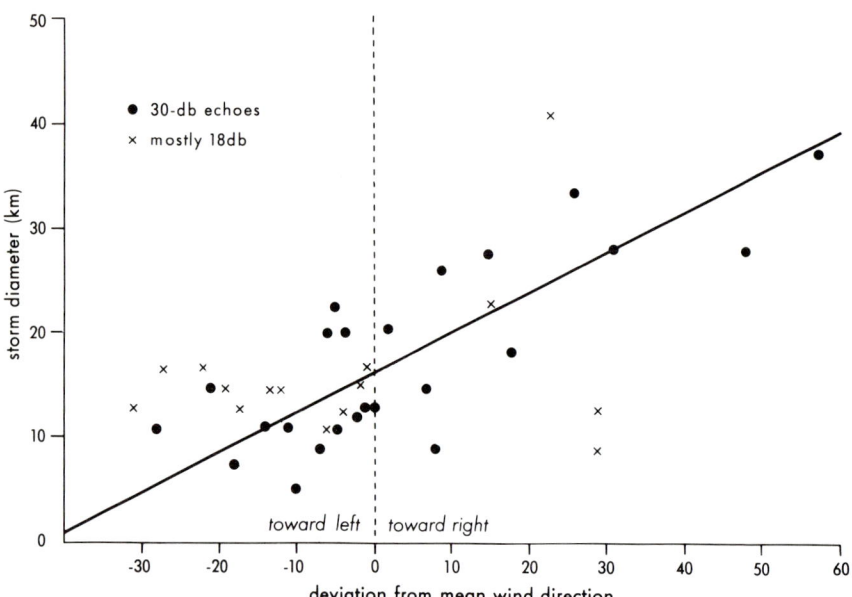

FIG. 127. Deviation of storm motion from direction of mean wind as a function of echo diameter. Regression line based on radar echo observations as shown. (After Newton and Fankhauser, 1964a,b.)

moved 25° to the right of the mean wind. Newton and Fankhauser (1975) analysed a multi-cellular storm some 80 km across and found it to move 55° to the right of the mean wind and with half its speed.

The "cross-wind" motion of storms is primarily due to propagation. In multi-cell storms this usually means that new cells form on one side of the storm, move through the storm, dissipating on the other side (Fig. 128(a)). Browning (1962) first suggested this, emphasizing that new cell growth tended to occur on the right forward flank of the storm and that the cells moved with the wind. Only the whole storm moved across the wind. In a later review, Newton and Fankhauser (1975) suggested that new cells tend to form in the rear, right quadrant, move through the storm, dissipating on approach to its forward, left flank. Uncertainty about the mechanism was increased somewhat by Marwitz's (1972b) suggestion of two additional modes of propagation (Fig. 128(b), (c)). In Fig. 128(b) Marwitz suggested that not only does the storm propagate discretely through its cells, but that the cells themselves are also continuously propagating to the right. Consequently the storm motion is substantially to the right of the wind direction. In contrast (Fig. 128(c)) the individual cells propagate continuously to the left of the mean winds whereas discrete propagation of cells occurred on the right flank

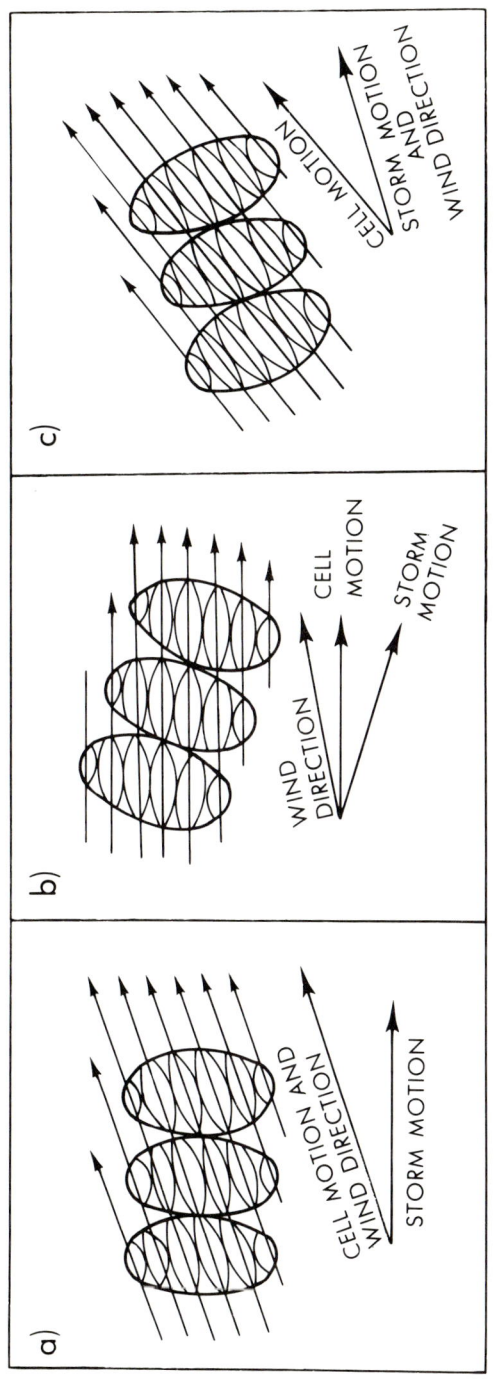

FIG. 128. Schematic diagrams of storm movement by propagation (see text). (After Marwitz, 1972b.)

of the storm. Thus this discrete propagation was acting to cause the storm complex to move to the right of the mean winds while continuous propagation of the cells was acting to move the storm to the left. As a result, the storm itself travelled along the mean wind direction. Attractive though this idea is, it has not as yet had widespread observational confirmation and the exact nature of the "continuous" as opposed to the "discrete" propagation is still unclear. Newton and Fankhauser (1975) still preferred the idea of discrete propagation by cells, suggesting that the occasionally observed rapid movement of large multi-cellular storms to the left of the mean wind is partly accounted for by a left-forward flank pattern of propagation. Whichever form of propagation is preferred, it is no doubt encouraged by the existence of "meso-lows" ahead of the cumulonimbus system. These lows, noted by both Hoxit et al. (1976) and Sanders and Emanuel (1977), foster a convergence of near-surface air immediately prior to its lifting into the storm updraught. Hoxit et al. suggested that high level subsidence warming of air ahead of and at roughly the same height as the anvil could cause such a surface low, but Sanders (1977) was not totally convinced by their argument. A lack of consensus about both pre-storm meso-lows and propagation remains, requiring both observational and theoretical studies for provision of satisfactory explanations.

Closely associated with the movement of storms are their propensities to split and to rotate. Splitting of storm echoes has been clearly documented by, among others, Achtemeier (1969), Charba and Sasaki (1971), Fujita and Grandoso (1968) and Harrold (1966). After splitting, the right-hand echo (looking along the general direction of storm movement) deviates to the right and the left-hand one to the left. In both cases, echo speeds are lower than mean wind speeds by up to $11 \, \mathrm{m\,s^{-1}}$. The angle between the direction of motion of the diverging echoes may be as large as 60°. It has been suggested, particularly by Fujita and Grandoso (1968) and Goldman (1966), that the continuous deviation of the split echoes may be due to storm rotation, principally through the effect of the Magnus force. Whilst such an effect may be real, it is difficult to assess its importance relative to normal propagation effects. Charba and Sasaki (1971) provided a useful review. Whereas the Magnus effect is a little speculative, there is less doubt that its supposed cause, storm rotation, does indeed occur. Byers (1942) was probably the first to recognize "rotating thunderstorms" and Fujita (1960) and Fujita and Grandoso (1968) have provided substantive observational evidence. The cause of the rotation is again rather uncertain. Fujita (1965) suggested that the vorticity necessary for rotation is acquired from that of the subcloud air as it converges into the storm. Alternatively Barnes (1968) outlined a way in which the strong vertical wind shear in the inflow may provide an additional source of cyclonic vorticity through a mechanism involving the twisting term of the vorticity equation.

(c) *Supercell storms* The term "supercell" was first used by Browning (1962) to describe what appeared then to be a particular form of the mature stage of a multi-cell storm. The "supercell" was far larger, more persistent and gave more severe weather than the normal mature cell. In addition it appeared to have a highly organized internal circulation in a nearly steady state, to be intimately related to the vertical shear in the ambient winds and to propagate continuously rather than discretely. In subsequent years Browning's persuasive writings have led to the fairly widespread use of the supercell idea in the studies of both the Alberta Council and the National Hail Research Experiment. The model has been modified several times since its inception,

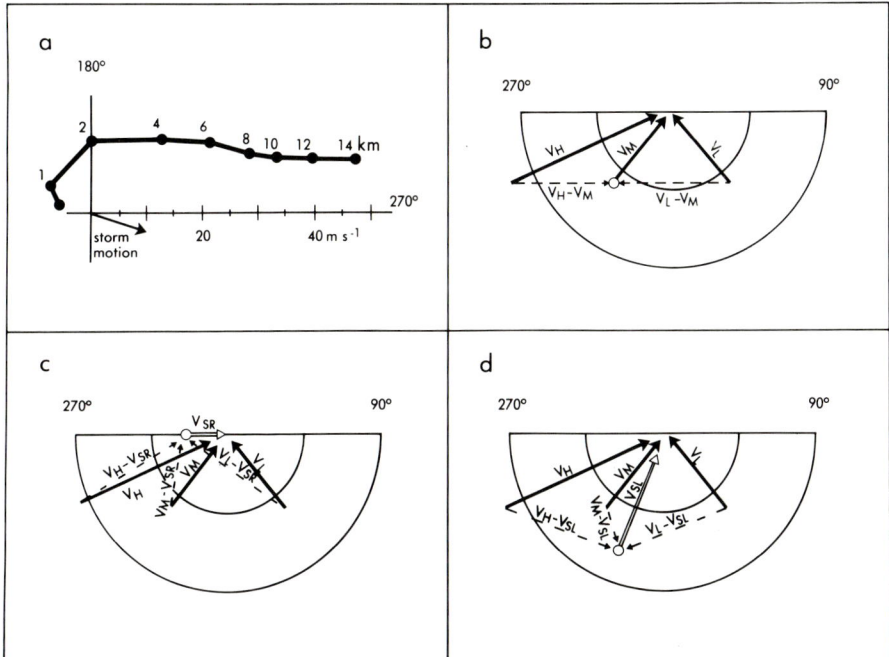

FIG. 129. (a) Typical wind hodograph for a supercell storm, showing low level directional shear. (After Chisholm and Renick, 1972). In parts (b), (c), (d) the wind arrows are in the opposite sense to those in (a). The diagrams show winds representative of low, middle and high levels in a typical supercell storm. The solid arrows show winds relative to the ground at these three levels, i.e. V_L, V_M, V_H. The dotted arrows in (b) show winds relative to a two-dimensional storm travelling at the velocity of the middle level winds. The double arrows in (c) and (d) represent the velocities V_{SR} and V_{SL} of severe right-moving and severe left-moving storms respectively. The dashed arrows in (c) and (d) show wind velocities relative to these storms; the winds at the three levels relative to a severe, right-moving storm, for example, are labelled $(V_L - V_{SR})$, $(V_M - V_{SR})$ and $(V_H - V_{SR})$. (After Browning, 1968.)

principally by Browning himself, but in its essentials it appears to account for the most significant kinematical characteristics of very severe local storms.

Supercell storms typically occur in the following type of synoptic scale environment: (1) strong instability, with parcel theory indicating a thermal buoyancy at 500 mb in excess of $+4\,°C$; (2) strong ($10\,\text{m s}^{-1}$) mean subcloud environmental winds; (3) strong environmental wind shear through the cloud layer, with values from 2.5×10^{-3} to $4.5 \times 10^{-3}\,\text{s}^{-1}$; (4) strong veer of winds with height, up to 90°. These observations support the earlier views (Dessens, 1960; Das, 1962; Newton, 1960a) based upon less data, that the occurrence and destructiveness of hail was related to strong winds and strong shear. Figure 129 illustrates a typical hodograph (Fig. 129(a)) for a supercell storm together with Browning's suggestions for relative flow for storms moving with (Fig. 129(b)), to the right (Fig. 129(c)) or to the left (Fig. 129(d)) of the middle level winds.

As noted earlier in thic chapter, the bulk of our knowledge about severe storms derives from the use of radar. Chisholm and Renick (1972) have admirably summarized a welter of information in their schematic PPI sections of a supercell storm. They also suggested that typical supercells exhibit the following radar characteristics:

(1) In plan view the supercell exhibits a single cellular structure which is circular to elliptical in shape with a characteristic horizontal dimension of 20–30 km and a depth of 12–15 km (Figs 130 and 131).

(2) A persistent bounded weak echo region (BWER) on the right-hand side (looking along the direction of motion) of the storm with horizontal dimension of 5–12 km. This BWER is frequently conical in shape, extending to a height of one-half to two-thirds of the distance through the storm depth. As noted earlier, WERs denote the storm updraught, in this case having speeds of $25–40\,\text{m s}^{-1}$.

(3) The strongest echo occurs on the left-hand side of the BWER, comprising precipitation with large hail on the side closest to the BWER.

(4) An extensive plume (60–150 km in length) observable by radar and accompanied by a visible anvil 100–300 km in length extends downstream from the main storm core.

Figure 131 illustrates the vertical "radar" structure of supercell storms, clearly showing the WER and the strong forward and rightward leaning of the echo, the degree of leaning increasing with the wind shear. In the absence of doppler measurements until quite recently, it was the shrewd intepretation of conventional radar data (as illustrated in Fig. 120) which led to the widely accepted kinematic models reviewed below. Figure 131 also illustrates the widely accepted relationship between echo configuration and inferred updraught; the downdraught is not as well documented, as we shall see.

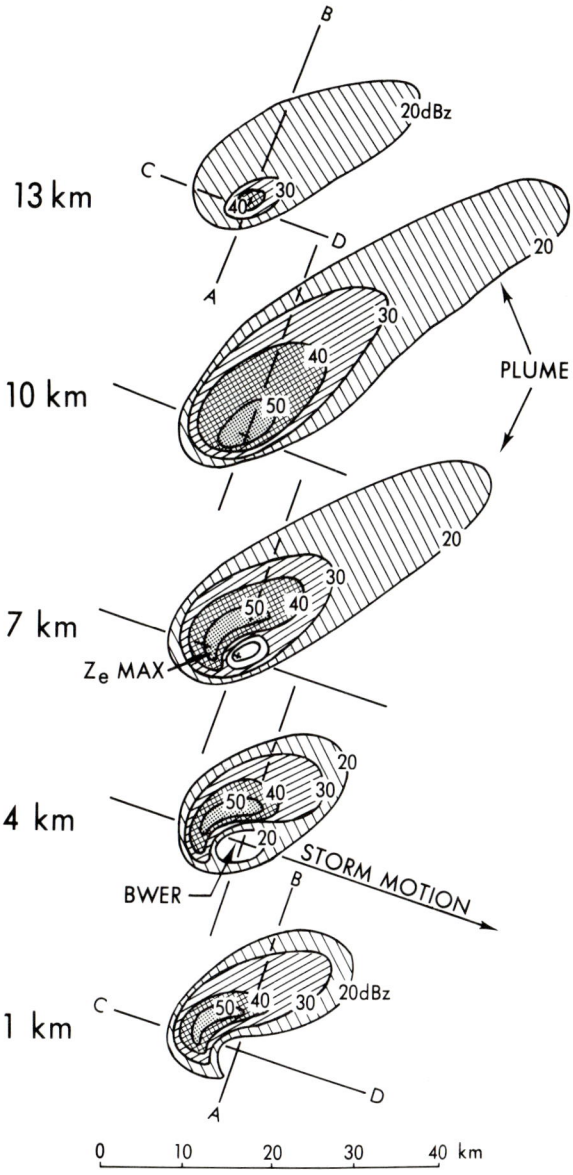

FIG. 130. Schematic plan position indicator (PPI) sections of radar structure of a supercell storm at heights of 1, 4, 7, 10 and 13 km. Contours of reflectivity are labelled in dBz. (After Chisholm and Renick, 1972.)

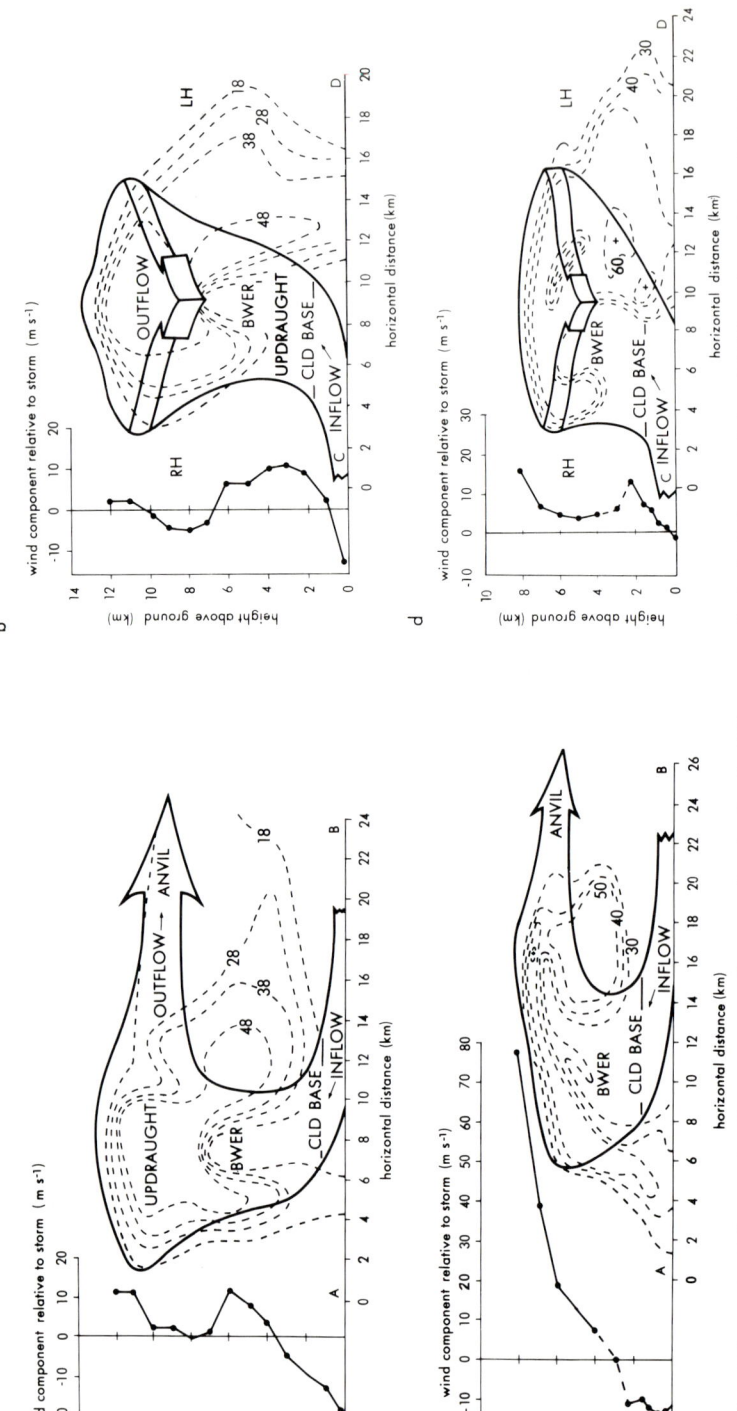

FIG. 131. Schematic airflow and radar structure of storms developing in different vertical wind shears—(a) and (b) low shear; (c) and (d) high shear. Reflectivity shown by dashed lines, labelled in dBz. (a) Section along plane of storm motion. Inflow air enters on downwind side, ascends almost vertically through the bounded weak echo region (BWER) to leave on the downwind side. (b) As (a) except airflow normal to plane of storm motion. Inflow air enters on right-hand flank, ascends through BWER and leaves out of the plane of the paper towards the reader. (c) As (a) but in stronger shear. (d) As (b) but in stronger shear. (After Chisholm, 1973.)

Various models of internal airflow of a supercell storm (all primarily due to Browning) are summarized in Fig. 132. Common themes exist in this series of models, but improved observations continually reveal tremendous variety in storm behaviour and restrict valid generalizations. Despite being two-dimensional, the Browning and Ludlam (1962) model (Fig. 132(a)) was a breakthrough in recognizing the persistent co-existence of up- and downdraughts in a strongly sheared environment. Browning and Donaldson (1963) added the third dimension, thus showing explicitly for the first time that the updraught approaches the storm from the front right and in its upper reaches twists cyclonically to stream out into the anvil. A similar picture emerged (Fig. 132(d)) from Browning's (1965) analysis of severe, right-moving storms. A year earlier he had made an important improvement to the Browning–Ludlam model by showing that the downdraught entered the storm from the middle-level right of the storm rather than the rear (Fig. 132(c)) (Browning, 1964). Further modifications came over a decade later (Fig. 132(e) and (f)) (Browning and Foote, 1976) with the appreciation that only a part of the updraught twisted cyclonically to help form the important source of embryos necessary for hailstone growth over the summit of the updraught itself. As Fig. 132(e) and (f) show, much of the updraught air moves directly to the cloud top and is then swept downstream into the anvil. The low level inflow once more comes from directly ahead of the storm rather than from the right. A notable omission from the model is the downdraught. Rather ironically, recent very comprehensive observations by doppler radars of the three-dimensional airflow in a storm (Kropfli and Miller, 1976) strongly support the early Browning and Ludlam (1962) two-dimensional model. This suggests that factors common to both studies are fairly typical of such storms:

(1) Both storms were persistently fed from opposite sides; the updraughts sloped in the upshear direction from near the surface and the downdraughts entered the backside at middle levels. Updraught speeds measured by Kropfli and Miller averaged 3.8 m s^{-1}.

(2) The downdraughts descended from middle levels with little curvature and most of them left the rear of the storms at the surface. The storm-average downdraught speed in Kropfli and Miller's case was 2.3 m s^{-1}.

(3) Environmental air in the middle and upper levels overtook the storms and flowed to either side as if around an obstacle.

(4) The updraught source was on the right forward flank of the storms.

In addition to this complexity for storms which all moved toward the right of the wind, Hammond (1967) showed that a storm with a mirror image of the Browning (1964) model (Fig. 132(c)) moved to the left (see Fig. 129). Fortunately he did not complicate matters further by suggesting that the storm had split and was rotating anticyclonically. This would have represented

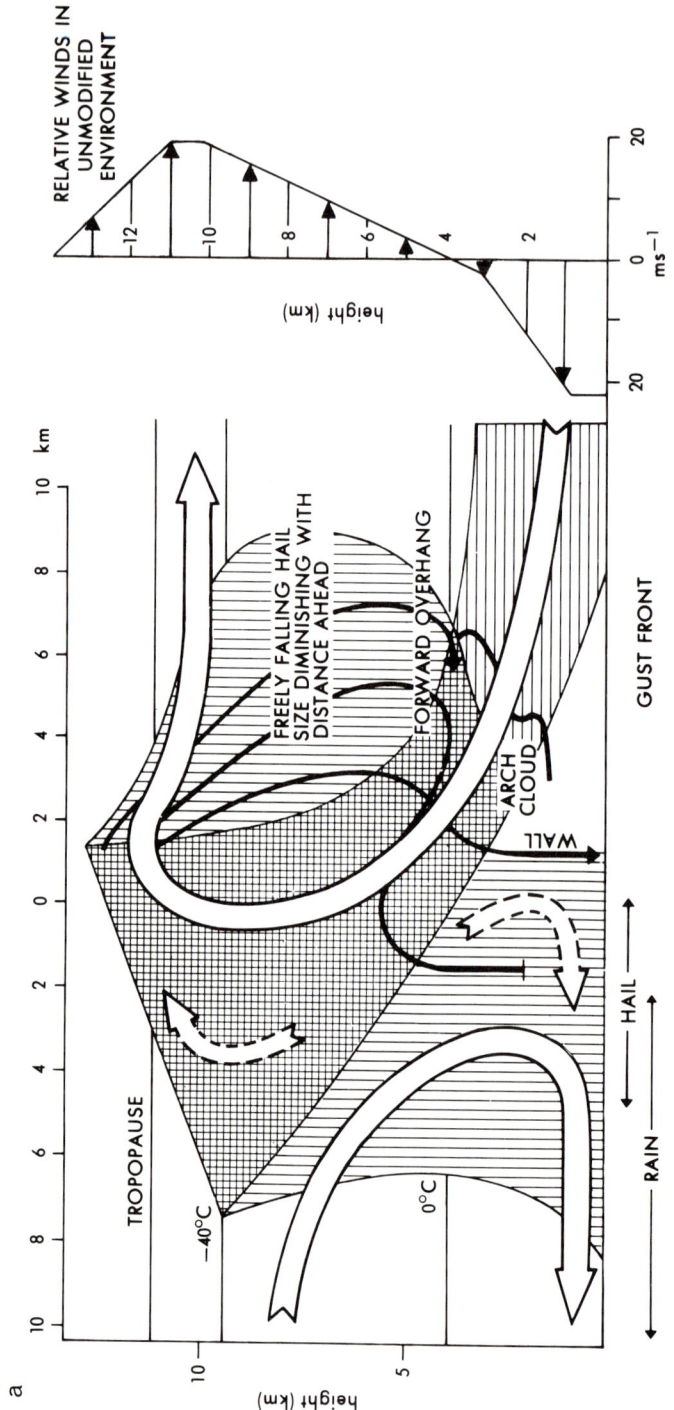

Fig. 132. *Various models of airflow within severe storms.* (a) Vertical section through centre of storm along direction of motion. Horizontal shading shows updraught and vertical shading the bulk of the radar echo. (After Browning and Ludlam, 1962.)

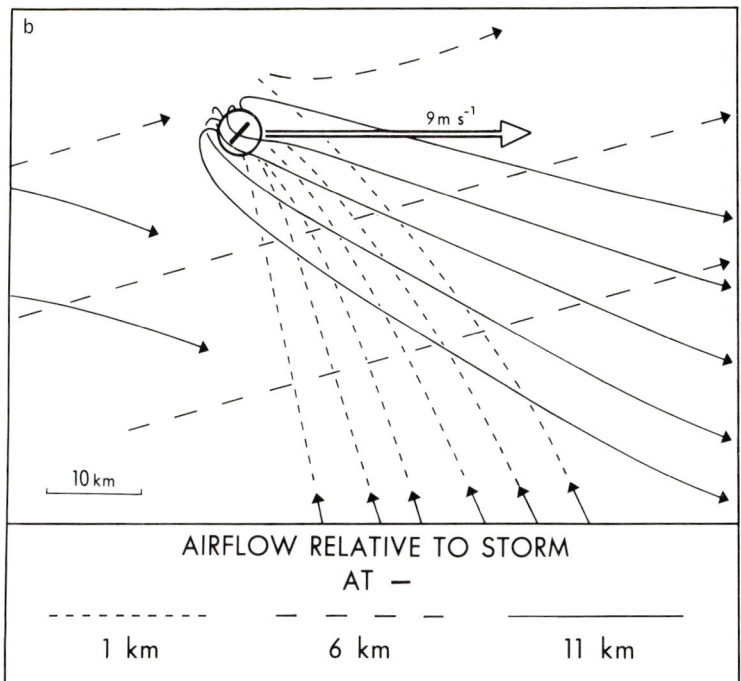

FIG. 132. *Various models of airflow within severe storms.* (b) Plan of airflows at three levels in a severe storm. The three levels depicted characterize the levels of inflow toward the updraught and downdraught and the level of outflow from the updraught. Closed contour shows position of updraught and thick line the position of the vault. (After Browning and Donaldson, 1963.)

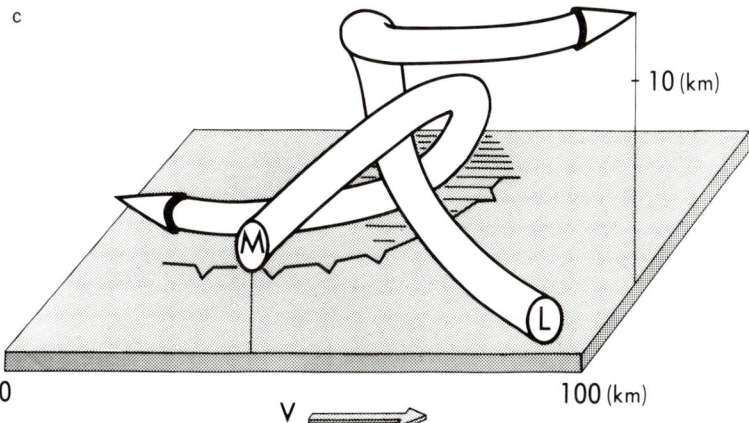

FIG. 132. *Various models of airflow within severe storms.* (c) Three-dimensional airflow in a severe storm that travels to the right of the winds in the middle troposphere. L (low) and M (middle) refer to the main levels of origin of updraught and downdraught respectively. Hatched area shows extent of surface precipitation; barbed line the position of surface gust front. V is storm velocity. (After Browning, 1964.)

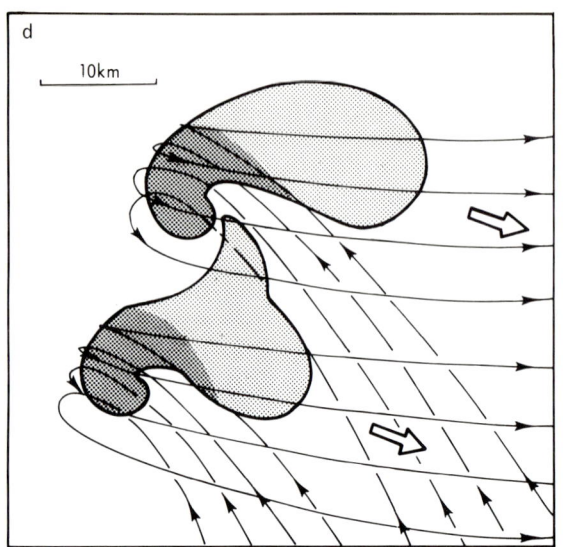

Fig. 132. *Various models of airflow within severe storms.* (d) Schematic plan of relative streamlines in two storms showing air approaching the updraughts at low levels and leaving them at high levels. Parts of the echo believed to have been associated mainly with updraughts and downdraughts respectively are heavily and lightly stippled. Storms move in direction of bold arrows. (After Browning, 1965.)

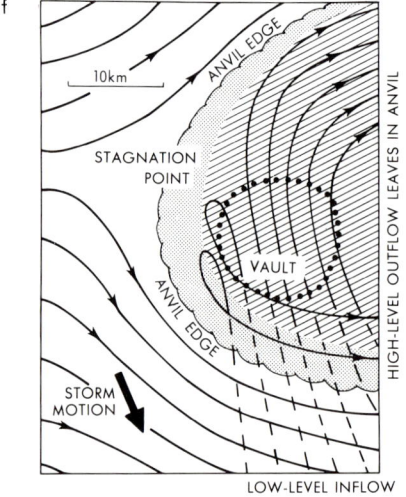

Fig. 132. *Various models of airflow within severe storms.* (f) Schematic plan of relative streamlines in a severe storm. Regions of radar echo are shown hatched; areas of cloud devoid of detectable echo are stippled. The dotted circle represents the extent of intense updraught in the middle atmosphere. Some of the streamlines represent environmental flow at the middle levels diverted around the main updraught. Others represent the low level inflow toward the updraught (dashed lines) and also the high level outflow. (After Browning and Foote, 1976.)

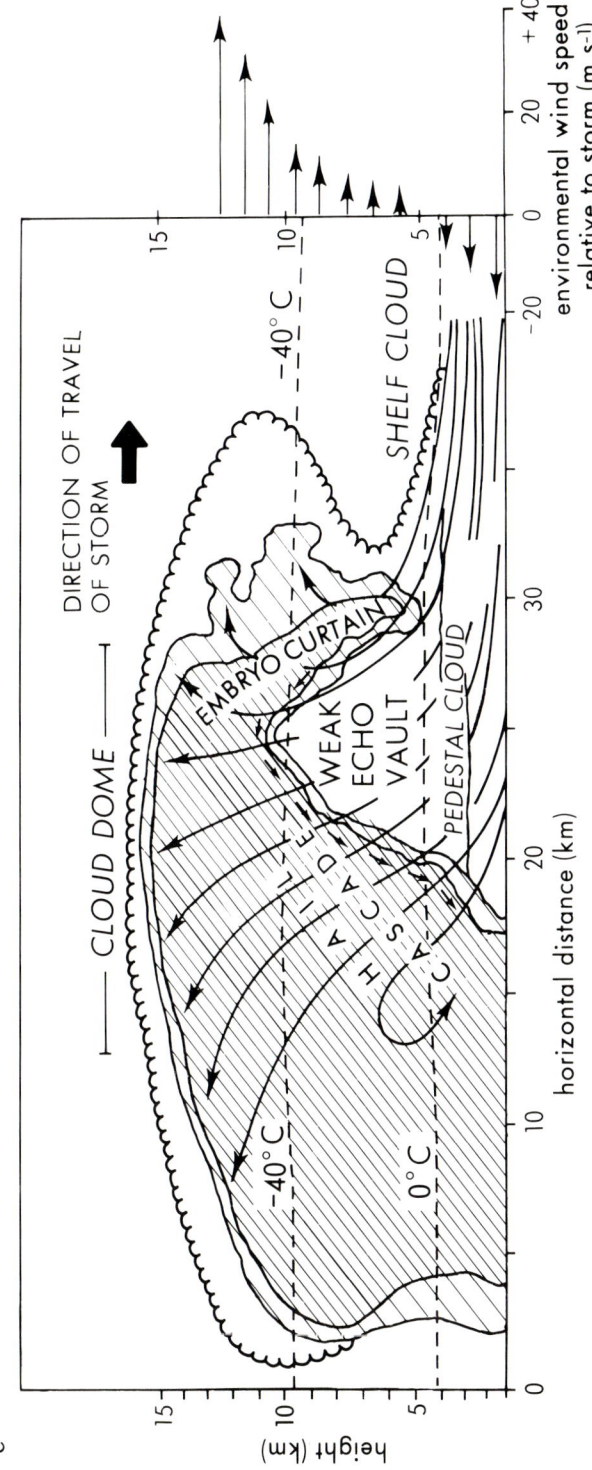

FIG. 132. *Various models of airflow within severe storms.* (e) Schematic vertical section along direction of travel of storm showing visual cloud boundaries, radar echo and streamlines of relative airflow. Two levels of radar reflectivity are represented by different densities of shading. To the right is a profile of the wind component along the storm's direction of travel derived from a nearby sounding. (After Browning and Foote, 1976.)

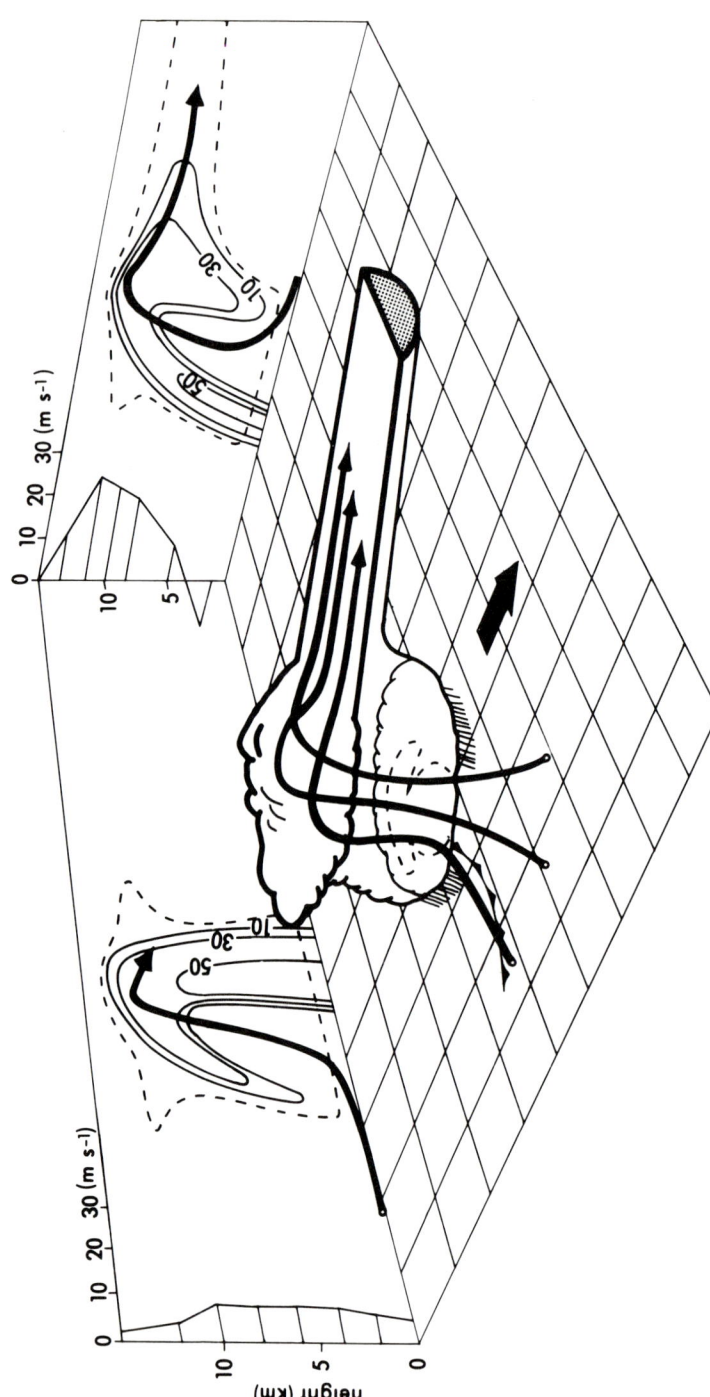

FIG. 133. Schematic three-dimensional view of a supercell storm depicting relative airflow and reflectivity. (After Chisholm and Renick, 1972.)

a particularly awkward combination of the then currently fashionable interpretations. Clearly our knowledge of internal airflow is still incomplete and we await more detailed observations from doppler radar. Meanwhile, Fig. 133 is Chisholm and Renick's (1972) most creditable attempt to summarize schematically the airflow and echo structure of supercell storms.

The clear qualitative differences in airflow between multi-cell and supercell storms are complemented by quantitative differences in common characteristics. Thus, supercell updraughts are larger (see section on radar characteristics) and more intense than in most multi-cell storms. Both Chisholm and Renick (1972) and Brandes (1977a) quoted values of 25–40 m s^{-1}. In addition supercell updraughts are comparatively steady state, notably in their lower and middle reaches. Indeed this particular feature constitutes the major characteristic of the storm as a whole. However, in its uppermost parts the updraught is complex and unsteady (Nelson and Braham, 1975), possibly due to periodic loading with precipitation followed by a new upper level updraught core on the inflow side of the previous one. With a recurrence interval of 9–12 min, such behaviour would explain surface hailstreaks (Changnon, 1970) and Phillips's (1973) observation of a precipitation pattern that was "decidedly cellular" despite the "quasi-steady nature of the radar echo structure". Supercell storms usually travel to the right (by as much as 50°)

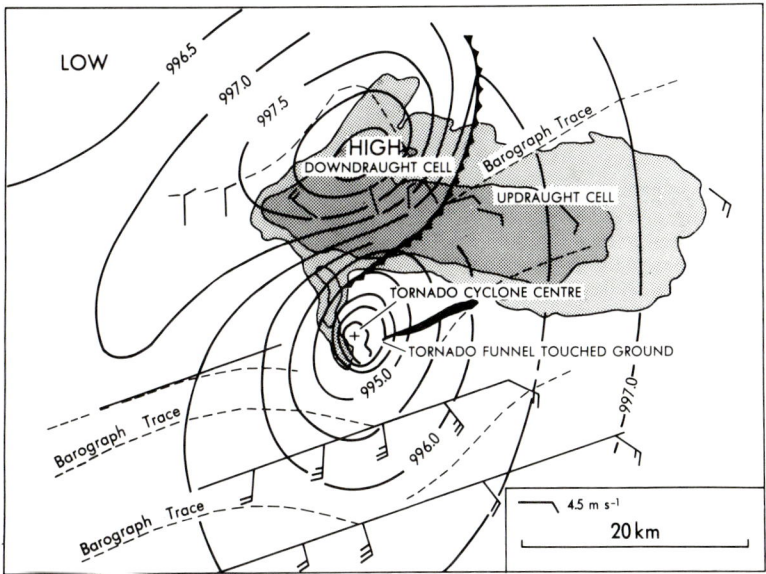

FIG. 134. Tornado cyclone within a severe storm. Solid lines represent isobars (in millibars). Shaded areas represent storm echoes of varying intensity. Frontal symbol represents gust front at head of outflow from downdraught. (After Fujita, 1958.)

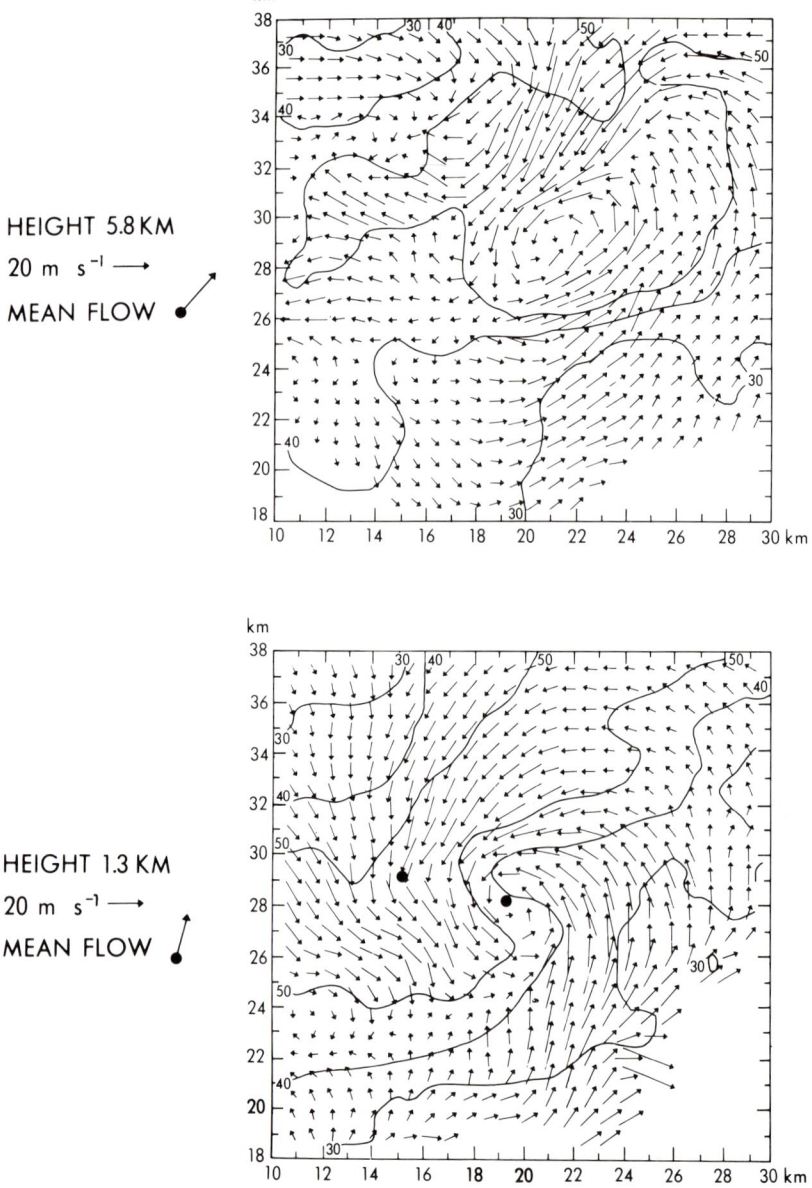

FIG. 135. Four horizontal sections of a tornado vortex forming in a severe storm. The sections are at heights of 5.8, 2.8, 1.3 and 0.3 km. Arrows show horizontal velocities at about every 0.5 km horizontal distance. Radar reflectivities (in dBz)

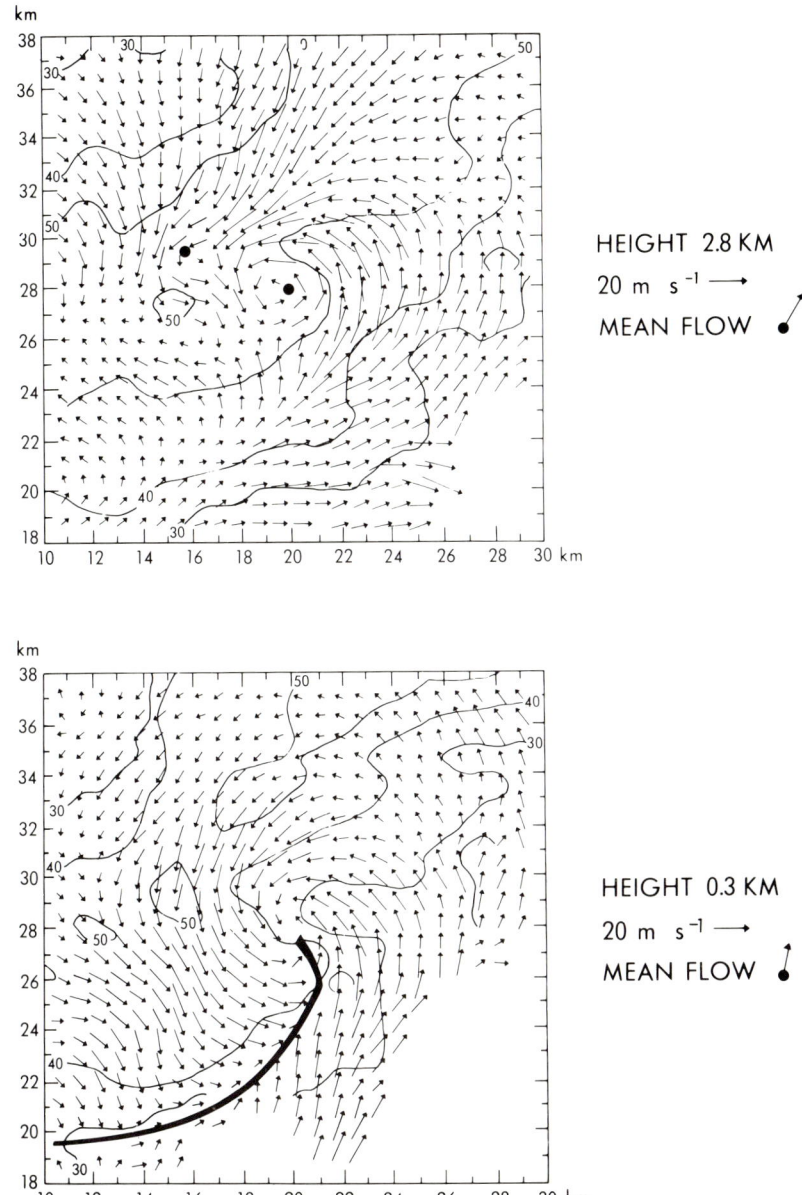

shown by solid lines. Major surface wind discontinuity shown by bold line and anomalous shear zones ($\geq 2 \times 10^{-2}$ s^{-1}) by a bold dot. (After Brandes, 1977b.)

and slower than the mean winds, the deviation being greater the larger the storm in accord with Newton and Fankhauser's (1964b) results. Continuous, as opposed to discrete, propagation is held to account for this strong movement to the right. Supercell storms have been observed to rotate but not to split.

3. *Associated circulations*

Closely associated with the essential elements of cumulonimbus airflow, namely the up- and downdraughts, are other circulations which at present cannot be attributed solely to either multi-cellular or supercell storms. There appear to be three such circulations that are worthy of attention: the tornado cyclone; the thunderstorm high and "wake" low couplet; and wake vortices.

(a) *Tornado cyclone*   Three decades ago Brooks's (1949) analysis of microbarograph traces in the vicinity of tornado paths led him to conclude that a tornado is surrounded by a low pressure system much larger than the tornado itself, but one or two orders of magnitude smaller than ordinary cyclones. He called the low surrounding the tornado a "tornado cyclone". These cyclones are frequently 5–10 km in horizontal extent and, as the name implies, have a cyclonic circulation. They have been most frequently observed in the right rear quadrant of a storm. The radar signature of such a feature is a "hook" echo, first photographed and analysed by Stout and Huff (1953). Later Fujita (1958) clearly demonstrated that this originally observed hook echo was associated with a tornado cyclone some 50 km in diameter. Figure 134 illustrates a typical tornado cyclone situation.

The advent of doppler radar has allowed direct observation of the cyclonic flow in both the surface "tornado cyclone" and the body of the parent severe storm (Ray et al., 1975). Brandes (1977b,c) has produced ample evidence of strong cyclonic circulation throughout a depth of nearly 5 km within a storm. There was some suggestion that the circulation started at the upper levels and moved downwards, eventually resulting in a marked gust front at the surface. Brandes was most careful not to call the "meso-cyclone" a tornado cyclone, but the circulation was in fact closely associated with the formation of a tornado and in all probability is the same feature that Fujita and other earlier investigators would call a tornado cyclone. Figure 135 shows the structure of the circulation at the time of the appearance of the tornado vortex. Clearly this type of circulation is a far from trivial characteristic of severe local storms. Indeed these doppler measurements show clearly that the updraught does in fact rotate and that this rotation is closely linked to both the "tornado" cyclone and the gust front that was discussed earlier in this chapter. Whereas the tornado cyclone is but a part, albeit an important one, of the parent storm, it is itself the "parent" to yet smaller vortex features that

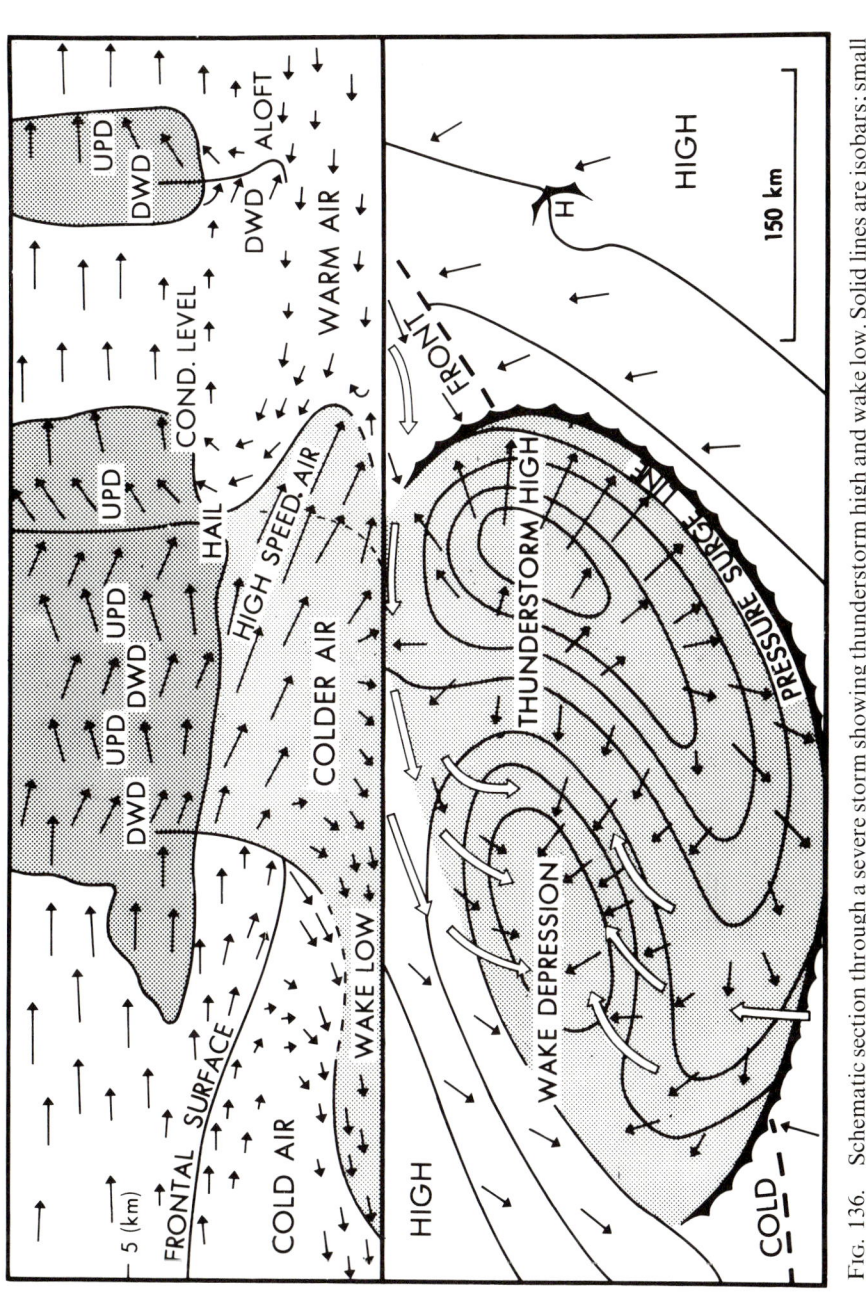

FIG. 136. Schematic section through a severe storm showing thunderstorm high and wake low. Solid lines are isobars; small arrows show airflow due to meso-scale pressure field; large arrows show disturbed flow due to presence of whole meso-system. (After Fujita, 1955.)

may be the direct driving forces of tornadoes. These vortices are not only too small to be of relevance to this book but also are very difficult to observe and analyse. Nevertheless attempts are being made (Agee *et al.*, 1976).

(b) *Thunderstorm high and wake low* When the airflow in the vicinity of severe storms is analysed indirectly in terms of the perturbation of pressure resulting from the storm passage, two clear circulations appear to emerge: the thunderstorm high and the "wake" low (Fig. 136). Clear examples have been provided by Fujita (1955) and Pedgley (1962). Both features may have pressure amplitudes of up to 5 mb relative to the undisturbed pressure field. In horizontal dimensions they are a function of the size of the parent storm. For a large multi-cell storm the high may be 300 km along its largest axis and the low 120 km. Each cell of the storm probably produces its own high, which merges with its neighbour to produce the overall storm high. Within such highs flow is strongly divergent (10^{-3} s^{-1}) near the ground, less so (10^{-4} s^{-1}) at a height of 1–2 km (Fujita, 1959*a*). The detailed mechanism of the high is considered in the section on theory, but it is clear at this stage that it is closely related to the downdraught and outflow. The wake low is far less well documented than the high. Fujita (1955) argued that it is to be expected if the high moves through the atmosphere faster than the winds. If this is the case, wake lows may be closely related to or even be identical with wake vortices.

(c) *Wake vortices* Careful observation by radar of bundles of chaff released upwind of a severe storm has revealed that indeed the storm—in particular its middle level part—may act rather like a cylinder in a stream (Jessup, 1972). This being so, the ambient airflow bifurcates and leads to the shedding of small vortices on the leeward side, some cyclonic, others anticyclonic (see Fig. 137). Jessup suggested that this behaviour is consistent with the shedding of a Kármán vortex train (see Chapter 4). Lemon (1976) has further shown that many such vortices are consistently anticyclonic, at variance with von Kármán theory. He suggests that these anticyclonic airflows are "starting" vortices which could develop during the transition of a storm updraught from non-rotational to rotational state.

4. *Mass and water budgets*

Our knowledge of the mass and water budgets together with the associated precipitation efficiency of storms results largely from American projects suggested by C. W. Newton. Based upon soundings in the neighbourhood of a squall line, the air and water fluxes in different branches of a single storm circulation are shown in Fig. 138. Of the 700 kton s^{-1} of air entering the updraught at low levels, about 20 per cent may circulate through the stratosphere tower of a pulsating storm whereas in a "steady state" storm, up

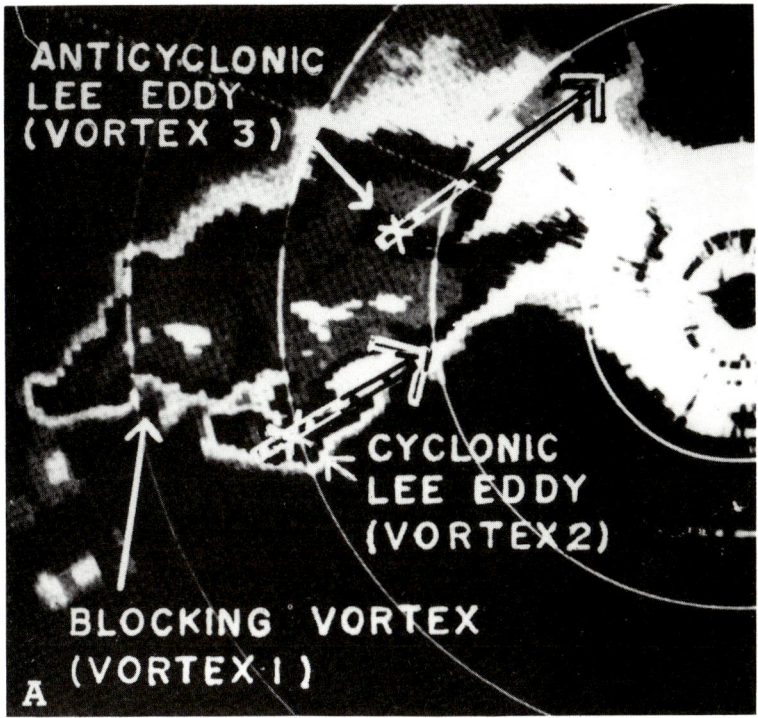

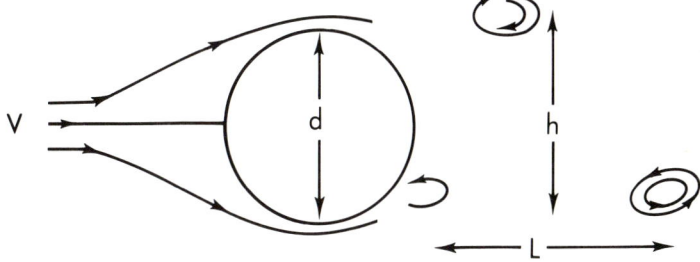

Fig. 137. Radar echo patterns of vortices in the wake of a severe storm. The patterns resemble a Kármán vortex train if vortices 2 and 3 are considered as lee eddies downstream from vortex 1. Arrows represent the direction of relative motion of the lee eddies over a period of 40 min. Dots along the arrows indicate approximate 10 min positions of vortex centres. (After Jessup, 1972.)

to 60 per cent may do so. To accommodate this massive upward flux of air the anvil expands downwind at a rate of $2 \, \text{km}^3 \, \text{s}^{-1}$. The lower tropospheric downdraught, with a mass flux of about $400 \, \text{kton} \, \text{s}^{-1}$, is about 60 per cent of the updraught flux.

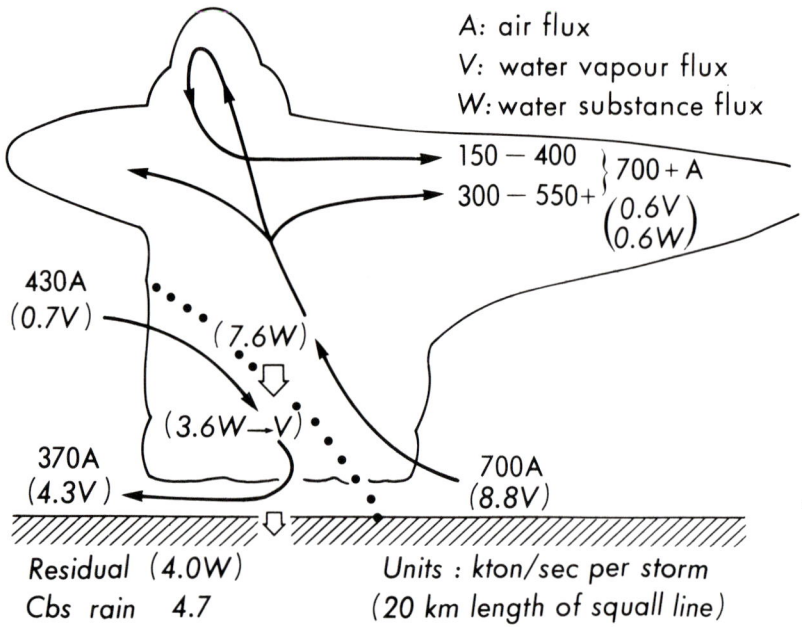

Fig. 138. Schematic diagram of airflow, and fluxes of air, water vapour and water within a single thunderstorm in a squall line. Fluxes are in kilotons per second per storm (20 km length of a squall line). (After Newton, 1966.)

The mass of water vapour entering the updraught is 8.8 kton s^{-1}. With a mixing ratio of 0.9 g kg^{-1} and an assumed water density of 0.3 g m^{-3}, 0.6 kton s^{-1} of vapour and 0.6 kton s^{-1} of water are exported in the anvil expansion, leaving 7.6 kton s^{-1} and the outflow to the rear at low levels is 4.3 kton s^{-1}, thus requiring 3.6 kton s^{-1} of this rain to be evaporated in the downdraught. This leaves a residual of 4.0 kton s^{-1} water available for surface precipitation. Newton's (1966) observations showed that the mean precipitation for the storm that was the object of analysis was 4.7 kton s^{-1}.

From the above it is clear that 80 per cent of the water vapour entering the updraught condenses and falls out aloft: but roughly half of that amount evaporates into the downdraught leaving the remainder to fall to the surface. Thus the "precipitation efficiency", or ratio of precipitation to vapour flux, is 53 per cent. Other observations (Table 45) support this figure, with the exception of Braham (1952) whose data were gathered from a small, single-cell air mass thunderstorm. Braham attributed his low values to the facts that most of the water condensed in a short-lived cell remained aloft as cloud droplets or was evaporated from the sides. In the large storms, of the kind treated in this chapter, the amount of precipitation deposited is essentially equivalent to the moisture convergence in low levels. Below

700 mb the vapour accumulation in the current case was $4.5\,\mathrm{kton\,s^{-1}}$ compared with $4.7\,\mathrm{kton\,s^{-1}}$ rainfall. Such a result is consistent with earlier comparisons of rainfall with water vapour convergence over broader areas encompassing convective precipitation regions.

5. *Squall lines*

As hinted earlier in this chapter scale concepts may be profitably applied to the study of severe local storms. Retaining our focus upon meso-scale circulations, only brief attention was given to single cell storms, but they were included to provide some context for their larger counterparts. For the same reason we conclude this observational half of the chapter with a brief look at the larger convective systems within which severe local storms frequently form.

Squall lines, or instability lines (Fulks, 1951) as they are alternatively known, are essentially synoptic scale lines of severe convective storms usually occurring 200–300 km ahead of a cold front in the warm sector of a mid-latitude cyclone. They are frequently hundreds of kilometres long, 50–100 km wide and lie roughly parallel to the front. Composites of radar PPI scopes gave us our first complete pictures of squall lines but now they are regularly visible on satellite photographs. Although these lines no doubt still exist, the terms "squall line" and "instability line" appear rarely if at all in the literature published since about 1960, the time, it will be noted, when research emphasis shifted to the individual storm. Between 1945 and 1960 the specially mounted Thunderstorm Project and the increasing operational use of radar spawned a series of articles concerned with the nature (Brunk, 1953; Fujita, 1955) and origin (Crawford, 1950; Newton, 1950; MacDonald, 1952) of squall lines. Some of them, for example that by Tepper (1950), produced interesting speculations about the mechanism of squall lines—speculations that subsequent observations did not support.

Of more direct interest to our current treatment are the analyses of "squall lines" by Fankhauser (1964) and Pedgley (1962). On the one hand Fankhauser recognized the squall line to comprise smaller, meso-scale "line segments", whereas Pedgley identified "clusters" of storms which moved along the "instability line".

The line segments were observed to be over 100 km long at times, about 30 km wide, to comprise several large, isolated and widely separated storms and to have a life history of about 5 h. In their lifetime, new cells tended to form 20–25 km from the right-hand end of the existing line segment, to move through the line segment and die out on the "left-hand" end—the right and left being relative to the direction of the path of movement of the line segment. The line segments as a whole appeared to move in a direction perpendicular to their orientation and in so doing cause the whole squall line

Table 45 Water content and water fluxes in some convective clouds

Source	System	Air flux (kton s^{-1})	(A) Vapour flux (kton s^{-1})	(B) Surface rain (kton s^{-1})	Precipitation efficiency, B/A (%)	Remarks
Auer and Marwitz (1968)	8 hailstorms, NE Colorado	310	3.0	—	—	
	8 hailstorms, NE Colorado	—	3.3	1.7	53	A different sample
	9 hailstorms, S Dakota	130	1.1	—	—	
	1 hailstorm, Oklahoma	440	6.6	—	—	
Austin et al. (1961)	8 squall lines, New England	—	—	4.1	—	Per 20 km length of squall line, in some cases roughly comprising one large thunderstorm
Braham (1952)	Average thunderstorm cell, Ohio and Florida	50	0.64	0.07	11	Mostly cumulus congestus and thunderstorms in early stage of development. Air fluxes at 500–600 mb; value uncertain and probably too large
Fankhauser (1965)	Squall line, Texas–Iowa	—	—	5.6	—	Per 20 km length of squall line, in some cases roughly comprising one large thunderstorm
Fankhauser (1967)	Large thunderstorm (50 km diameter)	—	24–32	13–16	52–66	

Reference	Description				Notes
Fankhauser (1971)	Large hailstorm	3000	—	—	Sub-cloud inflow
Foote and Fankhauser (1973)	Large hailstorm, Colorado	{ 2000 { 700	13 4.5	2 —	Inflow to updraught Outflow from downdraught
Fujita and Byers (1962)	Large hailstorm NE Colorado	550	—	—	Based on volume of fully-developed storm and its estimated past lifetime
Kropfli and Miller (1976)	Decaying supercell, Colorado	{ 230 { 220	1.4 —	—	At base of the updraught Downdraught at height of 3 km
Newton (1963)	Squall line, Ohio	200	3.4	—	Per 20 km length of squall line, in some cases roughly comprising one large thunderstorm
Newton (1966)	Squall line, Ohio	700	8.8	4.7	Per 20 km length of squall line, in some cases roughly comprising one large thunderstorm
Ulanski and Garstang (1978)	Thunderstorms, Florida	—	0.18	0.09	
Weickmann (1964)	Excessive hailstorm, Kansas	4800	43	8.6	Air and vapour fluxes estimated from observed surface hailfall and assumed hail production efficiency of 20%

Partly after Newton (1969).

to move in the same direction. Pedgley's "clusters", bearing some resemblance to the systems analysed by Sanders and Paine (1975), also comprised cells which formed on the right-hand side, moved through, and died on the left-hand side. But in contrast to the line segments, they moved along the length of the quasi-stationary instability line rather than in a direction perpendicular to the line orientation. Nevertheless, as Newton and Fankhauser (1964b) noted, in both types of meso-system, line segment or irregular cluster, the pattern of new echo formation is such as to cause the echo complex as a whole to move farther to the right of the upper winds than do the individual storms comprising the complex. The differences in morphology and movement of the two types of convective meso-system remain unexplained.

C. Tropical severe local storms

In common with most tropical circulations our knowledge of meso-scale, tropical, severe convective storms is meagre when compared with that of their mid-latitude counterparts. Indeed it was really only in the 1970s that such tropical systems came under close scrutiny. In the 1960s data from the Line Islands Experiment provided a first glimpse of these systems but substantial data arrived only with the Venezuelan International Meteorological and Hydrological Experiment (VIMHEX) and the GARP Atlantic Tropical Experiment (GATE) (GARP: Global Atmospheric Research Programme). Particularly as a result of the latter we now have a reasonable description of what has been called the "tropical squall (or disturbance) line". This term is generally agreed to mean the following: "squall line" refers to the line of cumulonimbus clouds and heavy precipitation forming along the downdraught squall front. The term "squall-line system" refers to the entire disturbance consisting of the squall line, its trailing anvil cloud, the squall front, downdraughts and all associated precipitation.

From the point of view of horizontal size, mid-latitude and tropical squall-line systems are quite similar. Each system can profitably be considered to comprise three scales of motion. At the largest scale is the whole squall-line system itself, which, as noted in the section on mid-latitude storms, is virtually a synoptic scale feature. Within this system, Houze (1977) has recognized what he calls "line elements" which are immediately compared in size to the mid-latitude "line segments" identified by Fankhauser (1964). These line elements are of the order of 60–100 km long, about 30 km wide, 14 km high and last for about $2\frac{1}{2}$ h (Houze, 1977). They are the primary building blocks of the tropical squall-line system. At the third and smallest scale are the individual cumulonimbus clouds within the line elements. In fact good observations of these third level systems are sparse and much of the literature, both observational and theoretical, is necessarily concerned with the two-

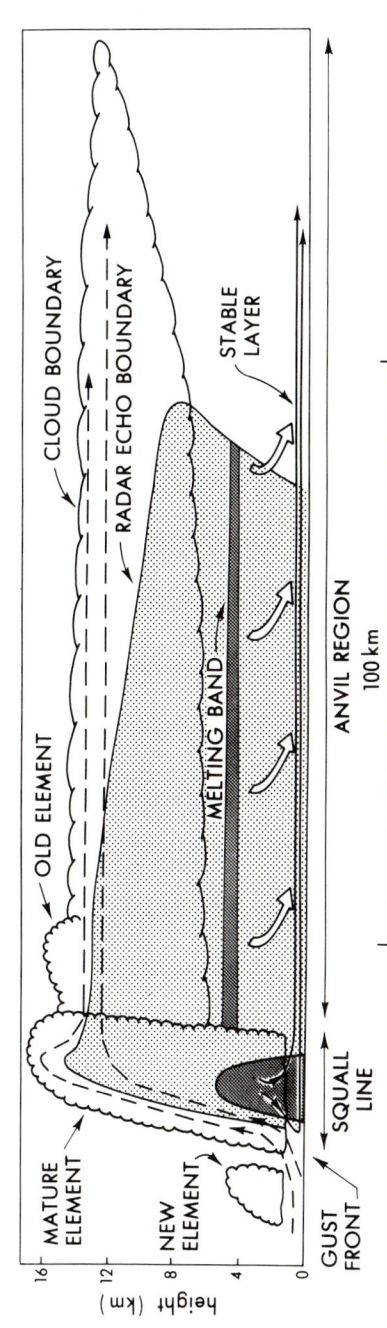

FIG. 139. Schematic cross-section through a tropical squall-line system. Streamlines show flow relative to the squall line. Dashed streamlines show updraught and anvil flow; thin solid streamlines show the convective scale downdraught and outflow; wide arrows show the meso-scale downdraught below the base of the anvil cloud. Dark shading shows strong radar echo in the melting band and in the heavy precipitation zone of the mature squall-line element. Light shading shows weaker radar echoes and scalloped line shows the visible cloud boundaries. (After Houze, 1977.)

dimensional structure of line elements, assuming no variation along their length. Although the two-dimensional approach has limitations (as revealed by Moncrieff and Miller (1976)), it is a useful preliminary step and the remainder of this section outlines the most important results.

1. *Storm circulation*

As a result primarily of the efforts of Zipser (1969, 1977), Betts (1976), Miller and Betts (1977), Betts *et al.* (1976) and Houze (1977), the two-dimensional structure of the tropical squall line has been sketched out (Fig. 139). At first sight it resembles the mid-latitude model morphologically, but it differs in at least two important ways. First, the anvil streams off *behind* the storm as opposed to in front of it; second, both up- and downdraught air originate

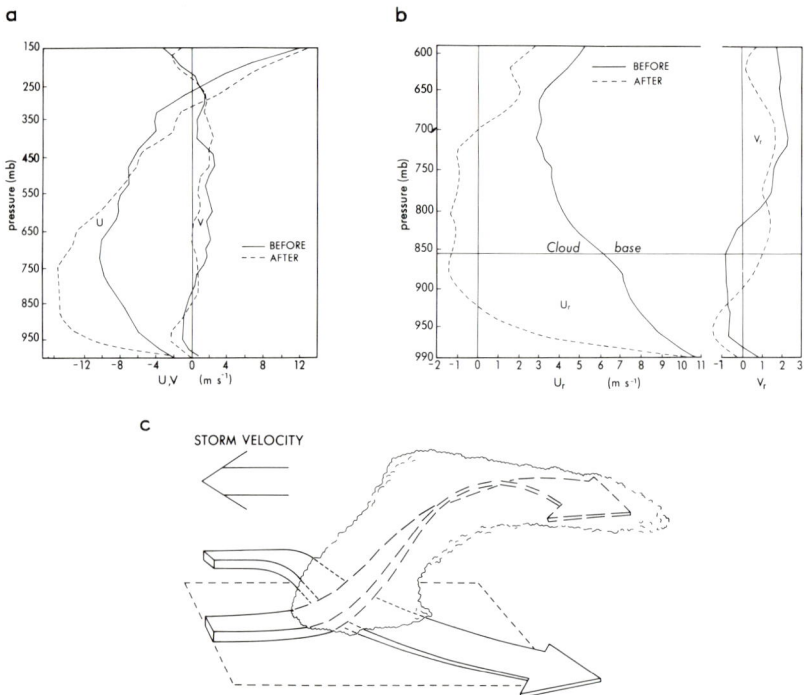

FIG. 140. (a) Mean tropospheric profiles of wind components along (u) and normal to (v) storm direction before and after the passage of a travelling convective storm. Positive values in direction of storm movement. (After Miller and Betts, 1977.) (b) Mean lower tropospheric profiles of wind components (u_r, v_r) relative to the storm before and after the passage of the storm. Horizontal line near 850 mb corresponds to cloud base. (After Miller and Betts, 1977.) (c) Schematic three-dimensional airflow in a tropical storm that propagates relative to all winds in the cloud layer as shown in (b). (After Moncrieff and Miller, 1976.)

from the front of the system. This second difference results from the ambient wind field (Fig. 140(a)) and necessitates a three-dimensional flow (Fig. 140(c)). In turn, it means that the storm propagates relative to all winds in the cloud layer (Fig. 140(b)) in contrast to mid-latitude systems which tend to move at a velocity equal to that of the ambient wind at some level within the cloud layer. Clearly the gross circulations of tropical and mid-latitude squall-line elements/segments are quite different.

Figure 139 shows that the squall-line system consists of a squall line forming the leading edge of the system and a trailing anvil region. New line elements form ahead of the squall line. Old line elements weaken toward the rear of the line and blend into the trailing anvil region as they dissipate. Each line element progresses through a period of rapid growth, with echo tops penetrating the tropopause to maximum heights of 16–17 km, then decreasing to heights of 13–14 km, which corresponds to the height of the anvil cloud with which the line elements merge at the end of their lifetimes. The squall line is located along the leading edge of a meso-scale downdraught, first noted by Zipser (1969), which spreads out in the middle and lower troposphere below the anvil cloud. Within the squall line, individual line elements contain smaller, convective (or cloud) scale downdraughts which penetrate to the Earth's surface. This cold, convective scale downdraught air also spreads out in a layer 200–400 m deep toward the rear of the system. The top of this layer of cold air is bounded by a stable layer. Houze (1977) showed that precipitation falling from the trailing anvil of a particular line element observed in GATE was stratiform and accounted for 40 per cent of the total rain from the squall-line system. He concluded that much of the anvil cloud comprises successive weakened, but precipitation-laden, old line elements from the back edge of the squall line, possibly aided by upward air motion within the upper level anvil cloud.

III. Theory

Despite enormous efforts in the last four decades a sound theoretical knowledge of atmospheric convection still lies beyond our grasp. This is perhaps not surprising in view of our still limited observational knowledge of the internal mechanisms of both cumulus and cumulonimbus circulations. The first half of this chapter has outlined the current state of our empirical knowledge of cumulonimbus and stressed that it is within only the last few years that doppler radar has allowed us to "see" inside these storms. Notwithstanding these difficulties a welter of literature exists on the theory of cloud- (as opposed to planetary-) scale convection, much of it concerned with the thermals and plumes that may comprise cumulus convection. But this material is of no concern here, as we are interested in those mechanisms of primary importance only to severe convective storms. Given that atmospheric instability is an important prerequisite for severe storms (as outlined in the first part of this chapter), the important question is: how does the convection subsequently become severe, organized and long-lasting, resulting in a self-propagating cloud system that produces at the Earth's surface very heavy precipitation (frequently including large hail) and high wind speeds? This question provides the central theoretical problem for analysts of both extra-tropical and tropical severe local storms. Within this broad topic, and in both types of storm, specific attention has been given to evaluating the effects upon storm behaviour of two main factors: first, vertical shear of the ambient wind—in both speed and direction; second, the cloud microphysics. In addition some attention has been given to the effects of a rotating updraught and the speed of the gust front.

Theoretical approaches to the problem of severe local convection may be considered to be either "diagnostic" or "prognostic". In the former category are those studies that seek to explain a set of observations and, in so doing, provide a theory (or at least hypothesis) that can be tested against later observations. In the latter category are the analytical and numerical studies that predict from the hydrodynamical and thermodynamical equations certain aspects of storm behaviour which again must be compared to reality. Among the leading exponents of the diagnostic type, in addition to F. H. Ludlam, were F. C. Bates and C. W. Newton. Their primary concerns were, respectively, with the internal structure of the cumulonimbus and with the movement of such storms. Unfortunately Bates wrote little in the open literature, but we do have his doctoral thesis (Bates, 1961) and an appreciation of his work by his colleagues (Severe Storms Research Group of St Louis University, 1970). In contrast, Newton has been prolific and is generous in his acknowledgement of Bates's contribution to his own work. Their work is reviewed below.

Within the "prognostic" approach, numerical modelling is far more popular than analytical methods. The numerical models can be conveniently divided into two classes. First are the zero- and one-dimensional models, which involve rather gross simplification of the fluid dynamics but may give careful attention to the microphysical processes. In a zero-dimensional cloud model, the growth of cloud and precipitation particles is studied by integrating the time-dependent equations of condensation, coalescence and other microphysical processes with the vertical velocity of the air specified as constant or as a known function of time. In one-dimensional clouds, height and time are independent variables allowing the simulation through time of the vertical development (the one dimension) of the cloud. Although the basic assumptions that the motion is axially symmetric, that horizontal pressure gradients are ignored and that entrainment through the sides can be estimated only from gross properties of the plume are important limitations, this type of model allows almost any degree of complexity of microphysics to be simulated within a plausible fluid dynamical framework. The assumption of axial symmetry nullifies any value of one-dimensional models to the simulation of severe local storms. In essence, it generally prevents the possibility of modelling simultaneous up- and downdraughts.

The second class of models includes those with two and three dimensions and it is with these that studies of severe local storms have proceeded. Several two-dimensional models are now well-established and initial experiments in three dimensions look most promising. The principal problems to be overcome in the development of a three-dimensional model of a severe local storm appear to lie in the area of modelling approximations and numerical methods. As Lilly (1975a, p. 723) noted: "It appears possible in principle to incorporate all the dynamic and microphysical processes which have been previously studied in isolation into a more complex model, but the problem is to fit such a model into any available computer and integrate it in a reasonable time at a reasonable cost".

The remainder of this chapter is concerned with these theoretical developments in the following framework. Within each of two sections on extra-tropical and tropical severe local storms, the "diagnostic" and "prognostic" theories are reviewed. In essence the reviews will be chronological because this has the advantage of progressing from the simpler, two-dimensional models of squall lines, through two- to three-dimensional models of the individual cumulonimbus as a whole, to the simulation of its component parts, such as up- and downdraught and gust front. Note that with increasing experience and computing power, it is proving possible to simulate the smaller, more complicated systems that are the essence of the meso-scale severe convective phenomenon. Throughout the whole treatment the emphasis lies on cloud dynamics rather than cloud microphysics, except in so far as the latter influences the former.

A. Extra-tropical severe local storms

1. *Diagnostic studies*

Although his name appears only infrequently in the literature, it is clear from his contributions that F. C. Bates was a pioneer in severe storm studies. Before the appearance of the seminal paper by Browning and Ludlam in 1962, Bates already believed that the storm updraught was steady state, sloped toward the rear of the cloud, and in the more severe storms, frequently rotated. All three insights were ahead of their time in that they were based on comparatively little observational evidence. Bates developed a steady state, non-entraining, numerical model for hail- and tornado-producing storms. He assumed that inside the updraught the vertical and horizontal wind components, the moisture distribution and the buoyancy terms were invariant in any given horizontal slice. The storm comprised discrete horizontal slices of updraught air, like a stack of infinitesimally thin wafers of air. Each wafer would enter the base of the storm with some initial vertical velocity. As the wafer moved through the updraught region, it would be acted upon by the aerodynamic drag force and the Magnus effect in the horizontal and by gravity, the precipitation drag force and the pressure gradient force in the vertical. To maintain mass continuity, the thickness of the wafer was allowed to vary with height along its path through the storm. As the wafer neared the top of the updraught, the vertical dimension of the wafer would approach zero and the area of the wafer would increase infinitely. Thus, as the wafer traversed the storm, it would change its diameter and thickness according to the variations in the geometry of the storm figure, that is, the shape of the storm at any one time. Figure 141 shows the paths of eight such wafers in a storm moving at a constant speed. As the storm moves, wafers enter the base of the storm and move upward through it—so that, in Fig. 141, the wafer following trajectory number 1 has risen to the top of the storm while wafer number 8 is just entering the foot of the storm. The intermediate wafers are shown at their appropriate positions relative to the Earth. The geometry of the entire ensemble of wafers at any one given time defines the profile of the moving storm—a profile comparable with several in Fig. 132.

Complementing Bates's work and setting a trend for many later studies,

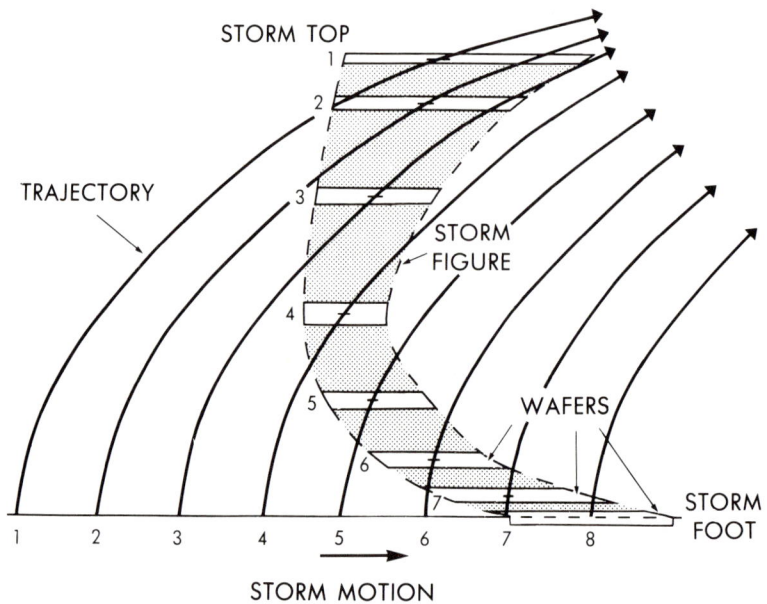

FIG. 141. Schematic vertical cross-section to show trajectories of eight "air wafers" relative to the ground together with the final storm figure. (After Severe Storm Group of St Louis University, 1970).

Newton and Newton (1959) made a preliminary analysis of the effects of vertical shear in the winds around severe storms. The stimulus for this work was the feeling that the observed shear that frequently (if not always) accompanied severe storms (noted in the first part of the chapter) should, in some way, affect the growth, propagation and thus movement of the storms. In addition storm resistance to deformation by strong shear, observed by Hitschfeld (1960), required explanation. In resisting, or indeed even thriving in vertical wind shear, severe cumulonimbus contrast markedly with cumulus, the growth of which is markedly inhibited by shear (Byers and Braham, 1949). But even cumulonimbus are inhibited by *severe* shear. Bhaskara Rao and Dekate (1967) concluded that deep convection was confined to layers in which relative top shear did not exceed 5 m s^{-1}: relative top shear was defined as that due to the difference between cloud velocity and the wind at the next level above cloud top. In their analysis of the effects of shear Newton and Newton (1959) simplified the problem by considering the vertical variation of wind to be undirectional: they then analysed the creation and effects of hydrodynamical pressures around the storm.

Figure 142(a) schematically illustrates a cumulonimbus in vertical shear. Two countering effects influence the horizontal velocity distribution within the cloud. First, elements of rising or descending air tend to carry with them the horizontal momentum they had at their levels of origin. This process tends to destroy the vertical wind shear within the cloud and to maintain mean in-cloud velocities ($\bar{V}_c$) different from those in the environment (V_e). Second, due to these differential motions (V_r), external boundary forces are set up which tend to shear the cloud away from the vertical.

Newton and Newton (1959) considered that, at any particular level, a convective cloud may be considered as an obstacle in motion relative to the ambient winds. As a result, an excess of pressure on the upstream sides and a deficit on the downstream sides of the obstacle occur. The resulting hydrodynamic pressure (defined as $b = p - p_h$, the departure of actual pressure p from the hydrostatic value p_h in the undisturbed environment) is distributed as shown by the plus and minus signs in Fig. 142(a). The horizontal accelerations due to these hydrodynamic pressure gradients lead to a shearing of the cloud. Vertical accelerations also occur and Newton and Newton showed that

$$\frac{dw}{dt} = g\left(\frac{\Delta T}{T_e} - \frac{b}{p} + \frac{\partial b}{\partial p_h}\right), \tag{195}$$

where ΔT is the excess of virtual temperature over that in the environment (T_e) at a given level. The second term on the right-hand side is very small, so equation (195) was simplified to

$$\frac{dw}{dt} \simeq g\left(\frac{\Delta T}{T_e} + \frac{\partial b}{\partial p_h}\right). \tag{196}$$

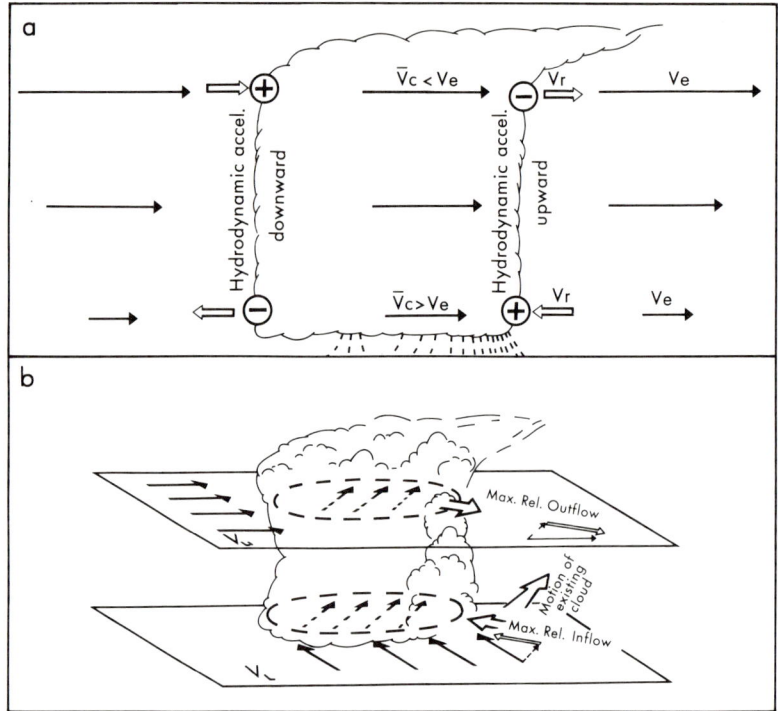

Fig. 142. (a) Schematic vertical cross-section of velocities of cloud (V_c), of the environment (V_e) and the resultant relative velocity (V_r). Plus and minus signs show areas of relative excess or deficits of pressure that result in upward or downward motion. (b) Schematic three-dimensional picture of ambient winds (upper level, V_u; lower V_l), in-cloud winds (dashed) and relative motion of the ambient with respect to in-cloud air (double arrows.) (After Newton, 1960b.)

If the hydrodynamic pressure decreases upward, an upward-acting force exists independent of the ordinary "parcel" buoyancy force associated with density anomaly (the first term). A net upward acceleration will exist as long as

$$\frac{\partial b}{\partial p_h} > -\frac{\Delta T}{T_e}; \qquad (197)$$

that is, potentially cool air could be lifted if the hydrodynamical pressure variation is strong enough. Such lifting is generally required for "triggering" unless the air mass is absolutely unstable. In Fig. 142(a) the distribution of b would favour lifting of the air adjacent to the existing cloud system on the downshear side, and inhibit it on the upshear side. Thus the same forces that tend to deform a cloud by shearing it out of the vertical, tend to promote new convection on a favoured flank of the cloud. Newton and Newton estimated that the maximum hydrodynamic pressures at the four points indicated in Fig. 142(a) would be about 0.5 mb. As negative pressure overlies positive at the front of the storm, the total vertical decrement would be about 1 mb. With a cloud 500 mb deep, the effect upon vertical acceleration would be equivalent to ordinary buoyancy with $\Delta T = 0.5\,°C$. If the downdraught outflow is incorporated, the maximum b at lower levels is increased by about four times and the fall of b throughout the depth of the cloud can be 2–3 mb, with associated increases in vertical acceleration.

It was noted in the first part of this chapter that, in association with speed changes, winds also frequently veer with height, particularly in organized convective situations in North America. In that case, the motions of the ambient air relative to in-cloud air are as shown in Fig. 142(b), where for simplicity the in-cloud velocity is shown as the mean through the depth of the cloud layer. In the lower levels, the relative motions are into the right-hand side of the storm with respect to mean motion of the in-cloud air. In upper levels, the opposite is true. Correspondingly the greatest positive pressure in lower levels is found on the right-hand side, beneath negative pressure in upper levels. Generation of new clouds is therefore most favoured on the right flank with respect to the mean wind in the cloud layer.

Newton and Newton's (1959) pioneering analysis of the effects of wind shear upon severe storms has been followed by several analytical and numerical investigations seeking deeper insights. Alberty (1969) provided a comparatively simple, specific test of their results using a quasi-steady state cumulonimbus model and he confirmed that vertical shear in unidirectional environmental flow led to enhanced vertical motion. Later models, reviewed below, were far more elaborate and provided useful information on facets other than wind-shear effects.

2. Prognostic studies

(a) *Two-dimensional models*   These will be considered in terms of analytical and then numerical results.

(i) *Analytical results*   In the last two decades both analytical and numerical two-dimensional models have been studied as a complement to the observational picture that was sketched with increasing confidence in the 1960s. Some of the earliest analytical work was undertaken by members of Imperial College, London, who attempted to explain the major features of the observational model produced by their colleagues Browning and Ludlam (1962). In particular Green and Pearce (1962) examined the relationship between, on the one hand, the convective overturning with a sloping updraught and, on the other, both vertical wind shear and the distribution of heat sources and sinks required for the maintenance of the convection. Pearce computed the temperature field required to maintain a flow pattern similar to that inferred in the storm studied by Browning and Ludlam (1962). He found it to be broadly consistent with saturated adiabatic ascent and descent with conservation of wet bulb potential temperature in the up- and downdraught respectively, except close to the interface between them. Green considered the thermodynamics, disregarding the details of flow inside the storm, and found some problems connected with the maintenance and slope of the interface between the up- and downdraught. It appeared that a steady state, two-dimensional flow pattern with conservation of wet bulb potential temperature could not exist and that the circulation in real storms must be essentially three-dimensional, involving motions which develop across the direction of the general winds. This early conclusion has amply been borne out by later observational and theoretical work (Moncrieff, 1978).

A decade later Moncrieff and Green (1972) re-appraised the problem of cumulonimbus convection in shear. Their treatment is more comprehensive than that by Haman (1976) but because of the similarity of approach and the relative simplicity of Haman's treatment, his analysis is used for illustration. In essence, a two-dimensional convective overturning as illustrated in Fig. 143 was analysed to establish the values of steering height and storm velocity. The effects of stability and shear were also examined. Using the Bernouilli equation as a basis, Haman showed that the difference in kinetic energy

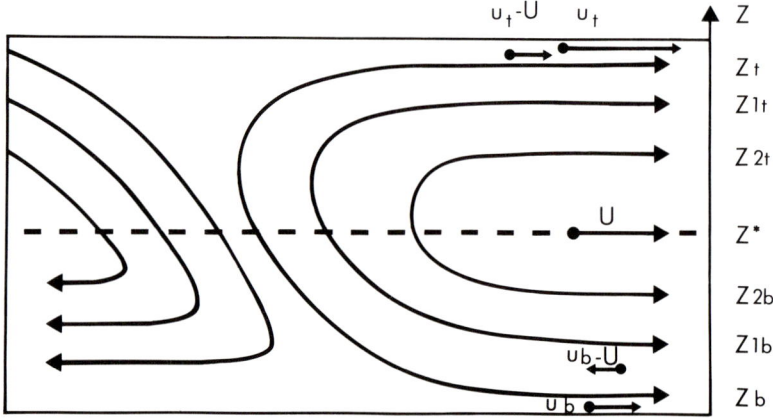

FIG. 143. Schematic vertical cross-section of airflow relative to a steady storm moving at the propagation speed. The undisturbed vertical wind shear is constant from left to right. u denotes the air speed relative to the ground far from the storm where the motion is purely horizontal. Indices "b" and "t" refer to the inflow and outflow levels along the same relative streamline far from the storm. U is the speed of the storm relative to the Earth—the steering, or propagation speed at the steering level z^*. (After Haman, 1976.)

between air at the top and bottom of the storm was due to buoyancy and pressure forces. In particular he showed that

$$\tfrac{1}{2}V_t^2 - \tfrac{1}{2}V_b^2 = \int_{\Psi_b}^{\Psi_t} \left(\frac{\rho_0}{\rho} - 1\right) d\Psi - \int_{p'_b}^{p'_t} \frac{dp'}{\rho}, \qquad (198)$$

where V_b and V_t are, respectively, the velocities of inflow and outflow at distances far from the storm, Ψ is the geopotential, ρ_0 is the density in environmental air undisturbed by the storm and p' is the deviation of pressure from the hydrostatic value in the unaltered environment. The first term on the right-hand side of equation (198) represents the work against the buoyancy forces whereas the second represents the work of non-buoyant pressure forces that are generated by the storm in the manner considered by Newton and Newton (1959).

The role of these perturbation pressure gradients is, as yet, unresolved, but Haman circumvented the problem by noting that the two terms on the right-hand side of equation (198) could be re-written as

$$\int_{p_b}^{p_t} \left(\frac{1}{\rho_0} - \frac{1}{\rho}\right) dp,$$

which defines the area between the in-cloud and environmental stratification curves on a thermodynamic diagram in $(p, 1/\rho)$ coordinates. This area is easily measured with the aid of aerological data plotted on such a diagram: Haman called this quantity B. Thus equation (198) reduces to

$$V_t^2 = V_b^2 + 2B. \qquad (199)$$

Aplying this result to the convective overturning illustrated in Fig. 143 we see that

$$(u_t - U)^2 = (u_b - U)^2 + 2B, \qquad (200)$$

where u_b and u_t are the air speeds far from the storm at heights equal to the bottom and top of the storm respectively and U is the steering speed of the storm. Haman then defined $S = u_t - u_b$ as the integrated shear in the convective layer and noted that the geometry of the system in Fig. 143 meant that S must be positive. Thus, a solution of equation (200) is

$$U = \frac{u_t + u_b}{2} - \frac{B}{S}, \qquad (201)$$

which shows that buoyancy and wind shear depress the steering speed below the mean velocity of the convection layer. Haman went on to show that a necessary condition for the possibility of two-dimensional, quasi-steady overturning is

$$-\tfrac{1}{2} < B/S^2 < \tfrac{1}{2}. \qquad (202)$$

The term $-B/S^2$ is usually interpreted as a kind of bulk Richardson number for this type of overturning of the convective layer.

Equation (201) is an attractively simple predictor of the storm velocity, as all the terms on the right-hand side are easily evaluated from operational data. Haman tested his model against two real storms and his predicted velocities were within 5 per cent of those observed.

Although also using the Bernouilli approach, the main tool employed by Moncrieff and Green (1972) in their examination of essentially the same problem was the vorticity equation. By manipulating the y-component of this equation they derived the following equation relating flow to stability and vertical shear:

$$\frac{1}{\rho}\frac{\partial}{\partial z}\left(\frac{1}{\rho}\frac{\partial \psi}{\partial z}\right) \approx \frac{g(\gamma - B)(z - z_0)}{2A\rho(z_0)(z_* - z_0)} + \frac{2A}{\rho(z_0)}, \qquad (203)$$

where $\rho(z_0)$ is air density at height z_0; z_0 is the height of an inflow streamline far from the storm; ψ is a stream function such that $\rho u = \partial \psi/\partial z$ and $\rho w = -\partial \psi/\partial x$; B is $\partial \phi/\partial z_0$; ϕ is $\ln \theta$, where θ is potential temperature; A is $\partial U_0/\partial z$, where U_0 is the inflow speed of a particle at a height z_0 on the streamline $\psi = $ constant; z_* is the height of the steering level of the storm: and the symbol $\approx$ means asymptotically equal to. The general configuration of flow is as shown in Fig. 143. Solution of equation (203) gave the height of the steering level in terms of the height (H) of the storm. As the vertical wind shear was initially specified, a knowledge of the steering height allowed easy calculation of the storm propagation speed relative to the ground. Moncrieff and Green expressed their results as a relationship between the steering height, a bulk Richardson number and a density-scaling parameter, and compared their predictions with at least six real storms. In most of the comparisons predictions were good: only in the case of a "severe right-moving" storm did the theory fail badly and this was probably due to its lack of the third dimension—an important facet of such storms, as noted earlier in this chapter. In addition to these important results Moncrieff and Green (1972) evaluated the effects of the perturbed pressure field upon vertical velocity and general cloud growth. They found that the error in updraught speeds due to neglect of the horizontal pressure gradient could range from 13 to 42 per cent. In their own words (Moncrieff and Green, 1972, p. 349):

> The horizontal pressure gradient therefore exerts an important control on the updraught speed and probably the growth rate as well. Moreover, since the horizontal gradient of non-hydrostatic pressure is large in cumulonimbus clouds (of the order of 1 mb/10 km) its inclusion in numerical models of these clouds is essential. The fact that non-hydrostatic pressure will inhibit the convection should be noted because it is sometimes quoted as having an enhancing effect.

(ii) *Numerical results* Probably the first attempt to analyse severe meso-scale convection by numerical methods was that by Sasaki (1959). Using the equations of motion in a vertical plane perpendicular to a squall line together with the temperature and humidity equations, Sasaki ran a simulation for 10 time steps of 6 min within a domain 278 km wide and 600 mb deep. The grid sizes were 18 km and 100 mb respectively. Starting with a sharp front extending to 700 mb, the model generated upward motion and condensation after 30 min ahead of the cold front, thus providing the first crude simulation of squall-line convection. Sasaki was well aware of the limitations of his analysis, largely due to limited computing capacity; nevertheless he had produced a pioneering study a decade ahead of those who followed. But the vast increases in computing power and progress in both theory and numerical methods in the 1960s eventually led to the far more detailed and realistic models of the 1970s which are reviewed below.

Several attempts at two-dimensional simulation of severe convective storms appeared in the early 1970s. Their aims were twofold: some concentrated on storm dynamics; others dealt primarily with the cloud microphysics. The first aim is of main interest here.

Efforts to fulfil this aim have taken two main forms. On the one hand are the models comprising analytically expressed "initial" conditions of vertical velocity, temperature, mixing ratio and updraught cross-section shape. These conditions are appropriate to an already well-developed storm. On the other hand are the more conventional models which initiate and develop the storm in the vertical plane. Both types are reviewed below.

Of the first type, Lin and Martin (1971) specified a non-entraining, axially symmetric, steady state storm with a sloping updraught. They showed that the vertical velocity of the updraught was sensitive to its radius, shape and slope. In general large (radius 4.5 km) and sloping (up to 30° from the vertical) updraughts had the largest vertical velocities. A later version of the model included a rotating updraught (Lin and Martin, 1972) and showed that temperatures at heights of 6–9 km within the rising air could be nearly 4 °C warmer than their environment. Associated with these changes the calculated perturbed pressure field in the centre of the storm ranged from a 9 mb deficit at heights of 1–3 km to +0.2 mb at 13–15 km. Whilst recognizing that 9 mb is a large figure, Lin and Martin claim that pressure jumps of this magnitude were reported by Fujita (1963) in association with the passage of a severe storm.

The second group of models had several common characteristics, including time-dependence, some recognition of basic cloud microphysics, but the exclusion of the ice phase, exclusion of coriolis and earth-friction effects, and the use of the so-called "anelastic equations" (Ogura and Phillips, 1962). In these equations the local time changes of air density are neglected in the continuity equation with the effect of filtering out acoustic waves which are of no meteorological importance to the "deep" convection problem. The anelastic equations form the basis of the models by Takeda (1966a,b, 1971),

Schlesinger (1973a,b), Hane (1973), Orville and Kopp (1977) and Orville and Sloan (1970) but the latter two are primarily interested in cloud microphysics. The studies by Takeda (1971), Schlesinger (1973a) and Hane (1973) are conceptually very similar and their essentials are illustrated with Schlesinger's model.

The model comprised approximate forms of the horizontal and vertical equations of motion, equation of continuity, first law of thermodynamics, water vapour equation and liquid water equation. The basic equations of motion as used were

$$\frac{\partial u}{\partial t} = -u\frac{\partial u}{\partial x} - w\frac{\partial u}{\partial z} - \frac{1}{\rho}\frac{\partial p}{\partial x} + F_x, \tag{204}$$

$$\frac{\partial w}{\partial t} = -u\frac{\partial w}{\partial x} - w\frac{\partial w}{\partial z} - \frac{1}{\rho}\frac{\partial p}{\partial z} - g - \frac{q_w g}{\rho} + F_z, \tag{205}$$

where q_w is liquid water content. Takeda's (1971) equations were identical except for the use of a diffusion term instead of friction. In equation (205) the fifth term on the right-hand side represents the drag force of water drops. The continuity equation was

$$\frac{\partial}{\partial x}(\rho_0 u) + \frac{\partial}{\partial z}(\rho_0 w) = 0, \tag{206}$$

where ρ_0 is the horizontal average of density in the undisturbed air. The temperature equation was

$$\frac{\partial T}{\partial t} = -u\frac{\partial T}{\partial x} - w\left(\frac{\partial T}{\partial z} - \Gamma\right), \tag{207}$$

where Γ is the adiabatic lapse rate, dry or saturated as appropriate.

The water vapour equation was

$$\frac{\partial q_v}{\partial t} = -u\frac{\partial q_v}{\partial x} - w\frac{\partial q_v}{\partial z} - \frac{c_p}{L_{vw}}(\Gamma_d - \Gamma)w, \tag{208}$$

where q_v is the mixing ratio for water vapour and L_{vw} is the latent heat of vaporization.

The liquid water equation was

$$\frac{\partial q_w}{\partial t} = -u\frac{\partial q_w}{\partial x} - w\frac{\partial q_w}{\partial z} + \frac{c_p}{L_{vw}}\rho_0(\Gamma_d - \Gamma)w + \frac{\partial}{\partial z}(VL_p) + \frac{q_w w}{\rho_0}\frac{\partial \rho_0}{\partial z}, \tag{209}$$

where V is the relative fall velocity of water drops and L_p is the precipitation content. The latent heat release terms in equations (208) and (209) are clearly non-zero only when Γ represents the saturated adiabatic lapse rate. The continuity equation (equation (206)) implies the existence of a stream function ψ such that

$$u = \frac{1}{\rho_0}\frac{\partial \psi}{\partial z} \quad \text{and} \quad w = -\frac{1}{\rho_0}\frac{\partial \psi}{\partial x}.$$

A mass-weighted vorticity η was defined in terms of ψ, i.e.

$$\eta = \nabla^2 \psi = \frac{\partial}{\partial z}(\rho_0 u) - \frac{\partial}{\partial x}(\rho_0 w). \tag{210}$$

In the majority of the two-dimensional models the equations of motion are manipulated into a vorticity equation applicable to the (x, z) plane. Equations (204) and (205) were initially re-written as follows:

$$\frac{\partial u}{\partial t} = -u\frac{\partial u}{\partial x} - w\frac{\partial u}{\partial z} - \frac{1}{\rho_0}\frac{\partial p}{\partial x} + F_x, \tag{211}$$

$$\frac{\partial w}{\partial t} = -u\frac{\partial w}{\partial x} - w\frac{\partial w}{\partial z} - \frac{1}{\rho_0}\frac{\partial p'}{\partial z} - \left(\frac{p'}{p_0} - \frac{T_v'}{T_{v0}}\right)g - \frac{q_w g}{\rho_0} + F_z, \tag{212}$$

where $p = p_0 + p'$, p_0 is the initial (hydrostatic) pressure, p' is the deviation from p_0, $T_v = T_{v0} + T_v'$, T_v is the virtual temperature, T_{v0} is the horizontal average of the initial virtual temperature and T_v' is the deviation of T_v from T_{v0}. After multiplying equations (211) and (212) by ρ_0 and cross-differentiating them, the following vorticity equation resulted:

$$\frac{\partial \eta}{\partial t} = -u\frac{\partial \eta}{\partial x} - w\frac{\partial \eta}{\partial z} + g\rho_0\left(\frac{1}{p_0}\frac{\partial p'}{\partial x} - \frac{1}{T_{v0}}\frac{\partial T_v'}{\partial x}\right) + g\frac{\partial q_w}{\partial x}$$

$$+ \frac{2w}{\rho_0}\frac{\partial \rho_0}{\partial z}\left(\eta - u\frac{\partial \rho_0}{\partial z}\right) + uw\frac{\partial^2 \rho_0}{\partial z^2} + \frac{\partial(\rho_0 F_x)}{\partial z} - \frac{\partial(\rho_0 F_z)}{\partial x}. \tag{213}$$

The term $(g\rho_0/p_0)(\partial p'/\partial x)$ represents the horizontal gradient of the perturbed pressure, a problematical component of all cumulonimbus modelling. Whilst recognizing its importance Takeda (1971) ignored this term. Schlesinger (1973a) evaluated it by substitution from equation (211).

Using these, or similar equations, Schlesinger, Takeda and Hane ran models in domains varying in dimensions from 25 to 173 km in the horizontal and from 10 to 14 km in the vertical. Horizontal grid distances ranged from 400 m to 3.2 km and in the vertical from 400 to 700 m. Time steps were typically less than 20 s. In general the domains were closed, particularly at the top and bottom. Lateral boundary conditions were necessarily more artificial. The initial conditions most frequently employed by each author are illustrated in Fig. 144, clearly showing that the effects of stability and, more importantly, vertical wind shear were being assessed in these studies. In all three cases an initial perturbation was introduced into the model to "trigger" the convection. In fact, in Hane's case, the initial perturbation was a well-developed thunder-

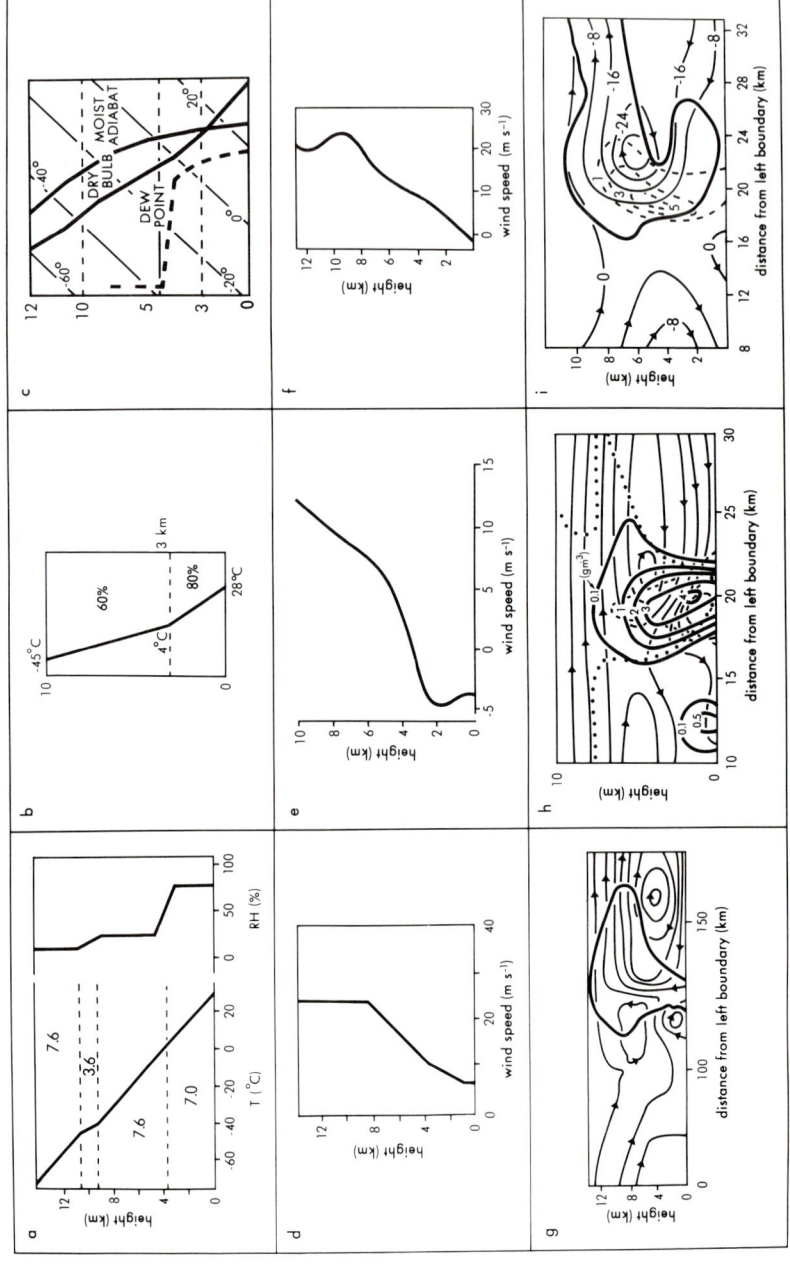

storm as he was primarily concerned with the maintenance of the convection rather than its initiation and subsequent development. Several variations of stability and shear were investigated and Fig. 144 illustrates the airflows (appropriate to the given initial conditions also shown in the figure) which each author considered to be his best simulation of reality. In general these first two-dimensional models perform remarkably well. All three authors were in broad agreement in their main conclusions. Strong instability, abundant low level moisture and weak shear led to intense but relatively short-lived updraughts. The main rainfall and downdraught occurred downshear of the updraught, isolating it from its source. Moderate or strong shear was found generally to favour both intense and long-lasting convection. But some differences of opinion existed. Takeda found that shear of constant sign inclined the cloud considerably downshear with an associated downdraught eventually cutting off inflow and isolating the updraught. The result was a short-lived cloud. When shear reversed sign a long-lived cloud developed, its existence being a function of the height of the first reversal of the shear. In contrast, Schlesinger (1973a) and Hane (1973) found that the updraught tilted over the downdraught, allowing the former to exist for long periods. Important as the shear has been shown to be, with insufficient moisture it cannot support an intense or long-lived storm. Summarizing in Hane's words (Hane, 1973, p. 1672): "The implication is that the squall-line thunderstorm, once initiated, maintains itself by interaction with its synoptic environment as long as it remains within an environment containing convectively unstable air whose motion is characterized by moderate-to-strong vertical shear".

These encouraging initial results led Schlesinger (1973b) to experiment further with his successful model. In this later work he assessed, among other

---

FIG. 144. Initial conditions and computed airflows in numerical models of severe storms. (a) Initial state of temperature and relative humidity in Schlesinger's model. The numbers in the layers separated by dashed lines indicate the lapse rate $-\partial T/\partial z$ (in kelvins per kilometre)—in each layer. (After Schlesinger, 1973a.) (b) Initial state of temperature and humidity in Takeda's model. (After Takeda, 1971.) (c) Initial state of dry bulb temperature (solid line) and dew-point temperature (dashed line) in Hane's model. (After Hane, 1973.) (d) Initial state of wind velocity in Schlesinger's model. (After Schlesinger, 1973a.) (e) Initial state of wind velocity in Takeda's model. (After Takeda, 1971.) (f) Initial state of wind velocity in Hane's model. (After Hane, 1973.) (g) Airflow in Schlesinger's model after 50 min simulated time. The heavy solid curve represents the cloud boundary. Thinner solid curves with arrows are streamlines at intervals of $10^7 \mathrm{g\,m^{-1}\,s^{-1}}$. (After Schlesinger, 1973a). (h) Airflow and rainwater content in Takeda's model after 60 min simulated time. Thin solid curves with arrows are streamlines. Heavy solid lines represent rainwater content ($\mathrm{g\,m^{-3}}$). Dotted line indicates the cloud boundary. (After Takeda, 1971.) (i) Airflow and rainwater content in Hane's model after about 42 min simulated time. Thin solid lines with arrows are streamlines at intervals of $10^6 \mathrm{g\,m^{-1}\,s^{-1}}$. Dashed lines indicate rainwater mixing ratio (in grams per kilogram); thick solid line indicates cloudwater mixing ratio outline. (After Hane, 1973.)

things, the role of liquid water drag and the effects of pressure perturbations upon the buoyancy.

Appreciation of the effects of pressure perturbations is helped by subdivision of the more familiar total vertical pressure gradient force,

$$F_1 = -\frac{1}{\rho}\frac{\partial p}{\partial z} - g. \tag{214}$$

Splitting p and ρ into a hydrostatic base state (p_0 and ρ_0) and deviations (p' and ρ') therefrom gave

$$p = p_0 + p' \quad \text{and} \quad \rho = \rho_0 + \rho'.$$

Since $|p'| \ll p_0$ and $|\rho'| \ll \rho_0$ the right-hand side of equation (214) may be linearized with the aid of the hydrostatic equation

$$\frac{\partial p_0}{\partial z} = -g\rho_0 \tag{215}$$

to give

$$F_1 = -\frac{1}{\rho_0}\frac{\partial p'}{\partial z} - \frac{g\rho'}{\rho_0}. \tag{216}$$

The first term on the right-hand side of equation (216) is the perturbed vertical pressure gradient force and the second term the gravitational buoyancy force. Schlesinger (1975) showed that this latter force comprises three parts to give the following expression for F_1:

$$F_1 = -\frac{1}{\rho_0}\frac{\partial p'}{\partial z} + g\left(\frac{T'}{T_0} + aq'_v - \frac{p'}{p_0}\right), \tag{217}$$

where T_0 and T' are base and deviation temperatures respectively, a is R_v/R_d where R_v and R_d are respectively gas constants for water vapour and dry air and q'_v is the deviation from base state of water vapour mixing ratio. Thus three factors (in parentheses on right-hand side of equation (217)) contribute to upward buoyancy in an air parcel: temperature excess, water vapour excess and pressure deficit. In the well-known parcel theory the first two of these are treated together as a thermal buoyancy while the third contribution is neglected since horizontal pressure differences are themselves neglected. Schlesinger called this less well-known contribution the "pressure buoyancy", partly because of its appearance as part of the gravitational buoyancy $-g\rho'/\rho_0$ and partly to distinguish it from the perturbed vertical pressure gradient force $-(1/\rho_0)(\partial p'/\partial z)$.

Schlesinger's primary findings were as follows. The overall storm circulation was very similar to those in Fig. 132. It was quasi-steady state with potentially warm, low level air feeding the updraught from downshear while potentially cool, middle level air fed the downdraught from upshear. During maturity thermal buoyancy was the dominant vertical force, but was strongly opposed by the vertical perturbed pressure gradient force. The buoyancy due to pressure (as opposed to temperature) perturbations was appreciable, with a maximum value about one-quarter that of the thermal buoyancy. Liquid water drag was intermediate in importance between the two buoyancy components. The vertical and horizontal net accelerations were comparable to one another and to the pressure buoyancy. Latent heat release ranged from 8.3×10^{12} J s^{-1} to 3.4×10^{13} J s^{-1} while the liquid water mass was between 1×10^8 and 5×10^8 kg (Sikdar et al., 1974). Dynamic entrainment of potentially cool air into the sides of the cloud eventually contributed to its dissipation as the downdraught spread laterally and isolated the updraught. The liquid water drag limited the updraught intensity but did not appear to be necessary for the formation of the downdraught, which was due instead to evaporative cooling. The model also indicated that fallout of precipitation was essential for storm dissipation. Without fallout, the liquid water accumulation at low levels was insufficient for significant downdraught development and the cloud core evolved to a steady state. Finally, Schlesinger (1973b) noted that negative pressure buoyancy in the upper portion of the cloud slightly limited the intensity of the developing updraught but the positive pressure buoyancy at and near the foot of the updraught reinforced its intensity during maturity.

Despite the success of the above experiments in simulating the overall cumulonimbus circulation, scope remains for the more detailed analysis of components such as the up- and downdraughts and the gust front. In fact, such studies as exist have concentrated upon the downdraught and the closely associated "thunderstorm high", cold outflow and gust front.

It is now generally acknowledged that the downdraught of cold air is primarily a function of the drag force exerted on the air by precipitation together with the cooling effect of evaporation of rain and cloud drops (Spillane and McCarthy, 1969). In his analysis of the latter effect Hookings (1965, 1967) showed that downdraught speed increased with decreasing-sized drops, decreasing initial relative humidity and increasing initial liquid water content. Kamburova and Ludlam (1966) built upon this result, confirming that small raindrop size favours strong downdraughts and further showing that more important factors were an intense rainfall rate and a general lapse-rate close to the dry adiabatic. Further refinements by Haman (1973) included the suggestion that downdraughts can be maintained by rapid evaporation of small droplets supplied by entrainment of cloudy air from neighbouring updraughts. Haman's comparatively simplified quantitative analysis showed that his proposed mechanism could maintain fairly strong, cold downdraughts even when hydrostatic instability is weak and evaporation of precipitation is unable to provide a sufficiently large rate of cooling. Haman concluded with the following plausible picture as a basis for further work (Haman, 1973, p. 232):

A downdraught starts in the upper part of the cloud in the zone in which precipitation is formed, under the drag of precipitation, being cooled there mainly by evaporation of the remainder of the small-droplet fraction of cloud water. After exhausting the resources of this easily-evaporated water, in the middle parts of the cloud the proposed mechanism of cooling...becomes dominant. Then at the lowest levels, coolness of the downdraught is due to the melting of precipitation and to the relatively steep lapse rate of environmental temperature, which permits maintenance of a negative temperature deviation even by relatively weak evaporation of large droplets.

Regardless of the details of the downdraught mechanism, its effects upon the surface pressure and flow fields and upon the subsequent development of convection (Takeda, 1965) are widely accepted, even if not yet fully understood. A rise in surface pressure, apparently associated with the onset of precipitation, had been frequently observed prior to 1945. This local feature became known as the "rain-" or "thunderstorm-high". The earlier attempted explanations concentrated upon the effects of vertical accelerations. Despite Mal and Rao's (1945) criticism of Levine's (1942) and Buell's (1943a,b) suggestion that downward accelerating air would cause a pressure rise at the surface, Schaffer (1947, 1952) remained convinced that there must be some effect even if it were not a major one. Sawyer (1946), noting that Mal and Rao offered no alternative mechanism, suggested that evaporative cooling was more than adequate to cause a near-surface rise in pressure. He outlined the mechanism as follows. Cooling by evaporation of precipitation increases the density of the air column which thus contracts; pressure therefore falls at the top of the cooled column and air flows into a new, low pressure centre at that level. This causes a rise of pressure at the ground. Were it not for friction in the layers near the surface, the air flowing out from the rising pressure at the ground would soon balance that flowing in at higher levels; but as the outflow is retarded near the ground more air enters the column than leaves the air column and surface pressure continues to rise. Sawyer calculated that, for an air column about 3 km deep with an initially dry adiabatic lapse rate, a surface temperature of 32 °C, and a uniform water content corresponding to a dew point of 7 °C, evaporation to saturation would cause a rise of surface pressure of 9 mb. Whilst recognizing that his calculations were simple and the result probably too large, Sawyer felt that, in the absence of other ideas, his suggested mechanism was on the right lines. The general result that the higher the level from which the rain falls and the drier the air, the greater the possible rise in pressure, has been amply confirmed by subsequent observations. In particular Fujita (1959b) established that the "excess" mass of cold air in the thunderstorm high was directly proportional to the evaporated rain.

As Sawyer indicated, and as noted in the first part of this chapter, divergence of the cold air in the high pressure area results in an outflow, frequently headed by a gust front. Morphologically this outflow strongly resembles a density, or gravity current, and Charba (1974) argued that dynamical considerations support this idea. A more comprehensive, fine scale, two-dimensional, numerical simulation by Mitchell and Hovermale (1977) appears to confirm this viewpoint. After the initiation of a downdraught in the model, the cold outflow rapidly achieved a constant configuration that represented a dynamic balance between the pressure gradient forces and the surface frictional drag forces. Simulation of this feature required grid lengths of about 200 m and a time step of about 0.35 s—far smaller than those employed in the two-dimensional simulation of the storm circulations as a whole. Notwithstanding these beginnings of the analysis of storm components, much remains to be done on the satisfactory simulation of the storm circulation as a whole. The following section considers recent efforts.

(b) *Three-dimensional models* It was perhaps inevitable that the development of three-dimensional models should follow hard on the heels of the two-dimensional ones. Schlesinger (1975), among others, has noted the limitations of the two-dimensional approach, particularly that precipitation must fall in the same plane as the inflow, since the updraught cannot turn with height, that the model storm cannot propagate at an angle to the mean cloud-layer ambient wind and that no air can flow around the model cloud. Whilst these would seem to be powerful arguments for an unrestricted three-dimensional approach, Miller (1978) sounded a note of caution about the construction of ever more complicated models that endeavour to simulate every physical process or effect that might play a role in determining the precise details of a particular convective situation. He noted that "...the inclusion of a large number of parameterizations into a fully three-dimensional model results in a cumbersome, expensive research tool which does not readily allow the modeller to gain insight into the sensitivity of the results to various parameters by re-running the simulations many times" (Miller, 1978, p. 415). In fact the sheer complexity and the massive computing requirements of the problem have forced researchers to simplify their models in reasonable accord with Miller's opinion.

The use of the vorticity equation in three-dimensional modelling is an almost impossible task so the major efforts have involved more basic forms of the equations of motion. Schlesinger (1975) continued to use the anelastic equations, as did Wilhelmson and Ogura (1972), Wilhelmson (1974) and Lin and Chang (1977). Miller and Pearce (1974) and Pastushkov (1975) used the primitive equations, the former in pressure coordinate. These exploratory papers again examined the interplay between instability, wind shear and cloud microphysics—with particular emphasis on the latter two. Pastushkov (1975) found that "strong" convective clouds were intensified when they occurred in shear. In contrast, and perhaps more surprisingly, the simulations by Schlesinger (1975) and Miller and Pearce (1974) revealed no encouragement

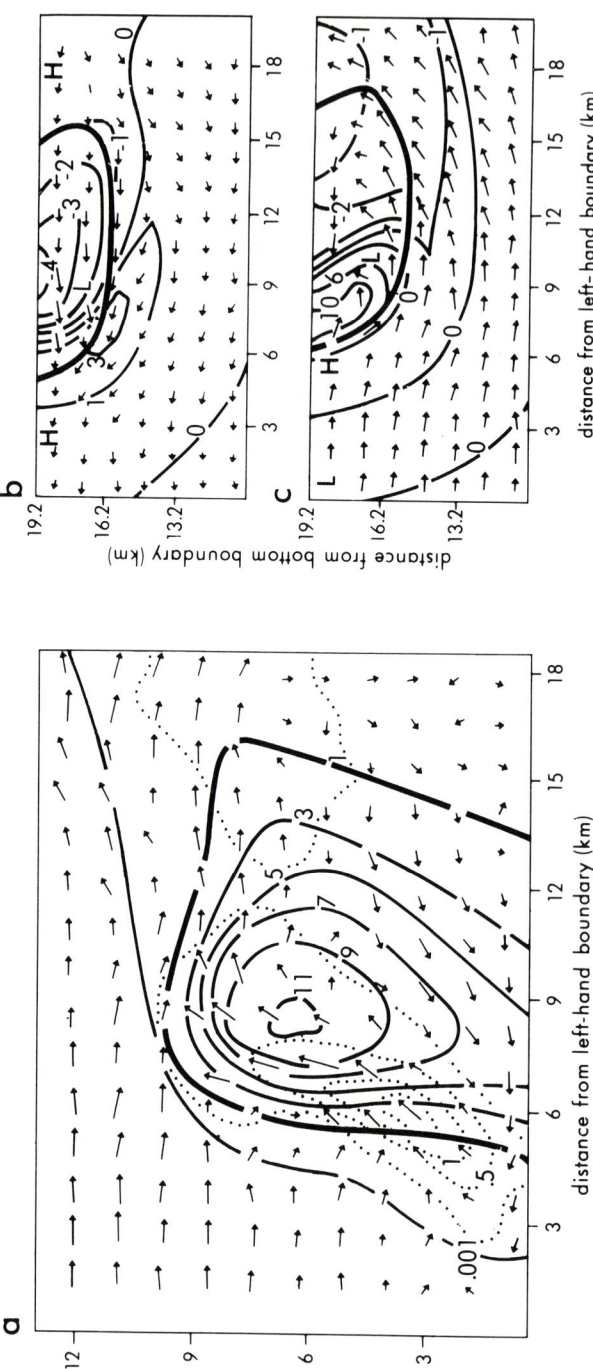

FIG. 145. Airflow and water content in Wilhelmson's model after 45 min simulated time. Small arrows are wind vectors; the magnitude of the largest vector is about 20 m s^{-1}. The thick solid line encloses an area with a mixing ratio of rainwater ≥ 1 g kg^{-1}. Dotted lines represent the mixing ratio of cloud droplets (g kg^{-1}). Solid lines represent the mixing ratio of rainwater (g kg^{-1}). (a) Central vertical section. The lower boundary is at a height of 0.3 km. (b) Horizontal section of vertical velocity (thin solid lines, in metres per second) and horizontal wind vectors (arrows) at a height of 0.9 km after 45 min simulated time. The magnitude of the largest vector is 8.2 m s^{-1}. Thick solid line shows mixing ratio of rainwater of 1 g kg^{-1}. (c) As (b) except for a height of 3.9 km. The magnitude of largest vector is 7.0 m s^{-1}. (After Wilhelmson, 1974.)

of severe convection by the presence of vertical shear. Both studies did, however, produce the up- and downdraught configuration. The essential results of Wilhelmson's (1974) pioneering study are illustrated in Fig. 145. The model storm developed in an initial shear of $1.54\,\text{m}\,\text{s}^{-1}\,\text{km}^{-1}$ in the x direction and symmetry was assumed in the y direction about the central (x, z) plane. An erect cloud developed from an initial 3.7 km radius impulse as the updraught core was fed from all directions by moist and warm low level air. After some time rain began to fall and a downdraught developed downshear of the updraught core cutting off the major low level supply of warm, moist air. Eventually, the updraught core began to tilt downshear and the centre of the lower part of the updraught core shifted away from the central (x, z) plane. Figure 145(a) shows the downshear tilt of the updraught after 45 min and Fig. 145(b) and (c) shows how the low level downdraught is overlain by the updraught. At middle cloud levels the general airflow was similar to that around a blocking cylinder and both the hydrostatic and dynamic pressures appeared to be important in decelerating the flow upwind of the cloud. Wilhelmson compared his three-dimensional model with one of two dimensions and found that the three-dimensional cloud developed faster, grew taller, lasted longer and travelled further and faster. He attributed these characteristics to differences in geometry which in turn were related to low level moisture supply for cloud growth and to subsidence in the cloud environment.

In large measure Wilhelmson's results are supported by those of Lin and Chang (1977) who used a model with initial conditions specified by analytical expression that described the distribution of observed horizontal wind and temperature around a mature severe storm. They found that a sheared and veering wind favoured upward motion on the right flank of the storm and downward motion on the left flank. The intensity and position of these up- and downdraughts were closely related to the degree of shearing and veering. Stronger shears significantly increased the intensity of the draughts, an opposite result to Takeda's (1966*b*) earlier studies. Lin and Chang also found well-developed horizontal and vertical perturbation pressure gradients. Pressure excesses of about 2 mb were consistently found on the leading and trailing parts of the storm core, whereas pressure deficits of about 3 mb lay on both flanks. The magnitude of the pressure gradient forces was directly proportional to environmental wind shears which in turn promoted accelerations and decelerations of air motion around the main updraught core.

More recently three-dimensional numerical models have provided greater insight into the effects of both uni- and multi-directional wind shear upon severe storms, in particular the configurations of updraught and downdraught, the longevity of the storms, their propensity to rotate, to split and to move to the left or right of the ambient winds. Schlesinger's (1978) simulation of a mature, isolated convective storm in three different vertical profiles of the ambient wind—no wind, unidirectional shear and multi-directional shear—produced the following major results. Both sheared storms had a virtually vertical, high speed updraught, a deep cyclonic–anticyclonic vortex couplet aloft, a middle level barrier flow around the updraught and a gradual splitting into cyclonic and anticyclonic cells moving to the right and left of the mean winds. The presence of shear favoured a stronger, more persistent mature storm with the main downdraught developing upshear of the updraught. A further interesting result, to some extent in conflict with Newton and Newton's (1959) findings, was that thermal buoyancy and the perturbed vertical pressure gradient force opposed each other, thus enabling parcels to accelerate upwards against negative buoyancy at the bases of the sheared updraughts.

Additional important results were forthcoming from the model derived by Klemp and Wilhelmson (1978a). This three-dimensional model was based on the compressible equations of motion, had parameterized microphysical and sub-grid turbulence processes and boundary conditions that were designed to allow major gravity waves to propagate out through the lateral boundaries without significant reflection. Klemp and Wilhelmson used the model to assess the effects of both uni- and multi-directional windshear upon storm splitting and movement. In the case of one-directional shear, of maximum magnitude $4 \text{ m s}^{-1} \text{ km}^{-1}$ (Wilhelmson and Klemp, 1978) the process was initiated by the formation of a precipitation-induced downdraught that split the low level updraught. The low level split updraughts which were located on the right and left flanks of the storm were supplied with moisture within a convergence zone that developed along a downdraught-induced gust front. The two split storms were organized to sustain themselves with precipitation falling out of the updraught into the downdraught which in turn undercut the updraught along the gust front, similar to that of left- and right-moving supercells. The updraught of the right-moving storm rotated cyclonically and the downdraught anticyclonically. Various simulations suggested that strong shear at and just above cloud base was important for successful splitting. For splitting to occur in the model the low level flow from the front of the storm had to be sufficiently strong to inhibit the propagation of the gust front ahead of the storm. If the gust front could propagate away from the storm, the region of low level convergence moved away from the storm and the splitting of the lower updraught could not be sustained. In general, without the precipitation-induced downdraught and associated low level outflow splitting did not occur.

Multi-directional shear was found to have an important influence on storm movement after splitting (Klemp and Wilhelmson, 1978b). By altering the direction of the environmental shear at low and middle levels, either a right- or left-moving storms could be selectively enhanced. If the wind hodograph turned clockwise with height, a single right-moving storm evolved from the splitting process. Conversely backing of the wind with height favoured development of a left-moving storm.

Numerical modelling of extra-tropical severe convective storms is still in its early stages—particularly so in the case of three-dimensional simulation. The next decade will no doubt witness major developments in this area, hopefully together with an ability to insert these models into those of larger scale circulations.

B. Tropical severe local storms

We noted earlier in this chapter our comparative ignorance of the observed structure of tropical cumulonimbus. It is thus understandable that theoretical work is still in its infancy. Nevertheless the few observations (Fig. 140(c)) have shown the necessity of three-dimensional flow in this type of convection and an adequate theory must explain this feature. The major attack on this problem came from Moncrieff and Miller (1976) who combined both analytical and numerical approaches. Figure 140(c) shows that the cumulonimbus propagates against the wind at all levels in the cloud layer such that inflows (of both up- and downdraught) enter the storm from the front and outflows leave from the rear. Moncrieff and Miller isolated the dynamical problem as one of finding the propagation speed and the modified outflow given the kinetic and thermodynamic state of the inflow.

In the analytical part of the paper a steady state storm was investigated using a Boussinesq system of equations. Along each air trajectory (streamlines in steady flow) the values of three quantities—the Bernoulli integral, the entropy in an adiabatic system and the mass flux—remained constant. The propagation speed of the cumulonimbus relative to the group (C) was found to be determined by a so-called convective available potential energy (CAPE) and the mid-level wind speed in the cumulonimbus layer (U_m):

$$C = U_m + 0.3(\text{CAPE})^{1/2}. \tag{218}$$

In practical terms the definition of convective available potential energy is

$$\text{CAPE} = \int_{z_b}^{H} g\left(\frac{\Delta\theta}{\theta}\right) dz, \tag{219}$$

where $\Delta\theta$ is the potential temperature difference at height z between the appropriate saturated adiabat on a thermodynamic diagram and the environment at the same level, θ is the ambient potential temperature at that level and z_b and H are cloud base and the equilibrium level of the adiabatic parcel respectively. Betts et al. (1976) compared theoretical values calculated from equation (218) with values observed in Venezuela and found a mean value of $C_{\text{obs}}/C_{\text{pred}}$ of 0.91.

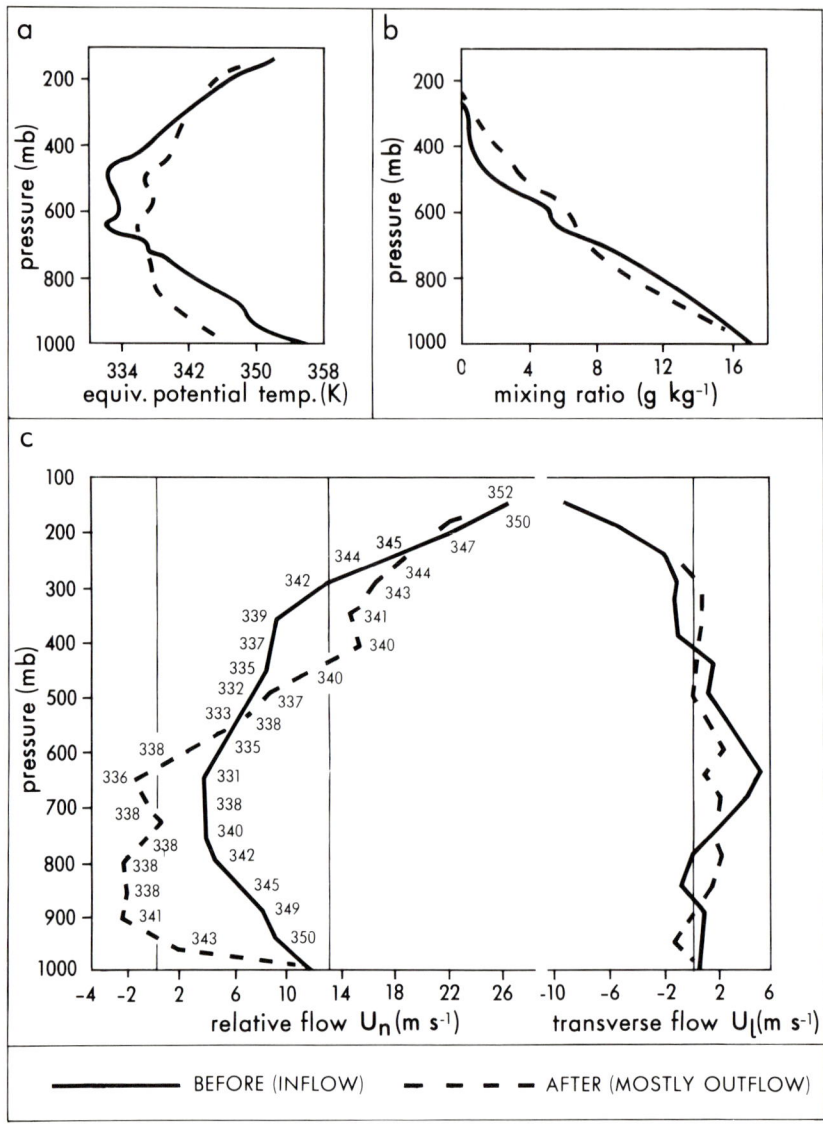

Fig. 146. Modification of environment by tropical squall-line storms as shown by conditions before and after passage of storm. (a) Equivalent potential temperature. (b) Mixing ratio. (c) Along and transverse flow relative to the average squall-line motion. The dash-dotted vertical line represents the average propagation speed. Corresponding value of equivalent potential temperature are shown alongside the normal relative flow curves. (After Betts et al., 1976.)

Table 46 Mean modification of wind, temperature and moisture computed for an area of 25 km² behind the squall line at 88 min simulation time

Pressure level (mb)	Δu (m s^{-1})	Δv (m s^{-1})	$\Delta \theta$ (K)	Δq (g kg^{-1})	$\Delta \theta_e$ (K)
200	0.0	0.0			
			−0.0	0.0	0.0
300	0.1	0.1			
			−0.4	0.2	0.3
400	1.2	0.9			
			−0.7	1.2	3.2
500	1.3	2.0			
			1.4	3.2	11.2
600	−0.2	2.8			
			1.2	2.2	7.7
700	−0.6	−0.3			
			1.0	−0.8	−1.4
800	−2.5	0.2			
			0.6	−1.6	−4.1
900	−4.5	0.9			
			−1.9	−0.9	−4.5
1000	1.7	−2.9			

Source: Moncrieff and Miller (1976).

In addition to providing a propagation speed Moncrieff and Miller's theory predicted how the atmosphere would be transformed by the passage of a convective system. Table 46 shows the mean modification of wind, temperature and moisture derived from the theory. The basic modification to the east–west wind component is to increase both upper and lower layer shears and hence to enhance the low level wind maximum (Fig. 140(a)). Thus the convection achieves an up-gradient momentum transfer to the dominant wind component. The north–south wind component loses some shear, probably due to a down-gradient transfer. The thermodynamic fields imply substantial upward heat and moisture transfers. Strong support for these theoretical results was provided by Betts et al. (1976) (Fig. 146).

References

Achtemeier, G. L. (1969). Some observations of splitting thunderstorms over Iowa on August 25–26, 1965. Report no. 69-4, prepared under Grant E22-49-68 (G) with the ESSA, Tallahassee, Florida.

Agee, E. M., Snow, J. T. and Clare, P. R. (1976). Multiple vortex features in the tornado cyclone and the occurrence of tornado families. *Mon. Weath. Rev.*, **104**, 552–563.

Alberty, R. L. (1969). A proposed mechanism for cumulonimbus persistence in the presence of strong vertical shear. *Mon. Weath. Rev.*, **97**, 590–596.
Appleman, H. (1959). An investigation into the formation of hail. *Nubila*, **2**, 28–37.
Atlas, D. (1976). Overview: the prediction, detection and warning of severe storms. *Bull. Am. met. Soc.*, **57**, 398–401.
Atlas, D., Booker, D. R., Byers, H., Douglas, R. H., Fujita, T., House, D. C., Ludlam, F. H., Malkus, J. S., Newton, C. W., Ogura, Y., Schleusener, R. A., Vonnegut, B. and Williams, R. T. (1963). Severe local storms. *Met. Monog.*, **5** (27).
Auer, A. H., Jr and Marwitz, J. D. (1968). Estimates of air and moisture flux into hailstorms on the high plains. *Proc. 5th Conf. Severe Local Storms, St Louis, 19–20 Oct. 1967*, 303–306.
Auer, A. H., Jr and Marwitz, J. D. (1968). Estimates of air and moisture flux into hailstorms on the high plains. *J. appl. Met.*, **7**, 196–198.
Auer, A. H., Jr and Sand, W. (1966). Updraught measurements beneath the base of cumulus and cumulonimbus clouds. *J. appl. Met.*, **5**, 461–466.
Auer, A. H., Veal, D. L. and Marwitz, J. D. (1969). Updraught deterioration below cloud base. *Preprints 6th Conf. Severe Local Storms, Chicago*, 16–19.
Auer, A. H., Jr, Veal, D. L. and Marwitz, J. D. (1970). The identification of organized cloud base updraughts. *J. Rech. Atmos.*, **4**, 1–6.
Austin, P. M., Cochran, H. B. and Patrick, G. O. (1961). Investigations concerning the internal structure of New England squall lines. *Proc. 9th Weath. Rad. Conf., Kansas City*, 193–198.
Barnes, S. L. (1968). On the source of thunderstorms rotation. ESSA Technical Memorandum RLTM-NSSL 38.
Barnes, S. L. (1976). Severe local storms: concepts and understanding. *Bull. Am. met. Soc.*, **57**, 412–419.
Bates, F. C. (1961). The Great Plains squall-line thunderstorm—a model. Ph.D. thesis, St Louis University. (University Microfilms, Ann Arbor. Mich., No. 61-6455.)
Battan, L. J. (1973). *Radar Observation of the Atmosphere*, Chicago University Press, Chicago.
Battan, L. J. (1975). Doppler radar observations of a hailstorm. *J. appl. Met.*, **14**, 98–108.
Battan, L. J. and Theiss, J. B. (1966). Observations of vertical motions and particle sizes in a thunderstorm. *J. atmos. Sci.*, **23**, 78–87.
Beebe, R. G. (1958). Tornado proximity soundings. *Bull. Am. met. Soc.*, **39**, 195–201.
Beebe, R. G. and Bates, F. C. (1955). A mechanism for assisting in the release of convective instability. *Mon. Weath. Rev.*, **83**, 1–10.
Betts, A. K. (1976). The thermodynamic transformation of the tropical subcloud layer by precipitation and downdraughts. *J. atmos. Sci.*, **33**, 1008–1020.
Betts, A. K., Grover, R. W. and Moncrieff, M. W. (1976). Structure and motion of tropical squall lines over Venezuela. *Q. Jl R. met. Soc.*, **102**, 395–404.
Bhaskara Rao, N. S. and Dekate, M. V. (1967). Effect of vertical wind shear on the growth of convective clouds. *Q. Jl R. met. Soc.*, **93**, 363–367.
Bonner, W. D. (1966). Case study of thunderstorm activity in relation to the low-level jet. *Mon. Weath. Rev.*, **94**, 167–178.
Braham, R. R. (1952). Water and energy budgets of the thunderstorm and their relation to thunderstorm development. *J. Met.*, **9**, 227–242.
Brandes, E. A. (1977a). Flow in severe thunderstorms observed by dual-doppler radar. *Mon. Weath. Rev.*, **105**, 113–120.
Brandes, E. A. (1977b). Mesocyclone evolution and tornado generation within the Harrah, Oklahoma storm. NOAA Technical Memorandum ERL NSSL-81.

Brandes, E. A. (1977c). Gust front evolution and tornado genesis as viewed by doppler radar. *J. appl. Met.*, **16**, 333–338.
Brooks, E. M. (1949). The tornado cyclone. *Weatherwise*, **2**, 32–33.
Brooks, H. B. (1946). A summary of some radar thunderstorm observations. *Bull. Am. met. Soc.*, **27**, 557–563.
Browning, K. A. (1962). Cellular structure of convective storms. *Met. Mag.*, **91**, 341–350.
Browning, K. A. (1964). Airflow and precipitation trajectories within severe local storms which travel to the right of the winds. *J. atmos. Sci.*, **21**, 634–639.
Browning, K. A. (1965). Some inferences about the updraught within a severe local storm. *J. atmos. Sci.*, **22**, 669–677.
Browning, K. A. (1968). The organization of severe local storms. *Weather*, **23**, 429–434.
Browning, K. A. and Donaldson, R. J. (1963). Airflow and structure of a tornadic storm. *J. atmos. Sci.*, **20**, 533–545.
Browning, K. A. and Foote, G. B. (1976). Airflow and hail growth in supercell storms and some implications for hail suppression. *Q. Jl R. met. Soc.*, **102**, 499–534.
Browning, K. A. and Ludlam, F. H. (1960). Radar analysis of a hailstorm. Technical Note no. 5, Contract AF61(052)-254, Department of Meteorology, Imperial College, London.
Browning, K. A. and Ludlam, F. H. (1962). Airflow in convective storms. *Q. Jl R. met. Soc.*, **88**, 117–135.
Browning, K. A., Fankhauser, J. C., Chalon, J.-P., Eccles, P. J., Strauch, R. G., Merrem, F. H., Musil, D. J., May, E. L. and Sand, W. R. (1976). Structure of an evolving hailstorm. V. Synthesis and implications for hail growth and hail suppression. *Mon. Weath. Rev.*, **104**, 603–610.
Brunk, I. W. (1953). Squall lines. *Bull. Am. met. Soc.*, **34**, 1–9.
Buell, C. E. (1943a). The determination of vertical velocities in thunderstorms. *Bull. Am. met. Soc.*, **24**, 94–95.
Buell, C. E. (1943b). The determination of vertical velocities in thunderstorms. *Bull. Am. met. Soc.*, **24**, 211–212.
Burley, M. W., Pfleger, R. and Wang, J. Y. (1964). Hailstorms in Wisconsin. *Mon. Weath. Rev.*, **92**, 121–127.
Byers, H. R. (1942). Non-frontal thunderstorms. Miscellaneous Report 3, Department of Meteorology, University of Chicago.
Byers, H. R. and Braham, R. R. (1949). *The Thunderstorm*, US Government Printing Office, Washington, D.C.
Carlson, T. N. and Ludlam, F. H. (1968). Conditions for the occurrence of severe local storms. *Tellus*, **20**, 203–226.
Chalker, W. R. (1949). Vertical stability in regions of air mass showers. *Bull. Am. met. Soc.*, **30**, 145–147.
Chalon, J. P., Fankhauser, J. C. and Eccles, P. J. (1976). Structure of an evolving hailstorm. I. General characteristics and cellular structure. *Mon. Weath. Rev.*, **104**, 564–575.
Champion, D. L. (1948). Seasonal distribution of hail in Gt Britain. *Weather*, **3**, 201–205.
Chandrashekhar Aiya, S. V. and Sonde, B. S. (1963). Number of cells developed during the lifetime of a thunderstorm. *Nature, Lond'.* **200**, 562–563.
Changnon, S. A., Jr (1970). Hailstreaks. *J. atmos. Sci.*, **27**, 109–125.
Charba, J. (1974). Application of gravity current model to analysis of squall-line gust front. *Mon. Weath. Rev.*, **102**, 140–156.

Charba, J. and Sasaki, Y. (1971). Structure and movement of the severe thunderstorm of 3 April 1964 as revealed from radar and surface mesonetwork data analysis. *J. met. Soc. Jap.* **49**, 191–214.

Chisholm, A. J. (1970a). Alberta hailstorms: a radar study and model. Ph.D. thesis, Department of Meteorology, McGill University, Montreal.

Chisholm, A. J. (1970b). The radar and airflow structure of Alberta hailstorms. *Preprints 14th Conf. Radar Met., Tucson, Ariz.*, 35–42.

Chisholm, A. J. (1973). Alberta hailstorms. I. Radar case studies and airflow models. *Met. Monog.*, **14** (36), 1–36.

Chisholm, A. J. and Renick, J. H. (1972). The kinematics of multicell and supercell Alberta hailstorms. Alberta Hail Studies. Research Council of Alberta Hail Studies, Report 72-2, pp. 24–31.

Chisholm, A. J. and Warner, C. (1969). Radar and stereo photo measurements. Part II. The hailstorm of 29 June 1967. Scientific Report MW-59, Stormy Weather Group, McGill University, Montreal, pp. 8–16.

Crawford, M. E. (1950). A synoptic study of instability lines. *Bull. Am. met. Soc.*, **31**, 349–357.

Das, P. (1962). Influence of wind shear on growth of hail. *J. atmos. Sci.*, **19**, 407–414.

Davies-Jones, R. P. and Henderson, J. H. (1974). Updraught properties deduced from rawin soundings. NOAA Technical Memorandum ERL NSSL-72, Norman, Oklahoma.

Davis, W. M. (1894). *Elementary Meteorology*, Ginn and Co., Boston.

Dennis, A. S., Schock, C. A. and Koscielski, A. (1970). Characteristics of hailstorms of western S. Dakota. *J. appl. Met.*, **9**, 127–135.

Derome, J. F. (1965). Large-scale vertical motions and the occurrence of severe storms. Scientific Report MW-42, Stormy Weather Group, McGill University, Montreal, pp. 1–41.

Dessens, H. (1960). Severe hailstorms are associated with very strong winds between 6,000 and 12,000 m. *Geophys. Monogr.*, **5**, 333–338.

Donaldson, R. J., Jr (1958). Analysis of severe convective storms observed by radar. *J. Met.*, **15**, 44–50.

Donaldson, R. J. (1959). Analysis of severe convective storms observed by radar (II). *J. Met.*, **16**, 281–287.

Easterbrook, C. C. (1967). Some Doppler radar measurement of circulation patterns in convective storms. *J. appl. Met.*, **6**, 882–888.

Ellrod, G. P. and Marwitz, J. D. (1976). Structure and interaction in the subcloud region of thunderstorms. *J. appl. Met.*, **15**, 1083–1091.

Fankhauser, J. C. (1964). On the motion and predictability of convective systems. Report no. 21, NSSP.

Fankhauser, J. C. (1965). A comparison of kinematically computed precipitation with observed convective rainfall. Technical Note 4-NSSL-25, Weather Bureau, Washington, D.C.

Fankhauser, J. C. (1967). Some physical and dynamical aspects of a severe right moving cumulonimbus. Technical Memorandum IERTM-NSSL 32, NSSL, Norman, Oklahoma, pp. 11–32.

Fankhauser, J. C. (1971). Thunderstorm–environment interactions determined from aircraft and radar observations. *Mon. Weath. Rev.*, **99**, 171–192.

Fawbush, E. J. and Miller, R. C. (1954). Types of airmasses in which North American tornadoes form. *Bull. Am. met. Soc.*, **35**, 154–165.

Fenner, J. H. (1976). The motion of thunderstorm cells in relation to the mean wind and mean wind shear. *Q. Jl R. met. Soc.*, **102**, 459–461.

Foote, G. B. and Fankhauser, J. C. (1973). Airflow and moisture budget beneath a northeast Colorado hailstorm. *J. appl. Met.*, **12**, 1330–1353.
Fujita, T. (1955). Results of detailed synoptic studies of squall lines. *Tellus*, **7**, 405–436.
Fujita, T. (1958). Mesoanalysis of the Illinois tornadoes of April 9 1953. *J. Met.*, **15**, 288–296.
Fujita, T. (1959a). Study of mesosystems associated with stationary radar echoes. *J. Met.*, **16**, 38–52.
Fujita, T. (1959b). Precipitation and cold air production in mesoscale thunderstorm systems. *J. Met.*, **16**, 454–466.
Fujita, T. (1960). A detailed analysis of the Fargo tornadoes of 20 June 1956. US Weather Bureau Research Paper no. 42.
Fujita, T. (1963). Analytical mesometeorology. A Review. *Met. Monogr.*, **5** (27), 77–125.
Fujita, T. (1965). On the formation and steering mechanisms of tornado cyclones and associated hook echoes. *Mon. Weath. Rev.*, **93**, 67–78.
Fujita, T. and Arnold, J. (1963). Preliminary result of analysis of the cumulonimbus cloud of 21 April 1961. Research Paper no. 16, Mesometeorology Project, Department of Geophysical Science, University of Chicago.
Fujita, T. and Byers, H. R. (1962). Model of a hail cloud as revealed by photogrammetric analysis. *Nubila*, **5**, 85–105.
Fujita, T. T. and Byers, H. R. (1977). Spearhead echo and downburst in the crash of an airliner. *Mon. Weath. Rev.*, **105**, 129–146.
Fujita, T. and Grandoso, H. (1968). Split of a thunderstorm into anticyclonic and cyclonic storms and their motion as determined from numerical model experiments. *J. atmos. Sci.*, **25**, 416–439.
Fujita, T., Newstein, H. and Tepper, M. (1956). Mesoanalysis—an important scale in the analysis of weather data. US Weather Bureau Research Paper no. 39.
Fulks, J. R. (1951). The instability line. In *Compendium of Meteorology* (T. F. Malone, ed.), American Meteorological Society, Boston, Mass., pp. 647–654.
Goff, R. C. (1976a). Some observations of thunderstorm induced low-level wind variations. *Proc. AIAA 9th Fluid and Plasma Dynamics Conf., San Diego, Calif., 14–16 July*, AIAA Paper no. 76-388.
Goff, R. C. (1976b). Vertical structure of thunderstorm outflows. *Mon. Weath. Rev.*, **104**, 1429–1440.
Goldman, J. L. (1966). The role of the Kutta–Joukowski force in cloud systems with circulation. US Department of Commerce, ESSA, Technical Note 48-NSSL 27, National Severe Storms Laboratory Report no. 27, pp. 21–34.
Green, J. S. A. and Pearce, R. P. (1962). Cumulonimbus convection in shear. Technical (Scientific) Note 12, Contract No. AF 61(052)-254, Imperial College, London. Prepared for GRD AFCRL, Bedford, Mass.
Haman, K. (1973). On the updraught-downdraught interaction in convective clouds. *Acta geophys. pol.*, **21**, 215–233.
Haman, K. E. (1976). On the airflow and motion of quasi-steady convective storms. *Mon. Weath. Rev.*, **104**, 49–56.
Hammond, G. R. (1967). Study of a left moving thunderstorm of 23 April 1964. Technical Memorandum IERTM-NSSL 31, Institute for Environmental Research, National Severe Storms Laboratory, Norman, Oklahoma.
Hane, C. E. (1973). The squall line thunderstorm: numerical experimentation. *J. atmos. Sci.*, **30**, 1672–1690.
Harrold, T. W. (1966). A note on the development and movement of storms over Oklahoma on 27 May 1965. US Department of Commerce, ESSA, Institute for

Environmental Research, Technical Memorandum NSSL-29, Norman, Oklahoma, pp. 1–8.

Hart, H. E. and Cooper, L. W. (1968). Thunderstorm airflow studies using radar transponders and superpressure balloons. *Preprints 13th Radar Met. Conf., Montreal*, 196–201.

Hitschfeld, W. (1960). The motion and erosion of convective storms in vertical wind shear. *J. Met.*, **17**, 270–282.

Hookings, G. A. (1965). Precipitation-maintained downdraughts. *J. appl. Met.*, **4**, 190–195.

Hookings, G. A. (1967). Hail maintained downdraughts. *J. appl. Met.*, **6**, 589–591.

Houze, R. A., Jr (1977). Structure and dynamics of a tropical squall-line system. *Mon. Weath. Rev.*, **105**, 1540–1567.

Hoxit, L. R., Chappell, C. F. and Fritsch, J. M. (1976). Formation of mesolows or pressure troughs in advance of cumulonimbus clouds. *Mon. Weath. Rev.*, **104**, 1419–1428.

Huff, F. A. and Changnon, S. A. (1959). Hail climatology of Illinois. Report of Investigation no. 38, Illinois State Water Survey, Urbana.

Jessup, E. A. (1972). Interpretations of chaff trajectories near a severe thunderstorm. *Mon. Weath. Rev.*, **100**, 653–661.

Kamburova, P. L. and Ludlam, R. H. (1966). Rainfall evaporation in thunderstorm downdraughts. *Q. Jl R. met. Soc.*, **92**, 510–518.

Klemp, J. B. and Wilhelmson, R. B. (1978a). The simulation of three-dimensional convective storm dynamics. *J. atmos. Sci.*, **35**, 1070–1096.

Klemp, J. B. and Wilhelmson, R. B. (1978b). Simulations of right- and left-moving storms produced through storm splitting. *J. atmos. Sci.*, **35**, 1097–110.

Kropfli, R. A. and Miller, L. J. (1976). Kinematic structure and flux quantities in a convective storm from dual-doppler radar observations. *J. atmos. Sci.*, **33**, 520–529.

Lemon, L. R. (1976). Wake vortex structure and aerodynamic origin in severe thunderstorms. *J.atmos. Sci.*, **33**, 678–685.

Levine, J. (1942). The effect of vertical accelerations on pressure during thunderstorms. *Bull. Am. met. Soc.*, **23**, 52–61.

Lhermitte, R. M. (1964). Doppler radars as severe storm sensors. *Bull. Am. met. Soc.*, **45**, 587–596.

Ligda, M. G. H. (1951). Radar storm observation. In *Compendium of Meteorology* (T. F. Malone, ed.), American Meteorological Society, Boston, Mass., pp. 1265–1282.

Lilly, D. K. (1975a). Severe storms and storm systems: scientific background, methods and critical questions. *Pageophysics* **113**, 713–734.

Lilly, D. K. (ed.) (1975b). *Open SESAME (Severe Environmental Storms and Mesoscale Experiment): Proceedings of SESAME Opening Meeting.* Prepared by the NOAA, Environmental Research Laboratories, Boulder, Colorado, for the proposed SESAME Project.

Lin, Y. J. and Chang, P. T. (1977). Some effects of the shearing and veering environmental wind on the internal dynamics and structure of a rotating supercell thunderstorm. *Mon. Weath. Rev.*, **105**, 987–997.

Lin, Y. J. and Martin, D. E. (1971). Some numerical aspects of a steady non-entraining severe storm with sloping updraughts. *J. atmos. Sci.*, **28**, 1472–1478.

Lin, Y. and Martin, D. E. (1972). Further study of a steady severe model storm with rotating updraught configuration. *Tellus*, **24**, 216–229.

Ludlam, F. H. (1958). The hail problem. *Nubila*, **1**, 12–96.

Ludlam, F. H. (1959). Hailstorm studies. *Nubila*, **2**, 38–50.

Ludlam, F. H. (1961). The hailstorm. *Weather*, **16**, 152–161.

Ludlam, F. H. (1963). Severe local storms: a review. *Met. Monogr.*, **5** (27), 1–30.

Ludlam, F. H. (1966). Cumulus and cumulonimbus convection. *Tellus*, **18**, 687–698.
Ludlam, F. H. (1976). Aspects of cumulonimbus study. *Bull. Am. met. Soc.*, **57**, 774–779.
MacDonald, J. D. (1952). On the formation of squall lines. *Bull. Am. met. Soc.*, **33**, 237–239.
Mal, S. and Rao, Y. P. (1945). Effect of vertical acceleration on pressure during thunderstorms. *Q. Jl R. met. Soc.*, **71**, 419–421.
Marwitz, J. D. (1972a). The structure and motion of severe hailstorms. I. Supercell storms. *J. appl. Met.*, **11**, 166–179.
Marwitz, J. D. (1972b). The structure and motion of severe hailstorms. II. Multi-cell storms. *J. appl. Met.*, **11**, 180–188.
Marwitz, J. D. (1972c). The structure and motion of severe hailstorms. III. Severely sheared storms. *J. appl. Met.*, **11**, 189–201.
Marwitz, J. D., Chisholm, A. J. and Auer, A. H. (1969). The kinematics of severe thunderstorms sheared in the direction of motion. *Preprints, 6th Conf. Severe Local Storms, Chicago*, 6–12.
Means, L. L. (1952). On thunderstorm forecasting in central USA. *Mon. Weath. Rev.*, **80**, 165–188.
Miller, J. E. (1955). Intensification of precipitation by differential advection. *J. Met.*, **12**, 472–477.
Miller, M. J. (1978). The Hampstead storm: a numerical simulation of a quasi-stationary cumulonimbus system. *Q. Jl R. met. Soc.*, **104**, 413–428.
Miller, M. J. and Betts, A. K. (1977). Travelling convective storms over Venezuela. *Mon. Weath. Rev.*, **105**, 833–848.
Miller, M. J. and Pearce, R. P. (1974). A three-dimensional primitive equation model of cumulonimbus convection. *Q. Jl R. met. Soc.*, **100**, 133–154.
Mitchell, K. E. and Hovermale, J. B. (1977). A numerical investigation of the severe thunderstorm gust front. *Mon. Weath. Rev.*, **105**, 657–675.
Moncrieff, M. W. (1978). The dynamical structure of two-dimensional steady convection in constant vertical shear. *Q. Jl R. met. Soc.*, **104**, 543–568.
Moncrieff, M. W. and Green, J. S. A. (1972). The propagation and transfer properties of steady convective overturning in shear. *Q. Jl R. met. Soc.*, **98**, 336–352.
Moncrieff, M. W. and Miller, M. J. (1976). The dynamics and simulation of tropical cumulonimbus and squall lines. *Q. Jl R. met. Soc.*, **102**, 373–394.
Musil, D. J., Sand, W. R. and Schleusener, R. A. (1973). Analysis of data from T-28 aircraft penetrations of a Colorado hailstorm. *J. appl. Met.*, **12**, 1364–1370.
Namias, J. (1940). Synoptic aspects of thunderstorms. In *Air Mass and Isentropic Analysis* by J. Namias, American Meteorological Society, Boston, Mass., pp. 60–67.
Nelson, S. P. and Braham, R. R. (1975). Detailed observational study of a weak echo region. *Pageophysics* **113**, 735–746.
Newton, C. W. (1950). Structure and mechanism of pre-frontal squall line. *J. Met.*, **7**, 210–222.
Newton, C. W. (1960a). Morphology of thunderstorms and hailstorms as affected by vertical wind shear. *Geophys. Monogr.*, **5** (27), 339–347.
Newton, C. W. (1960b). Hydrodynamic interactions with ambient wind field as a factor in cumulus development. In *Cumulus Dynamics* (C. E. Anderson, ed.), Pergamon Press, New York, pp. 135–144.
Newton, C. W. (1963). Dynamics of severe convective storms. *Met. Monogr.*, **5** (27), 33–58.
Newton, C. W. (1966). Circulations in large sheared cumulonimbus. *Tellus*, **18**, 699–712.
Newton, C. W. (1967). Severe convective storms. *Adv. Geophys.*, **12**, 257–308.

Newton, C. W. (1969). Convective cloud dynamics—a synopsis. *Proc. int. Conf. Cloud Phys., Toronto*, 487–498.
Newton, C. W. and Fankhauser, J. C. (1964a). On the movements of convective storms, with emphasis on size discrimination in relation to water-budget requirements. *J. appl. Met.*, **3**, 651–668.
Newton, C. W. and Fankhauser, J. C. (1964b). Movement and development patterns of convective storms, and forecasting the probability of storm passage at a given location. NSSP Report no. 22, US Department of Commerce, Weather Bureau, Washington, D.C.
Newton, C. W. and Fankhauser, J. C. (1975). Movement and propagation of multicellular convective storms. *Pageophysics*, **113**, 747–764.
Newton, C. W. and Katz, S. (1958). Movement of large convective rain storms in relation to winds aloft. *Bull. Am. met. Soc.*, **39**, 129–136.
Newton, C. W. and Newton, H. R. (1959). Dynamical interactions between large convective clouds and environment with vertical shear. *J. Met.*, **16**, 483–496.
Ninomiya, K. (1971). Mesoscale modification of synoptic situations from thunderstorm development as revealed by ATS III and aerological data. *J. appl. Met.*, **10**, 1103–1121.
Normand, C. (1946). Energy in the atmosphere. *Q. Jl R. met. Soc.*, **72**, 145–167.
Ogura, Y. and Chen, Y.-L. (1977). A life history of an intense mesoscale convective storm in Oklahoma. *J. atmos. Sci.*, **34**, 1458–1476.
Ogura, Y. and Phillips, N. A. (1962). Scale analysis of deep and shallow convection in the atmosphere. *J. atmos. Sci.*, **19**, 173–179.
Orville, H. D. and Kopp, F. J. (1977). Numerical simulation of the life history of a hailstorm. *J. atmos. Sci.*, **34**, 1596–1618.
Orville, H. D. and Sloan, L. J. (1970). A numerical simulation of the life history of a rainstorm. *J. atmos. Sci.*, **27**, 1148–1159.
Pastushkov, R. S. (1975). The effects of vertical wind shear on the evolution of convective clouds. *Q. Jl R. met. Soc.*, **101**, 281–292.
Pedgley, D. E. (1962). A meso-synoptic analysis of the thunderstorms on 28 August 1958. *Geophysical Memoirs*, **14**, no. 106, London, Meteorological Office.
Phillips, B. B. (1973). Precipitation characteristics of a sheared, moderate intensity, supercell-type Colorado thunderstorm. *J. appl. Met.*, **12**, 1354–1363.
Pitchford, K. L. and London, J. (1962). Low level jet as related to nocturnal thunderstorms in mid-west of US. *J. appl. Met.*, **1**, 43–47.
Porter, J. M., Means, L. L., Hovde, J. E. and Chappell, W. B. (1955). A synoptic study on formation of squall lines in northern central US. *Bull. Am. met. Soc.*, **36**, 390–96.
Ragette, G. (1973). Mesoscale circulations associated with Alberta hailstorms. *Mon. Weath. Rev.*, **101**, 150–159.
Ray, P. S., Doviak, R. J., Walker, G. B., Sirmans, D., Carter, J. and Bumgarner, B. (1975). Dual-doppler observation of a tornadic storm. *J. appl. Met.*, **14**, 1521–1530.
Rhea, J. O. (1966). A study of thunderstorm formation along dry lines. *J. appl. Met.*, **5**, 58–63.
Roach, W. T. (1967). On the nature of the summit areas of severe storms in Oklahoma. *Q. Jl R. met. Soc.*, **93**, 318–336.
Sanders, F. (1977). Comments on "Formation of mesolows or pressure troughs in advance of cumulonimbus clouds". *Mon. Weath. Rev.*, **105**, 1061–1063.
Sanders, F. and Emanuel, K. A. (1977). The momentum budget and temporal evolution of a mesoscale convective system. *J. atmos. Sci.*, **34**, 322–330.
Sanders, F. and Paine, R. J. (1975). The structure and thermodynamics of an intense mesoscale convective storm in Oklahoma. *J. atmos. Sci.*, **32**, 1563–1579.

Sansom, H. W. (1966). Occurrence and distribution of hail in Africa. *Met. Mag.*, **95**, 212–218.
Sartor, J. D. (1962). Essential factors of thunderstorm forecasting. Memorandum RM-3049-PR, USAF Project Rand, Rand Corp., Santa Monica, Calif.
Sasaki, Y. (1959). A numerical experiment of squall-line formation. *J. Met.*, **16**, 347–353.
Saunders, P. M. (1962). Penetrative convection in stably stratified fluids. *Tellus*, **14**, 177–194.
Sawyer, J. S. (1946). Cooling by rain as a cause of the pressure rise in convectional squalls. *Q. Jl R. met. Soc.*, **72**, 168.
Schaffer, W. (1947). The thunderstorm high. *Bull. Am. met. Soc.*, **28**, 351–355.
Schaffer, W. (1952). Second note on the thunderstorm high. *Bull. Am. met. Soc.*, **33**, 150–152.
Schlesinger, R. E. (1973a). A numerical model of deep moist convection. I. Comparative experiments for variable ambient moisture and wind shear. *J. atmos. Sci.*, **30**, 835–856.
Schlesinger, R. E. (1973b). A numerical model of deep moist convection. II. A prototype experiment and variations upon it. *J. atmos. Sci.*, **30**, 1374–1391.
Schlesinger, R. E. (1975). A three-dimensional numerical model of an isolated deep convective cloud: preliminary results. *J. atmos. Sci.*, **32**, 934–957.
Schlesinger, R. E. (1978). A three-dimensional numerical model of an isolated thunderstorm. I. Comparative experiments for variable ambient wind shear. *J. atmos. Sci.*, **35**, 690–713.
Severe Storms Research Group of St Louis University (1970). F. C. Bates' conceptual thoughts on severe thunderstorms. *Bull. Am. met. Soc.*, **51**, 481–489.
Showalter, A. K. (1953). A stability index for thunderstorm forecasting. *Bull. Am. met. Soc.*, **34**, 250–252.
Sikdar, D. N., Schlesinger, R. E. and Anderson, C. E. (1974). Severe storm latent heat release: comparison of radar estimate versus a numerical experiment. *Mon. Weath. Rev.*, **102**, 455–465.
Sinclair, P. C. (1973). Severe storm air velocity and temperature structure deduced from penetrating aircraft. *Preprints, 8th Conf. Severe Local Storms, Denver, Col.*, 25–32.
Spillane, K. T. and McCarthy, M. J. (1969). Downdraught of the organised thunderstorm. *Aust. met. Mag.*, **17**, 1–24.
Steiner, R. and Rhyne, R. H. (1962). Some measured characteristics of severe storm turbulence. NSSP Report no. 10, USWB.
Stout, G. E. and Huff, F. A. (1953). Radar records Illinois tornado genesis. *Bull. Am. met. Soc.*, **34**, 281–284.
Takeda, T. (1965). The downdraught in convective shower cloud under the vertical wind shear and its significance for the maintenance of convective system. *J. met. Soc. Jap.*, **43**, 302–309.
Takeda, T. (1966a). The downdraught in the convective cloud and raindrops: a numerical simulation. *J. met. Soc. Jap.*, **44**, 1–11.
Takeda, T. (1966b). Effects of the prevailing wind with vertical shear on the convective cloud accompanied with heavy rainfall. *J. met. Soc. Jap.*, **44**, 129–143.
Takeda, T. (1971). Numerical simulation of a precipitating convective cloud: The formation of a "long-lasting" cloud. *J. atmos. Sci.*, **28**, 350–376.
Tepper, M. (1950). A proposed mechanism of squall lines: the pressure jump line. *J. Met.*, **7**, 21–29.
Ulanski, S. L. and Garstang, M. (1978). The role of surface divergence and vorticity

in the life cycle of convective rainfall. I. Observations and analysis. *J. atmos. Sci.*, **35**, 1047–1062.

Weickmann, H. K. (1964). The language of hailstorms and hailstones. *Nubila*, **6**, 7–51.

Weston, K. J. (1972). The dry-line of Northern India and its role in cumulonimbus convection. *Q. Jl R. met. Soc.*, **98**, 519–531.

Wichmann, M. (1951). Uber das Vorkommen und Verhalten des Hagels in Gewittervolken. *Annln Met.*, **4**, 218–225.

Wilhelmson, R. (1974). The life cycle of a thunderstorm in three dimensions. *J. atmos. Sci.*, **31**, 1629–1651.

Wilhelmson, R. B. and Klemp, J. B. (1978). A numerical study of storm splitting that leads to long-lived storms. *J. atmos. Sci.*, **35**, 1974–1986.

Wilhelmson, R. and Ogura, Y. (1972). The pressure perturbation and the numerical modelling of a cloud. *J. atmos. Sci.*, **29**, 1295–1307.

Winston, J. S. (ed.) (1956). Forecasting tornadoes and severe thunderstorms. Forecasting Guide no. 1, US Weather Bureau, Washington, D.C.

Wojtiw, L. (1975). Climatic summaries of hailfall in Central Alberta 1957–1973. Alberta Research Atmospheric Science Report 75-1.

Zipser, E. J. (1969). The role of organized unsaturated convective downdraughts in the structure and rapid decay of an equatorial disturbance. *J. appl. Met.*, **8**, 799–814.

Zipser, E. J. (1977). Mesoscale and convective scale downdraughts as distinct components of squall-line structure. *Mon. Weath. Rev.*, **105**, 1568–1589.

9

Shallow Cellular Circulations

I. Introduction

The post-war era of high-flying aircraft, radar and satellite has witnessed the identification of a suite of shallow meso-scale convective systems. These systems, which may be cloudy or dry, vary in both form and horizontal size. In form they are usually considered to comprise "cells", but this term has meanings other than that employed in the chapter on severe storms. In this chapter the word "cell" is employed to describe a system comprising many individual cumulus towers linearly linked so as to form distinctive shapes, the most striking perhaps being polygons and cloud streets (Fig. 147). Whilst these shapes appear to be quite different they may profitably be linked, for both convenience and dynamical reasons, by considering the polygons (usually hexagons) to be a three-dimensional and the streets to be a two-dimensional form of cellular convection, the latter essentially comprising a transverse circulation with little or no circulation along the street. In size the three-dimensional cells vary from 5 to 10 km in diameter (seen by radar; Hardy and Ottersten, 1969), with a diameter to height ratio of about 5 : 1, to cells of up to 100 km diameter and a ratio of 30 : 1. Cloud streets (or longitudinal rolls) have been known for much longer but only scantily studied. A range of sizes may also exist here. Angell *et al.* (1968) observed horizontal helical structures with both diameters and depths of about 2 km. They suggested that such circulations would favour the formation of the type of cloud streets so clearly seen in satellite photographs (Fig. 147). This chapter considers our observational and theoretical knowledge of these meso-scale cellular convective systems. At this stage we should note that the

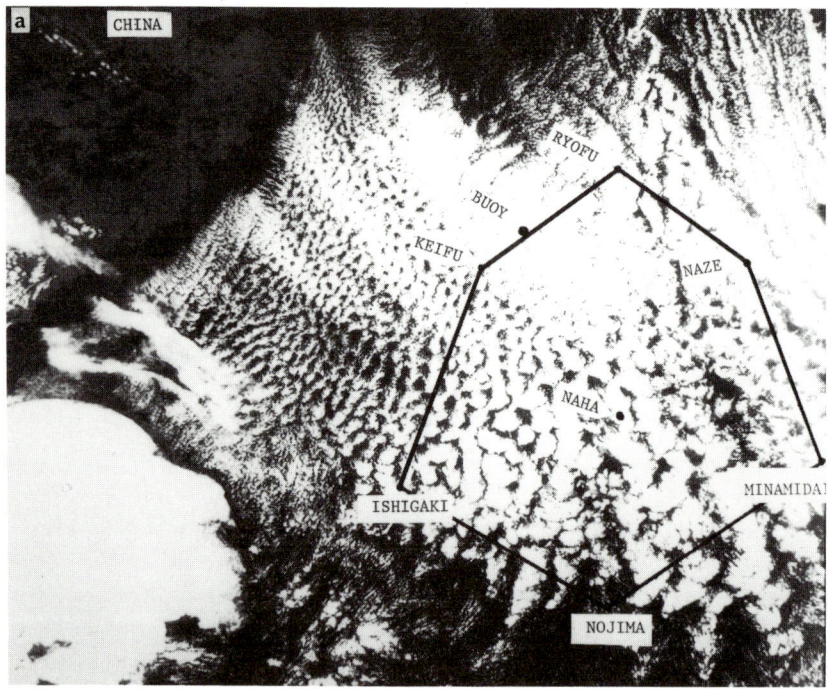

FIG. 147. (a) Co-existing open and closed cells of shallow convection over the East China Sea, 1155 JST, 16 February 1975. (Courtesy of E. M. Agee.) (b) Well-developed cloud streets over land surface. (After Finley Holiday Films.)

cellular convection discussed in this chapter is not that deep meso-scale convection identified by Mason (1970) as comprising the main component of the tropical "cloud clusters" which are one subject of the GARP Atlantic Tropical Experiment (GATE). Whilst these tropical systems are no doubt important, they have not yet been either fully observed or explained and, as such, perforce receive only this mention in this book.

II. Observation

The repercussions of the comparatively recent "discovery" of shallow atmospheric cellular convection are that neither observational nor theoretical knowledge are well developed. Although cloud streets have been known to exist for many decades, their nature and full extent have become clearer only with observations from high-flying aircraft and both manned and unmanned satellites. Similarly, three-dimensional convective cells were fully appreciated only on the receipt of the first satellite pictures in the early 1960s. We now consider the observational structure of each type of cell.

A. Two-dimensional cells

In this type of cell the essential convective overturning exists in a plane perpendicular to the main axis of the "street"—cloudy or dry. It is in this sense that the cell is "two-dimensional". We owe the bulk of our knowledge on these "band structures" in the atmosphere to Kuettner (1959, 1971).

1. *Description*

For convenience cloudy as opposed to dry streets are considered. Most cloud streets are easily recognized to comprise individual cumuli lined up like pearls on a string: the most spectacular cases resemble continuous roll clouds. In both cases the streets vary diurnally as do normal cumuli. Kuettner (1959) noted that although individual cumulus streets sometimes originate over and downwind from a heat source, such as an industrial area, a fire, a heat island or mountain range, the phenomenon discussed here is, in general, independent of orography. In fact, orography tends to disturb rather than create cloud streets and probably for that reason they are best developed over uniform flat terrain and the oceans. Together with data from aircraft and satellites (Weston, 1977; Kamiko and Okano, 1971; Streten, 1975), observations from the Barbados Oceanographic and Meteorological Experiment (BOMEX) suggest the following typical charactcristics of convective cloud streets: length ranging from 20 to 500 km; spacing from 2 to 8 km; layer height from 0.8 to 2 km; width to height ratio from 2 to 4; alignment along the mean wind of the convective layer.

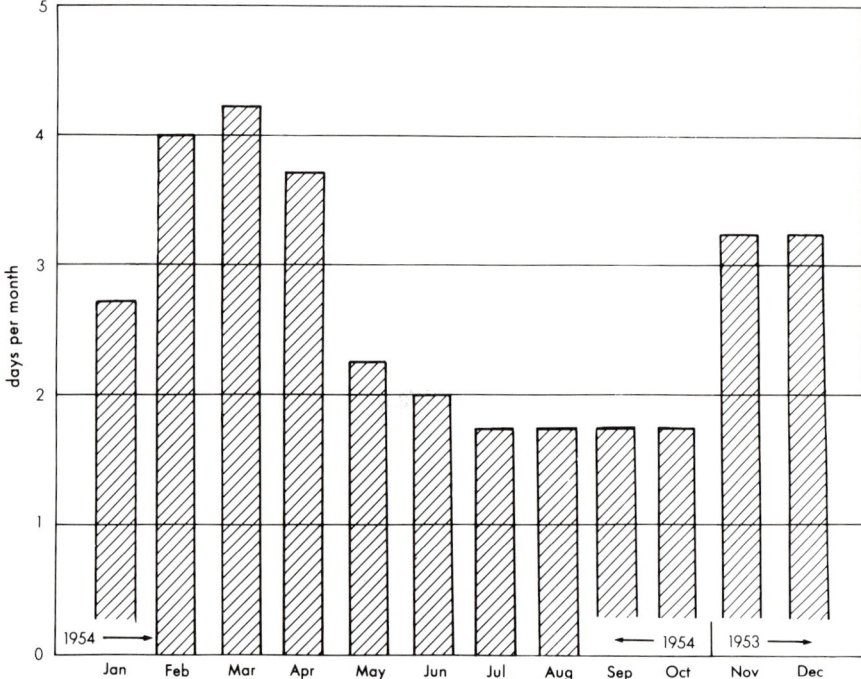

Fig. 148. Seasonal variation of cloud street occurrence at Boston, USA during 1953 and 1945. (After Kuettner, 1959.)

2. Climatology

Their comparatively transient nature, contributing to and combined with the sparsity of observation, hinders a sensible climatological analysis of cloud streets. Kuettner (1959) attempted such a study near Boston and produced the results shown in Fig. 148. Despite the recognized limitations in the data, a real seasonal variation was suggested, the higher frequencies being in the winter half of the year. The overall number of days with streets (35) was considered by Kuettner to be an underestimate—a figure of 40–50 days being more likely. He also noted that cloud streets are far more common in other latitudes. At present, to the author's knowledge, no climatological maps of cloud street occurrence exist.

3. Synoptic background

Kuettner found that most extra-tropical cloud streets occurred in vigorous outbreaks of polar air that is heated from below during its progress towards warmer regions. The formation of a street was favoured by the presence of an inversion, usually at a height of about 2 km above the ground. The convection

9 SHALLOW CELLULAR CIRCULATIONS

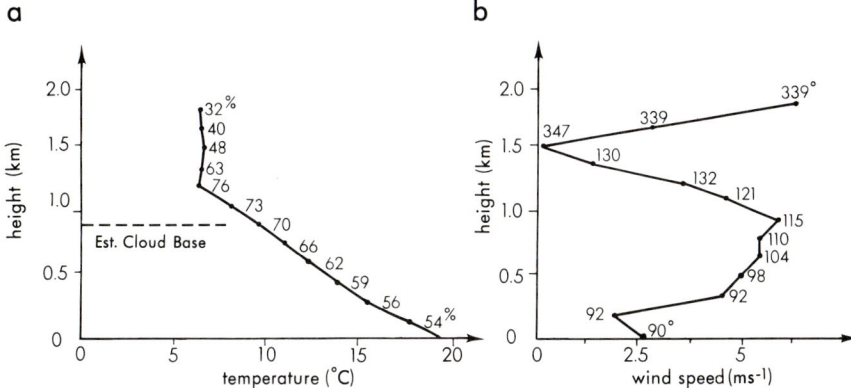

FIG. 149. (a) Vertical distribution of temperature and humidity associated with cloud streets. Numbers on the temperature curve are relative humidities (in per cent). (After Kuettner, 1971.) (b) Vertical distribution of horizontal winds associated with cloud streets. Numbers on the curve indicate wind direction.

occurred beneath the inversion in a layer where the lapse rate was greater than the dry adiabatic value (Fig. 149(a)). In the convective layer the vertical wind speed profile showed strong curvature containing in most cases a definite wind maximum within the layer. The vertical gradient of the wind speed shear (called the "profile curvature" by Kuettner) ranged from 10^{-7} to 10^{-6} cm^{-1} s^{-1}. There was little indication of a linear wind shear or uniform flow (Fig. 149(b)). Similarly, variations in wind direction with height were quite small.

B. Three-dimensional cells

Nearly two decades ago Krueger and Fritz (1961) presented the first pictures of the honeycomb structure that lay over thousands of square kilometres of ocean but not over land. Since then thousands of satellite photographs have confirmed the existence of this form of convection and have also revealed variations on the basic theme (Kuettner and Soules, 1966). The frequent and widespread occurrence of three-dimensional cellular convection posed an apparently new meteorological problem that warranted inclusion in the schedule of the Global Atmospheric Research Programme (GARP). Indeed it occupied a major role in the Air Mass Transformation Experiment (AMTEX)—a major observational experiment undertaken in the north-west Pacific in the early 1970s (Lenschow and Agee, 1976). Figure 147 shows that there are two types of cellular convection (first noted by Hubert, 1966) with horizontal dimensions of 20–100 km: on the one hand open cells; and on the other hand closed cells. Both types of cell are visible of course only because of the presence of cloud. On any one occasion cells may change from one form to the other: for example Sekihara et al. (1978) noted a change from closed to open cells associated with a thickening of the mixing layer in which they are formed. The inferred circulations are as follows: in the open cells air rises in convective towers on the borders and sinks to give clear skies in the centre; in the closed cells air rises in the centre and sinks on the borders. These circulations provide the central theoretical problem of these meteorological phenomena—a topic considered in Section III of this chapter.

1. Description

Within the last decade a descriptive picture of meso-scale cellular convection has been constructed with the aid of both satellite and surface observations. Following in the tradition of Krueger and Fritz (1961) and Hubert (1966), Agee and Dowell (1974) combined conventional rawinsonde data and surface synoptic reports with satellite cloud photography in an analysis of 38 cases of meso-scale cellular convection. In particular they considered cell diameter, depth and aspect ratio—the last-mentioned being the diameter divided by the depth. They also noted the wind shear in the cell layer and the energy flux between air and sea. Table 47 summarizes some of their results. In these early

Table 47 Summary of average statistics for 38 cases of meso-scale shallow cellular convection

	Open (25)	Closed (13)	All (38)
Diameter, D (km)	30.48	32.09	31.25
Depth, h (km)	2.26	1.25	**
D/h	15.12	27.57	**
Winds:			
Magnitude of shear (m s^{-1} km^{-1})	2.25	1.25	1.82
Absolute value†	3.26	3.30	3.27
Directional shear (deg km^{-1})	−5.73	6.67	**
Absolute value†	6.75	7.19	6.89
Gradient (m s^{-1} km^{-2})	1.39	1.25	1.37
Absolute value†	1.73	2.16	1.92
Lapse rate (°C km^{-1})	8.30	7.92	8.17
$T_s - T_a$ (°C)	2.1 (6)	−0.4 (5)	**

Numbers in parentheses indicate how many cases were used in averaging. Double asterisk indicates that open and closed cells are statistically different at the 1 per cent level and averages are inappropriate. T_s is sea surface temperature; T_a is air temperature. Negative directional shear for open cells indicates a backing wind with height or cold air advection. Positive directional shear for closed cells indicates a veering wind or warm air advection.
† Averages computed using only absolute values.
Source: Agee and Dowell (1974).

results open cells were found to be smaller in diameter but deeper than closed ones, with concomitant differences in the aspect ratio. Later measurements of depth of the cells by Agee and Lomax (1978) showed the opposite result: mean convective depths for open and closed cells were 1529 m and 2066 m respectively. The gradient of the wind shear was small in both cases, in fact much less than the 10^{-7}–10^{-6} cm^{-1} s^{-1} values reported by Kuettner (1971) for two-dimensional convection or cloud streets. The much smaller gradient was interpreted by Agee and Dowell as being necessary to allow the three-dimensional convective mode in the atmosphere. The negative directional shear for open cells indicated a backing wind with height or cold air advection. The positive directional shear for closed cells indicated a veering wind or warm air advection. As suggested in Table 47, and shown more fully by Agee and Dowell, the aspect ratio is inversely proportional to the convective depth. This may be due to eddy anisotropy, a feature discussed in the second part of this chapter.

These results derived from satellite data were broadly confirmed and complemented by those from surface buoy data gathered in AMTEX. Burt and Agee (1977) found that closed cells forming in an outbreak of cold polar air over the much warmer Kuroshio had a diameter of 28 km. They also added the first quantitative information on the internal flow temperature and humidity. Within a closed cell air near the sea surface moved from the edge

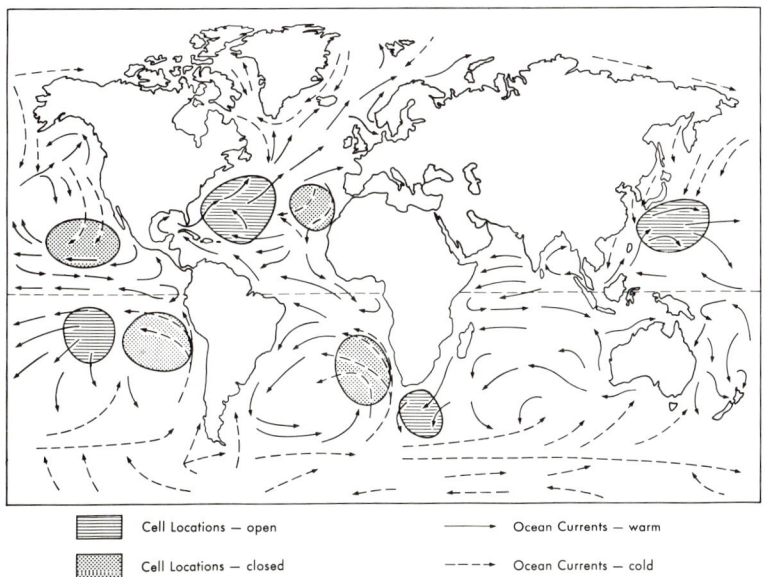

FIG. 150. Semi-schematic map of favoured regions of open and closed cellular convection with respect to warm and cool ocean currents. (After Agee *et al.*, 1973.)

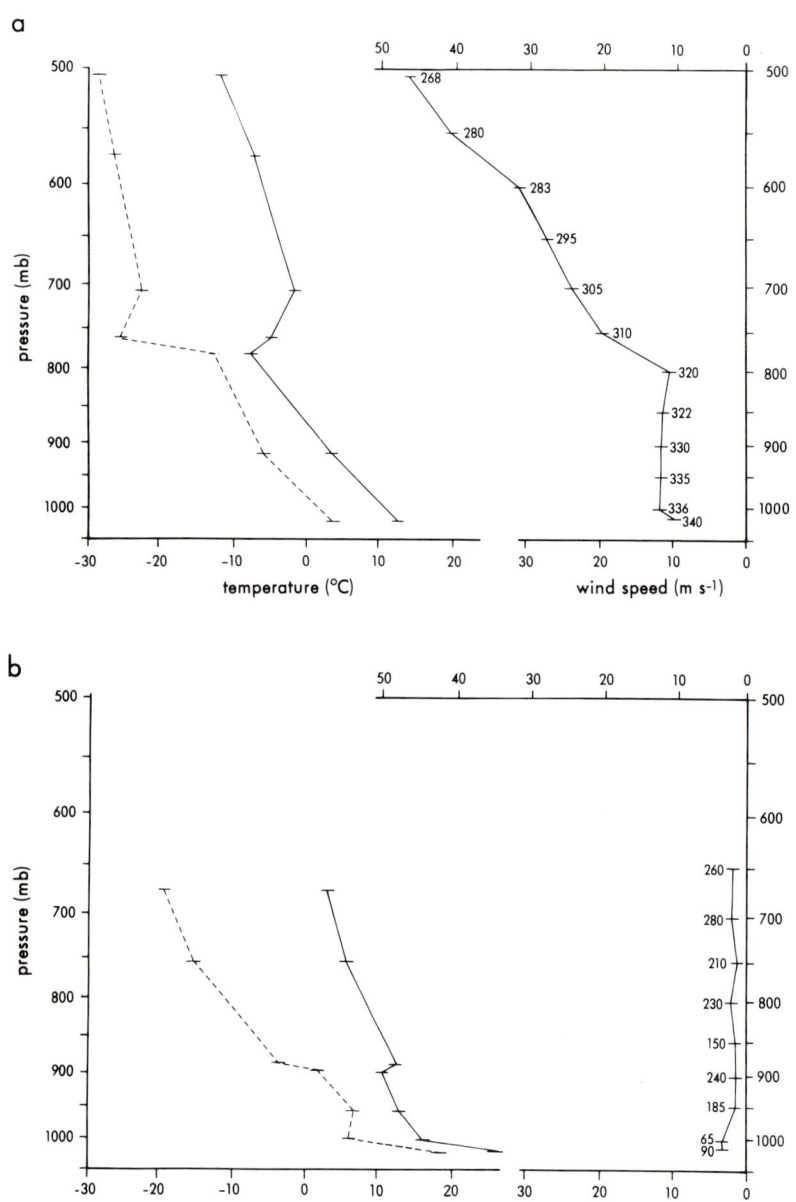

FIG. 151. Vertical distribution of temperature (solid line), dew point temperature (dashed line) and horizontal wind speed and direction in degrees (numbers against wind speed line) typical of a situation giving cellular convection: (a) for open cells; (b) for closed cells. (After Agee and Dowell, 1974.)

toward the centre at a speed of $0.6\,\mathrm{m\,s^{-1}}$. The temperature difference near the sea surface between the relatively cold descending branch and the warm ascending branch was $0.2\,°C$ and the specific humidity difference between the less moist descending air near the edge and the moist ascending air near the cell centre was 9 per cent ($c.\ 0.4\,\mathrm{g\,kg^{-1}}$).

2. Climatology

Both the nature and recent discovery of meso-scale cellular convection inhibit full climatological study. Nevertheless, experience has grown and Agee et al. (1973) presented a tentative climatological picture (Fig. 150) in which open convection cells preferably occur to the east of continents over warm ocean currents, while closed cells tend to occur west of the continents over cooler ocean currents.

3. Synoptic background

The qualitative climatology noted above is essentially a composite of the synoptic conditions favouring the existence of meso-scale cellular convection. Figure 151 shows that the convection is capped by an inversion, both stronger and higher in the case of open cells (Fig. 151(a)). Below the inversion lapse rates are nearly dry adiabatic (see Table 47) or even greater in a shallow, near-surface layer (Fig. 151(b). Figure 150 also shows the marked absence of wind shear in the convective layer. The combined effects of lapse rate and wind were, in essence, summarized by Agee and Sheu (1978). In fact they analysed the relationship of meso-scale cellular convection of wind speed (V) and the temperature difference between sea surface water and air (ΔT) (Fig. 152). They found that:

(1) If $V > 9\,\mathrm{m\,s^{-1}}$ a direct relationship existed between V and the minimum T required to produce the meso-scale convective pattern.

(2) At $V \simeq 9\text{–}5\,\mathrm{m\,s^{-1}}$, if T was between $c.\,4$ and $7\,°C$, the minimum positive ΔT values at which meso-scale cellular convective clouds were observed was inversely related to V. In this range of V the initiation efficiency was apparently reduced so that higher positive ΔT values, i.e. $7\text{–}10\,°C$ were required to induce the convection.

(3) V between 8 and $10\,\mathrm{m\,s^{-1}}$ was the optimal range for the formation of the convection at the minimum $\Delta T \simeq 5\,°C$.

These results make no distinction between closed and open cells and include values of ΔT about twice those found by Agee and Dowell (1974). In this latter study the authors found that sea surface temperature exceeded the air temperature by only $2.1\,°C$ in open cell cases but was $0.4\,°C$ cooler in closed cell cases (Table 47). In view of the fact that Agee and Dowell used only

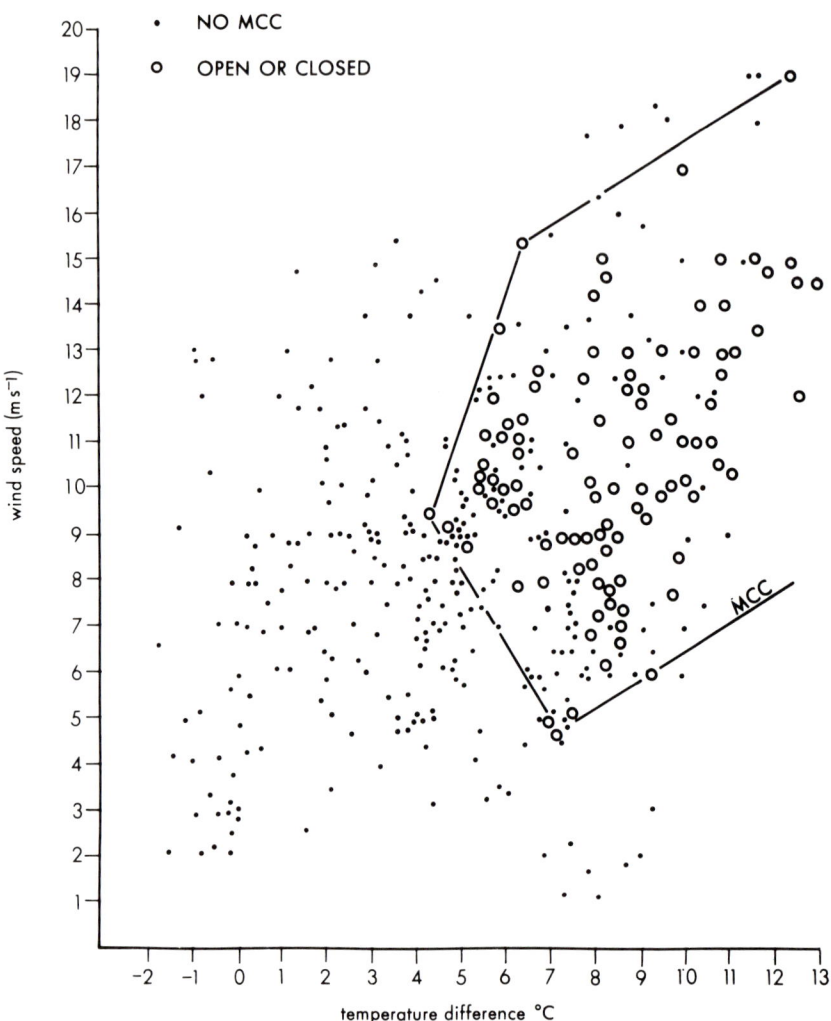

Fig. 152. Occurrence of meso-scale cellular convection (MCC) in relation to wind speed and air–sea temperature difference as observed at 3 hourly intervals over a 3 week period in 1975 south of Japan. Circles represent conditions in the presence of MCC and dots represent conditions in the absence of MCC. The solid line marked MCC separates an absolute non-MCC region from a likely MCC region. (After Agee and Sheu, 1978.)

operationally gathered data whilst Agee and Sheu used especially gathered data in AMTEX, the latter's results are probably more reliable.

The interplay between cells and large scale flow has been preliminarily studied by Sheu and Agee (1977) who calculated the horizontal divergence,

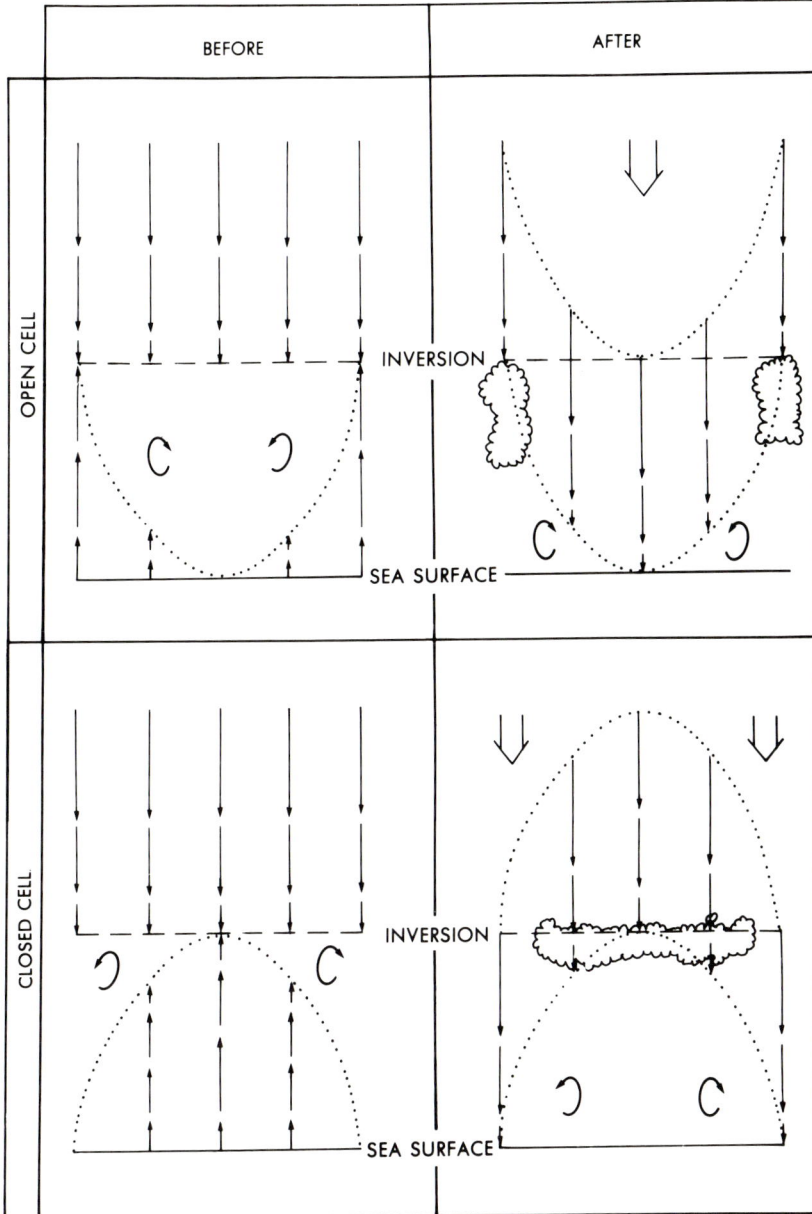

FIG. 153. Schematic vertical cross-sections of vertical motion associated with open and closed cells before and after their formation. (After Sheu and Agee, 1977.)

relative vorticity and vertical motion of the synoptic scale flow for cases of both open and closed cells. They found that both types of cell would form in an environment with large scale divergence and downward motion. The distribution of vorticity showed no characteristic sign for either type of cell. Sheu and Agee envisaged the following relationship between synoptic scale flow and meso-scale cellular convection. Both types of cell form in cold, polar outflows characterized by both large scale divergence and downward motion. In such a region the divergence and sinking are likely to exist above an inversion, with weak ascending motion in a narrow column below the inversion during formation of the wall of an open cell. They also envisaged, with no physical argument for support, a column of strong ascending motion during the formation of the cloudy centre of a closed cell. Though the mixing process due to eddies of various sizes may be in the neighbourhood of the major vertical motion columns, the net circulation within a convecting cell over a meso-scale cellular convection region remains as divergence and downward motion as depicted schematically in Fig. 153. Clearly much remains to be clarified in the analysis of airflow at both the meso- and larger scales, but Sheu and Agee were of the opinion that the direction of the large scale vertical motion does not determine the direction of the circulation in the convection cells.

4. *Heat fluxes*

As in the case of all atmospheric convective phenomena the ultimate *raison d'être* of the cellular structure is the necessity for vertical heat transfer. In their preliminary study Agee and Dowell (1974) suggested that in open cells a total heat transfer from sea to air of over 600 W m^{-2} took place whereas in closed cells a flux of about 200 W m^{-2} existed from atmosphere to sea. These early results, in which open and closed cells were distinguished, have found little support and in fact Burt and Agee (1977) have subsequently shown that closed as well as open cells can exist with strong surface heating. Further analyses by the AMTEX team showed that the threshold value (where ΔT is small) of total (i.e. sensible plus latent) heat transfer for the production of meso-scale cellular convection was 200 W m^{-2}. Once the convection was well developed in cold air, sea to air heat flux rose to about 750 W m^{-2} (Agee and Howley, 1977) or even 1200 W m^{-2} (Sheu and Agee, 1977).

This part of the chapter may be summarized as follows. Convection cells may be two- or three-dimensional, the latter being open or closed, with diameters of about 30 km and aspect ratios of 20–30 : 1, with open cells being less flat then closed cells. Both two- and three-dimensional cells form when the Earth's surface temperature is at least 5 °C higher than that of the air and wind speeds are greater than 5 m s^{-1}. In both cases the mean lapse rate of the convective layer is approximately dry adiabatic. A vertical heat flux of at least

200 W m^{-2} is necessary to drive the three-dimensional convection. The major difference between the situations favouring two- or three-dimensional cellular convection lies in the vertical distribution of wind speed. In the former case, shear is marked with a speed maximum in the convective layer; in the latter shear is virtually non-existent. This difference has important dynamical consequences that are considered in the following section of this chapter.

III. Theory

Theoretical examination of cellular convection has been both comparatively sparse and beset with some confusion. It is probably fair to say that, until Kuettner's (1959) preliminary examination, cloud streets had received little or no serious theoretical attention. Somewhat ironically the more complicated three-dimensional convective cells appeared to have an already developed theoretical explanation when they were first observed in the early 1960s. The apparency of this "explanation" is clarified below.

Krueger and Fritz's (1961) observations of a honeycomb of cells immediately triggered thoughts of the convective pattern resulting from Benard cells. Such cells were first noted by Thomson (1882) but only thoroughly observed by Benard (1900, 1901). As a result of his experiments with liquids only about 1 mm deep, Benard found that the cooling of the upper free surface rapidly led to the development of cells in which the motion was upward in the middle and downward at the common boundary between a cell and its neighbours. The convective motion tended towards a regular hexagonal pattern. In attempting to explain these observations Lord Rayleigh (1916) produced a classic paper in fluid mechanics and also unwittingly initiated some confusion about the origins of both Benard and atmospheric convective cells. In essence Rayleigh showed that in a horizontal layer of fluid at rest convection would occur only if the upward increase in density (resulting in an upward buoyancy force) was sufficiently large to overcome the effects of viscosity. He expressed this in terms of what has come to be known as the Rayleigh number:

$$\frac{\rho_2 - \rho_1}{\rho_1} > \frac{27\pi^4 \kappa \nu}{4gh^3},$$

where ρ_2 and ρ_1 are respectively the densities at the top and bottom of the fluid, κ is the diffusivity of heat and h is the depth of the fluid. The more conventional way of expressing the result is as follows:

$$\text{Ra} = \frac{g\alpha \Delta T h^3}{\kappa \nu} > A,$$

where Ra is the Rayleigh number, ΔT is the temperature difference across the layer and A is a constant for a particular convective situation.

In Rayleigh's original experiment he assumed that the liquid had two free surfaces and A was evaluated at $27\pi^4/4$ i.e. 657.5. Later, Jeffreys (1926, 1928) showed that if the fluid were bounded by two rigid surfaces then $A = 1708$ and if by one free and one rigid surface than $A = 1101$, but the essential mechanism as outlined by Rayleigh remained valid. What Rayleigh did not show conclusively was that the form of the convection should be hexagonal, as found by Benard. This is not surprising because Rayleigh had not in fact explained Benard cells, but a completely different type in which buoyancy forces played a major role. Although this important difference was hinted at in the subsequent literature (Brunt, 1937), it is only comparatively recently that Agee and Chen (1973) have suggested that cells driven by buoyancy be called Rayleigh cells and that the term Benard cell be restricted to those driven by some other force. This other force has been shown by Pearson (1958) to be surface tension. Thus, although Benard cells continue to be investigated from a meteorological viewpoint (Somerville, 1967; Somerville and Lipps, 1973), it is the Rayleigh type of cell that is of concern here.

The main problems to be addressed by theory are the preferred mode of convection (two- or three-dimensional?), the closely associated direction of circulation (particularly in the three-dimensional case) and the aspect ratio. In his analysis of cloud bands, Kuettner (1971) gave a valuable overview of both two- and three-dimensional convection which, in turn, provided a useful framework for the more specific studies of three-dimensional convection encouraged by the satellite imagery of the 1960s. Somewhat paradoxically the apparently simpler atmospheric phenomena—the cloud streets—require a more elaborate explanation than the more complex three-dimensional cells. The latter may be considered to be a special case of the former in the absence of vertical shear of wind speed. Consequently, in the treatment below, the more general case is considered. Before looking at the theory in detail it is profitable to dwell a little on the physical mechanisms involved, as seen by Kuettner.

The observed wind speed profile confirms the existence of vorticity around a horizontal axis normal to the basic flow direction (Fig. 154). It is known that the restoring forces experienced by a displaced parcel conserving its vorticity in an environment of varying vorticity tend to return the parcel to its original level. Kuettner reminded us that this mechanism underlies the Rossby waves in the westerlies and drew upon Lin (1955) to illustrate the idea in the present context. If, in a configuration such as shown in Fig. 154, a parcel rises from the layer below the wind maximum, it conserves its vorticity and constitutes a "relative vortex" having an excess vorticity over its new environment. Kuettner then noted, with no explanation, that the resulting distortion of the basic vorticity field causes fluid elements on the left side of the vortex to be replaced by elements from lower levels (having an excess

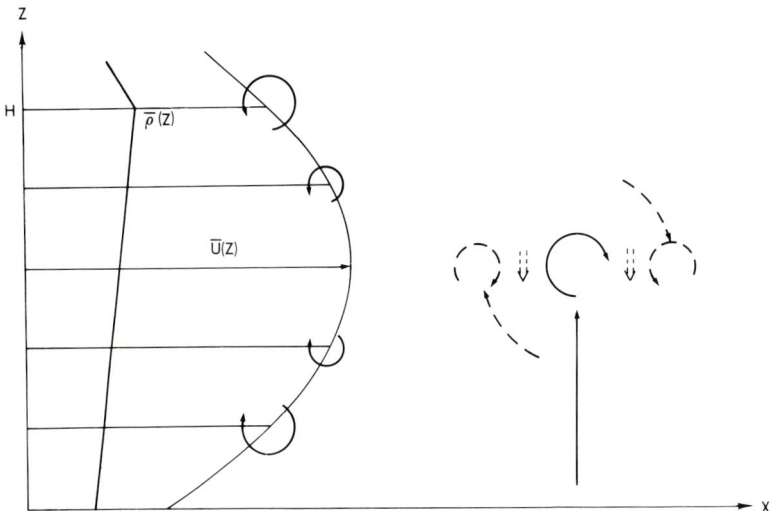

FIG. 154. Schematic vertical distribution of velocity and density of the basic flow (left) and the relative vorticity of a vertically displaced convective element with a resulting restoring force (right). (After Kuettner, 1971.)

verticity) and those on the right side by elements from higher levels (having a vorticity deficiency). This redistribution induces a downward acceleration of the "vortex" returning the displaced element toward the level where it belonged and where it finds no differential vorticity.

In such a mechanism fluid elements displaced upwards or downwards under the action of buoyancy forces will therefore have to overcome the restoring forces resulting from the vorticity gradient in an airstream in which vertical shear exists. We can do no better than quote Kuettner (1971, p. 412) on the implications of this result:

> Nature can circumvent this restriction of convective motion by selecting a specific mode of convection in which such restoring forces cannot arise. If all fluid elements along a horizontal line in the direction of flow (x) organize themselves to move upwards simultaneously and complete their circulation in a plane normal to the x, z plane (z-vertical), no differential vorticities can develop between them in the x, z plane and convection is uninhibited by restoring forces. This mode of organized convection represents helical motions in longitudinal roles, the precise dimensions of which are regulated by other factors such as viscosity. It should be noted that in the y, z plane the convective circulations do not encounter restoring forces either, because in this place there exists no basic flow and therefore no vorticity gradient.

Kuettner went on to point out that the well-known critical Rayleigh number (noted above) which characterizes the onset of convection in a fluid at rest holds also for two dimensional rolls. All other convection modes require higher Rayleigh numbers, that is higher vertical density gradients. This is true for three-dimensional cells and for two-dimensional cells of different orientation, the highest numbers being required for transverse rolls. Because of the requirement for higher Rayleigh numbers of three-dimensional convection, Kuettner suggested that longitudinal bands represent the prevailing mode of convection.

Kuettner suggested that theoretical analysis is facilitated by considering an incompressible, viscous and conductive fluid of height H and infinite horizontal extension flowing steadily and under gravity in the horizontal x direction. Its speed, but not its direction, varies with height (Fig. 154). The vertical density gradient is constant and positive, setting the stage for convection. If this gradient is small relative to the density itself, the ratio of the two can be neglected everywhere except in connection with gravity. This is the Boussinesq approximation first applied to the problem of cellular convection by Rayleigh (1916). Theoretical analysis is further simplified by the assumptions of invariance of kinematic viscosity and heat diffusivity throughout the fluid, that a linear thermal expansion function holds for the small density changes under consideration and that the coriolis force can be neglected. In using small perturbation methods, symbols with bars indicate properties of the basic flow and the perturbed quantities have no bars.

In the conventional Cartesian framework, the above assumptions may be expressed as follows:

$$\frac{\partial}{\partial t}, \quad \frac{\partial}{\partial y}(\bar{\rho}, \bar{p}, \bar{u}) = \frac{\partial}{\partial x}(\bar{\rho}, \bar{u}) = \bar{v} = \bar{w} = 0,$$

$$\frac{\partial \bar{\rho}}{\partial z}, \quad \frac{\partial \rho}{\partial T}, v, \kappa = \text{constant}.$$

In the atmosphere the diffusivity is represented by the eddy diffusivity K_z^H. The perturbation equations of motion as used by Kuettner were

$$\frac{\partial u}{\partial t} + \bar{u}\frac{\partial u}{\partial x} - v\nabla^2 u + \frac{\partial \bar{u}}{\partial z}w - \frac{\rho v}{\bar{\rho}}\frac{\partial^2 \bar{u}}{\partial z^2} = -\frac{1}{\rho}\frac{\partial p}{\partial x}, \quad (220)$$

$$\frac{\partial v}{\partial t} + \bar{u}\frac{\partial v}{\partial x} - v\nabla^2 v = -\frac{1}{\bar{\rho}}\frac{\partial p}{\partial y}, \quad (221)$$

$$\frac{\partial w}{\partial t} + \bar{u}\frac{\partial w}{\partial x} - v\nabla^2 w + \frac{g\rho}{\bar{\rho}} = -\frac{1}{\bar{\rho}}\frac{\partial p}{\partial z}, \quad (222)$$

where ∇^2 is the three-dimensional Laplacian operator. The continuity equation in incompressible flow was

$$\frac{\partial u}{\partial x} + \frac{\partial v}{\partial y} + \frac{\partial w}{\partial z} + \frac{w}{\bar{\rho}}\frac{\partial \bar{\rho}}{\partial z} = 0, \quad (223)$$

where the last term can be neglected according to the Boussinesq approximation. The energy equation was

$$\frac{\partial T}{\partial t} + \bar{u}\frac{\partial T}{\partial x} - K^H \nabla^2 T + \frac{\partial \bar{T}}{\partial z}w = 0. \tag{224}$$

Kuettner manipulated equations (220)–(224) into a sixth-order differential equation for the vertical velocity component w, eliminating density (ρ) and introducing a stability parameter

$$\beta = \frac{1}{\bar{\rho}}\frac{\partial \bar{\rho}}{\partial z}.$$

The basic equation thus became

$$\left\{F\left[f(\nabla^2) - \frac{\partial^2 \bar{u}}{\partial z^2}\frac{\partial}{\partial x}\right] - g\beta\nabla_h^2\right\}w = 0, \tag{225}$$

where ∇_h^2 is the two-dimensional Laplacian operator in the horizontal plane,

$$f = \frac{\partial}{\partial t} + u\frac{\partial}{\partial x} - \nu\nabla^2$$

and

$$F = \frac{\partial}{\partial t} + \bar{u}\frac{\partial}{\partial x} - K^H \nabla^2.$$

In passing, Kuettner noted that for steady ($\partial/\partial t = 0$), inviscous ($\nu = 0$), non-conductive ($K^H = 0$), two-dimensional flow ($\partial^2/\partial y^2 = 0$), the operator $f = F = \bar{u}(\partial/\partial x)$ and equation (225) reduces to

$$\frac{\partial^2 w}{\partial x^2} + \frac{\partial^2 w}{\partial z^2} - \left(\frac{g\beta}{\bar{u}^2} + \frac{1}{\bar{u}}\frac{\partial^2 \bar{u}}{\partial z^2}\right)w = 0, \tag{226}$$

which is the equation governing the lee-wave problem.

In treating equation (225) Kuettner approximated the height-dependent flow velocity $\bar{u}(z)$ by a characteristic mean velocity $\bar{u}$ and the curvature $\partial^2 \bar{u}/\partial z^2$ by a characteristic mean value $\overline{\partial^2 \bar{u}/\partial z^2}$ and introduced three-dimensional harmonic perturbations. If l and m are wave numbers along x and y and if σ is the exponential time constant, a complex quantity, all perturbations contain the factor

$$e^{[i(lx+my)+\sigma t]}.$$

Assuming that w is proportional to $\sin(nz)$ and with Rayleigh's boundary conditions that $w = \partial^2 w/\partial z^2 = 0$ at $z = 0, H$, Kuettner derived the following expression for the growth of the disturbance:

$$(\sigma_r + K^H d^2 + is)[(\sigma_r + \nu d^2 + is)d^2 + il\overline{\partial^2 \bar{u}/\partial z^2}] - g\beta(l^2 + m^2) = 0, \tag{227}$$

where σ_r is the real part of σ; $s = \bar{u}l + \sigma_i$, where σ_i is the imaginary part of σ;

and $d^2 = l^2 + m^2 + n^2$, characterizing the dimensions of a convective cell. Equation (227) allowed an examination of convection in a medium both at rest and flowing. Kuettner considered first, the medium at rest, the case studied by Rayleigh and, at first sight, that which supports three-dimensional cells.

A. Medium at rest

In a resting medium $\bar{u} = \overline{\partial^2 \bar{u}/\partial z^2} = 0$ and in stationary conditions ($\sigma_i = 0$), equation (227) is reduced to

$$(\sigma_r + K^H d^2)(\sigma_r + v d^2)d^2 - g\beta(l^2 + m^2) = 0. \tag{228}$$

Equation (228) corresponds to equation (37) in Rayleigh (1916), who derived the condition for convection (Ra $\geq$ 658) for unlimited values of the sum $(l^2 + m^2)$. Clearly l^2 and m^2 can vary whilst still retaining the same sum and this means that two-dimensional cells (or rolls, lines, strips, streets) could be equally amplified as three-dimensional cells but their spacing would be different. Bands would have a wavelength of $2\sqrt{2}H$ and symmetrical cells a wavelength of $4H$, where H is the depth of the convective layer. Thus in an air mass with an inversion at a height of 2 km (say), bands would be about 5.5 km apart and cells would have a diameter of 8 km. The theoretical band spacing accords well with observation, but predicted cell diameters are rather small. The diameter and flatness of three-dimensional cells, along with their direction of circulation, have been central themes in theoretical analysis and the main results are reviewed below.

Since the pioneering study by Krueger and Fritz (1961) the geometry of three-dimensional convective cells has been under increasing scrutiny. Casting a wide net Agee (1976) related the aspect ratio (diameter divided by height) to the height of the convective layers (Fig. 155) within which five types of cellular convection occurred. Two clear messages emerged: first, the aspect ratio decreases with increasing height of convective layer, large, "flat" cells being very shallow; and second, the geometry of radar-observed and laboratory cells is so different from the rest as to suggest a separate physical mechanism. Drawing on the work of Priestley (1962) and Ray (1965), Agee suggested that the differences in cell geometry are primarily due to the anisotropy of the eddy viscosity in the convective layer. Priestley speculated that the stable layer that acts as a lid to the cellular convection would result in the suppression of the vertical eddy viscosity (K_z^M) relative to the horizontal coefficients (K_x^M, K_y^M), which in turn would encourage "flat" cells due to the relatively favoured horizontal transfers. Ray (1965) followed up this idea and agreed that, in general, a horizontal transfer coefficient will flatten the cells. Agee (1976) enthusiastically took up the idea and claimed that the "physical"

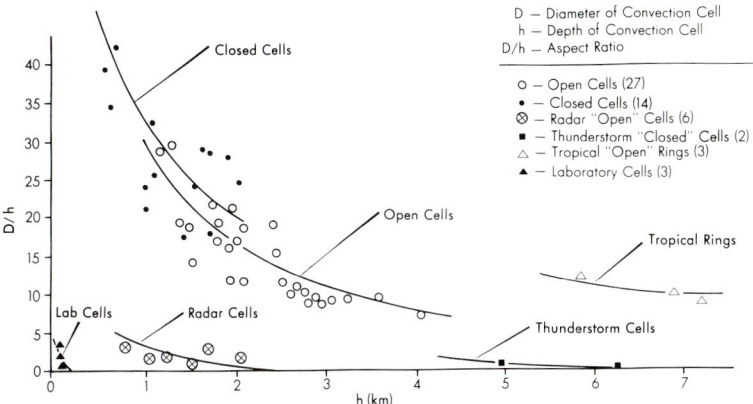

FIG. 155. Relationship between aspect ratio (D/h) (or cell flatness) and convective depth, based on observational case studies of various convective phenomena. (After Agee, 1976.)

differences in Fig. 155 are due to eddy anisotropy. In particular he felt that the "clear air" cells observed by radar occur in an environment that is more isotropic than that containing meso-scale cellular convection and also that the laboratory cells, being largely driven by molecular viscosity and conductivity, are also free of effects of eddy anisotropy. His general conclusion was that for a given convective depth the aspect ratio increases (i.e. flatter cells) as the degree of eddy anisotropy increases.

Agee and his colleagues have also used variations in eddy viscosity in an examination of cell circulation (Agee and Chen, 1973). In this case they concluded that the sign of the change of eddy viscosity with height determined the direction of circulation in cellular convection. In an unstable layer the cells should be open (closed) if the convective layer has an eddy viscosity profile that decreased (increased) with height. They also concluded, however, that the vertical variation of eddy viscosity had little effect on cell flatness. In a two-pronged attack on the problem, Krishnamurti (1975a,b,c) refuted Agee and Chen's main result, claiming that observed vertical variations of viscosity were inadequate to account fully for differences in cell circulation. As an alternative she suggested that regions of large scale descending motion favoured open cells and that regions of large scale ascending motion favoured close cells. Both mathematical and laboratory models supported her conclusions.

It is indicative of the state of the art that very few attempts at a fairly comprehensive theory of three-dimensional cellular convection appear in the literature. Painting on a broader canvas than either Agee and Chen or Krishnamurti, Rosmond (1973) presented a linear stability model that was forced by a parameterization of the net heating due to small scale cumulus convection. He used a stable atmosphere to give physical justification for the assumption of anisotropic mixing coefficients. He stressed, however, that the assumption of anisotropy was not essential to the production of the observed flattened cells. The main direct results of Rosmond's experiment were that relative humidity and large scale static stability were the most important factors affecting the cellular convection. The former represented the positive buoyant energy in the model and the latter the negative buoyant energy. Humidities of about 80 per cent and a conditionally unstable lapse rate of 7.3 K km^{-1} gave cells of about 40 km diameter and similarly realistic growth rates. Of equal importance to the above results was the general conclusion that the meso-scale cellular convection was the consequence of the cumulative, large scale heating effect of the cumulus clouds present in a conditionally unstable atmosphere. As Rosmond (1973, p. 1408) noted: "...cellular patterns...represent an important intermediate step in the transfer of heat, moisture and momentum from the cumulus to the synoptic scale..." and no doubt his pioneering linearized analysis will stimulate production of the necessary non-linear models. Enough has been said to show that theoretical analysis of the internal structure and dynamics of three-dimensional cellular convection is still in its infancy.

B. Flowing medium

In a flowing medium, the more general case of equation (227), clearly $\bar{u}$ and $\overline{\partial^2 \bar{u}/\partial z^2}$ are not equal to zero. By neglecting the difference between the diffusivities for momentum and heat and assuming the Prandtl number

$$\mathrm{Pr} = v/k_c = 1,$$

where k_c is the thermal conductivity, Kuettner (1971) showed that the amplification constant (σ_r) was as follows:

$$\sigma_r = \mp [g\beta(l^2+m^2)/d^2 - (\overline{l\partial^2 \bar{u}/\partial z^2}/2d^2)^2]^{1/2} - v^*d^2, \qquad (229)$$

where $v^* = (k_c v)^{1/2}$. The choice of sign in front of the bracket indicated the possibility of a positive growth rate.

The first term in the bracket containing gravity and the density gradient represents the buoyancy. Its magnitude depends on the cell dimensions such that, with decreasing horizontal cell size (increasing l and m), the term tends towards a finite maximum value, namely $g\beta$, while it tends toward zero with increasing cell size. The second term in the bracket is the vorticity term. Regardless of the sign of the shear gradient, this term always subtracts from the buoyancy. If larger than the buoyancy term, the vorticity term is responsible for waves rather than the convection currently under discussion. The last term in equation (229), the viscosity/conductivity term, is always negative and tends to suppress or dampen buoyant motions. Kuettner went on to show that shear in the flow reduces the spacing of transverse bands in comparison with that of longitudinal bands ($2.8H$) but increases the critical Rayleigh number according to the fourth power of the wind profile curvature, thus strongly penalizing the transverse against the longitudinal mode of convection. By relating the critical Rayleigh numbers for longitudinal, transverse and symmetrical cells to a Froude number (comprising the inertia and buoyancy forces in the flow) Kuettner revealed that a given Rayleigh number may well be supercritical for the longitudinal mode, but subcritical for the transverse or symmetrical mode. In this way longitudinal bands are first amplified and emerge as the prevailing convection pattern in a flowing medium such as an airstream.

References

Agee, E. M. (1976). Observational evidence of cell flatness as a function of convective depth and eddy anisotropy. *J. met. Soc. Jap.*, **54**, 68–71.
Agee, E. M. and Chen, T. S. (1973). A model for investigating eddy viscosity effects on mesoscale cellular convection. *J. atmos. Sci.*, **30**, 180–189.
Agee, E. M. and Dowell, K. E. (1974). Observational studies of mesoscale cellular convection. *J. appl. Met.*, **13**, 46–53.
Agee, E. M. and Howley, R. P. (1977). Latent and sensible heat flux calculations at the air–sea interface during AMTEX 74. *J. appl. Met.*, **16**, 443–447.
Agee, E. M. and Lomax, F. E. (1978). Structure of the mixed layer and inversion layer associated with patterns of mesoscale cellular convection during AMTEX 75. *J. atmos. Sci.*, **35**, 2281–2301.
Agee, E. M. and Sheu, P. J. (1978). MCC and gull flight behaviour. *Boundary Layer Met.*, **14**, 247–252.
Agee, E. M., Chen, T. and Dowell, K. E. (1973). A review of mesoscale cellular convection. *Bull. Am. met. Soc.*, **54**, 1004–1012.
Angell, J. K., Pack, D. H. and Dickson, C. R. (1968). A lagrangian study of helical circulations in the planetary boundary layer. *J. atmos. Sci.*, **25**, 707–717.
Benard, H. (1900). Les tourbillons cellulaires dans une nappe liquide. *Rev. gen. Sci. pures appl.*, **11**, 1261–1271, 1309–1328.
Benard, H. (1901). Les tourbillons cellulaires dans une nappe liquide transportante de la chaleur par convection en regime permanent. *Annls Chim. Phys.*, **23**, 62–144.
Brunt, D. (1937). Natural and artificial clouds. *Q. Jl R. met. Soc.*, **63**, 277–288.
Burt, W. V. and Agee, E. M. (1977). Buoy and satellite observations of meso-scale cellular convection during AMTEX 75. *Boundary Layer Met.*, **12**, 3–24.
Hardy, K. R. and Ottersten, H. (1969). Radar investigations of convective patterns in the clear atmosphere. *J. atmos. Sci.*, **26**, 666–672.
Hubert, L. F. (1966). Mesoscale cellular convection. Meteorological Satellite Laboratory Report no. 37, National Environmental Satellite Centre, ESSA, Washington.
Jeffreys, H. (1926). The stability of a layer of fluid heated below. *Phil. Mag.*, **2**, 833–844.

Jeffreys, H. (1928). Some cases of instability in fluid motion. *Proc. R. Soc. A*, **118**, 195–208.
Kamiko, T. and Okano, M. (1971). Cellular cloud pattern. *Geophys. Mag.*, **35**, 275–292.
Krishnamurti, R. (1975a). On cellular cloud patterns. I. Mathematical model. *J. atmos. Sci.*, **32**, 1353–1363.
Krishnamurti, R. (1975b). On cellular cloud patterns. II. Laboratory model. *J. atmos. Sci.*, **32**, 1364–1372.
Krishnamurti, R. (1975c). On cellular cloud patterns. III. Applicability of the mathematical and laboratory models. *J. atmos. Sci.*, **32**, 1373–1383.
Krueger, A. F. and Fritz, S. (1961). Cellular cloud patterns revealed by TIROS I. *Tellus*, **13**, 1–7.
Kuettner, J. (1959). The band structure of the atmosphere. *Tellus*, **11**, 267–294.
Kuettner, J. P. (1971). Cloud bands in the earth's atmosphere: observation and theory. *Tellus*, **23**, 404–425.
Kuettner, J. P. and Soules, S. D. (1966). Organized convection as seen from space. *Bull. Am. met. Soc.*, **47**, 364–370.
Lenschow, D. H. and Agee, E. M. (1976). Preliminary results from the Air Mass Transformation Experiment (AMTEX). *Bull. Am. met. Soc.*, **57**, 1346–1355.
Lin, C. (1955). *The Theory of Hydrodynamic Stability*, Cambridge University Press, London.
Mason, B. J. (1970). Future developments in meteorology: an outlook to the year 2000. *Q. Jl R. met. Soc.*, **96**, 349–368.
Pearson, J. R. A. (1958). On convection cells induced by surface tension. *J. Fluid Mech.*, **4**, 489–500.
Priestley, C. H. B. (1962). The width–height ratio of large convective cells. *Tellus*, **14**, 123–124.
Ray, D. (1965). Cellular convection with non-isotropic eddies. *Tellus*, **17**, 434–439.
Rayleigh, Lord (1916). On convection currents in a horizontal layer of fluid when the higher temperature is on the under side. *Phil. Mag.*, **32**, 529–546.
Rosmond, T. E. (1973). Mesoscale cellular convection. *J. atmos. Sci.*, **30**, 1392–1409.
Sekihara, K., Suzuki, Y. and Murai, K. (1978). On the formation of convective cloud cells during cold air outbreaks in AMTEX 75. *Pap. Met. Geophys.*, **29**, 65–73.
Sheu, P. J. and Agee, E. M. (1977). Kinematic analysis and air–sea heat flux associated with mesoscale cellular convection during AMTEX 75*. *J. atmos. Sci.*, **34**, 793–801.
Somerville, R. C. J. (1967). A non-linear spectral model of convection in a fluid unevenly heated from below. *J. atmos. Sci.*, **24**, 665–676.
Somerville, R. C. J. and Lipps, F. B. (1973). A numerical study in three space dimensions of Bénard convection in a rotating fluid. *J. Atmos. Sci.*, **30**, 590–596.
Streten, N. A. (1975). Cloud cell size and pattern evolution in Arctic air advection over the North Pacific. *Arch. Met. Geophys. Bioklim. A*, **24**, 213–228.
Thomson, J. (1882). On a changing tesselated structure in certain liquids. *Proc. phil. Soc. Glasg.*, **13**, 464–468.
Weston, K. J. (1977). Cellular cloud patterns. *Weather*, **32**, 446–450.

10

Circulations in Cyclones

I. Introduction

Probably the most familiar meteorological circulation, to both layman and professional alike, is the synoptic scale, extra-tropical, frontal cyclone. The well-known Norwegian model of this type of cyclone still retains much validity after more than half a century of widespread and intensive use. Perhaps the excellence of this classic model discouraged further scrutiny of the extra-tropical cyclone. Whatever the reasons, serious examination of the meso-scale structure of such cyclones is essentially a feature of the last two decades in the history of meteorological analysis. Even more neglected from a meso-scale viewpoint are tropical cyclones, for the simple reason that, until recently, observation of the tropical atmosphere as a whole has been very limited in comparison to that in extra-tropical areas. Despite these disadvantages we do have some knowledge of meso-scale circulations within both tropical and extra-tropical cyclones and this chapter reviews the current state of our understanding.

The bulk of our knowledge of these particular meso-scale circulations is derived from three data sources: radar (both conventional and doppler), autographic raingauges and special radiosonde or balloon ascents. Of the three data sources, radar provides the basis for the majority of studies (see Browning, 1971a,b; Donaldson and Atlas, 1963) and by its very nature leads to analysis of echo behaviour and growth of precipitation particles, to some extent at the expense of analysis of airflow. Autographic raingauge records have the advantage of allowing detailed examination of the evolution of meso-scale precipitation signatures over a larger area than can be covered by one radar: but, in a direct sense they tell us nothing about the airflows which ultimately cause those signatures. Direct evidence is available from sequential balloon ascents. Clearly, any project combining all three sources, perhaps together with aircraft and satellite data, would give the most complete picture of cyclonic meso-scale circulations.

As in some previous chapters a brief reappraisal of scale is beneficial at this stage. Following the study by Matsumoto (1968) and indeed the framework outlined in Chapter 1, Browning (1974) recognized three scales of motion systems in the atmosphere—large, meso- and convective—having characteristic horizontal dimensions of order 10^3, 10^2 and 10 km respectively. In association with these dimensions, vertical velocities are about 1 cm s^{-1}, 10 cm s^{-1} and 1 m s^{-1} respectively, together with characteristic divergence fields of order 10^{-5}, 10^{-4} and 10^{-3} s^{-1} respectively. Largely as a result of the work by Professor P. M. Austin and her collaborators it is now fairly clear that the meso-scale in this particular context can be sensibly divided into two parts giving, quite simply, a large and a small meso-scale respectively. Two decades ago Austin (1960) initially subdivided the meso-scale, thus turning the three-part scheme outlined by Browning (1974) into a four-part one. Working with radar echoes in extra-tropical cyclones Austin recognized the following: a synoptic scale; a large meso-scale; a small meso-scale; and a cell scale. Table 48 summarizes some of their characteristics. Despite this early statement, the presence of which was rediscovered over a decade later, several subsequent studies (e.g. Wallington, 1963; Bodolai and Jakus, 1969;

Table 48 Summary of characteristics of synoptic and subsynoptic scale precipitation areas

Synoptic	Large meso-scale area (LMSA)	Small meso-scale area (SMSA)	Cells
Occurrence			
Ahead of both warm and cold fronts	Several within synoptic precipitation area	3–6 within LMSA	1–7 within SMSA
Area (km^2)			
Over 10^4	1300–2600	250–400	5–10
Duration (h)			
Over 12	2–5	1	0.1–0.5
Rainfall rate ($mm\,h^{-1}$)			
c. 1–2	2–4	4–8	8–80

Source: Austin and Houze (1970).

Browning and Harrold, 1969) chose to ignore its potentiality as an analytical framework. This was probably because Austin's classification was based upon an instantaneous picture (radar echoes) whereas both Wallington and Browning and Harrold used hourly amounts of surface precipitation, albeit, in the latter case, only as support for more comprehensive radar data. In later studies (e.g. Harrold, 1973), the four-part classification re-emerged, possibly as a result of support by Atkinson and Smithson (1972), and it is now fairly widely accepted. We should note, however, that several Japanese meteorologists (e.g. Matsumoto *et al.*, 1970; Ninomiya and Akiyama, 1972) have found what they call "medium" scale disturbances, with characteristic lengths of about 1000 km. Whereas some authors consider such features to be of meso-scale, Ninomiya and Akiyama (1974) and Akiyama (1974) reject that idea, retaining the description for the smaller systems ($c.$ 100 + km) that they have frequently found in the cyclones which visit their islands. This chapter is concerned with the nature, mechanism and interrelationships of large and small meso-scale circulations in cyclones and their interaction with synoptic scale flows.

II. Observation

The following review of our empirical knowledge of meso-systems within cyclones has two main parts: the first covers extra-tropical systems; the second tropical systems. Within each section, as far as information allows, the climatology and nature of the meso-scale systems are elucidated. Mechanism and theory are covered in Section III of the chapter.

A. Extra-tropical circulations

1. *Radar and precipitation evidence*

Our description of these systems relies heavily upon both radar and raingauge data, to the extent that "precipitation areas" (PA) are frequently used as an indirect indication of the existence, and a measure of the size, shape, intensity and movement of circulation systems. The validity of this procedure has been shown in the detailed analyses by Atkinson and Smithson (1972, 1974*a*, 1978). The surface precipitation is largely due to the existence of "generator cells" (Marshall, 1953; Gunn *et al.*, 1954) which have horizontal and vertical dimensions of 1–2 km and occupy about 5 per cent (Langleben, 1956) of the precipitating area at the generator level within the upper reaches of clouds. Relationships between surface precipitation intensities and those aloft in the generators are, however, by no means clear. The roles of accretion, aggregation, shattering, evaporation, melting and wind shear upon the ice crystal or raindrop spectrum, and therefore upon precipitation rates at both the surface and aloft, have been subject to marked disagreement in the literature. Earlier studies, such as those by Mason and Ramanadham (1954), Gunn and Marshall (1958), Wexler and Atlas (1958) and more recently Cornford (1966) suggested that the drop spectra in the low levels of clouds, at or below the 0 °C level, are substantially modified by one or more of the processes mentioned above. In contrast, Rigby et al. (1954) considered the effects of coalescence, accretion and evaporation to be negligible, a view partially supported by Caton (1966) for the case of accretion below the melting level. Even if the latter school of thought is believed it does not exclude the operation of relevant processes above the metling level, and

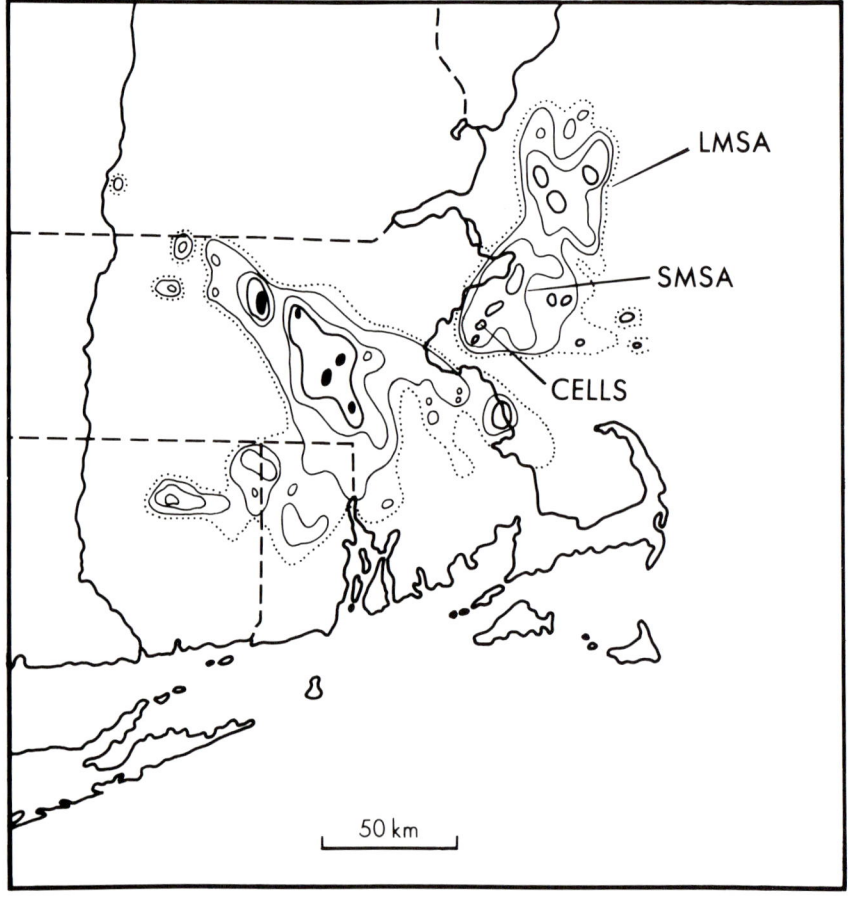

FIG. 156. Radar echoes showing large meso-scale areas (LMSA), small meso-scale areas (SMSA) and cells over New England, 8 July 1963. Contours indicate equivalent rainfall rates: 1 (dotted), 2, 4, 8 and 16 (solid) mm h^{-1}. (After Austin and Houze, 1972.)

indeed it permits large production rates of precipitation-sized particles in generators at high levels. The resultant ice crystal spectrum, and upon melting, raindrop spectrum, would, if the second school of thought is correct, be modified more by vertical wind shear above the freezing level than by any other process. These effects of wind shear upon precipitation have been examined by Gunn and Marshall (1955) on the assumption of little non-shear modification of the raindrop spectrum. Even if the spectrum is modified by factors other than wind shear, they are of the opinion that the velocity of precipitation areas on the ground is related to the velocity of the source of particles aloft (i.e. the generators). It has been suggested by several writers (e.g. Hobbs and Locatelli, 1978) that meso-scale precipitation areas (MPAs) result primarily from meso-scale clusters of generator cells aloft. Observations of clusters of the required size have been made by Langleben (1956) and Douglas et al. (1957). On the basis of the above work it is now fairly widely agreed that both MPAs, and the meso-scale echo areas, such as analysed by Austin and Houze (1972), are closely related to vertical circulations of similar size, although the exact nature of the relationship remains obscure. Following up Austin's (1960) earlier work, Austin and Houze (1972) provided further clear exemplification of the large and small meso-scale echo areas (LMSA and SMSA respectively), with the yet smaller scale cells within the SMSAs (Fig. 156). This nomenclature has since been slightly modified to cover precipitation areas observed by either radar or raingauges and the accepted terms are now large and small meso-scale precipitation area (LMPA and SMPA).

(a) *Climatology* The nearest thing to a climatology of meso-scale circulations within extra-tropical cyclones lies in the analyses of precipitation

Table 49 Mean areas and durations of large mesoscale precipitation areas in New England

Date	Area (km^2)	Duration (min)
18 May 1963	3000	200
	2200	240
8 July 1963	3000	80
	3300	>80
	2500	720
29 August 1962	4000	240
17 September 1963	4500	>210
	3300	>210

Source: Austin and Houze (1972).

patterns in New England by Austin and Houze (1972) and in Japan by Matsumoto et al. (1969). In the former study observations of eight cyclones provided information on eight LMSAs, 25 SMSAs and 125 cells. Table 49 shows that LMSAs varied from about 2000 to 4500 km² in area and from 80 to 720 min in duration. In many cases the LMSAs were band-shaped, their length being three to five times larger than their width. A typical LMSA would be 20–30 km wide and about 100 km long. In the Japanese study, seven PAs (defined as areas with a precipitation intensity of over 2 mm per 10 min period) ranged in size from 1000 to 7000 km² (average 2500 km²) and in duration from 130 min to 320 min (average 190 min), the larger systems lasting longer. Matsumoto et al. estimated that duration (T) and size (S) were related as follows:

$$T \propto S^{1.5}.$$

Figure 157 shows the frequency distribution of areas of SMSAs in New England. The majority were between 100 and 400 km², with occasional larger ones up to 800 km². Durations varied from 20 to 150 min, the median value being 50 min. An earlier synoptic climatological analysis of MSAs (Austin et al., 1966) proved rather inconclusive, except to show that band structures were frequently associated with fronts.

(b) *Organization of meso-scale precipitation areas* Whereas Austin and Houze (1972) found that SMPAs were irregularly distributed throughout a cyclone, Harrold (1973) devised a scheme whereby Austin's four types of circulation fitted together in a coherent way within an extra-tropical frontal cyclone. Fundamental to his scheme was the notion of a "conveyor belt", a well-defined flow (relative to the cyclone) of low level, moist, warm air in the

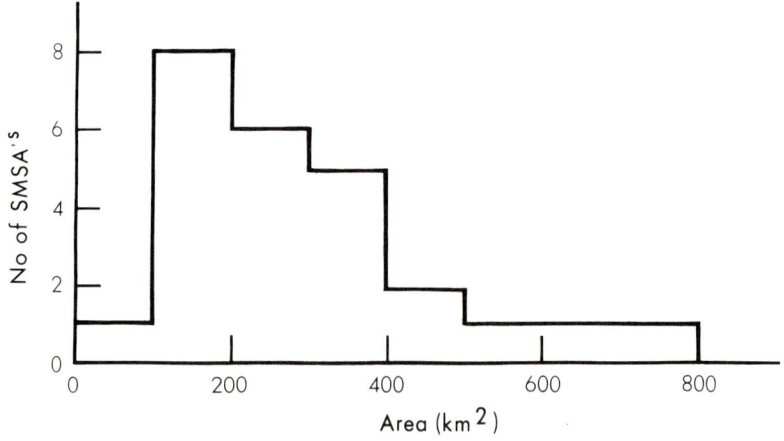

FIG. 157. Typical areas of small meso-scale precipitation areas as observed in New England (24 cases.) (After Austin and Houze, 1972.)

10 CIRCULATIONS IN CYCLONES   427

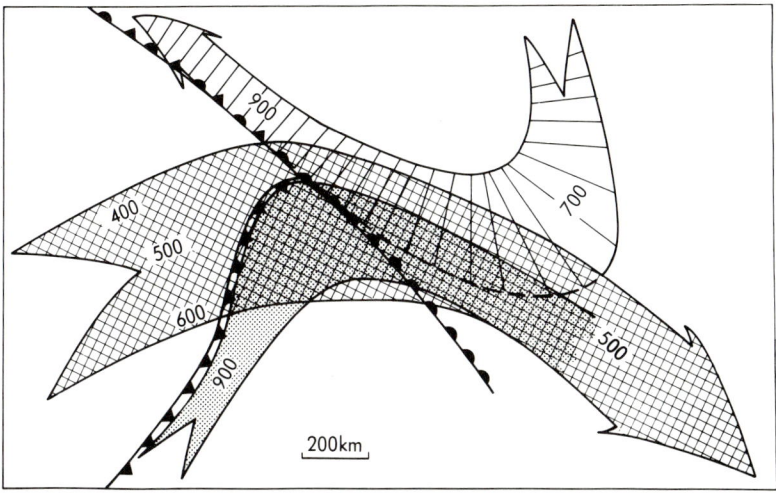

FIG. 158. Schematic representation of the features of the large scale flow which determine the distribution of precipitation at the surface. Most of the precipitation is produced within the stippled flow ascending from in advance of the surface cold front, but the leading portion of this precipitation is evaporated before it reaches the ground. The mid-tropospheric flow of potentially cold air from behind the cold front may lead to instability, and hence small scale convective overturning, within the large scale ascent. (After Harrold, 1973.)

warm sector, parallel to the cold front, typically a few hundred kilometres wide and 2 km deep. The flow ascends above the main warm frontal zone turning anticyclonically to run parallel to the surface warm front, but at a high level (Fig. 158). In Browning's (1974, p. 316) words: "The configuration of the flow, and the location of regions of ascent within it, largely determine the distribution of precipitation within mid-latitude depressions. The fronts are a secondary feature". In their several papers, Browning and Harrold, both individually and as co-authors, argued that the structure and mechanisms of MPAs are primarily dependent upon the nature of the conveyor belt. Figure 159(a) shows Harrold's (1973) schematic representation of the meso-scale structure of the precipitation within a partly occluded depression and its association with the conveyor belt. The parallels with Austin's 1960 classification are clear: the system band is roughly equivalent to the well-known, synoptic scale, warm frontal rain area; the cluster bands represent LMPAs; the clusters represent SMPAs; and although not shown in the figure, Harrold recognized the existence of smaller cells within the SMPAs. The model also shows warm sector cluster bands that may occur in the "upstream" part of the conveyor belt. Figure 159(b) shows Harrold's estimate of the meso-scale structure of the vertical velocity within the ascending conveyor belt. This picture was inferred from the distribution of different types of precipitation.

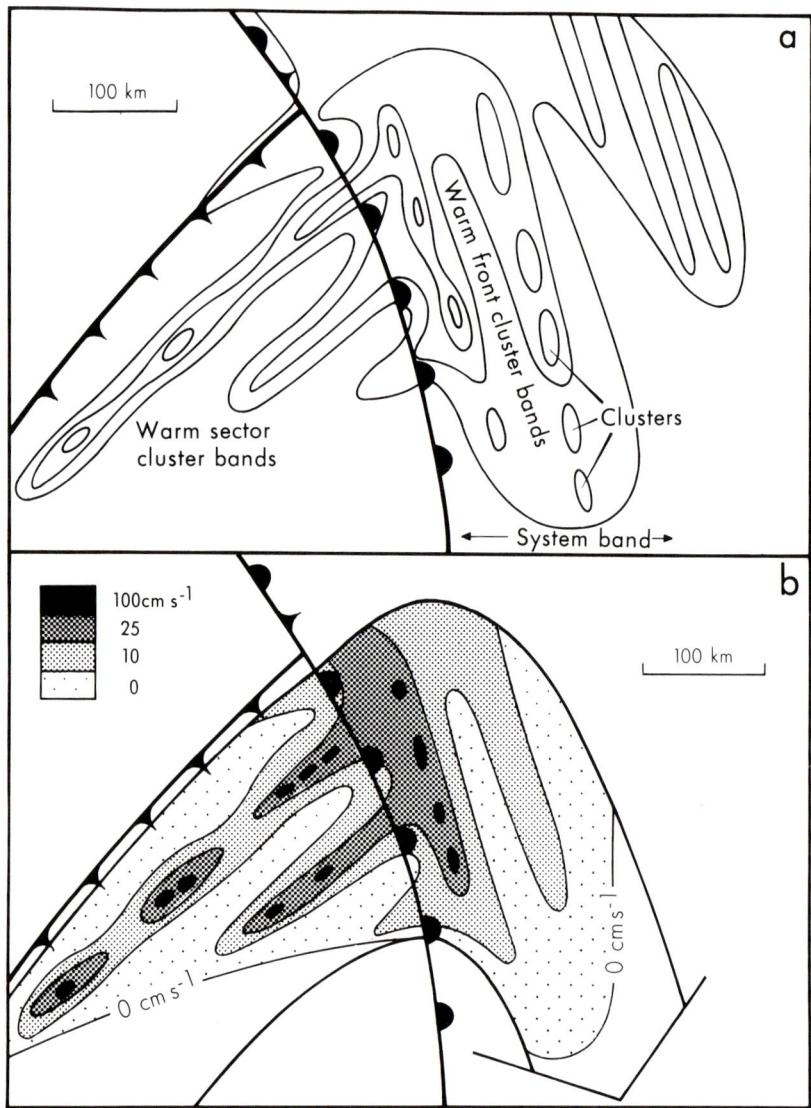

FIG. 159. (a) Schematic picture of the meso-scale structure of the precipitation associated with a partially occluded cyclone. (b) Schematic picture of the meso-scale structure of the vertical velocity within the ascending conveyor belt. (After Harrold, 1973.)

10 CIRCULATIONS IN CYCLONES 429

We are now in a position to look more closely at both the LMPAs and the SMPAs.

(i) *Large meso-scale precipitation areas* In the search for some order in Nature's apparent chaos we frequently succumb to the temptation to see regular geometrical shapes where perhaps none really exists. Although there may be an element of this in the investigation of LMPAs, on the whole many such systems do seem to take on a linear shape resulting in the "rainbands" that have recently proved popular objects of analysis. Although we should

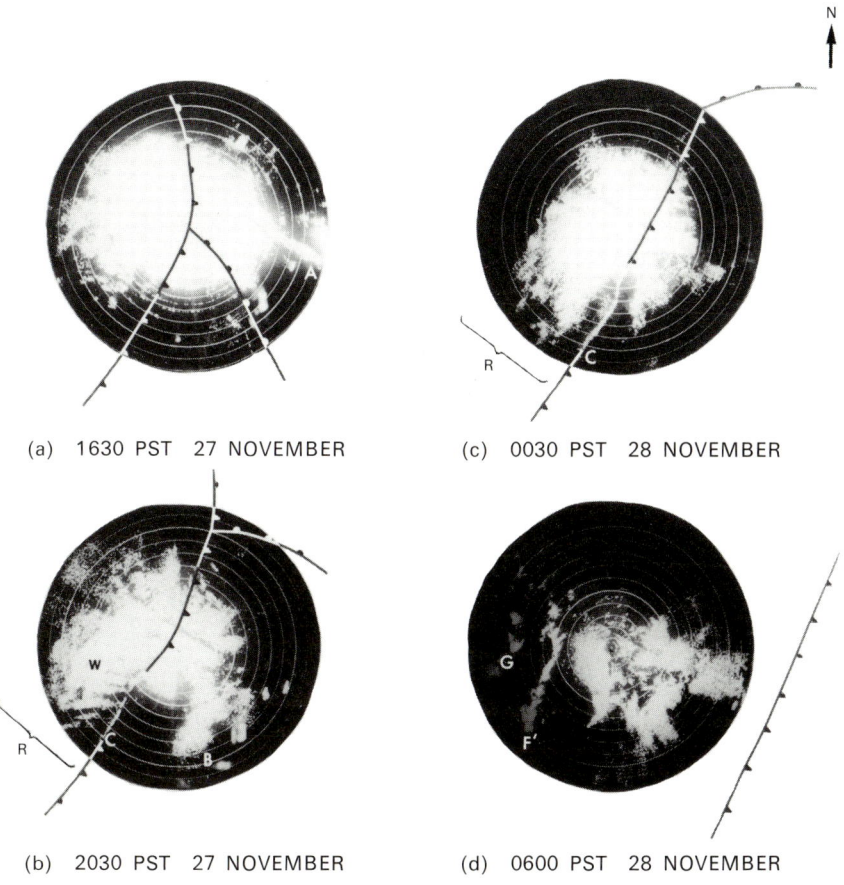

FIG. 160. Photographs of PPI displays and sea-level frontal positions illustrating various types of rainband. Band A is Type 1; Band B is Type 2; Band C is Type 4; W denotes Type 5; Bands G and F' are Type 6. Range markers are separated by intervals of 18.5 km. (After Houze *et al.*, 1976a.)

remember that not all LMPAs are necessarily rainbands it is true that non-linear LMPAs are difficult to find in both the atmosphere and the literature.

The systems found both within and near the conveyor belt can be conveniently analysed within the framework of a classification of rainbands by Houze et al. (1976a). They recognized six types of rainband as follows:

(1) *Warm frontal*. Bands approximately 50 km wide oriented parallel to the warm front and found toward the leading edge of a frontal cloud shield.

(2) *Warm sector*. Bands typically 50 km wide, found equatorward of the intersection of the surface warm and cold fronts and tending to be parallel to cold fronts. These could therefore of course be called pre-cold frontal bands. As such they could be construed to be pre-frontal squall lines—features dealt with in Chapter 8. In the current context Houze et al.'s (1976a) original nomenclature and meaning are retained.

(3) *Cold frontal—wide*. Bands approximately 50 km wide oriented parallel to the cold front and found toward the trailing edge of a frontal cloud shield.

(4) *Cold frontal—narrow*. Extremely narrow band (5 km wide) coinciding with the surface cold front.

(5) Wave bands occurring in a very regular pattern similar to waves, typically 10–20 km wide.

(6) *Post frontal*. Rainbands located in the convective cloud field behind a frontal cloud shield.

Figure 160 illustrates several of these types as seen on PPI scopes.

This classification is largely morphological and somewhat incomplete: for example, there is no mention of rainbands perpendicular to fronts, but in fairness, although such have been observed (Atkinson and Smithson, 1978), they are transient in comparison with the above six types. In the present context, type 5 is rather out of place, fitting more readily into the section on SMPAs. The classification also omits occlusions but this is not as serious as it may initially seem because such observations as are available suggest that

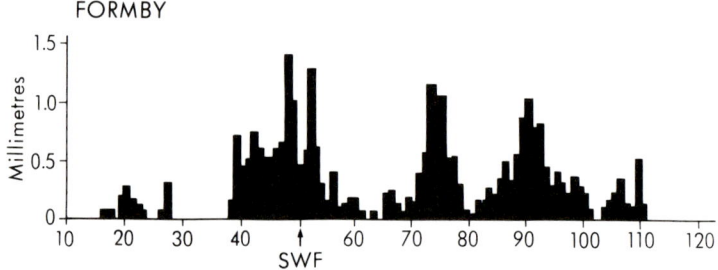

Fig. 161. Meso-scale precipitation areas revealed by variations through time at one station, Formby, UK. Histogram of precipitation amounts in 13 min periods over a 24 h period. Horizontal axis shows number of 13 min periods. SWF indicates passage of surface warm front. (After Atkinson and Smithson, 1972.)

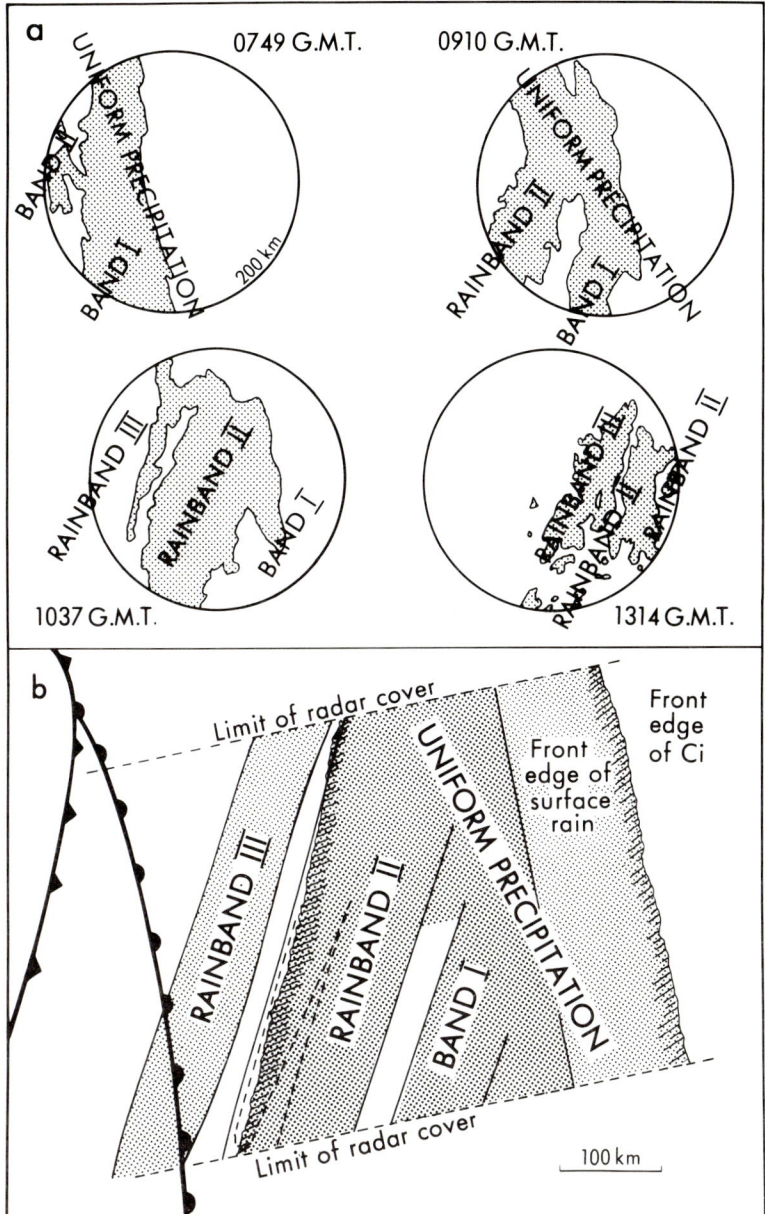

Fig. 162. Actual and schematic radar echoes of rainbands in extra-tropical cyclones. (a) PPI displays with 200 km range showing extent of precipitation echo at full gain for four times. (b) Schematic distribution of precipitation as inferred from qualitative, long range PPI data. The extent of surface rain is shown stippled. The further extent of precipitation aloft is lightly stippled. The leading and rear edges of the dense, high level cirrus canopy is indicated by hatched shading. (After Browning et al., 1973.)

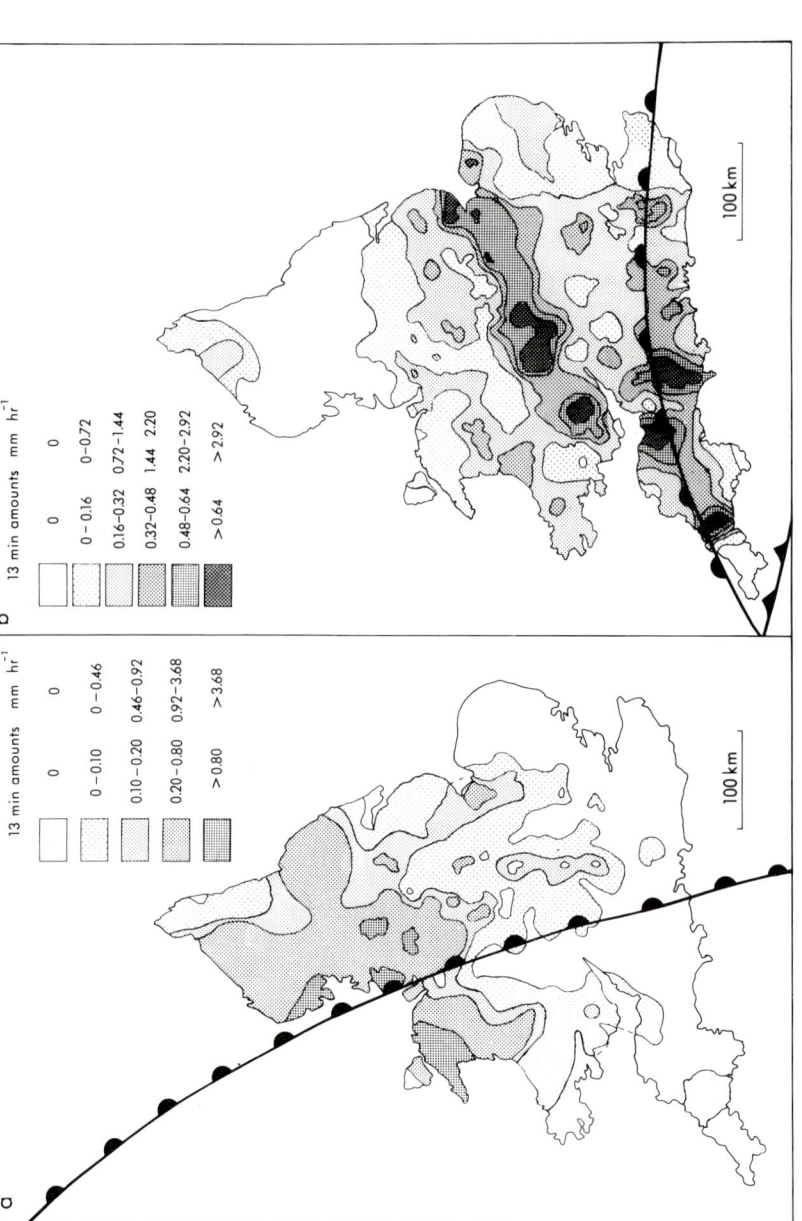

FIG. 163. Rainbands revealed by 13 min period—amounts of surface precipitation. (a) Two rainbands ahead of the surface warm front, most readily identifiable by the shading for 0.10–0.20 mm (0.46–0.92 mm h^{-1}). (After Atkinson and Smithson, 1972.) (b) Large rainband ahead of a surface occluded front, most readily identifiable by the shading over 0.32 mm (1.44 mm h^{-1}). Within this band three small meso-scale rain areas are most clearly identifiable by precipitation amounts of over 0.64 mm (2.92 mm h^{-1}). (After Atkinson and Smithson, 1974a.)

occlusions strongly resemble either warm or cold fronts, both in their synoptic scale structure and in their production of meso-scale circulations. This is elaborated below. These minor drawbacks apart, the classification provides a useful framework for the more detailed treatment that follows:

(1) *Warm-frontal bands* (including warm fronts in occlusions). Our understanding of warm frontal bands lies in careful interpretation of both radar and autographic raingauge data. In the following outline, both raw data from the two sources and schematic interpretations are employed: in this way the reader can evaluate the interpretation and also appreciate the less clear-cut realities of the atmospheric system. The bands are most clearly distinguished when viewed as a sequence of many rainfall maps of radar echoes, a time lapse movie film being the best way to identify the features. As Houze et al. (1976a, p. 869) noted: "The snapshots of the radar display... are not a completely satisfying way to depict the results of examining [a] movie which emphasises the time continuity exhibited by the bands". The same is true of a rainfall map series, as stressed by Atkinson and Smithson (1972).

The main characteristics of rainbands are illustrated in Figs 161–163 in terms of radar echoes and surface precipitation distribution. The meso-scale temporal variation in surface precipitation as measured at one point is shown in Fig. 161. A spatial view is provided by Fig. 162 which shows three bands, as seen by radar, associated with a warm front, together with a schematic composite (Fig. 162(b)). Figure 163(a) and (b) show distinct bands in a short-period precipitation distribution established from raingauge data. The vertical structure of typical rainbands is shown in Fig. 164 in terms of echo intensity and distribution. All the major studies of warm-frontal rainbands (Browning and Harrold, 1969; Kreitzberg and Brown, 1970; Browning *et al.*, 1973; Atkinson and Smithson, 1972, 1974a, 1978; Houze *et al.*, 1976a, b) agree that they tend to be 50–100 km wide, 100–300 km long, last for 3–6 h, lie roughly parallel to the surface warm front (except in the case of Browning *et al.*, where they were ahead of the warm front but parallel to the cold front) and that their movement is primarily governed by the interaction between the nature, evolution and movement of SMPAs within them and the larger, synoptic scale flow. This is considered in more detail below in our treatment of SMPAs.

(2) *Warm sector bands.* Bold schematic diagrams such as appear in Browning and Harrold (1969), Harrold (1973) (see Fig. 159(a)) and Nozumi and Arakawa (1968) lead us to believe in the evidence of rainbands in warm sectors. Whilst they do not appear to occur as frequently as most of the other types of band, their reality is confirmed by both radar (Fig. 160(b)) and surface precipitation distribution (Fig. 165). More than one band may exist (a point made most forcefully by Nozumi and Arakawa (1968)) and they tend to lie roughly parallel to both one another and the cold front, as shown by

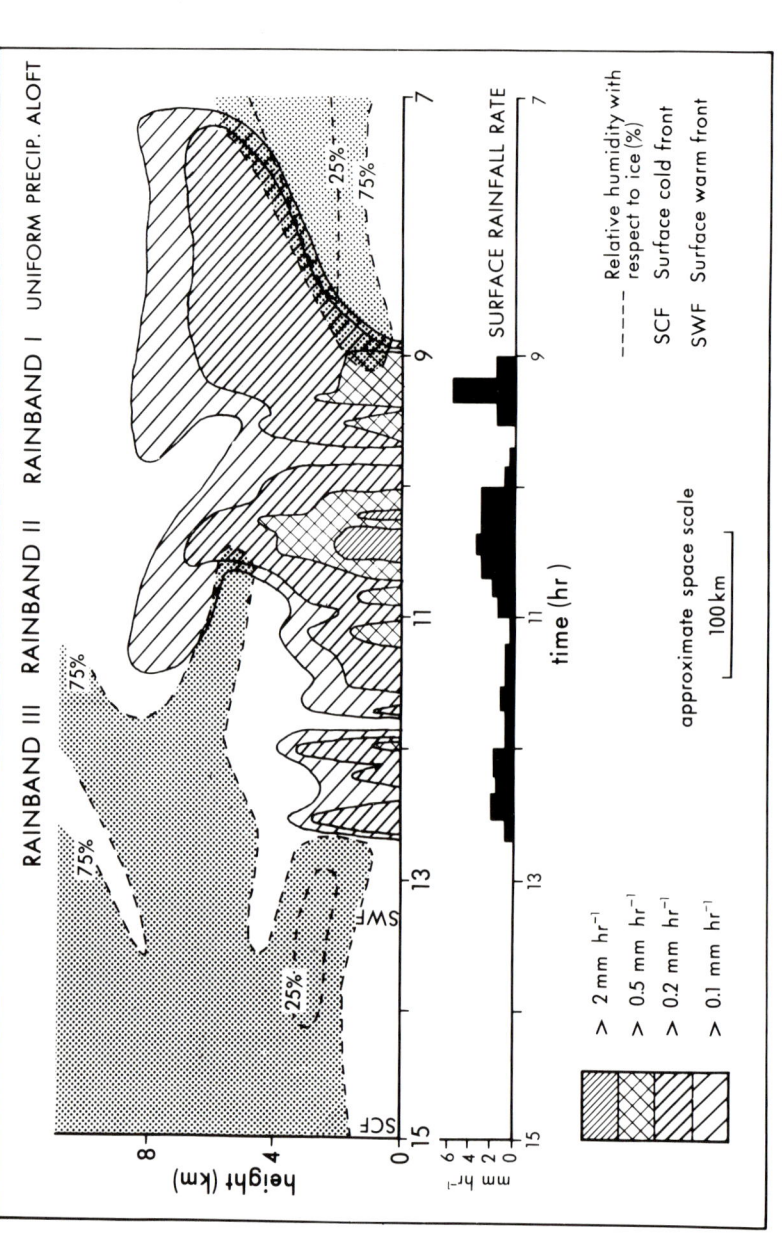

Fig. 164. Vertical structure of rainbands. Time–height cross-section showing the intensity of the precipitation echo and the distribution of relative humidity during the passage of a frontal system. The echo pattern has been averaged over a band normal to the fronts and 60 km wide. The hatched areas represent four levels of radar reflectivity roughly equivalent to the rainfall rates shown in the key. The dashed lines denote 25 and 75 per cent relative humidity with respect to ice. Regions of less than 75 per cent relative humidity are stippled. The positions of the surface cold front (SCF) and surface warm front (SWF) are indicated. Surface rainfall rate averaged over 10 min periods is plotted at the base of the diagram. (After Browning et al., 1973.)

Browning *et al.* (1974). In such a position they are of course in the "upstream" part of the conveyor belt. Their dimensions and lifetime are comparable with the warm frontal bands. Whilst fitting neatly into the currently used classification from a purely morphological viewpoint, the rainbands analysed

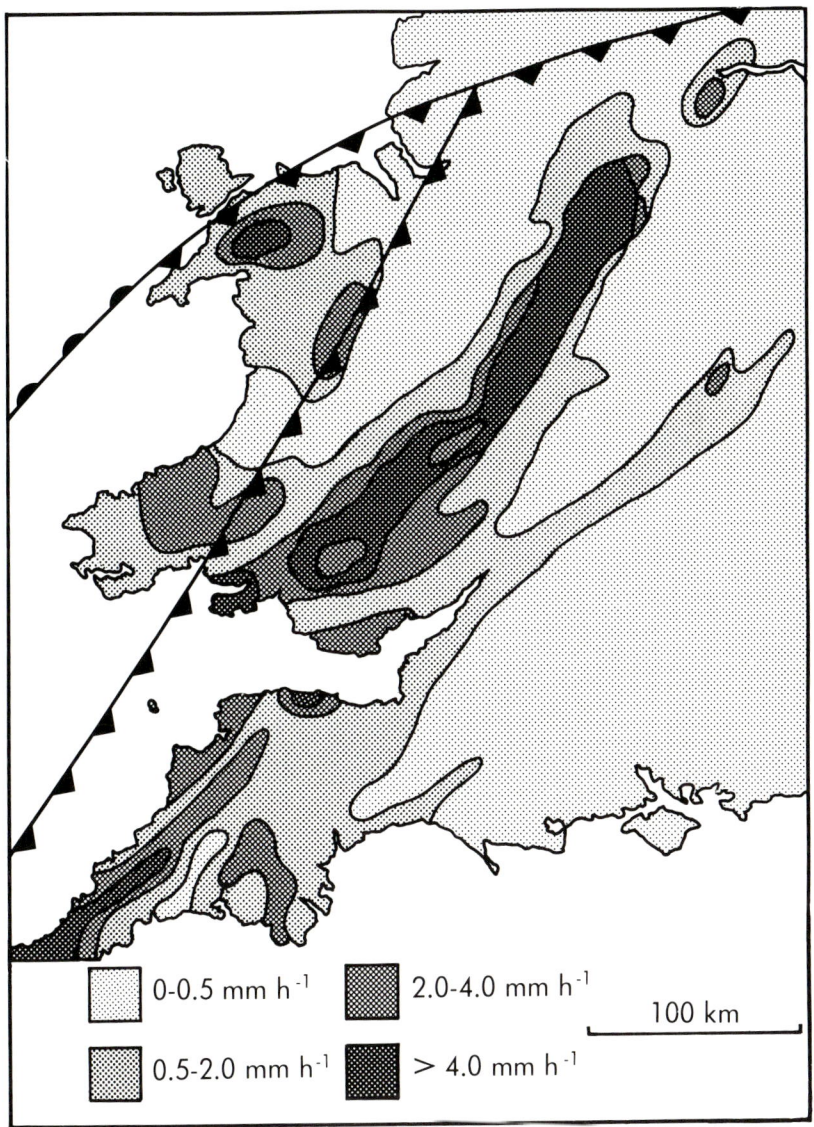

Fig. 165. Warm sector rainbands shown by surface rate of rainfall averaged over 10 min. (After Harrold, 1973.)

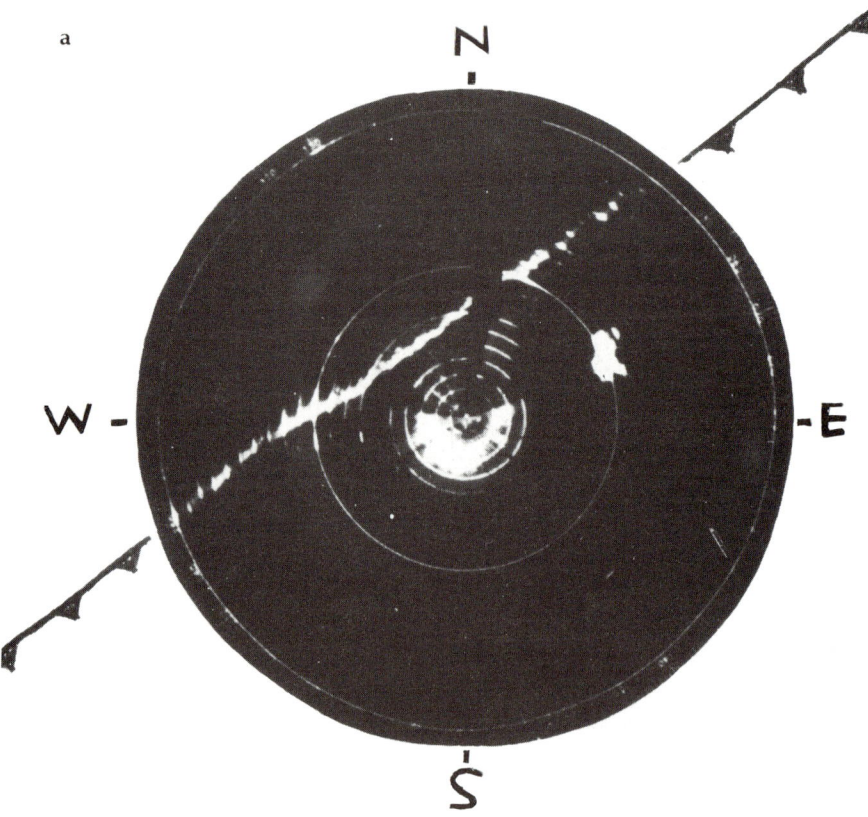

FIG. 166. Radar (a) and satellite (b) view of a typical very narrow line of cloud and precipitation. ((a) after Browning and Pardoe, 1973; (b) after Browning and Harrold, 1970.)

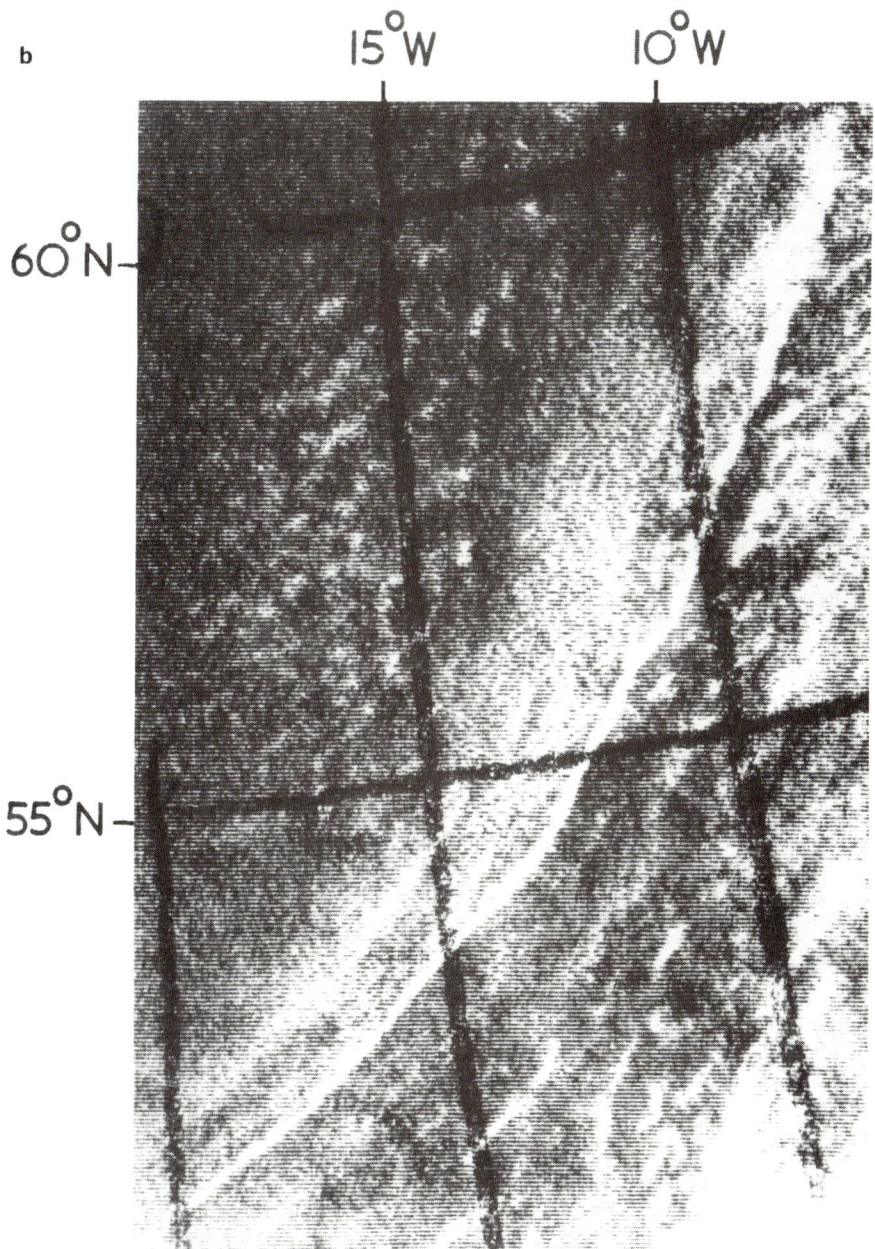

by Browning and Bryant (1975) were probably due to different mechanisms and, as such, would form a separate category. The authors argued that the longitudinal stationary bands were initiated and maintained in the planetary boundary layer downwind of hills. In particular they suggested that the bands were probably associated with the right-hand corkscrew circulations within the adiabatic boundary layer. The circulations were several kilometres wide and up to 100 km long with updraughts exceeding 30 cm s^{-1}.

(3) *Cold frontal—wide.* Houze *et al.* (1976a) were not alone in considering

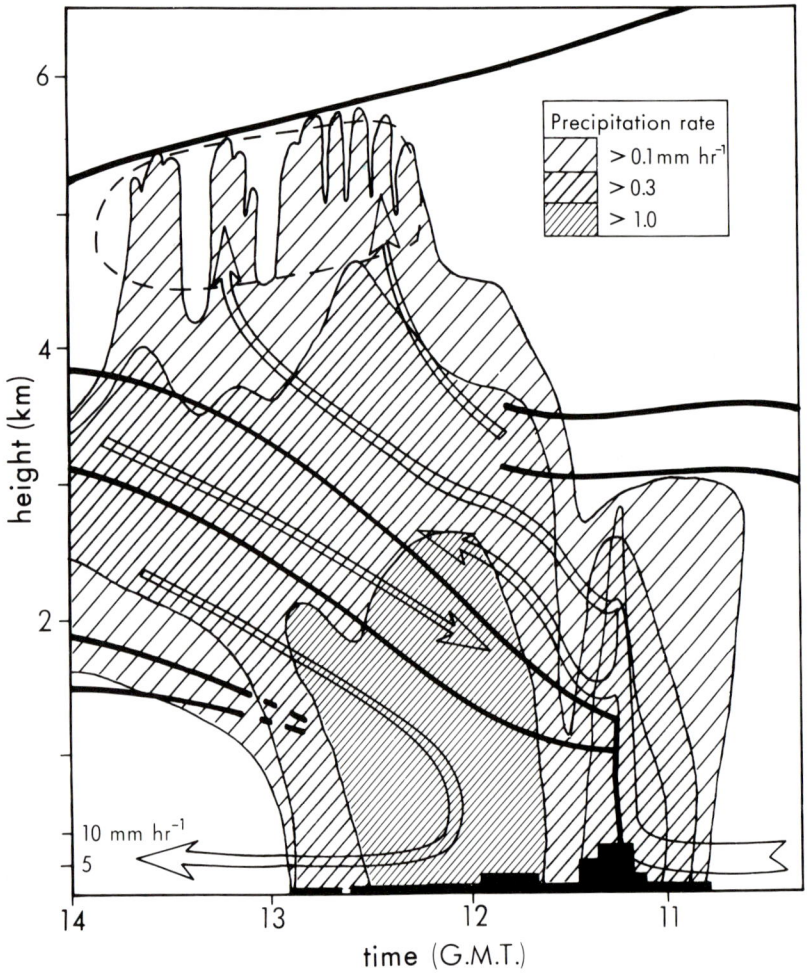

FIG. 167. Time–height section showing structure of a narrow cold front in terms of precipitation intensity and airflow. Surface rainfall rate is shown along the abscissa. Solid arrows are streamlines; thick lines denote boundaries of the frontal zone and other stable layers; and the dashed line delineates the extent of small scale convection at high levels. (After Browning and Harrold, 1970.)

this to be a separate category of rainband in a meso-scale context. Although few other authors appear to have analysed them both, Kreitzberg and Brown (1970) and Houze *et al.* (1976a) showed that a few such bands have been observed—their dimensions being approximately 50 km wide and over 100 km long or even larger.

(4) *Cold frontal—narrow*. Perhaps the best examples of this feature are those studied by Kessler and Wexler (1960), Browning and Harrold (1970) and Browning and Pardoe (1973). Figure 166 gives both a satellite and radar view of the typically very narrow line of cloud and precipitation. A characteristic vertical structure, in terms of precipitation rates and airflow, is shown in Fig. 167. These bands are associated with what Browning and

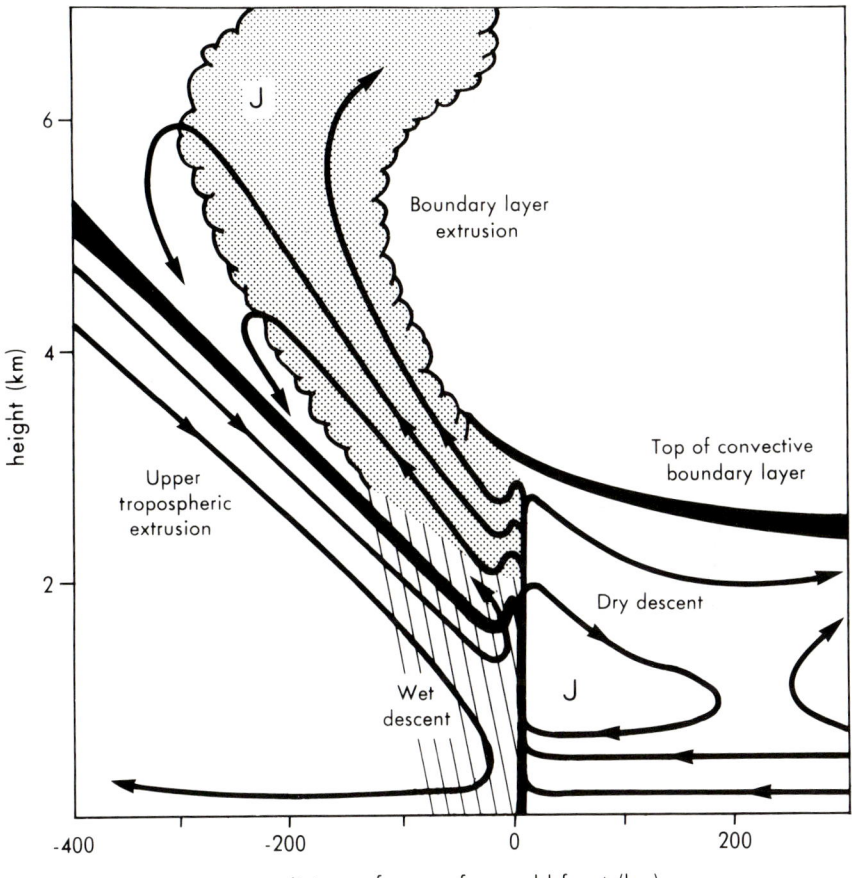

FIG. 168. Schematic model of airflow associated with an ana-cold front. Thin lines are streamlines relative to the moving system. Thick lines represent the cold frontal zone and the top of the convective boundary layer. Regions of saturated ascent are stippled. Jets shown by letter J. (After Browning and Pardoe, 1973.)

Harrold (1970) called "line convection", a fairly self-explanatory term for very narrow (2 km), shallow (2 km) but abrupt uplift on the frontal side of the synoptic scale, low level jet which frequently occurs just ahead of cold fronts. Figure 168 shows a schematic model of this configuration of airflow at a typical extra-tropical ana-front.

(5) *Wave*. Once again these features are at first sight peculiar to the systems analysed by Houze *et al.* (1976a). They observed such waves in the broad rain area behind the cold front (see W in Fig. 160(b)). These ripples were 15 km × 70 km in average dimension and moved with the low to mid-tropospheric wind parallel to the front at a speed of 32 m s^{-1}. Houze *et al.* (1976a) speculated that these bands were due to gravity waves but pursued the matter no further.

(6) *Post (cold) frontal*. Figure 160(d) illustrates a post-cold frontal band, about 20 km wide and over 100 km long. Similar bands were noted by Nagle and Serebreny (1962) in their schematic model of radar echoes in occluded cyclones over the eastern Pacific Ocean. Matsumoto *et al.* (1967) also observed meso-scale precipitation bands in cold air masses to the west of surface cyclonic systems during the winter over the Sea of Japan.

As hinted at above, not all LMPAs are obviously linear. In particular, Japanese meteorologists analysing the Baiu front have found that the very large rainbands associated with synoptic scale fronts, such as identified by Tatehira (1964, 1968) and Fujiwara (1958), may comprise either what they call "medium" or "intermediate" scale echoes (Matsumoto *et al.*, 1970;

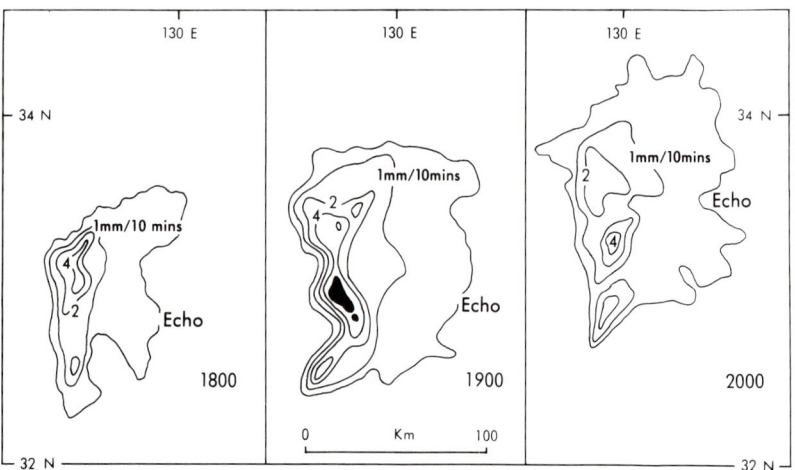

FIG. 169. Meso-scale echo cluster near Japan over a period of 2 h. Isolines show precipitation intensity (in millimetres per 10 min period). Outer line is limit of echo. (After Ninomiya and Akiyama, 1974.)

Murakami, 1972; Ninomiya and Akiyama, 1974), with characteristic horizontal dimensions of at least 1000 km, or meso-scale clusters, as confirmed by Ninomiya and Akiyama (1974) and Akiyama (1974), with characteristic dimensions of 200 km. Figure 169 clearly illustrates three such "mesoscale echo clusters" together with the distribution of precipitation intensity. Ninomiya and Akiyama (1974) noted that the bulk of the precipitation lay on the western side of the cluster and indeed, if intensities of less than 1 mm per 10 min period are ignored, one is left with a form very similar in shape and size to the rainbands outlined above. But non-linear LMPAs have been observed. For example Atkinson and Smithson (1978) found such systems with characteristic dimensions of 80–100 km (Fig. 170) and durations of about 4 h. It was uncertain as to whether they formed in a warm sector but they certainly moved across the sector perpendicular to the warm front, eventually passing through the surface front. In Japan Matsumoto and Akiyama (1970) have frequently found non-linear LMPAs with dimensions of 100–200 km within the Baiu front.

(ii) *Small meso-scale precipitation areas* Small meso-scale precipitation areas have received less attention than their larger counterparts, no doubt due to their size, duration and lack of distinctive shape discouraging observa-

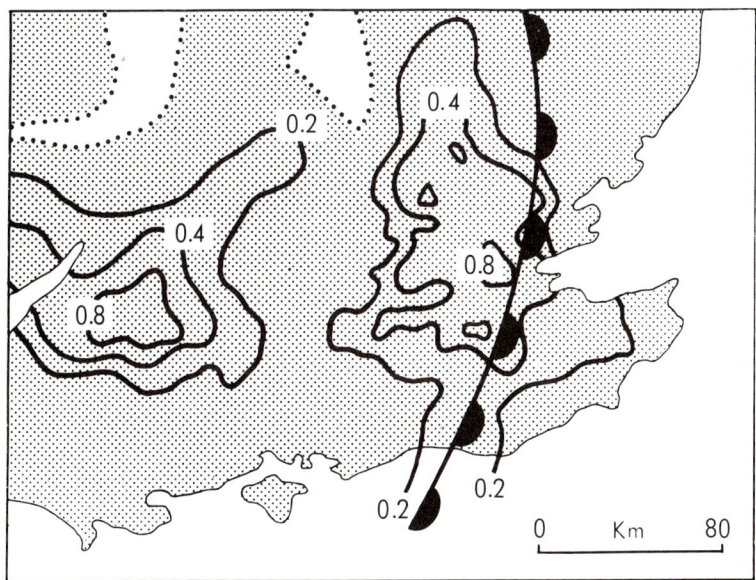

FIG. 170. Non-linear large meso-scale precipitation areas near a warm front. The two meso-scale areas are most easily identified by the 0.2 mm isohyet. Non-stippled areas received no rain. (After Atkinson and Smithson, 1978.)

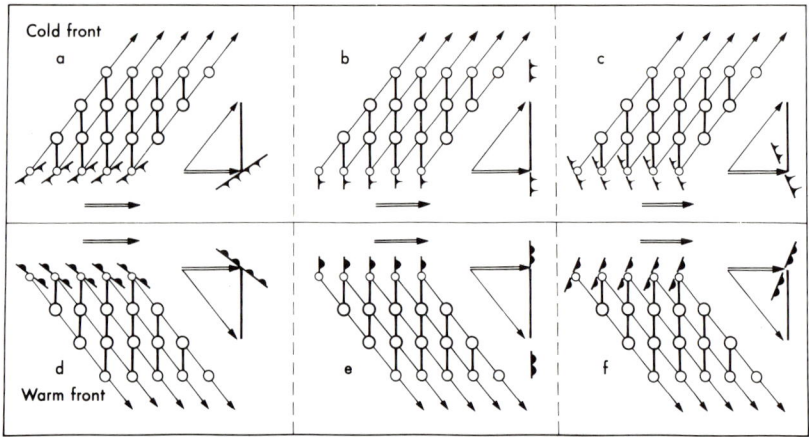

FIG. 171. Schematic diagram to show how the orientation of rainbands relative to that of fronts is related to frontal velocity and the air velocity at the height of generator source. Circles of varying size show how generator cluster varies throughout its lifetime. Bold arrow shows frontal (and therefore generator source) velocity; thin arrows show wind velocity at generator source level; bold lines punctuated by circles show the rainband at any particular time. (a) and (d) show rainband behind fronts; (b) and (e) show rainbands along fronts; (c) and (f) show rainbands ahead of fronts.

tional interest. Yet close examination (Atkinson and Smithson, 1972, 1974a, 1978) has revealed that SMPAs are the primary component parts of many, if not all LMPAs, as suggested in Austin's (1960) classification. Thus the majority, if not all, the rainbands considered above were but linear configurations of SMPAs. In this case the nature, size, evolution and movement of SMPAs clearly must largely determine the nature and behaviour of LMPAs, be they bands or otherwise. Whilst not stated explicitly by Tatehira (1968), his diagrams clearly revealed that he appreciated this particular relationship between LMPA and SMPA well before most other workers on this subject. The point was also made independently by Atkinson and Smithson (1972) and subsequently illustrated in later work.

As noted earlier in this chapter, SMPAs are due to clusters of generator cells that usually occur at the 500–600 mb level. These generators grow and die within 1–2 h and as they pass through their life cycle they are blown along by the winds at the level of their source. Figure 171 schematically illustrates how the relationship between the velocity of the generator source, which is in fact the velocity of the frontal system, and the velocity of the generator clusters, may determine the orientation of rainbands relative to both warm and cold fronts. In addition, the actual size and speed of evolution of the SMPAs also clearly affect the particular size, duration and orientation of linear LMPAs.

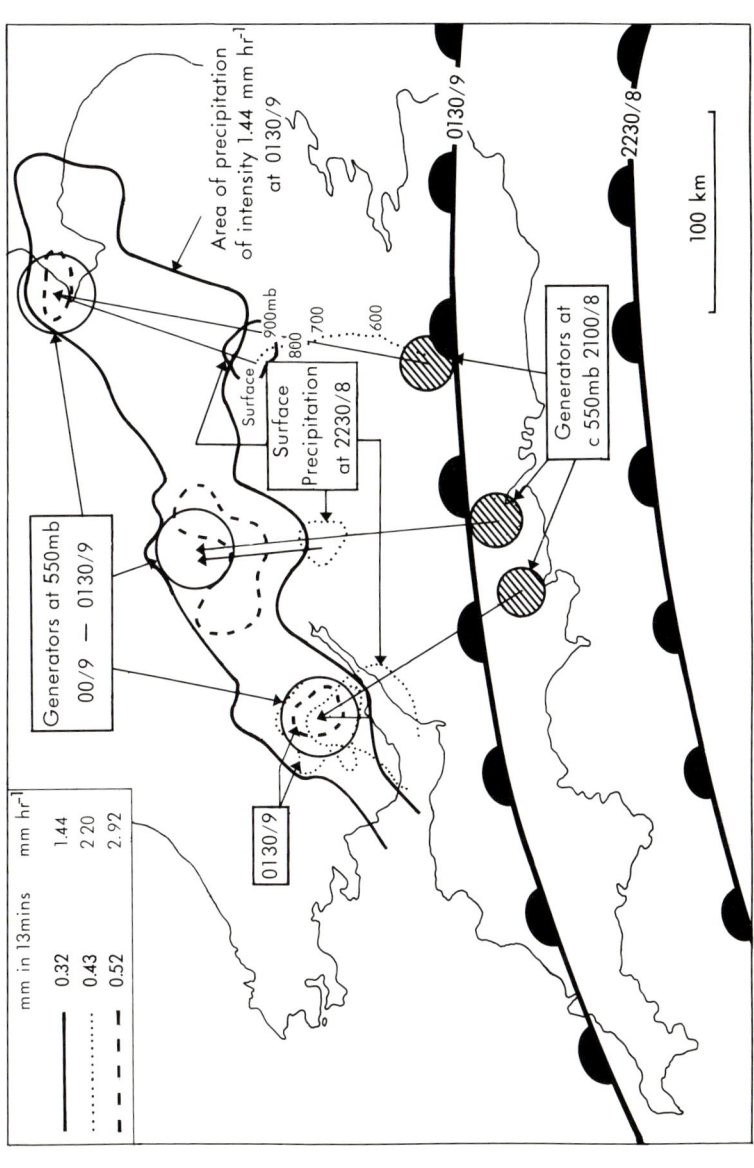

FIG. 172. Semi-schematic figure to illustrate the initial and final locations of small meso-scale precipitation areas from 2230 GMT 8 March to 0130 GMT 9 March 1967 and their relationships to locations of clusters of generators at a height of about 550 mb. The dashed line from the eastern cluster of generators at 2100 GMT represents a typical trajectory of a precipitation particle falling from these generators. (After Atkinson and Smithson, 1974a.)

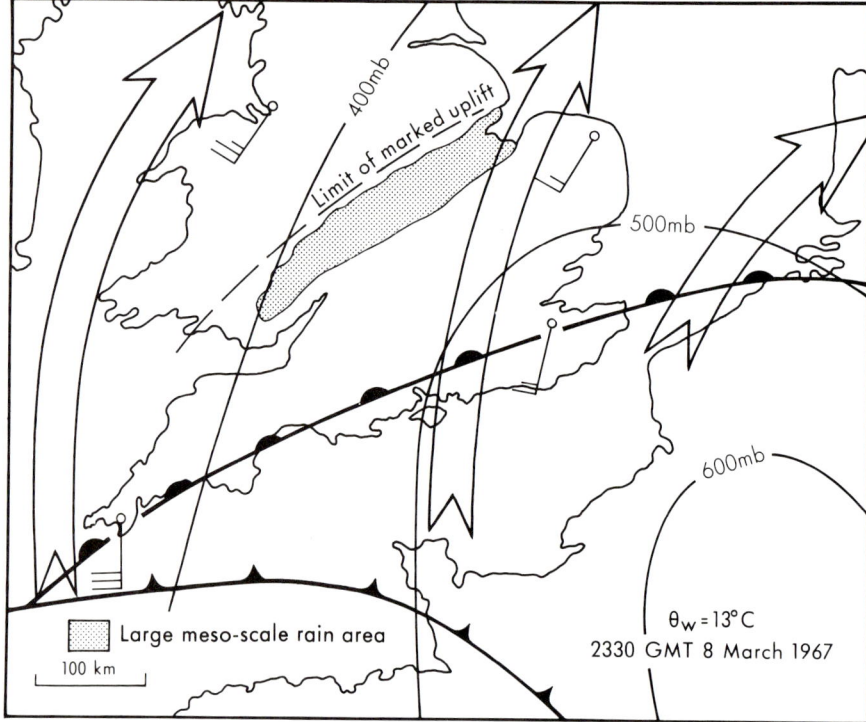

FIG. 173. Relative airflow on the $\theta_w = 13°C$ surface (as defined by isobars) at 2330 GMT 8 March 1967, showing the accordance between the northern limit of the large meso-scale rain area and the limit of uplift. (After Atkinson and Smithson, 1974a.)

A slight variation of this theme was exemplified by Atkinson and Smithson (1974a). On 8 March 1967 a pronounced rainband lay ahead of and roughly parallel to a warm front (Fig. 163(b)). Once more this band comprised SMPAs (three in this case) which themselves were due to generator clusters at a height of about 550 mb. Figure 172 shows how the movements of these clusters together with wind shear effects led to the linear configuration of three SMPAs. The resultant LMPA remained fairly stationary for about 3 h and this was due to the larger scale airflow. Figure 173 shows that the northern limit of the rainband coincided with the limit of the uplift area over the front. Thus, although the generator clusters existed to the north of the rainband, the lack of uplift meant that production of precipitation ceased or at least was reduced to such an amount that precipitation particles did not reach the ground because of evaporation.

In addition to the above mechanisms, which are quite capable of producing multiple rainbands of various sizes, durations and orientations, Browning

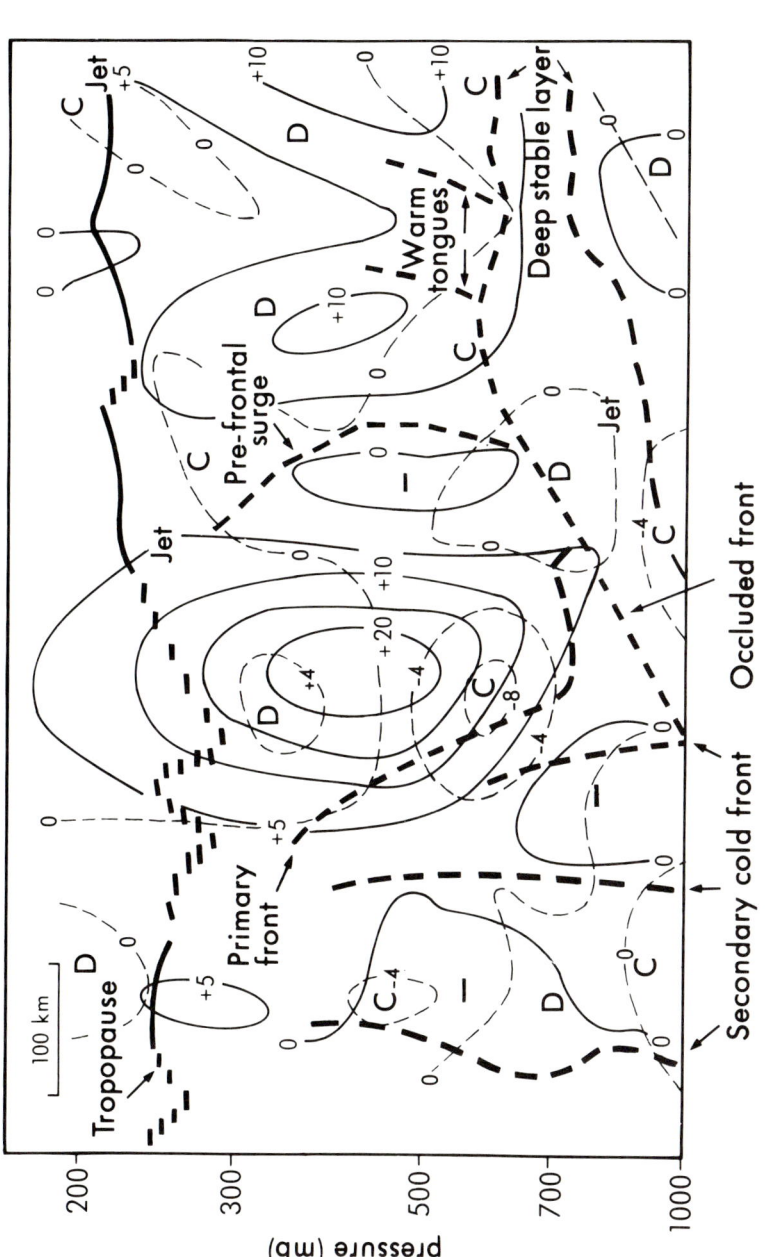

Fig. 174. Time–height section of vertical velocity and divergence in an occluded front. Solid lines are vertical velocity (in centimetres per second) and dashed lines divergence (10^{-5} per second). Centres of divergence and convergence are labelled D and C and thermal features are shown by thick dashed lines. (After Kreitzberg, 1968.)

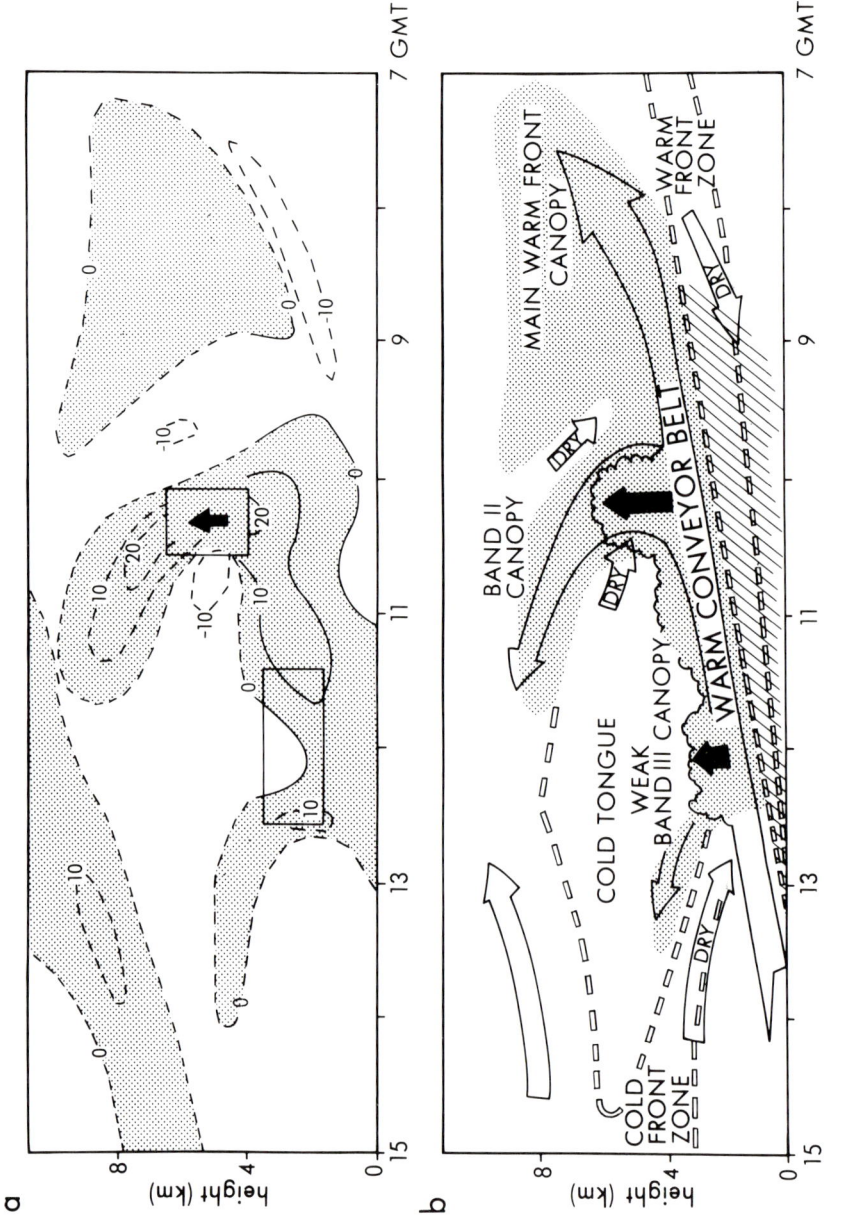

(1974) proposed that such features were due to a multiple, fingered structure to the leading edge of cold air which overran the conveyor belt. This is considered in more detail in the section on airflow.

Finally, in this section, we should note that not all SMPAs behave in the manner described above. In particular Atkinson and Smithson (1974b) observed warm sector SMPAs that moved outward from the cyclone centre across the wind directions at any level in the troposphere. These SMPAs appeared to be associated with isallobaric lows which moved contemporaneously with the same direction and speed as the PAs. With the aid of previous work by Brunt and Douglas (1928) and Pothecary (1954), Atkinson and Smithson (1974b) showed that vertical velocities adequate for production of cloud and precipitation were generated in the isallobaric lows. They concluded with a speculation that the movement of these lows was related to the transverse circulation in the frontal jet stream at higher levels.

2. *Direct observations of meso-scale airflow*

As shown above, radar and precipitation evidence has clearly revealed meso-scale structures within extra-tropical cyclones. Complementary observations of airflow which could possibly be associated with these structures are far fewer in number. As a result it is still not clear as to whether the MPAs result from meso-scale uplift *per se* or from convective scale uplift which happens to be clustered on the meso-scale. The weight of opinion (e.g. Austin and Houze, 1972; Elliott and Hovind, 1964; Browning and Harrold, 1969) appears to favour the latter view. Observational tests of this idea are both very difficult and expensive (Hardman *et al.*, 1972) and such direct evidence of airflow as is available is still at too large a scale finally to decide the matter.

Airflows associated with rainbands have been observed by Kreitzberg (1968) and Browning *et al.* (1973). Figure 174 clearly reveals fairly large meso-scale areas of uplift (up to $20\,\mathrm{cm\,s^{-1}}$) above the warm frontal zone. The distribution of convergence and divergence also has a definite meso-scale structure. On a similar scale, Browning *et al.* (1973) derived the vertical velocity field (Fig. 175(a)) and the schematic overall structure (Fig. 175(b)) of the circulations associated with the rainbands shown in Fig. 162. In the clear

---

FIG. 175. (a) Time–height section of vertical velocity associated with the rainbands shown in Fig. 162. Solid lines represent vertical velocity (in centimetres per second) as derived accurately from dropsonde data and doppler radar. Dashed lines represent vertical velocity inferred approximately by kinematical methods. Regions of ascent are stippled. The small rectangular frames represent regions of vertical convection with local updraughts of order $1\,\mathrm{m\,s^{-1}}$, as detected by doppler radar scanning vertically. (b) Schematic time–height cross-section showing some of the principal features of the convective rainbands shown in Fig. 162 and Fig. 175(a). Cloudy areas are stippled.
(After Browning *et al.*, 1973.)

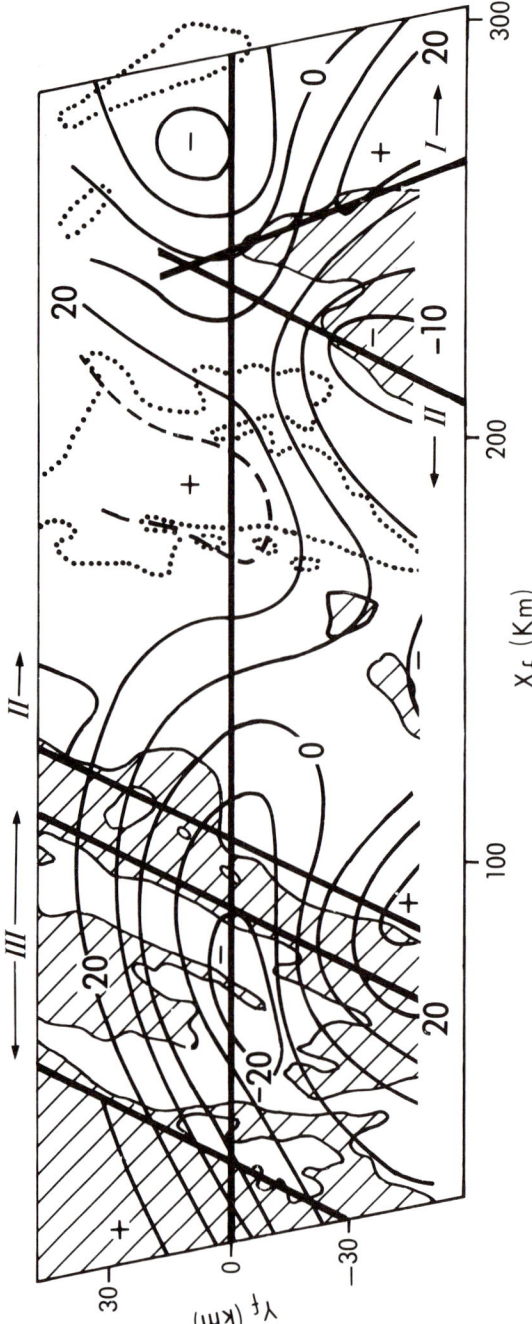

FIG. 176. Horizontal section of vertical velocity at a height of 3.5 km and two levels of radar reflectivity corresponding to rainfall rates of 0.2 and 3 mm h^{-1}. The heavy lines show the three main rainbands outlined in Fig. 162 (with some slight discrepancy in the orientation of band one). Thicker solid and dashed lines show vertical velocity (in centimetres per second) positive upward; thinner solid and dotted lines show reflectivities of 3 and 0.2 mm h^{-1} respectively. Reflectivities over 3 mm h^{-1} are shaded. (After Roach and Hardman, 1975.)

belief that these rainbands were really such, Browning et al. identified three thermally direct circulations, causing each rainband to have a rearward-sloping anvil cloud canopy, with moist ascending air riding over regions of colder and drier descending air. Figure 175(a) shows that Browning et al.'s (1973) estimates of meso-scale ascent were basically equal to Kreitzberg's. A closer examination of the same case study as Browning et al. revealed clear *small* meso-scale areas of lift (Roach and Hardman, 1975). Figure 176 shows vertical velocities at a height of 3.5 km within the three rainbands. Only Band II is clearly associated with uplift at that particular height. Cross-sections (Fig. 177) reveal a coherent vertical structure to these meso-scale vertical velocities. Although Roach and Hardman (1975, p. 443) noted that the vertical velocity field "reflects the rather large scale (200 km)" of the

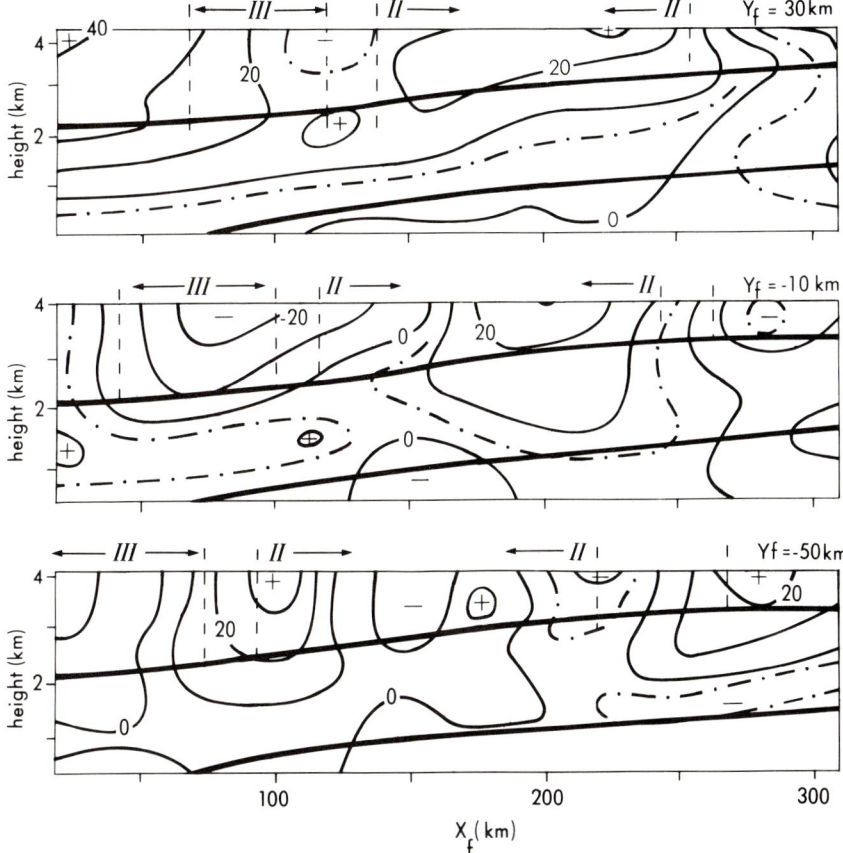

FIG. 177. Vertical cross-sections of vertical velocity (in centimetres per second) along the lines $y = 30$, -10 and -50 km of Fig. 176. Heavy lines delimit frontal zone. (After Roach and Hardman, 1975.)

horizontal windfield from which it was derived, their results are in fact some of the very few to cover the genuinely meso-scale and to be associated validly and directly to MPAs.

B. Tropical circulations

Meso-scale circulations within tropical cyclones are much less well known than their extra-tropical counterparts. As noted earlier, the primary reason for this is that observational capacity within the tropics has been and, in most areas, is still inadequate to record circulations of sub-synoptic size. Radar and satellites have gone some way towards solving this problem but, as pointed out by Fujita *et al.* (1967) and as shown below, their potentialities are as yet under-developed. Nevertheless it was radar observation of tropical cyclones which first clarified their major structural components (Ooeda, 1961; Yanagisawa, 1961; Imakado, 1966). In particular radar images revealed a spiral rain band structure (Fig. 178) that was soon identified as being typical

FIG. 178. Photograph of PPI display of Hurricane Betsy near Miami on 8 September 1965. The banded structure is quite apparent. (Courtesy of National Hurricane and Experimental Meteorology Laboratory, Coral Gables, Florida.)

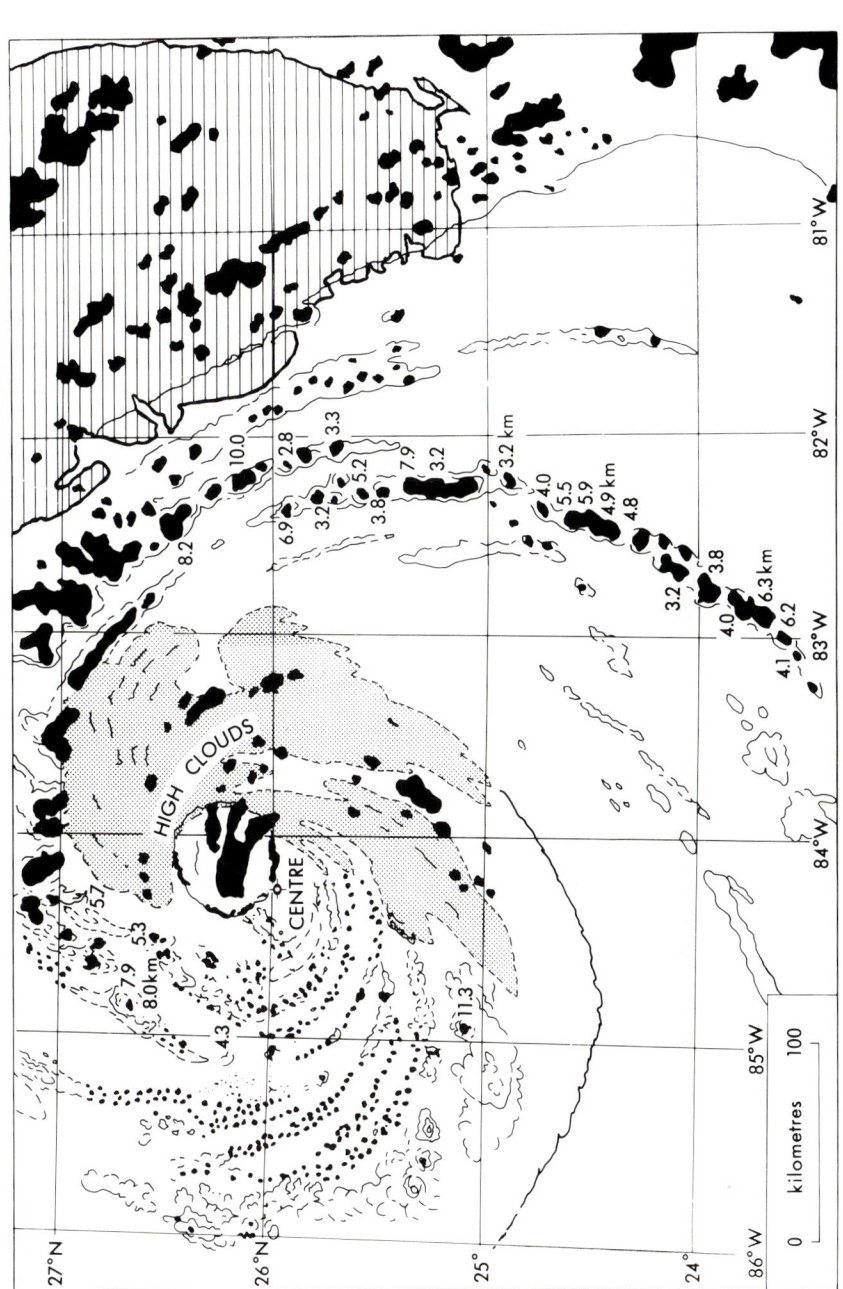

Fig. 179. Composite of radar PPI echoes (black) superimposed upon the cloud derived from the satellite Apollo 7 pictures. Cloud heights shown by the echoes are in kilometres above mean sea level. (After Gentry et al., 1970.)

of such storms (Wexler, 1947; Simpson, 1954). The many thousand of radar photographs now available have subsequently been supplemented by satellite pictures which, in some cases, reveal spiral cloud bands that are not shown on the radar. Figure 179 exemplifies a situation where photographs from the manned flight of Apollo 7 show patterns to the west of the eye of hurricane Gladys which were beyond the capability of radars. To the east of the eye, however, both satellite and radar images strikingly define bands that form spirals around that part of the storm.

As much of our knowledge of meso-scale rainbands within tropical cyclones derives from the use of radar it is helpful to have as background knowledge an overall view of the typical cyclone as seen by radar (Fig. 180). In advance of the storm are intense, sharply defined echoes which do not generally exhibit cyclonic motion about the storm centre and may form narrow "pre-hurricane squall lines". Wexler (1947) called these lines "precursor bands", whereas Ligda (1955), Kessler and Atlas (1956) and Fujita *et al.* (1967) referred to them as "outer bands". Although these squall lines are associated with the hurricane, they are usually found about 200–500 km from

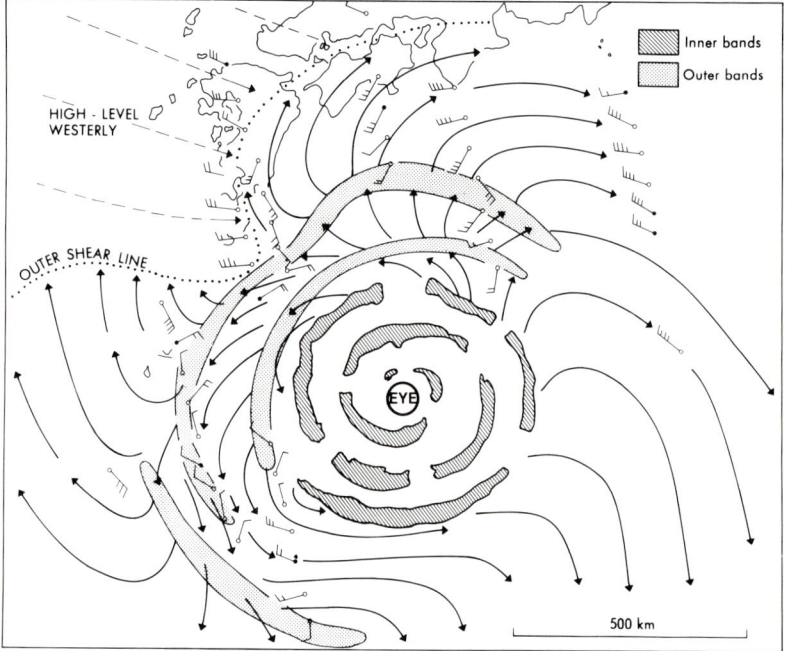

FIG. 180. Schematic plan of a typical hurricane showing the eye, inner (crosshatched) and outer (stippled) bands. Solid lines are streamlines of outflow from hurricane centre at high level. (After Fujita *et al.*, 1967.)

the storm centre. Senn *et al.* (1959) emphasized that these lines should not be identified as outer spiral bands. The pre-hurricane squall lines are generally separated by 80–100 km from echoes of the "rain shield", a ragged mass of spiralling weather that extends from 80 to 150 km ahead of the storm centre,

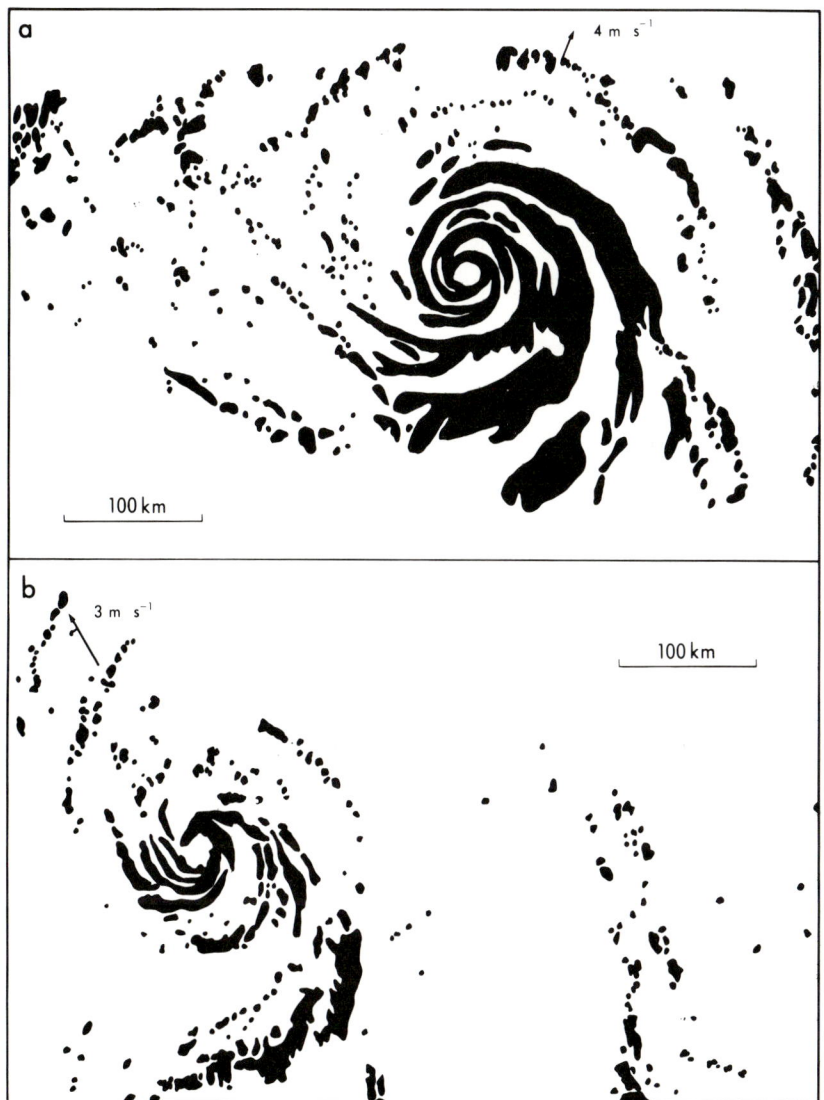

FIG. 181. Composite distributions of radar echoes in hurricane Daisy, August 1958, in the Caribbean. Arrow at the top indicates the motion of the storm centre: (a) at a height of about 4 km; (b) at a height of 10.6 km. (After Colon and staff, 1961.)

or there may also be scattered distinct echoes between the pre-hurricane squall lines and the lighter precipitation of the rain shield. A closer examination of the rain shield reveals many spiral bands (the outer bands) which apparently merged as they have been "thrown outward" from the centre of the storm, becoming wider, longer and less well defined in the process (Staff members, Division of Meteorology, University of Tokyo, 1969, 1970). The leading edge of this rain shield has often been observed to have the shape of a well-defined spiral band. The centre of the storm is usually not directly connected to this mass of spiralling weather but is separated from it by a space of 20–80 km that is occupied by one or several much more discrete spirals. The spiral bands nearest the storm centre are most clearly defined and continuous, and it is these which often define the eye of the hurricane. In what follows we deal with the well-defined spirals in the area between the storm centre and the rain shield.

1. *Band structure*

"...A lot of surmising has been done, and a lot of tentative conclusions have been drawn without sufficient data for verification. No detailed and complete description of the rainbands has been furnished" (Gentry, 1964, p. 6). Gentry's statements remain essentially true and we are able to present but a rough picture of the echo structure of rainbands together with some observations of temperature, wind and humidity, these latter, non-radar measurements being largely due to the efforts of Gentry himself.

Figure 181 presents two plan views of hurricane Daisy 1958 derived from airborne radars at heights of 4 km (Fig. 181(a)) and 10.6 km (Fig. 181(b)): at both heights the spiral structure is clearly evident. Within about 40 km of the centre of the storm the bands are 5–10 km in width, at least 100 km long (particularly at the lower levels) and they retain these dimensions throughout the depth of the troposphere. In the area between 40 and 180 km from the centre the bands vary markedly in the vertical. At the 4 km level they are major constituents of the storm, being 20–40 km wide and several hundred kilometres long, whereas at the 10 km level many of these bands do not exist, being replaced by the "string of pearls" form of individual echoes. Clearly these outer bands are low level features in this particular storm, a conclusion supported by Kessler and Atlas (1956) in their analysis of hurricane Edna 1954. Figure 182 shows the evolution of rainband structure within a typhoon over a 40 h period as seen by the radar on Mt Fuji in Japan. Notwithstanding problems of decreasing signal return with increasing range, the pictures reveal significant changes in the size and location of the rainbands relative to the storm centre. In their analyses of four Atlantic hurricanes in the mid 1950s Senn and Hiser (1959) found that individual bands had measurable lifetimes of about 20 min, although one band was recognizable for 2 h. It is important

10 CIRCULATIONS IN CYCLONES 455

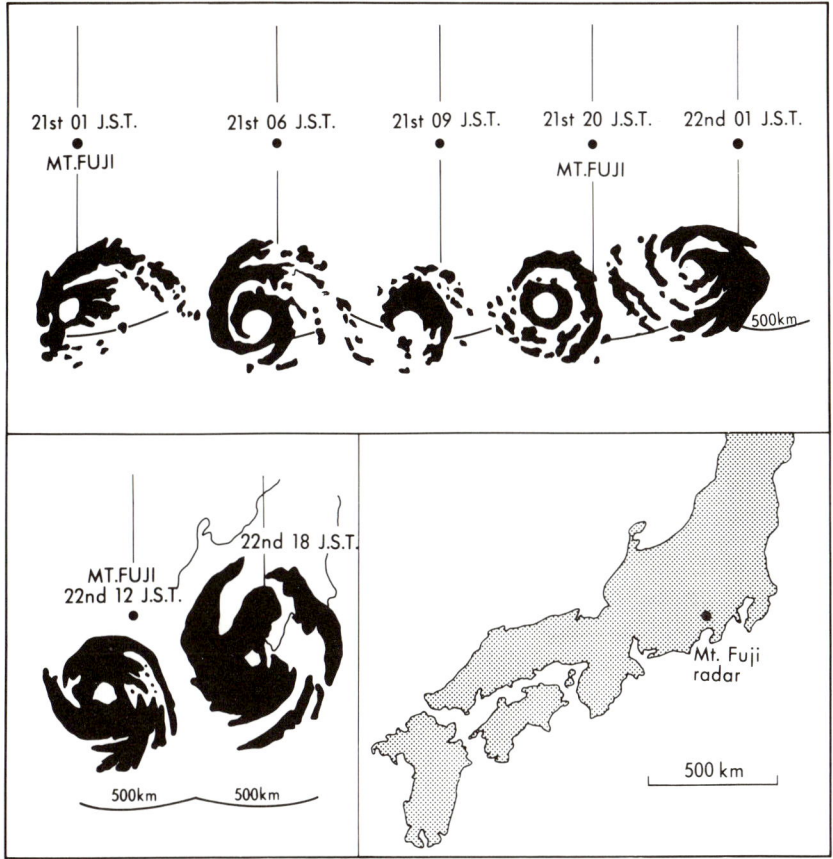

Fig. 182. Evolution of rainband structure around a hurricane near Japan over a period of 36 h. (After Tatehira and Itakura, 1966.)

to realize that bands are not always easy to see on radar pictures. Senn and Hiser (1959, p. 421) noted that "...the bands are normally fragmentary and...their shapes become increasingly significant only with research and familiarity with them". These authors also registered their surprise at the short durations they found, but emphasize that "...only five bands were found in the storms studied which could be consistently recognised and measured for over 20 min" (Senn and Hiser, 1959, p. 421).

Perhaps one of the most surprising aspects of the studies of spiral bands within tropical cyclones is the comparative lack of explicit attention to the component parts of these bands. In much of the literature the internal structure is implied rather than explicitly analysed, receiving attention only in the more detailed examination of rainband movement. Nevertheless, the bulk

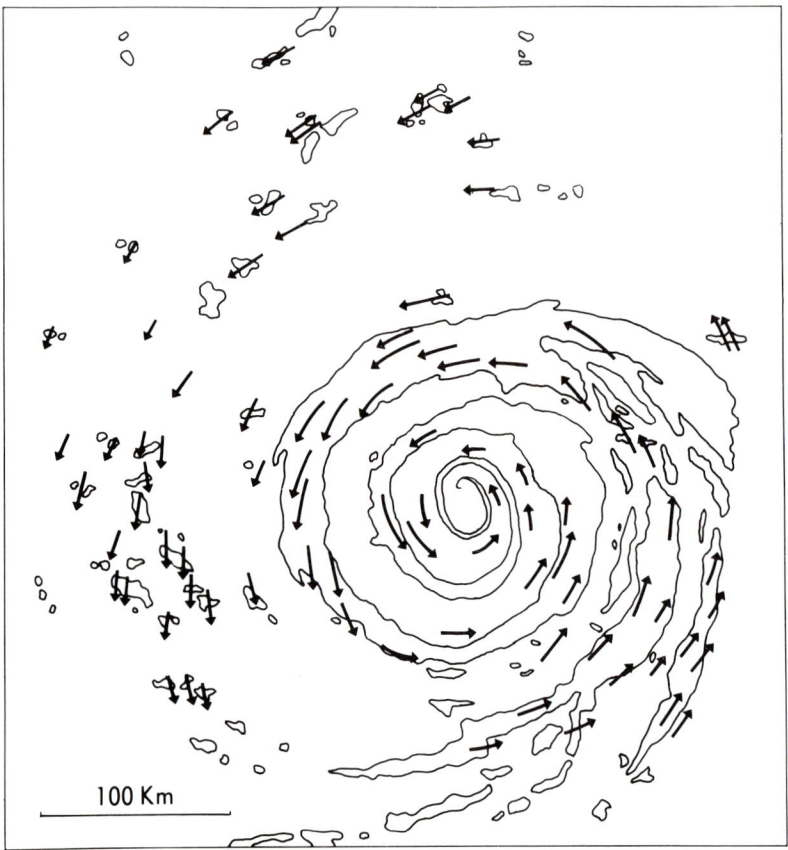

FIG. 183. Echo velocities relative to the centre of a hurricane. The length of each velocity vector represents the echo displacement during a 10 min period. (After Fujita and Black, 1970.)

of the literature suggests that the spiral bands comprise smaller echoes, probably from individual convective clouds—or, at most, from very few such clouds. Senn and Hiser (1959, p. 424) supported this conclusion noting that the "small, discrete rain cells" had lifetimes of 20–30 min. In their brief lives, the echoes move relative to the storm, but the direction of this relative flow is not yet yet agreed upon. Figure 183, resulting from studies by Fujita and Black (1970), reveals a systematic motion of radar echoes around the hurricane centre. Their detailed motion may be diagnosed with the aid of the echo "crossing angle" which is the angle between the tangent to the circle whose radius is the distance from the echo to the storm centre and the vector representing echo motion in relation to a stationary storm centre (Fig. 184).

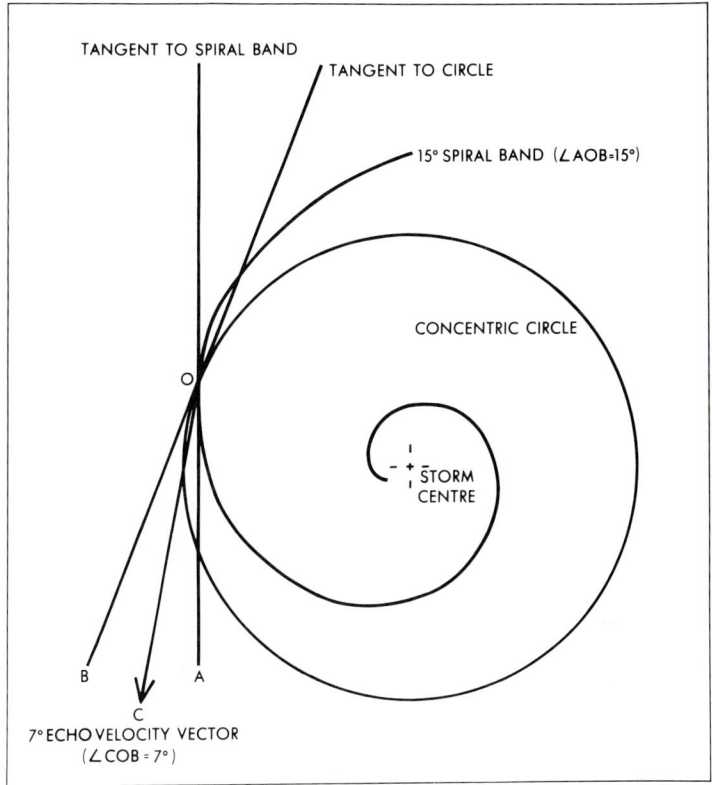

FIG. 184. Schematic diagram to illustrate the meaning of echo and spiral band crossing angles. (After Senn et al., 1959.)

Fujita and Black defined four sectors—left, right, front and rear—by drawing two 45° lines relative to the motion vector of the hurricane centre (Fig. 183). The echo velocity profiles (Fig. 185) are in good agreement with the mean echo velocity data computed by Senn (1960, 1963) and Senn et al. (1960), with the highest relative velocities near the storm eye. The pattern of inward and outward motion (Fig. 185) is, however, at variance with other studies. Fujita and Black found that the crossing angle was very small within 100 km of the storm centre, increasing to a maximum at a radius of about 180 km. Senn et al.'s (1960) data showed a maximum crossing angle at about 45 km decreasing outward. In addition, Fujita and Black noted outward moving echoes mainly in the left and front quadrants whereas Senn et al. (1960) found such motion in the left rear quadrant and in both the left rear and left front quadrants at small radii. Further complication is provided by Senn and Hiser (1959) who found inward motion everywhere except in the left front quadrant at distances

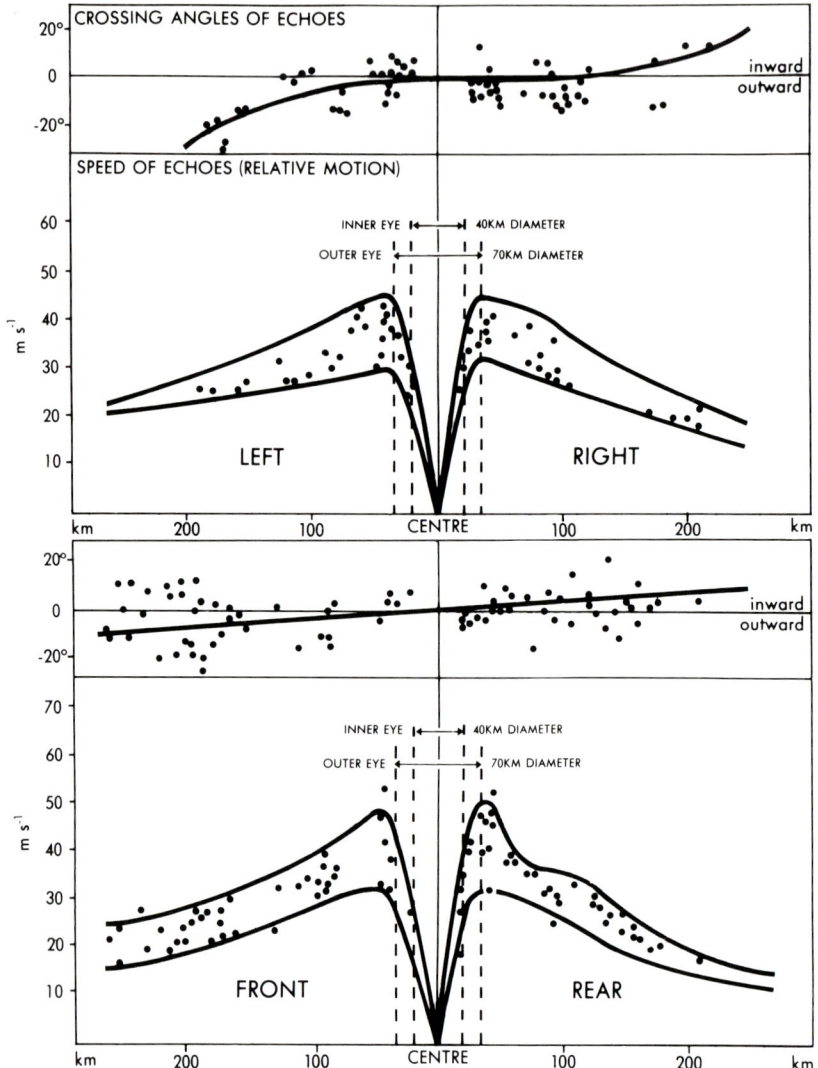

Fig. 185. The crossing angle and relative speed of radar echoes in the left, right, front and rear sectors of a hurricane. Data are the same as those used in Fig. 183. (After Fujita and Black, 1970.)

of over 180 km from the centre of the storm. Some of the discrepancy between all these results naturally stems from the differences in definition of the quadrants: in contrast to Fujita and Black, Senn and colleagues divided their storms into left front, left rear, right front and right rear quadrants.

Nevertheless, it is clear that we need more observations to elucidate the relative motion of the echoes within the rainbands of tropical cyclones.

2. Band movement

As a result of scrutinizing many photographs of radar scopes, Senn and Hiser (1959) showed that spiral bands moved outward but did not rotate relative to the storm centre. Figure 186(a) shows the consistent outward movement of a spiral until parts of it were almost 280 km from the centre. At this stage the tail of the band slowed and the head dissipated, broke up or fell beneath the radar beam. Figure 186(b) clearly illustrates that the bands remained in the same quadrant of the storm as they progressed outward.

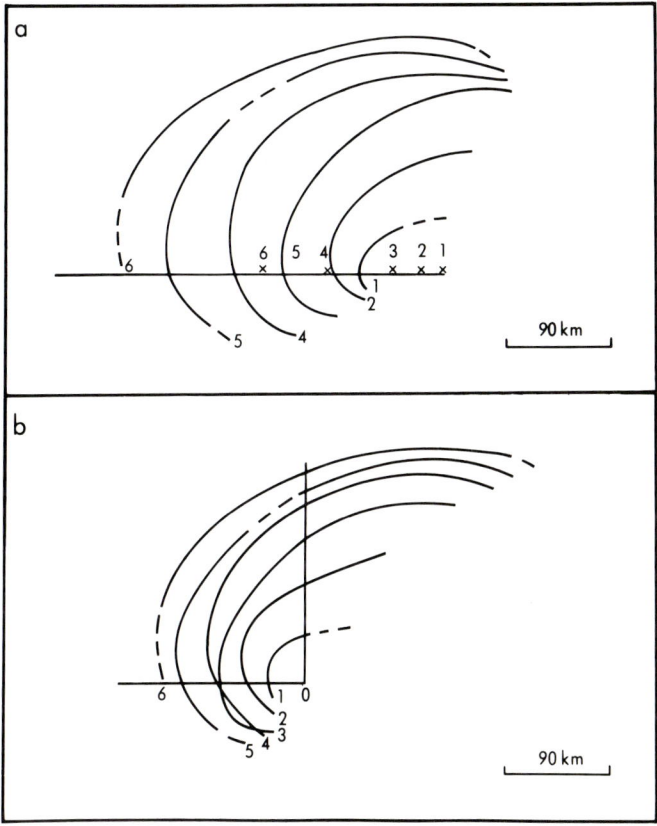

FIG. 186. Schematic diagrams of band movement. (a) Hourly positions of bands relative to a moving storm centre. Crosses represent eye positions. (b) Hourly positions of bands relative to a stationary storm centre. (After Senn and Hiser, 1959.)

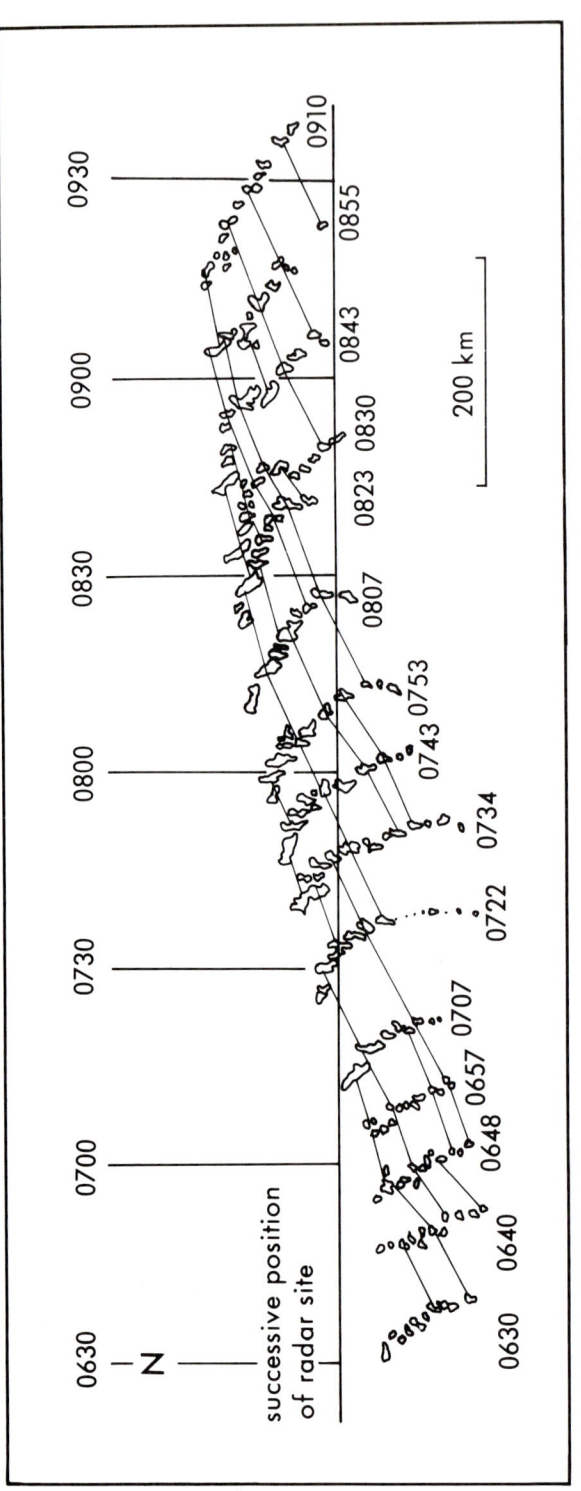

FIG. 187. Movement of cells within a rainband. The cells connected with a thin line are identical. The morphology of the whole rainband is shown at 15 different times. (After Tatehira, 1962.)

10 CIRCULATIONS IN CYCLONES

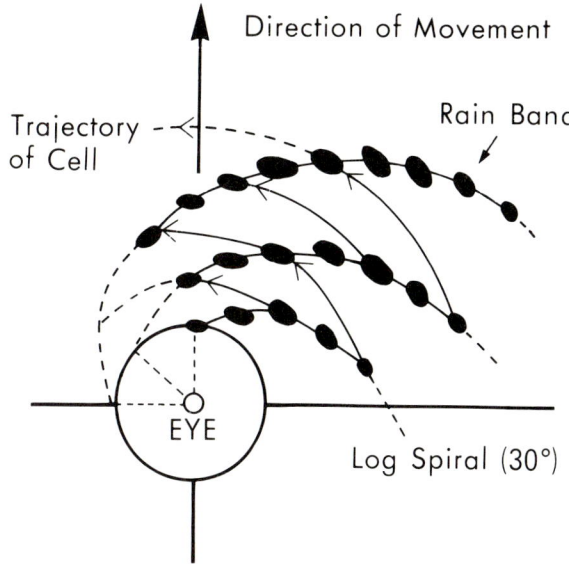

FIG. 188. Schematic picture to show displacement of rainbands relative to the eye and the movement of individual echoes as noted in Fig. 187. In this schematic diagram the crossing angle of the rainband is exaggerated. (After Tatehira, 1962.)

The explanation of band movement lies, as it does in the case of the extra-tropical bands, in the movement of the constituent cells. In one of the few studies to include an analysis of cell behaviour, Tatehira (1962) showed that individual cells originated at the outer end of the rainband and moved through to demise at the inner end. Figure 187 shows an actual case of this behaviour and Fig. 188 shows a schematic model. The relative flow outlined in Fig. 188 accords with the band and crossing angles as defined in Fig. 184. Fundamental to the cell behaviour shown in Fig. 188 is the necessity for an increase in cell speed with distance of the cell from the storm eye. Tatehira (1962) showed that indeed the cell speed increases linearly out to a distance of 150 km. This is in marked contrast to the results of Fujita and Black (1970) as shown in Fig. 185. Tatehira's results were in agreement with those of Ligda (1955) and Senn and Hiser (1959): at one and the same time they account for the apparent observation that echoes move cyclonically down the spiral bands towards the storm centre together with the outward movement of the bands. As yet it is unclear as to whether the cells simply move with the low level flow or have a propagation element. Similarly no serious work has yet been done on the initiation and demise of the cells. These problems await

3. Band temperatures and winds

Despite the value of all the work reported upon above, Gentry (1964) was dissatisfied:

> Published reports... tell us much about the weather at the surface associated with rainbands, about the geometry and kinematics of the bands, and to some extent about the liquid water distribution through the bands. They are entirely inadequate, however, for describing variations of temperature, wind, and pressure through the bands and for answering many of the pertinent questions about the bands. This lack of detailed and accurate knowledge may account in part for the failure of anyone to present a generally acceptable theory to account for the origin and maintenance of the bands even though many hypotheses concerning the bands have been advanced (Gentry, 1964, p. 5).

In an initial attempt to rectify the situation Gentry (1963) had collected data on the distribution of temperature and wind both along and normal to the long axis of rainbands. Typical results are shown in Fig. 189. In Fig. 189(a), the band was about 190 km south west of the storm centre and the temperature data were for a height of 3.9 km. Within the band (shown by the broken vertical lines), both temperature and wind varied markedly. Temperatures changed by 2.5 °C, which was about as much as the total variation from the outer edge of the eye wall to the outer edge of the strong cyclonic circulation for many hurricanes. Strong variations in wind speed led to convergence of about $10^{-3} s^{-1}$. Figure 189(b) shows the distribution of temperature and wind along the axis of a rainband, at a height of 3.9 km. Temperature gradients were as much as $0.6\,°C\,km^{-1}$ and wind speeds frequently changed more than $5\,m\,s^{-1}$ within 10 km. General conclusions from the work were as follows. First, rainbands in tropical storms of less than hurricane intensity tended to have a mean temperature lower than that of the ambient atmosphere. This was especially true for the outer bands. Second, in mature hurricanes rainbands tended to have mean temperatures warmer than those in the ambient atmosphere, especially when the band was located near the eye wall. Third, gradients of temperature, wind direction and speed were much greater inside rainbands than in areas between bands. Fourth, the anomaly of the mean temperature in the bands varied directly with the intensity of the storm, directly with altitude (at least to above the 5 km level) and inversely with radius. Fifth, exchange of air between various portions of the same band and between the various bands and their immediate environments took place very rapidly in both the outer bands and the eye walls. Lastly, all the evidence suggested the existence of convective cells embedded within the bands. Such a result provides further confirmation for the ideas used in the sub-sections on band structure and movement.

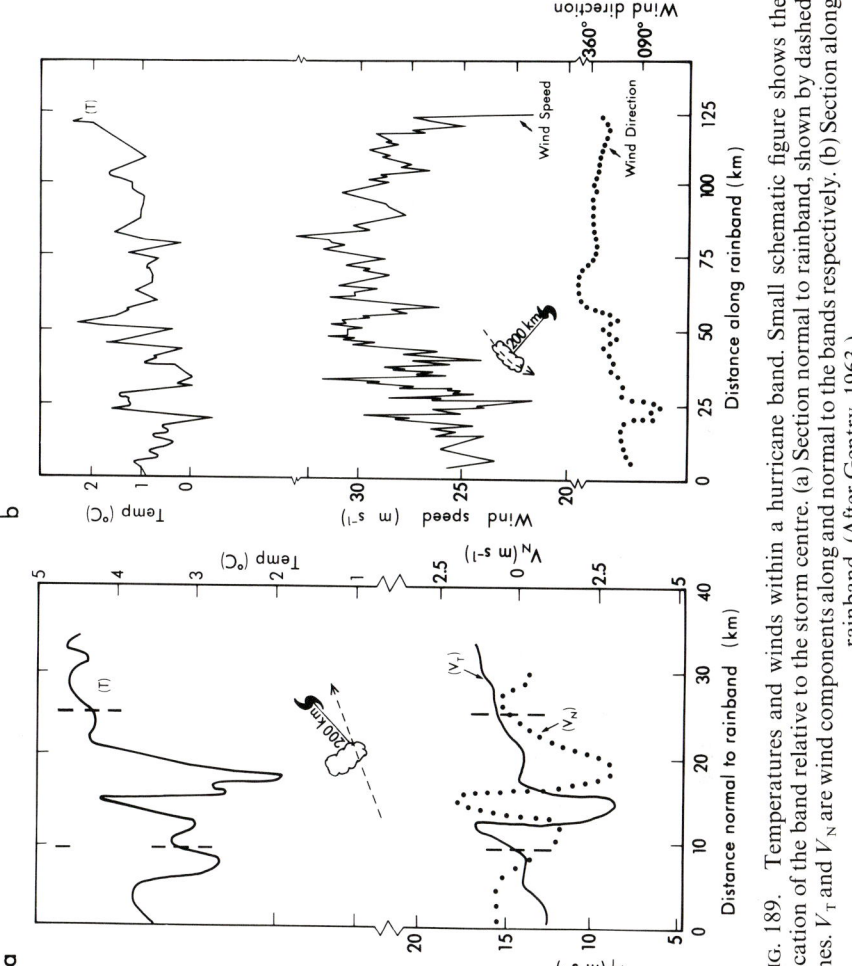

FIG. 189. Temperatures and winds within a hurricane band. Small schematic figure shows the location of the band relative to the storm centre. (a) Section normal to rainband, shown by dashed lines. V_T and V_N are wind components along and normal to the bands respectively. (b) Section along rainband. (After Gentry, 1963.)

III. Theory

The comparative recency of close observation of meso-scale circulations within cyclones has resulted in their limited theoretical analysis. Much of the theory derived within the last two decades was of the diagnostic, rather qualitative type, primarily concerned with the establishment of associations in space and time between, on the one hand, the circulations and, on the other, the instabilities that may have caused them. More recent theories have been of the prognostic type, largely consisting of numerical models. The disparity in observation of the tropical and extra-tropical cyclonic meso-systems is also evident in theoretical treatment. In the extra-tropical case enough is known about the kind of instability involved to warrant models of the meso-scale effects of cumulus convection. In tropical cyclones the internal structure of the meso-scale bands, so important to their size, shape and movement, is at present implicitly incorporated in models of whole hurricanes. Despite this drawback, the present models represent impressive progress over the last decade.

The remainder of this chapter reviews both the diagnostic and prognostic theoretical work on meso-scale circulations in extra-tropical and tropical cyclones.

A. Extra-tropical circulations

Virtually all studies of meso-scale circulations within extra-tropical systems have revealed the existence of potential instability in the region of origin of the meso-systems. This potential instability is primarily a result of differential flows at the synoptic scale within the cyclone. Figure 158 shows how the conveyor belt underlies a cooler, drier flow originating from behind the cold front. A vertical section through these flows (Fig. 190(a)) reveals the area in which the resultant potential instability occurs. Figure 190(b) illustrates a real, as opposed to a schematic, situation. Most, if not all, authors agree that it is the release of this potential instability by frontal uplift which is the ultimate cause of the meso-scale circulations—by which we mean those equivalent to SMPAs. The instability is manifest in the clusters of generators noted in the first part of the chapter. But the real problem is as outlined by Browning (1974, p. 316): "This convection occurs in clusters of meso-scale dimensions but it is not known what determines the size of these clusters". Although some progress has been made since those words were written a large measure of ignorance still remains.

Despite the lack of an explanation of the basic meso-scale circulation,

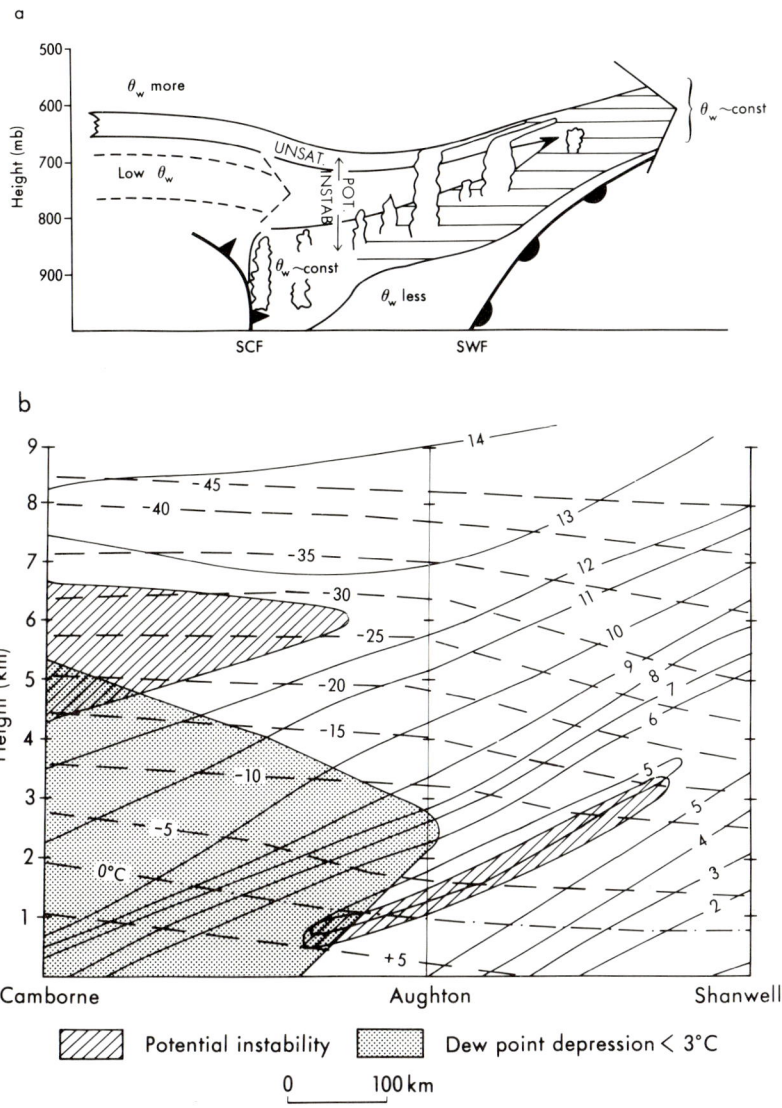

FIG. 190. Potential instability above a front, leading to development of meso-scale circulations. (a) Schematic vertical cross-section of airflow within an extra-tropical cyclone. Hatching shows layer cloud and convective overturning is depicted by the cumuli form shapes. Two initially separate flows, with lower θ_w between, mix to produce a lapse of nearly constant θ_w above the warm frontal zone. (After Harrold, 1973.) (b) An actual case of potential instability above the warm front on 8 March 1967. Vertical cross-section from Camborne to Shanwell, UK. Solid lines represent wet bulb potential temperature (in degrees Celsius) and dashed lines dry bulb temperature (in degrees Celsius.) (After Atkinson and Smithson, 1974a.)

Browning *et al.* (1973) provided a preliminary hypothesis for the origin of rainbands:

> It is not clear whether the organization of the convection into bands was the result of pre-existing banded structure within the potentially cold air aloft or to other (unknown) dynamical factors leading to a banded pattern of large scale ascent within the underlying conveyor belt flow. Of the two explanations, that involving pre-existing bandedness within the pattern of potential instability is favoured by the finding that the bands lost their identity when the potential instability became exhausted. Alternatively it is possible that the existence and spacing of the bands is some function of the cumulonimbus convection itself rather than the result of streakiness previously introduced.... If this is the case, then it would be necessary to think of the band development as leading to the disturbance in the field of wind and baroclinicity rather than vice versa (Browning *et al.*, 1973, p. 229).

In fact, subsequent work by Roach and Hardman (1975) favoured this latter view because a scale analysis of the equations of motion and continuity suggested that small differences in density between air in and between the bands would produce enough hydrostatic pressure difference to induce (through the resultant perturbation of the horizontal pressure field) wind perturbations of the characteristic dimensions and magnitudes observed. Further support for this general idea is forthcoming from numerical modelling techniques, which have only very recently been applied to this problem.

A substantial preliminary attack on the problem has been made by Kreitzberg and Perkey (1976, 1977). Their aim was to understand and predict the release of potential instability by large scale lifting and the subsequent interaction of the cumulus convection and the larger scale fields. Their approach was to use a non-hydrostatic cumulus model to provide the convective scale changes for use in a hydrostatic primitive equation model for the meso-scale and larger scale fields and to study the interaction of these two models. Thus the models attempted to simulate, first, the effects of cumulus convection on the temperature, pressure and humidity fields and then, second, how these fields generate meso-scale circulations.

Kreitzberg and Perkey (1976) stressed that the cumulus model was "tailor-made" for their particular problem: "The convection model is very simple as convection models go but it is complex, flexible and highly versatile as convective adjustments go" (Kreitzberg and Perkey, 1977, p. 1570). The one-dimensional, Lagrangian form cumulus model provided information on the base of the convection, the cumulus plume that built, the environmental subsidence and the mixing of the subsided environment with the cumulus plume after rainout. The cumulus model was "called" every 20 min in the running of the meso-scale hydrostatic model, building plumes sequentially on each cell. The convective changes resulting from this model entered smoothly into the hydrostatic primitive equations and responded appropriately to the larger scale vertical motions.

10 CIRCULATIONS IN CYCLONES

The meso-scale model was a two-dimensional one with grid sizes of 20 km in the horizontal and 1 km in the vertical. The primary features of the model were as follows:

(1) Height above a level surface was used as the vertical coordinate.

(2) East (u) and north (v) wind speed, virtual temperature (T_v), water vapour specific humidity (q), cloud water concentration (q_c), precipitation concentration (q_p), pressure tendency at the top $[(\partial p/\partial t)_{z_t}]$ and surface precipitation were predicted. Pressure was diagnosed hydrostatically using an effective temperature that accounted for cloud and precipitation loading. Vertical motion (w) was diagnosed from the continuity equation.

(3) Effects of sub-grid scale convection were evaluated from the cumulus model noted above and were added to the tendency equations for T_v, q, q_c and $(\partial p/\partial t)_{z_t}$.

(4) At the upper boundary z_t, $dp/dt \equiv \omega = 0$ was specified, where d/dt is the material derivative (change following the motion). At the level ground surface, $\omega = 0$ was specified. An x–z area average value of divergence normal to the cross-section was usually specified to keep the x-average surface pressure tendency zero.

The primitive equation set comprised seven prognostic equations of the form:

$$\frac{\partial \psi}{\partial t} = \frac{d\psi}{dt} - u\frac{\partial \psi}{\partial x} - v\frac{\partial \psi}{\partial x} - w\frac{\partial \psi}{\partial z} + \left(\frac{\partial \psi}{\partial t}\right)_s, \qquad (230)$$

where ψ is $u, v, T_v, p(z_t), q, q_c$, and q_p and $(\partial \psi/\partial t)_s$ are the sub-grid changes of $\psi = T_v, q, q_c$, and $p(z_t)$, which are supplied from the cumulus convection sub-routine. The material derivatives ($d\psi/dt$) were evaluated as follows:

conservation of momentum,

$$\frac{du}{dt} = fv - \frac{1}{\rho_e}\frac{\partial p}{\partial x}, \qquad (231)$$

$$\frac{dv}{dt} = fu - \frac{1}{\rho_e}\frac{\partial p}{\partial y}; \qquad (232)$$

conservation of energy,

$$\frac{dT_v}{dt} = \frac{\dot{Q}}{c_p} + \frac{1}{c_p \rho}\frac{dp}{dt}; \qquad (233)$$

conservation of moisture,

$$\frac{dq}{dt} = E_c + E_p - Cd, \qquad (234)$$

$$\frac{dq_c}{dt} = Cd + E_c - Cv - Cl, \tag{235}$$

$$\frac{dq_p}{dt} = Cv + Cl - E_p - D_p; \tag{236}$$

where ρ_e is the equivalent density, $\dot{Q}$ is the diabatic heating within the meso-scale model exclusive of convective scale heating, Cd is condensation of cloud water, E_c is evaporation of cloud water, Cv is the conversion from cloud water to precipitation, Cl is the collection of cloud water by precipitation, E_p is evaporation of precipitation and D_p is precipitation divergence.

All the experiments began with zero horizontal motion in the (x, z) plane and had a large scale, vertical motion field superimposed; this was constant in time and in the horizontal. The vertical motion came from specifying the divergence normal to the (x, z) plane. A passive boundary layer formulation was used and the disturbance arose from the release of potential instability due to large scale lifting. The initial humidity field decreased sinusoidally from the centre of the domain so that convection was initiated at the centre of the symmetrical model. Simulation tests were about 12 h long with the first half being primarily convective and the last half being primarily a steady meso-scale circulation in a hydrostatically neutral updraught.

The primary results of the experiment are illustrated in Fig. 191, which shows the initial distribution of vertical velocity, together with its evolution over a 12 h period. After 2 h a definite cell (or band, in three dimensions) appeared, with maximum vertical velocities of 14.4 cm s^{-1} at a height of about 6 km. The "band width" was about 300 km. After 4 h the band narrowed slightly but intensified so that maximum upward velocity was 17.8 cm s^{-1}, with compensatory subsidence of about 1 cm s^{-1}. The band widened and weakened a little after 6 h, but further intensified after 8 h, being about 200 km wide with maximum updraughts of 18.8 cm s^{-1} and sinking of 5.0 cm s^{-1}. After 12 h the band was still clearly evident at the end of the simulation. Together with these results Kreitzberg and Perkey (1977) also showed that both convective and condensation temperature changes ranged from about 1 to 4 °C $(10^4 \text{s})^{-1}$, complementing rather than duplicating each other. Kreitzberg and Perkey (1977, p. 1580) noted that:

> The convection produces upper troposphere [sic] rather than mid-troposphere warming and the subsidence drying near convection base (3 km) prevents saturation and latent heating from large scale ascent. The way in which convection avoids instability on the mesoscale in the atmosphere as well as in this model is by changing the vertical distribution of the warming resulting from condensation.

The authors further concluded that the meso-scale disturbance initiated by the convection produced most of its precipitation from stable updraughts in the hydrostatically neutral atmosphere after the potential energy had been

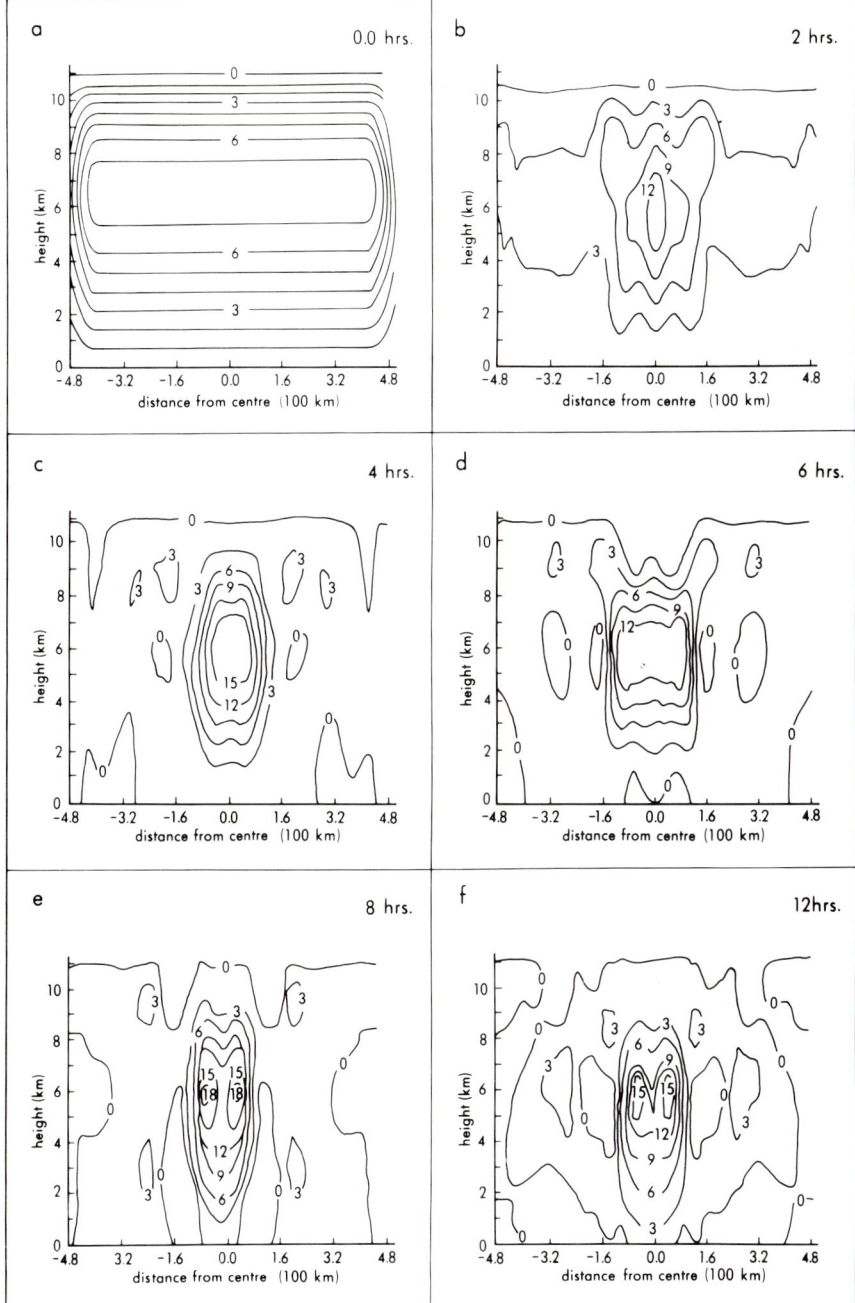

FIG. 191. Main results from Kreitzberg and Perkey's model in the form of vertical cross-sections of vertical velocity (in centimetres per second): (a) initial state; (b) after 2 h; (c) after 4 h; (d) after 6 h; (e) after 8 h; (f) after 12 h. (After Kreitzberg and Perkey, 1977.)

released. Therefore the release of the potential instability caused the heavy precipitation but the precipitation was produced for the most part in hydrostatic meso-scale updraughts after the cumulus convection. In a subsequent preliminary assessment of the behaviour of the model with varying governing physical factors, Kreitzberg and Perkey (1977) found that increasing the band width of the initial relative humidity distribution gave a wider resultant meso-scale precipitation band. An increase in large scale vertical motion had the same effect. Evaporation of precipitation above cloud base had only a minor effect on the lifetime of the meso-scale circulation. The effects of the initial cloud updraught radius remained somewhat uncertain: clearly more effort must be expended on this important initialization problem. In tests of the model's sensitivity to numerical factors, it was found that the rainband width was sensitive to the horizontal grid size, "so we have no way of knowing that narrow bands might not develop if it were not for the numerical constraint" (Kreitzberg and Perkey, 1977, p. 1594).

Despite the large size of these simulated bands (up to 300 km wide—far wider than those considered in the first part of this chapter) and the drawbacks of the model (clearly recognized by its creators), these simulations represented an important first step towards understanding the release of potential energy and thus the meso-scale systems within extra-tropical cyclones. The next few years will no doubt witness substantial progress in the analysis of this very difficult problem.

B. Tropical circulations

Our theoretical understanding of meso-scale rainbands in tropical cyclones is still in its infancy. It is not unreasonable to say that our present-day knowledge is roughly comparable with our insight into extra-tropical squall lines of some three decades ago. A major difference in the analysis of these two types of system has been the almost immediate application of numerical methods to the tropical circulations. In fact most of the numerical models are of tropical cyclones as a whole—only very recently have the spiral bands received particular attention. Before briefly reviewing the results of these numerical and other analyses, two points are worth making. First, in the "whole-storm" models, grid sizes were at least, and frequently greater than, 20 km in length. Consequently the models resolved only fairly large features, when viewed on the meso-scale. In this respect they were similar to Kreitzberg and Perkey's experiments. Thus the bands identified in tropical storm models were usually 1–200 km wide and hundreds of kilometres long—more akin to "outer" than "inner" hurricane bands. Very little, if any, internal structure of the type outlined in the first part of this chapter appeared in these models. The emphasis was still upon the reasonably realistic simulation of band shape, size and gross movement. The second point concerns our real understanding of bands that appear in such models. The appearance of these bands does not automatically mean that we understand their mechanism. As Diercks and Anthes (1976b, p. 1714) noted: "The formation of these bands in complicated multi-level models has called for an explanation of the generation mechanisms in these models". It is only the advent of diagnostic studies of the models that has allowed the closer examination of those factors influencing the existence of tropical cyclone bands.

As a result of both analytical and numerical work it is increasingly agreed that rainbands in tropical cyclones, and possibly other linear rain bands in the extra-tropics (Lindzen and Tung, 1976), are closely associated with gravity-inertia waves moving outward from the eye of the storm. Diercks and Anthes (1976a) claimed that, because of the small horizontal scale, the effect of the Earth's rotation is relatively minor and the travelling waves are nearly pure gravity waves. Such a mechanism was first speculated

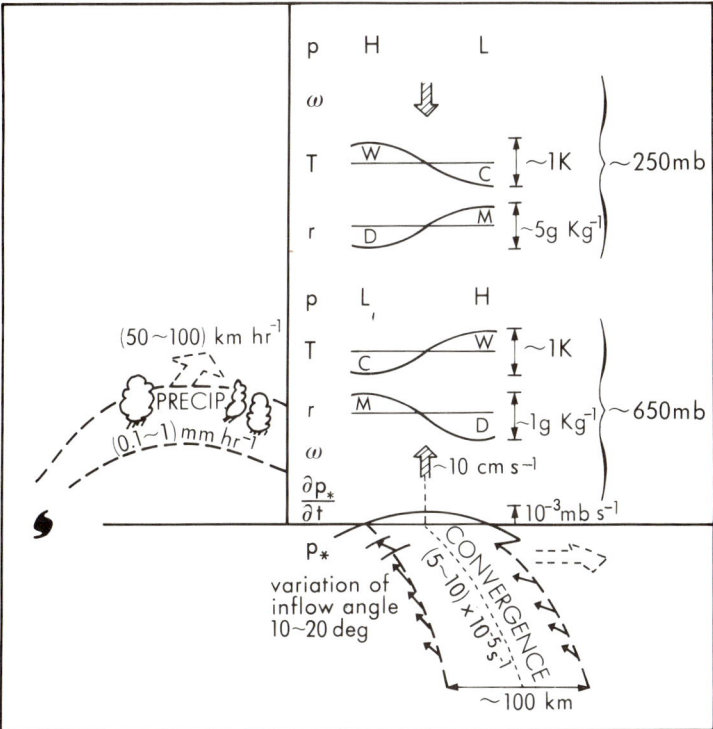

FIG. 192. Main results from Kurihara and Tuleya's model in schematic form. p is pressure, $\omega = dp/dt$ is the vertical velocity, T is temperature, r is the mixing ratio of water vapour, p_* is surface pressure. (After Kurihara and Tuleya, 1974.)

upon by Tepper (1958), soon to be followed by Yamamoto (1963) and Abdullah (1966), but the basic idea of these studies appeared to receive little credence until later numerical models began to produce similar results. Thus by the early 1970s a three-dimensional model generated bands with a width of 90 km, "...considered fairly acceptable compared to observations when the coarse resolution of the model is considered" (Anthes, 1972, p. 468). The author concluded that "...the spiral bands...are undoubtedly internal gravity waves modified by latent heat..." but also cautioned that "...the mechanism for their generation is unknown" (Anthes, 1972, p. 468).

Similar results were forthcoming from the elaborate 11 level, primitive equation model run by Kurihara and Tuleya (1974). Bands about 100 km wide formed near the centre of the cyclone and propagated outward at a speed of about 28 m s^{-1}. Diagnosis of the surface pressure tendency, wind speed and direction, divergence and vertical velocity resulted in the composite schematic picture shown in Fig. 192. Near the surface the windfield gave convergence of 5×10^{-5}–10×10^{-5} s^{-1} and the pressure field was such that the band propagated outward. Vertical velocities associated with the low level convergence reached about 10 cm s^{-1}. In the middle and high troposphere the pressure, temperature and humidity fields varied across the band in a systematic manner, but were out of phase between the 250 and 650 mb levels. Kurihara and Tuleya's (1974) preliminary interpretation of these features was that they could be due to the passage of internal gravity waves. Figure 193

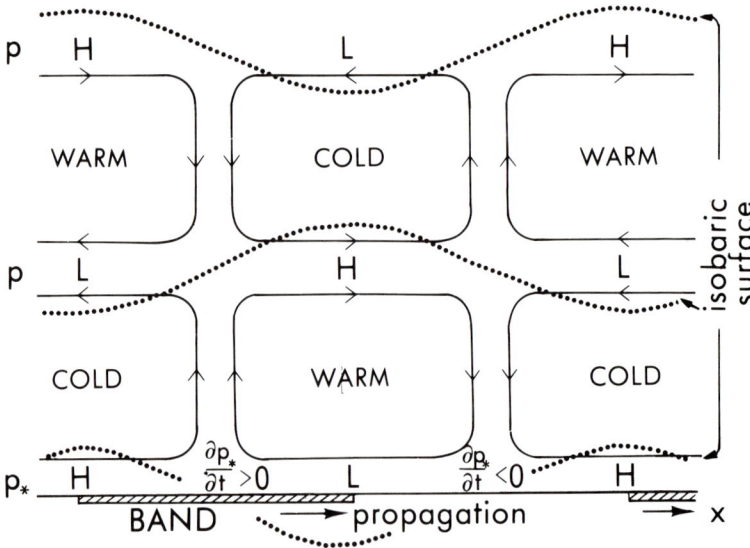

FIG. 193. Schematic diagram showing a typical structure of an internal gravity wave that may represent the band shown in Fig. 192. (After Kurihara and Tuleya, 1974.)

schematically illustrates a typical gravity wave structure that could represent the characteristics of the band shown in Fig. 192.

Although we are still unclear about the precise mechanisms of the effects of these gravity waves on the bands, some factors have been preliminarily assessed. First, rotation of flow appears to be a basic requirement. Diercks and Anthes (1976b) found that rotation organized initial random perturbations into a spiral form in experiments with unstable static stability. With stable static stability, significant growth of the bands was not observed, but random perturbations were still organized into a spiral pattern. Second, the same authors found that the coriolis parameter was relatively unimportant. Third, Kurihara (1976) found that a radial shear in the basic flow around a cyclone could cause an unstable, outward propagating wave and that this wave formed in the presence of stable static stability and without diabatic heating from cumulus convection. Fourth, the effects of release of latent heat are still uncertain. As just noted, Kurihara (1976) found growth without latent heating and Diercks and Anthes (1976a) also found that latent heating effects were small. In their diagnosis of a non-linear hurricane model they purposely suppressed latent heating from convective and non-convective processes in the bands. Nevertheless, bands of upward motion continued to form and propagate outward: without latent heating the bands propagated somewhat faster but did not rise as high into the troposphere as when the heating was included. The authors concluded that the latent heating in the bands was a result of the convergence pattern associated with the travelling wave and did not play an important role in the maintenance or propagation of the bands at large radii. In contrast to these results Mathur (1975) did find some effects of latent heat release. He divided the heating into two parts: that due to convection and that due to large scale ascent of moist air—the so-called non-convective release of latent heat. The latter was included in his model only in the upper troposphere and was about only one-third in magnitude of the convective latent heating. Nevertheless, Mathur (1975) showed that propagating bands did not form if the non-convective release of latent heat was omitted from the model. In contrast the inclusion of this release enhanced upper tropospheric outflow which induced strong zones of convergence in the boundary layer. The resulting increase in the upward motion at the top of the boundary layer augmented the convective release of latent heat and led to a rapid intensification of the disturbance.

IV. Hardware models

Numerical modelling is not the only approach employed in the analysis of hurricane rainbands. Faller (1962) argued that hardware modelling has much to offer through three elements: established fact, analogy, and conjecture.

Factually, a series of laboratory experiments have demonstrated the instability of a laminar Ekman boundary layer, the initial instability taking the form of spiral convective bands, whose spacing is proportional to the depth of the boundary layer. Arakawa (1964) and Arakawa and Manabe (1963, 1966) provided illustration of these features. Morphologically these laboratory bands are good analogues of those in hurricanes and Faller's conjecture covers discussion and rationalization of the differences between experimental results and hurricane phenomena.

Within a rotating fluid the boundary layer flow has been observed to accord with the Ekman spiral. Also the zonal speed of flow varied inversely with the radius. Therefore it was expected that at some radius approaching the centre the flow should become sufficiently strong to destabilize the boundary layer flow. This was found to be so and the nature of the instability was a spiral of waves in the fluid. Faller (1962) freely admitted that hurricane boundary layer flow is not that of an Ekman spiral but claimed that sufficient similarity existed to allow useful inference. In particular Faller concluded that the direct implication of the work was that the organization and maintenance of the convection in a banded structure is due to the boundary layer processes.

It is clear from this brief review of mathematical and hardware experiments that much remains to be discovered about the theory of meso-scale rainbands within tropical cyclones. The purist could claim with some justification that derivation of true meso-scale theory has yet to start, and there can be little doubt that the next decade will see major advances in this part of the subject. To only a slightly smaller extent these remarks are also true of meso-systems within extra-tropical cyclones. It is still too early to paint a coherent picture of the common themes of meso-scale circulations in cyclones worldwide—save perhaps to say that they are both probably convective in origin. Perhaps the major unifying feature of both tropical and extra-tropical systems is the question already hinted at earlier in this chapter: why does the convection manifest itself on the meso-scale? This rather fundamental puzzle still awaits a satisfactory answer.

References

Abdullah, A. J. (1966). The spiral bands of a hurricane: a possible dynamic explanation. *J. atmos. Sci.*, **23**, 367–375.

Akiyama, T. (1974). Mesoscale organization of cumulus convection in large-scale rainbands in the Baiu season. *J. met. Soc. Jap.*, **52**, 448–451.

Anthes, R. A. (1972). Developments of asymmetries in a three-dimensional numerical model of the tropical cyclone. *Mon. Weath. Rev.*, **100**, 461–476.

Arakawa, H. (1964). A model experiment on spiral rain bands—a dish pan experiment. *Proc. Symp. trop. Met. Rotorua*, 409–415.

Arakawa, H. and Manabe, D. (1963). Investigation of spiral rain bands and frontal structures in terms of shallow water waves. *Pap. Met. Geophys.*, **14**, 127–143.

Arakawa, H. and Manabe, D. (1966). Some considerations on spiral cloud patterns in typhoonic area by means of experiments using travelling disturbing element through a shallow water tank. *Geophys. Mag.*, **33**, 89–98.

Atkinson, B. W. and Smithson, P. A. (1972). An investigation into meso-scale precipitation distributions in a warm sector depression. *Q. Jl R. met. Soc.*, **98**, 353–368.

Atkinson, B. W. and Smithson, P. A. (1974a). Meso-scale circulations and rainfall patterns in an occluding depression. *Q. Jl R. met. Soc.*, **100**, 3–22.

Atkinson, B. W. and Smithson, P. A. (1974b). Rain pulses in a warm sector. *Weather*, **29**, 44–53.

Atkinson, B. W. and Smithson, P. A. (1978). Mesoscale precipitation areas in a warm frontal wave. *Mon. Weath. Rev.*, **106**, 211–222.

Austin, P. M. (1960). Microstructure of storms as described by quantitative radar data. *Geophys. Monogr.*, **5**, 86–92.

Austin, P. M. and Houze, R. A. (1970). Analysis of meso-scale precipitation areas. *Preprints 14th Weather Radar Conf.*, 329–334.

Austin, P. M. and Houze, R. A., Jr (1972). Analysis of the structure of precipitation patterns in New England. *J. appl. Met.*, **11**, 926–935.

Austin, P. M., Nason, C. K. and Kraus, M. J. (1966). Mesoscale precipitation patterns in New England and their relation to macroscale parameters. *Proc. 12th Radar Met. Conf.*, 234–240.

Bodolai, I. and Jakus, E. (1969). Some meso-scale particularities in the pattern of cyclonal precipitation. *Magyar Met. Tars. Foly., Időj.*, **73**, 257–264.

Browning, K. A. (1971a). Radar measurements of air motion near fronts. Part I. *Weather*, **26**, 293–304.

Browning, K. A. (1971b). Radar measurements of air motion near fronts. Part II. *Weather*, **26**, 320–340.

Browning, K. A. (1974). Mesoscale structure of rain systems in the British Isles. *J. met. Soc. Jap.*, **52**, 314–327.

Browning, K. A. and Bryant, G. W. (1975). An example of rainbands associated with stationary longitudinal circulations in the planetary boundary layer. *Q. Jl R. met. Soc.*, **101**, 893–900.

Browning, K. A. and Harrold, T. W. (1969). Air motion and precipitation growth in a wave depression. *Q. Jl R. met. Soc.*, **95**, 288–309.

Browning, K. A. and Harrold, T. W. (1970). Air motion and precipitation growth at a cold front. *Q. Jl R. met. Soc.*, **96**, 369–389.

Browning, K. A. and Pardoe, C. W. (1973). Structure of low-level jet streams ahead of mid-latitude cold fronts. *Q. Jl R. met. Soc.*, **99**, 619–638.

Browning, K. A., Hardman, M. E., Harrold, T. W. and Pardoe, C. W. (1973). The structure of rainbands within a mid-latitude depression. *Q. Jl R. met. Soc.*, **99**, 215–231.

Browning, K. A., Hill, F. F. and Pardoe, C. W. (1974). Structure and mechanism of precipitation and the effect of orography in a winter time warm sector. *Q. Jl R. met. Soc.*, **100**, 309–330.

Brunt, D. and Douglas, C. K. M. (1928). The modification of the strophic balance for changing pressure distribution and its effect on rainfall. *Mem. R. met. Soc.*, **3**, (22).

Caton, P. G. F. (1966). Raindrop size distributions in the free atmosphere. *Q. Jl R. met. Soc.*, **92**, 15–30.

Colon, J. A. and staff (1961). On the structure of hurrican Daisy (1958). National

Hurricane Research Program, Report no. 48, US Department of Commerce, Weather Bureau, Miami.

Cornford, S. G. (1966). A note on some measurements from aircraft of precipitation within frontal clouds. *Q. Jl R. met. Soc.*, **92**, 105–113.

Diercks, J. W. and Anthes, R. A. (1976a). Diagnostic studies of spiral rainbands in a non-linear hurricane model. *J. atmos. Sci.*, **33**, 959–975.

Diercks, J. W. and Anthes, R. A. (1976b). A study of spiral bands in a linear model of a cyclonic vortex. *J. atmos. Sci.*, **33**, 1714–1729.

Donaldson, R. J., Jr and Atlas, D. (1963). Radar in tropical meteorology. *Proc. Symp. trop. Met., Rotura*, 423–473.

Douglas, R. H., Gunn, K. L. S. and Marshall, J. S. (1957). Pattern in the vertical of snow generation. *J. Met.*, **14**, 95–114.

Elliott, R. D. and Hovind, E. L. (1964). On convection bands within Pacific coast storms and their relation to storm structure. *J. appl. Met.*, **3**, 143–154.

Faller, A. J. (1962). An experimental analogy to and proposed explanation of hurricane spiral bands. National Hurricane Research Program, Report no. 50, Part II, pp. 307–313, US Department of Commerce, Weather Bureau, Miami.

Fujita, T. T. and Black, P. G. (1970). In- and outflow field of hurricane Debbie as revealed by echo and cloud velocities from airborne radar and ATS-III pictures. *Preprints 14th Radar Met. Conf., Tucson*, 353–358.

Fujita, T., Izawa, T., Watanabe, K. and Imai, I. (1967). A model of typhoons accompanied by inner and outer rainbands. *J. appl. met.*, **6**, 3–19.

Fujiwara, M. (1958). Rainbands and a weather system around a wave crest in humid season. *Pap. Met. Geophys.*, **9**, 63–86.

Gentry, R. C. (1963). The role of the rainbands in hurricanes. *Geofisica Internacional*, **3**, 117–122.

Gentry, R. C. (1964). A study of hurricane rainbands. National Hurricane Research Project, Report no. 69, US Department of Commerce, Weather Bureau, Miami.

Gentry, R. C., Fujita, T. T. and Sheets, R. C. (1970). Aircraft, spacecraft, satellite and radar observations of Hurricane Gladys 1968. *J. appl. Met.*, **9**, 837–850.

Gunn, K. L. S., Langleben, M. P., Dennis, A. S. and Power, B. A. (1954). Radar evidence of a generating level for snow. *J. Met.*, **11**, 20–26.

Gunn, K. L. S. and Marshall, J. S. (1955). The effect of wind shear on falling precipitation. *J. Met.*, **12**, 339–349.

Gunn, K. L. S. and Marshall, J. S. (1958). The distribution with size of aggregate snowflakes. *J. Met.*, **12**, 452–461.

Hardman, M. E., James, D. G., Goldsmith, P. (1972). The measurement of mesoscale vertical motions in the atmosphere. *Q. Jl R. met. Soc.*, **98**, 38–47.

Harrold, T. W. (1973). Mechanisms influencing the distribution of precipitation within baroclinic disturbances. *Q. Jl R. met. Soc.*, **99**, 232–251.

Hobbs, P. V. and Locatelli, J. D. (1978). Rainbands, precipitation cores and generating cells in a cyclonic storm. *J. atmos. Sci.*, **35**, 230–241.

Houze, R. A., Jr, Hobbs, P. V., Biswas, K. R. and Davis, W. M. (1976a). Mesoscale rainbands in extratropical cyclones. *Mon. Weath. Rev.*, **104**, 868–878.

Houze, R. A., Jr, Locatelli, J. D. and Hobbs, P. V. (1976b). Dynamics and cloud microphysics of the rainbands in an occluded frontal system. *J. atmos. Sci.*, **33**, 1921–1936.

Imakado, M. (1966). Detailed analysis of typhoon structure by means of weather radars. *Geophys. Mag.*, **33**, 99–106.

Kessler, E., III and Atlas, D. (1956). Radar synoptic analysis of Hurricane Edna, 1954. Geophysical Research Paper no. 50 AFCRC, Bedford, Mass.

Kessler, E. and Wexler, R. (1960). Observations of a cold front, 1 October 1958. *Bull. Am. met. Soc.*, **41**, 253–257.

Kreitzberg, C. W. (1968). The mesoscale wind field in an occlusion. *J. appl. Met.*, **7**, 53–67.

Kreitzberg, C. W. and Brown, H. A. (1970). Mesoscale weather systems within an occlusion. *J. appl. Met.*, **9**, 417–432.

Kreitzberg, C. W. and Perkey, D. J. (1976). Release of potential instability. Part I. A sequential plume model within a hydrostatic primitive equation model. *J. atmos. Sci.*, **33**, 456–475.

Kreitzberg, C. W. and Perkey, D. J. (1977). Release of potential instability. Part II. The mechanism of convective/mesoscale interaction. *J. atmos. Sci.*, **34**, 1569–1595.

Kurihara, Y. (1976). On the development of spiral bands in a tropical cyclone. *J. atmos. Sci.*, **33**, 940–958.

Kurihara, Y. and Tuleya, R. E. (1974). Structure of a tropical cyclone developed in a three-dimensional simulation model. *J. atmos. Sci.*, **31**, 893–919.

Langleben, M. P. (1956). The plan pattern of snow echoes at the generating level. *J. Met.*, **13**, 554–560.

Ligda, M. G. H. (1955). Analysis of motion of small precipitation areas in the hurricane, August 23–28, 1949. Technical Note no. 3, Massachusetts Institute of Technology, Cambridge, Mass.

Lindzen, R. S. and Tung, K.-K. (1976). Banded convective activity and ducted gravity waves. *Mon. Weath. Rev.*, **104**, 1602–1617.

Marshall, J. S. (1953). Precipitation trajectories and patterns. *J. Met.*, **10**, 25–29.

Mason, B. J. and Ramanadham, R. (1954). Modification of size distribution of falling raindrops by coalescence. *Q. Jl R. met. Soc.*, **80**, 388–394.

Mathur, M. B. (1975). Development of banded structure in a numerically simulated hurricane. *J. atmos. Sci.*, **32**, 512–522.

Matsumoto, S. (1968). Smaller scale disturbance in the temperature field around a decaying typhoon with special emphasis on the severe precipitation. *J. met. Soc. Jap.*, **46**, 483–495.

Matsumoto, S. and Akiyama, T. (1970). Mesoscale disturbances and related rainfall cells embedded in the "Baiu front", with a proposal on the role of convective momentum transfer. *J. met. Soc. Jap.*, **48**, 91–102.

Matsumoto, S., Ninomiya, K. and Akiyama, T. (1967). A synoptic and dynamic study on the three-dimensional structure of mesoscale disturbances observed in the vicinity of a cold vortex centre. *J. met. Soc. Jap.*, **45**, 64–81.

Matsumoto, S., Akiyama, T. and Tsuneoka, Y. (1969). Some characteristic features of the heavy rainfalls observed over western Japan on July 9 1967. Part I. Mesoscale structure and short period pulsation. Part II. Displacement and life cycle of mesoscale rainfall cells. *J. met. Soc. Jap.*, **47**, 255–266, 267–278.

Matsumoto, S., Yoshizumi, S. and Takeuchi, M. (1970). On the structure of the "Baiu front" and the associated intermediate-scale disturbances in the lower atmosphere. *J. met. Soc. Jap.*, **48**, 479–491.

Murakami, M. (1972). Intermediate-scale disturbances appearing in the ITC zone in the tropical western Pacific. *J. met. Soc. Jap.*, **50**, 454–464.

Nagle, R. E. and Serebreny, S. M. (1962). Radar precipitation echo and satellite cloud observation of a maritime cyclone. *J. appl. Met.*, **1**, 279–295.

Ninomiya, K. and Akiyama, T. (1972). Medium-scale echo clusters in the Baiu front as revealed by multi-radar composite echo maps (Part I). *J. met. Soc. Jap.*, **50**, 558–569.

Ninomiya, K. and Akiyama, T. (1974). Band structure of mesoscale echo clusters associated with low-level jet stream. *J. met. Soc. Jap.*, **52**, 300–312.

Nozumi, Y. and Arakawa, H. (1968). Prefrontal rain bands located in the warm sector of sub tropical cyclones over the ocean. *J. geophys. Res.*, **73**, 487–492.

Ooeda, R. (1961). Patterns of the rainbands and their relation to the typhoon centre (1). *J. met. Soc. Jap.*, **39**, 310–312.

Pothecary, I. J. W. (1954). Short period variations in surface pressure and wind. *Q. Jl R. met. Soc.*, **80**, 395–401.

Rigby, E. C., Marshall, J. S. and Hitschfeld, W. (1954). The development of the size distribution of raindrops during their fall. *J. Met.*, **11**, 362–372.

Roach, W. T. and Hardman, M. E. (1975). Mesoscale air motions derived from wind finding dropsonde data: the warm front and rainbands of 18 January 1971. *Q. Jl R. met. Soc.*, **101**, 437–462.

Senn, H. V. (1960). The mean motion of radar echoes in the complete hurricane. *Proc. 8th Weath. Radar Conf.*, 427–434.

Senn, H. V. (1963). Radar precipitation echo motion in Hurricane Donna. *Proc. 3rd tech. Conf. Hurricanes trop. Met.*, 71–78.

Senn, H. V. and Hiser, H. W. (1959). On the origin of hurricane spiral rain bands. *J. Met.*, **16**, 419–426.

Senn, H. V., Hiser, H. W. and Low, E. F. (1959). Studies of the evolution and motion of radar echoes from hurricanes 1 July 1958 to 30 June 1959. Final Report, Contract no. Cwb-9480, University of Miami, Marine Laboratory, Coral Gables, Fla.

Senn, H. V., Hiser, H. W. and Nelson, R. D. (1960). Studies of the evolution and motion of radar echoes from hurricanes I July 1959 to 30 June 1960. Final Report, Contract Cwb-9727, Report no. 8944-1, University of Miami, Marine Laboratory, Coral Gables, Fla.

Simpson, R. H. (1954). Structure of an immature hurricane. *Bull. Am. met. Soc.*, **35**, 335–350.

Staff Members, Division of Meteorology, University of Tokyo (1969). Precipitation bands of Typhoon Vera in 1959 (Part I). *J. met. Soc. Jap.*, **47**, 298–309.

Staff Members, Division of Meteorology, University of Tokyo (1970). Precipitation bands of Typhoon Vera in 1959 (Part II). *J. met. Soc. Jap.*, **48**, 103–117.

Tatehira, R. (1962). A mesosynoptic and radar analysis of typhoon rain band (case study of typhoon "Helen" 1958). National Hurricane Research Program, Report no. 50, Part 1, pp. 115–126, US Department of Commerce, Weather Bureau, Miami.

Tatehira, R. (1964). Structure and mechanism of a huge radar rain band. *J. met. Soc. Jap.*, **42**, 362–371.

Tatehira, R. (1968). A study of rainband. *Geophys. Mag.*, **34**, 115–136.

Tatehira, R. and Itakura, H. (1966). Radar observation of typhoon Lucy by Mt. Fuji radar. *Proc. 12th Conf. Radar Met.*, 432–435.

Tepper, M. (1958). A theoretical model for hurricane radar bands. *Preprints 7th Weath. Radar Conf.*, K56–65.

Wallington, C. E. (1963). Meso-scale patterns of frontal rainfall and cloud. *Weather*, **18**, 171–181.

Wexler, H. (1947). Structure of hurricanes as determined by radar. *Ann. N. Y. Acad. Sci.*, **48**, 821–844.

Wexler, R. and Atlas, D. (1958). Moisture supply and growth of stratiform precipitation. *J. Met.*, **15**, 531–538.

Yamamoto, R. (1963). A dynamical theory of spiral rain band in tropical cyclones. *Tellus*, **15**, 155–161.

Yanagisawa, Z. (1961). An analysis of stationary rainbands as observed by radar (I). *Pap. Met. Geophys.*, **12**, 294–309.

Author Index

Aanensen, C. J. M., 31, 76, 110–112, 121
Abdullah, A. J., 472, 474
Abe, M., 31, 71, 76, 105, 106
Achtemeier, G. L., 340, 389
Agee, E. M., 356, 389, 400, 403, 404, 405, 406, 407, 408, 409, 410, 412, 416, 417, 419, 420
Air Ministry, 143, 145, 147, 209
Akiyama, T., 423, 440, 441, 474, 477
Alaka, M. A., 28, 42, 76
Alberty, R. L., 371, 390
Ali, B., 288, 307
Anderson, C. E., 381, 397
Anderson, I. I., 35, 36, 76
Angell, J. K., 163, 164, 207–208, 209, 212, 285, 308, 399, 419
Anthes, R. A., 471–473, 474, 476
Appleman, H., 318, 390
Arakawa, H., 158, 174, 175, 209, 433, 474, 475, 478
Arakawa, S., 84, 87, 97, 106
Arnold, J., 315, 320, 393
Asai, T., 188, 214
Ashwell, I. Y., 85, 106
Atkinson, B. W., 423, 430, 432, 433, 441, 442, 443, 444, 447, 465, 475
Atlas, D., 313, 315, 390, 421, 424, 452, 454, 476, 478
Atmanathan, S., 220, 222, 224, 225, 229, 232, 246, 257, 277
Auer, A. H., 325, 334, 335, 337, 360, 390, 395
Austin, P. M., 360, 390, 422, 424, 425, 426, 442, 447, 475
Axford, D. N., 292, 307
Ayer, H. S., 249, 277

Bailey, M., 36, 76
Balachandran, N. K., 289, 293, 308
Ball, F. K., 231, 276, 277
Baralt, G. L., 147, 209
Barnes, S. L., 313, 314, 340, 390
Barnett, K. M., 120–122
Barry, R. G., 90, 106
Bates, F. C., 323, 366, 367, 368, 390

Battan, L. J., 314, 315, 335, 390
Bean, B., 292, 308
Becker, R., 82, 106
Beebe, R. G., 319, 323, 390
Bemmelen, W. van, 143, 144, 145, 173, 214
Benard, H., 411, 412, 419
Beran, D. W., 85, 93, 94, 96, 97, 106, 292, 308
Berenger, M., 42, 78
Betts, A. K., 364, 387–389, 390, 395
Bhaskara, Rao, N. S., 369, 390
Biggs, W. G., 203, 209
Bilwiller, R., 93, 106
Binder, G. J., 71–73, 78
Birkhoff, G., 118, 122
Biswas, K. R., 429, 430, 433, 438, 439, 440, 476
Bjerknes, J., 4, 20, 117, 122
Black, P. G., 289, 293, 309, 456, 457, 458, 461, 476
Blackadar, A. D., 12, 22
Blanford, H. F., 147, 209
Bodolai, I., 422, 475
Bonner, W. D., 323, 390
Booker, D. R., 315, 390
Bosart, L. F., 289, 304, 308
Bowker, N. W., 114, 118, 119, 122
Bowley, C. J., 112, 122
Bradbury, T. A. M., 75, 76
Braham, R. R., 276, 277, 313, 327, 331, 335, 336, 351, 358, 360, 369, 390, 395
Brandes, E. A., 351–353, 354, 390
Bretherton, F. P., 10, 21, 76
Brinkmann, W. A. R., 81, 82, 83, 85, 87, 91, 92, 106
Brittain, O. W., 133, 156, 209
Brooks, E. M., 354
Brooks, H. B., 337
Brown, H. A., 228, 277, 433, 439, 477
Brown, R. A., 147, 209
Browning, K. A., 34, 36, 37, 38, 40, 79, 321, 324, 325, 327, 330, 331–337, 338, 341–349, 367, 371, 372, 391, 421, 422, 423, 427, 431, 433, 434, 435, 436, 438, 439, 444, 447, 449, 464, 466, 475

Brundidge, K. C., 8, 9, 21
Brunk, I. W., 359, 391
Brunt, D., 20, 21, 28, 76, 283, 293, 308, 412, 419, 447, 475
Bryant, G. W., 438, 475
Buell, C. E., 382, 391
Buettner, K. J. K., 216, 217, 218, 223, 225, 226, 235, 238, 240, 241, 244, 246, 247, 248, 274, 277
Bumgarner, B., 354, 396
Burger, A., 250, 277
Burley, M. W., 317, 318, 391
Burns, A., 8, 9, 22
Burt, W. V., 410, 419
Byers, H. R., 208, 209, 313, 315, 317, 331, 335, 336, 337, 340, 361, 369, 391, 393

Calheiros, R. V., 37, 58, 79
Capper, J., 146, 209
Carlson, T. N., 319, 323, 391
Carter, J., 354, 396
Caton, P. G. E., 424, 475
Chakravortty, K. C., 134, 135, 136, 137, 138, 141, 142, 213
Chalker, W. R., 319, 391
Chalon, J-P., 331–337, 391
Champion, D. L., 317, 391
Chandrashekhar Aiya, S. V., 331, 391
Chang, P. T., 383–385, 394
Changnon, S. A., 317, 351, 391
Chappell, C. F., 340, 394
Chappell, W. B., 319, 396
Charba, J., 336, 340, 383, 391
Charney, J. G., 5, 16–18, 21
Chisholm, A. J., 323–334, 341–343, 350, 351, 392, 395
Chen, T., 405, 407, 412, 417, 419
Chen, Y. L., 323, 396
Chopra, K. P., 109, 112, 113, 117, 118, 122
Claerbout, J. F., 304, 308, 309
Clare, P. R., 356, 389
Clark, T. L., 42, 76
Clarke, R. H., 9, 21, 146, 209
Cochran, H. B., 360, 390
Collins, G. F., 207, 209
Collis, R. T. H., 36, 77
Colon, J. A., 453, 475
Colson, de V., 36, 40, 41, 77
Conrad, V., 82, 106
Cook, R. H., 304, 308

Cooper, L. W., 335, 394
Corby, G. A., 28, 35, 36, 39, 42, 43, 51, 52, 53, 54, 55, 56, 57, 77
Cornett, J. S., 8, 9, 21
Cornford, C. E., 277
Cornford, S. G., 424, 476
Cotton, W. R., 206, 209
Craig, R. A., 143, 144, 209
Cramer, O. P., 143, 147, 152, 213
Crapper, G. D., 61, 77
Crary, A. P., 304, 308
Crawford, M. E., 359, 392
Cross, C. M., 227, 238, 277
Crowley, F. A., 304, 308
Cunning, J. B., 289, 308
Curry, M. J., 286, 289, 293, 308
Cussen, J. P., 289, 304, 308

Das, P., 342, 392
Davidson, B., 230, 234, 235, 237, 241, 244, 246, 249, 250, 251, 252, 277, 278
Davies-Jones, R. P., 334, 335, 392
Davis, W. M., 146, 210, 313, 392
Davis, W. M., 429, 430, 433, 438, 439, 440, 476
Day, S., 208, 210
De, U. S., 31, 64, 77
Defant, F., 159, 173–175, 210, 255, 278
de Felice, P., 143, 144, 145, 210
Dekate, M. V., 130, 131, 133, 134, 135, 136, 137, 139, 140, 210, 369, 390
Dennis, A. S., 329, 335, 392, 423, 476
Derome, J. F., 323, 392
Dessens, H., 342, 392
Dickson, C. R., 207–209, 399, 419
Diercks, J. W., 471–473, 476
Dixit, C. M., 143, 144, 145, 147, 151, 210
Donaldson, R. J., 315, 321, 325, 345, 347, 391, 392, 421, 476
Donn, W. L., 289, 293, 308
Döös, B. R., 58, 77
Douglas, C. K. M., 447, 475
Douglas, R. H., 315, 390, 425, 476
Doviak, R. J., 354, 396
Dowell, K. E., 404, 405, 406, 407, 410, 419
Draginis, M., 276, 277
Dutton, J. A., 10, 12, 22

Eady, E. T., 4, 21
Easterbrook, C. C., 335, 392
Eccles, P. J., 331–337, 391
Eddy, A., 144, 145, 210

AUTHOR INDEX

Edinger, J. G., 208, 210
Einaudi, F., 305, 309
Ekhart, E., 238, 243, 244, 247, 250, 251, 252, 278
Eldridge, R. H., 220, 223, 232, 278
Eliassen, A., 65, 66, 67, 68, 77
Elliott, A., 130, 132, 210
Elliott, R. D., 447, 476
Elliott, W. P., 8, 9, 21, 156, 212
Ellrod, G. P., 335, 392
Emanuel, K. A., 340, 392
Emmanuel, C. B., 292, 308
Eom, J., 285, 308
Erickson, C. O., 289, 290, 308
Estoque, M. A., 186–193, 196, 197, 204, 205, 210, 266, 278

Faller, A. J., 473–474, 476
Fankhauser, J. C., 331–338, 340, 354, 359–362, 391, 392, 393, 396
Farooqui, S. M. T., 64, 77
Fawbush, E. J., 319, 392
Feit, D. M., 127, 129, 137, 145, 162, 203, 210
Fenner, J. H., 337, 392
Fergusson, P., 147, 210
Ficker, H. von, 90, 95, 106
Fiedler, F., 6, 7, 12, 21
Findlater, J., 143, 147, 151, 156, 157, 210
Fisher, E. L., 143, 144, 145, 146, 147, 148, 150, 151, 152, 153, 163, 183–186, 210, 266, 278
Flaurand, E. A., 304, 308
Fleagle, R. E., 255–259, 278
Flohn, H., 87, 106
Foldvik, A., 36, 57, 61, 77
Foote, G. B., 345, 348, 349, 361, 391, 393
Förchgott, J., 29, 30, 77
Fosberg, M. A., 143, 147, 152, 213
Fournet, M. J., 217, 278
Frey, K., 82, 106
Fritsch, J. M., 340, 394
Fritz, S., 36, 77, 403, 404, 411, 416, 420
Frizzola, J. A., 143, 144, 146, 148, 150, 151, 152, 153, 210
Frye, D. E., 8, 9, 21
Fua, D., 305, 309
Fujita, T., 5, 21, 112, 115, 122, 315, 320, 329, 336, 340, 351, 354–355, 356, 359, 361, 375, 382, 390, 393, 450, 451, 452, 456, 457, 458, 461, 476

Fujiwara, M., 440, 476
Fulks, J. R., 359, 393

Gannon, P. T., 206, 209
Garstang, M., 361, 397
Gasne-Tabbagh, J., 143, 144, 145, 210
Gedzelman, S. D., 289, 291, 293, 297, 305–307, 308
Gentry, R. C., 451, 454, 462–463, 476
Gerbier, N., 42, 76
Gilet, M., 208, 211
Gill, D. S., 133, 210
Gill, G. C., 277, 278
Glaser, A. H., 112, 122
Gleeson, T. A., 267–270, 278
Glenn, C. L., 84, 93, 106
Goff, R. C., 336–337, 393
Goldman, J. L., 9, 10, 21, 340, 393
Goldsmith, P., 447, 476
Gossard, E. E., 283, 285, 288, 308
Gower, J. F. R., 114, 118, 119, 122
Grandoso, H., 340, 393
Graves, M. E., 203, 209
Green, J. S. A., 371, 372, 374, 393, 395
Grober, K. W., 90, 106
Gross, J., 204, 210
Grover, R. W., 364, 387–389, 390
Gunn, K. L. S., 423, 424, 425, 476

Hage, K. D., 36, 77
Haman, K. E., 372–374, 381, 393
Hamann, R. R., 84, 106
Hammond, G. R., 345, 393
Hane, C. E., 376–379, 393
Hann, J. V., 93, 106, 217, 251, 278
Hardman, M. E., 431, 433, 434, 447, 448, 449–451, 466, 475, 476, 478
Hardy, K. R., 287, 289, 293, 296, 307, 308, 309, 399, 419
Harney, P. J., 143, 144, 209
Harrold, T. W., 340, 393, 423, 426, 427, 428, 431, 433, 434, 435, 436, 438, 439, 440, 447, 449, 465, 466, 475, 476
Hart, H. E., 335, 394
Hatcher, R. W., 143, 210
Haurwitz, B., 42, 77, 147, 150, 159–167, 175–178, 210
Hawkes, H. B., 250, 278
Heffter, G. L., 118, 122
Helvey, R. A., 208, 210
Henderson, J. H., 334, 335, 392

Herron, T. J., 289, 304, 305, 308, 309
Hess, G. D., 9, 21
Hess, S. L., 13, 21, 116, 122, 159, 210
Hewson, E. W., 277, 278
Heywood, G. S. P., 220, 230, 278
Hicks, J. J., 285, 308
Hidy, G. M., 18, 21, 69, 77
Hill, F. F., 435, 475
Hines, C. O., 283, 303–304, 308
Hiser, H. W., 453, 454–459, 461, 478
Hitschfeld, W., 369, 394, 424, 478
Hobbs, P. V., 425, 429. 430, 433, 438, 439, 440, 476
Hočevar, A., 258, 279
Hoecker, W. H., 207–209
Hogström, U., 9, 22
Holmboe, J., 4, 20, 30, 31, 32, 36, 77, 117, 122
Holmes, R. M., 36, 77
Hooke, W. H., 283, 285, 287, 289, 292, 293, 307, 308
Hookings, G. A., 31, 78, 381, 394
Houghton, D. D., 97, 98, 106
House, D. C., 315, 390
Houze, R. A., 362, 364–365, 394, 422, 424, 425, 426, 429, 430, 433, 438, 439, 440, 447, 475, 476
Hovermale, J. B., 383, 395
Hovde, J. E., 319, 396
Hovind, E. L., 447, 476
Howley, R. P., 410, 419
Hoxit, L. R., 340, 394
Hseuh, Y., 10, 21
Hsu, S-A., 126, 143, 144, 145, 147, 149, 150, 156, 161, 162, 164, 212
Hubert, L. F., 112, 113, 117, 118, 122, 403, 404, 419
Huff, F. A., 317, 354, 394, 397
Huschke, R. E., 109, 122
Hwang, H. J., 10, 21

Imai, I., 450, 452, 476
Imakado, M., 450, 476
Isaacson, E., 98, 106
Itakura, H., 455, 478
Izawa, T., 450, 452, 476

Jaffe, S., 251, 278
Jakus, E., 422, 475
Jambunatham, R., 147, 213

James, D. G., 447, 476
Jedina, R., 90, 106
Jeffreys, H., 14–16, 21, 158, 162, 175, 211, 251, 252, 270, 278, 412, 420
Jehn, K. H., 147, 211
Jelinek, A., 229, 233, 234, 235, 278
Jessup, E. A., 356–357, 394
Jewell, C. J., 135, 211
Johnson, A., 143, 144, 145, 147, 211
Johnson, N., 285, 288, 292, 293, 308
Jordan, A. R., 286, 304, 309
Julian, L. T., 84, 106
Julian, P. R., 84, 106

Kakuta, M., 164, 212
Kamburova, P. L., 381, 394
Kamiko, T., 401, 420
Kao, S-K., 10, 21, 305, 309
Karman, T. von, 117, 122
Kasahara, A., 97, 98, 106
Katz, I., 143, 144, 209
Katz, S., 337, 396
Kauper, E. K., 207, 211
Keen, C. S., 143, 144, 145, 147, 211
Keliher, T. E., 293–295, 304, 309
Kennedy, P. J., 31, 32, 33, 75, 76, 78
Kessler, E., 439, 452, 454, 476, 477
Keulegan, G. H., 207, 211
Kimble, G. H. T., 143, 145, 147, 151, 152, 156, 211
Kistler, A. L., 156, 212
Klieforth, H., 30, 31, 32, 36, 68, 77, 79, 95, 96, 99, 107
Kleinschmidt, E., 251, 278
Klemp, J. B., 33, 78, 99, 100, 101, 102, 103, 104, 106, 107, 386, 394, 398
Kobayasi, T., 158, 211
Kolmogorov, A., 10, 21
Kopfmüller, A., 130, 211
Kopp, F. J., 376, 396
Koschmeider, H., 145, 152, 155, 211
Koscielski, A., 329, 335, 392
Koutnik, W., 88, 107
Kraft, D. W., 304, 308
Kraus, M. J., 426, 475
Kreitzberg, C. W., 433, 439, 445, 447, 449, 466–470, 477
Krishnamurti, R., 417, 420
Kropfli, R. A., 345, 361, 394
Krueger, A. F., 112, 122, 403, 404, 411, 416, 420

Kuettner, J. P., 401–403, 405, 411–419, 420
Kurihara, Y., 471–473, 477
Küttner, J. P., 31, 36, 69, 78, 97, 98, 107

Lai, H. W., 204, 210
Lalas, D. P., 305, 309
Lambert, S., 201–202, 211
Lamberth, R. L., 37, 79
Langleben, M. P., 423, 425, 476, 477
Lemon, L. R., 356, 394
Lenschow, D. H., 403, 420
Leopold, L. B., 208, 211
Lester, P. E., 37, 79
Lester, P. F., 99, 107
Lettau, H. H., 219, 278
Levine, J., 382, 394
Lhermitte, J. S., 208, 211
Lhermitte, R. M., 314, 394
Lied, N. T., 231, 278
Ligda, M. G. H., 5, 21, 313, 394, 452, 461, 477
Lilly, D. K., 31, 32, 33, 36, 37, 75, 76, 78, 81, 99, 100, 101, 102, 103, 104, 106, 107, 316, 367, 394
Lin, C., 412, 420
Lin, C. C., 118, 122
Lin, J. T., 71–73, 78
Lin, Y., 375, 394
Lin, Y. J., 383–385, 394
Lindsay, C. V., 31, 78
Lindzen, R. S., 471, 477
Lipps, F. B., 412, 420
Locatelli, J. D., 425, 433, 476
Lockwood, J. G., 96, 107
Lomax, F. E., 405, 419
London, J., 323, 396
Long, R. R., 70, 71, 96, 105, 107
Loomis, F. E., 146, 211
Low, E. F., 453, 457, 478
Ludlam, F. H., 313, 314, 315, 316, 318, 319, 323, 325, 330, 335, 345, 346, 366, 367, 371, 372, 381, 390, 391, 394, 395
Lyons, W. A., 112, 115, 122, 130, 133, 143, 144, 145, 147, 156, 211
Lyra, G., 42, 43, 48, 78

MacDonald, J. D., 359, 395
MacHattie, L. B., 222, 223, 224, 226, 250, 278
Machta, L., 207–209

MacPherson, J. I., 99, 107
Madden, T. R., 304, 308, 309
Magata, M., 193–196, 205, 211
Mahrer, Y., 197–201, 204, 211, 212
Mak, M. K., 203, 211
Mali, S., 382, 395
Malkus, J. S., 315, 390
Manabe, D., 474, 475
Manley, G., 31, 37, 78
Mansfield, D. A., 152–155, 213
Mantis, H. T., 8, 9, 21
Marshall, J. S., 423, 424, 425, 476, 477, 478
Marshall, W. A. L., 147, 212
Martin, D. E., 375, 394
Marvin, C. F., 257, 262, 278
Marwitz, J. D., 324, 325, 330, 331, 334, 335, 337, 338–339, 360, 390, 392, 395
Mason, B. J., 401, 420, 424, 477
Mastrantonio, G., 305, 309
Math, F. A., 84, 107
Mather, K. B., 216, 278
Mathur, M. B., 473, 477
Matsumoto, S., 422, 423, 426, 440, 441, 477
May, E. L., 331–337, 391
Mazelle, E., 86, 107
McAllister, L. E., 292, 308
McCaffery, W. D. S., 148, 212
McCarthy, M. J., 381, 397
McPherson, R. D., 188, 196–198, 204, 205, 212
Means, L. L., 319, 395, 396
Mears, A. H., 304, 308
Mendonca, B. G., 220–221, 230, 235, 236, 278
Merceret, F. J., 289, 293, 309
Meroney, R. N., 71, 79
Merrem, F. H., 331–337, 391
Meyer, J. H., 143, 145, 147, 212
Milford, J. R., 152–155, 213
Miller, E. S., 216, 278
Miller, J. E., 323, 395
Miller, L. J., 345, 361, 394
Miller, M. J., 364, 383, 387–389, 395
Miller, R. C., 319, 392
Mitchell, K. E., 383, 395
Mizumi, M., 164, 212
Moll, E., 235, 279
Moncrieff, M. W., 364, 372, 374, 387–389, 390, 395

Moroz, W. J., 143, 144, 145, 147, 150, 212
Munk, W., 288, 308
Murai, K., 404, 420
Murakami, M., 441, 477
Murty, R. C., 286, 289, 293, 308
Musil, D. J., 331–337, 391, 395

Nagle, R. E., 440, 477
Nakamura, K., 86, 108
Namias, J., 319, 395
Narayanan, V., 130, 143, 212
Nason, C. K., 426, 475
Nelson, R. D., 457, 478
Nelson, S. P., 351, 395
Neumann, C., 125, 212
Neumann, J., 178, 197–201, 209, 212
Newcomb, R. J., 112, 122
Newnham, E. V., 232, 279
Newstein, H., 5, 21, 315, 393
Newton, C. W., 315, 316, 323, 329, 337–338, 340, 342, 354, 356–359, 361, 362, 366, 369, 370, 371, 373, 386, 390, 395, 396
Newton, H. R., 329, 369, 371, 373, 386, 396
Nicholls, J. M., 25, 28, 37, 68, 78
Nicholson, J. R., 143, 144, 145, 147, 151, 210
Ninomiya, K., 316, 396, 423, 426, 440, 441, 477
Normand, C., 334, 396
Nozumi, Y., 433, 478

Obenland, E., 82, 107
O'Brien, J. J., 143, 144, 145, 147, 156, 211
O'Dell, C. A., 143, 147, 152, 213
Ogura, Y., 262, 279, 315, 323, 375, 383, 396, 398
Okano, M., 401, 420
Olsson, L. E., 156, 212
Ooeda, R., 450, 478
Oort, A. H., 8, 9, 21
Orlanski, I., 19, 20, 21
Orville, H. D., 262–266, 270, 274, 276, 279, 376, 396
Osmond, H. W., 82, 107
Ottersten, H., 399, 419

Pack, D. H., 163, 164, 207–209, 212, 399, 419
Paine, R. J., 362, 396

Palm, E., 57, 61, 78
Panofsky, H. A., 6, 7, 10, 12, 21, 42, 78
Paradiz, B., 90, 107
Pardoe, C. W., 431, 433, 434, 435, 436, 439, 447, 449, 466, 475
Partsch, J., 125, 212
Pasquill, F., 4, 21
Pastushkov, R. S., 383, 396
Patrick, G. O., 360, 390
Patrinos, A. A. N., 156, 212
Pearce, R. P., 156, 178–183, 186, 212, 371, 383, 393, 395
Pearson, J. R. A., 412, 420
Pearson, R. A., 201–203, 212
Pedgley, D. E., 143, 144, 145, 147, 151, 212, 315, 356, 359, 362, 396
Peltier, W. R., 42, 76
Perkey, D. J., 466–470, 477
Perry, A. H., 90, 106
Peters, S. P., 130, 132, 134, 143, 212
Petkovšek, Z., 258, 279
Pfleger, R., 317, 318, 391
Phillips, B. B., 351, 375, 396
Physik, W., 204, 212
Pielke, R. A., 204–206, 208, 209, 211, 213–214
Pierson, W. J., 159, 171–173, 213, 267, 279
Pinus, N. Z., 10, 21
Pitchford, K. L., 323, 396
Pohle, J., 12, 22
Poje, D., 90, 107
Pollak, L. W., 249, 279
Pollard, J. R., 292, 308
Pond, S., 8, 9, 21
Porter, J. M., 319, 396
Potapov, N. S., 90, 107
Pothecary, I. J. W., 288, 304, 309, 447, 478
Power, B. A., 423, 476
Prandtl, L., 219, 235, 243, 244, 246, 255, 259–262, 270, 279
Prestel, M. A. F., 146, 213
Preston-Whyte, R. A., 143, 145, 147, 209, 213, 215, 216, 250, 279
Priestley, C. H. B., 416, 420

Queney, P., 42, 43, 44, 45, 48, 49, 68, 78

Ragette, G., 336, 396
Ramakrishnan, K. P., 147, 213
Ramanadham, R., 143, 144, 145, 213, 424, 477

Ramanathan, K. R., 143, 144, 147, 213
Ramdas, L. A., 128, 130, 131, 138, 140, 141, 143, 144, 145, 213
Rao, D. V., 137, 138, 140, 141, 213
Rao, P. K., 230, 237, 241, 244, 246, 249, 250, 251, 252, 258, 259, 277, 278, 279
Rao, Y. P., 382, 395
Ray, D., 416, 420
Ray, P. S., 354, 396
Rayleigh, Lord, 411, 414, 415, 416, 420
Reed, R. J., 289, 296, 309
Reiter, E. R., 8, 9, 10, 21, 22
Rekustad, J. E., 66, 67, 68, 77
Renick, J. H., 323–334, 341–343, 350, 351, 392
Reynolds, R. D., 37, 79
Rhea, J. O., 323, 396
Rhyne, R. H., 335, 397
Richter, J. H., 285, 308
Riedel, A., 234, 278
Riehl, H., 85, 87, 107
Rigby, E. C., 424, 478
Rilling, R. A., 289, 291, 293, 297, 305–307, 308
Ringe, C., 146, 213
Roach, W. T., 329, 335, 396, 449–451, 466, 478
Robinson, G. D., 10, 12, 22
Rodebush, H. R., 208, 209
Rosenberg, C., 5, 22
Rosmond, T. E., 417–418, 420
Rossby, C-G., 4, 22
Roy, A. K., 130, 131, 139, 141, 143, 213
Rudder, B. de, 90, 106

Sand, W. R., 331–337, 390, 391, 395
Sanders, F., 340, 362, 396
Sansom, H. W., 317, 318, 397
Sarker, R. P., 37, 58, 79
Sartor, J. D., 319, 397
Sasaki, T., 158, 211
Sasaki, Y., 340, 374–375, 397
Saunders, P. M., 208, 213, 335, 397
Sawyer, J. S., 6, 22, 42, 43, 46, 57, 58, 59, 60, 61, 62, 63, 64, 65, 66, 75, 77, 79, 143, 210, 382, 397
Schaffer, W., 382, 397
Schlesinger, R. E., 376–385, 397
Schleusener, R. A., 315, 335, 390, 395
Schmauss, A., 251, 279
Schmidt, F. H., 159, 167–171, 213

Schmitt, W., 82, 107
Schock, C. A., 329, 335, 392
Schroeder, M. J., 143, 147, 152, 210, 213
Schuetz, J., 82, 107
Schultz, L. G., 146, 210
Schweitzer, H., 97, 107
Scorer, R. S., 28, 42, 46, 47, 48, 49, 51, 52, 57, 61, 63, 68, 79, 95, 96, 99, 107
Sekihara, K., 404, 420
Sen Gupta, P. K., 134, 135, 136, 137, 138, 141, 142, 213
Senn, H. V., 453, 454–459, 461, 478
Serebreny, S. M., 440, 477
Severe Storms Research Group of St. Louis University, 366, 367, 397
Shaw, N., 14
Sheets, R. C., 451, 476
Sherman, O. T., 125, 143, 144, 213
Sheu, P. J., 407, 408, 409, 410, 419, 420
Shieh, L. J., 249, 279
Showalter, A. K., 319, 397
Shur, G. N., 10, 21
Sikdar, D. N., 381, 397
Simpson, J. E., 152, 153, 154, 155, 207, 213
Simpson, R. H., 452, 478
Sinclair, P. C., 335, 397
Sirmans, D., 354, 396
Sloan, L. J., 376, 396
Smedman-Hogström, A. S., 9, 22
Smith, M. F., 143, 147, 213
Smith, R. B., 42, 68, 79
Smithson, P. A., 423, 430, 432, 433, 441, 442, 443, 444, 447, 465, 475
Snow, J. T., 356, 389
Solberg, H., 4, 20
Somerville, R. C. J., 412, 420
Sonde, B. S., 331, 391
Soules, S. D., 403, 420
South African Weather Bureau, 135, 213
Spillane, K. T., 381, 397
Staley, D. O., 150, 156, 213
Starr, J. R., 34, 36, 37, 38, 40, 79
Steiner, R., 335, 397
Steinhauser, F., 82, 107
Stepanova, N. A., 83, 107
Stephens, E. R. 207, 213
Stout, G. E., 354, 397
Strauch, R. G., 331–337, 391
Streten, N. A., 220, 223, 225, 228, 230, 231, 232, 279, 401, 420

Strouhal, V., 117, 122
Subbaramayya, I., 143, 144, 145, 213
Sutcliffe, R. C., 4, 5, 22, 143, 146, 147, 151, 214
Sutton, O. G., 4, 13, 22
Suzuki, S., 105, 107
Suzuki, Y., 404, 420
Swingle, D. M., 5, 22

Takeda, T., 375–379, 382, 385, 397
Takeuchi, M., 423, 477
Tamiya, H., 90, 108
Tang, W., 246, 279
Tatehira, R., 440, 442, 455, 460–461, 478
Taylor, A., 8, 9, 22
Taylor, A. D., 146, 214
Taylor, G. I., 305, 309
Tepper, M., 5, 21, 22, 315, 359, 393, 397, 472, 478
Theiss, J. B., 335, 390
Thorpe, S. A., 298–301, 309
Thompson, P. D., 283, 284, 301–303, 309
Thomson, J., 411, 420
Thomson, R. E., 114, 118, 119, 122
Thyer, N. H., 216, 217, 223, 225, 226, 235, 238, 240, 241, 244, 246, 247, 248, 270–275, 277, 279
Tollner, H., 234, 279
Tolstoy, I., 289, 304, 305, 308, 309
Tsuchiya, K., 112, 118, 119, 122
Tsuneoka, Y., 426, 477
Tuleya, R. E., 471–473, 477
Tung, K-K., 471, 477
Tyson, P. D., 213, 214, 215, 216, 217, 220, 221, 222, 224, 226, 241, 242, 243, 244, 246, 247, 249, 250, 279

Uccellini, L. W., 285, 304, 309
Ulanski, S. L., 361, 397
Ungeheuer, H., 82, 108
Utsugi, M., 158, 174, 175, 209

van der Hoven, I., 7, 8, 21, 22
Veal, D. L., 334, 337, 390
Vergeiner, I., 33, 37, 79
Vinnichenko, N. K., 10, 11, 12, 21, 22
Vonnegut, B., 315, 390

Wagner, A., 251, 252, 279
Wagner, A. J., 285, 309
Walker, G. B., 354, 396
Wallington, C. E., 41, 42, 52, 53, 54, 55, 56, 77, 79, 152, 153, 155, 156, 201, 214, 422, 423, 478
Walsh, J. E., 203, 209, 211, 214
Wang, J. Y., 317, 318, 391
Ward, R. De C., 146, 210
Warner, C., 325, 392
Watanabe, K., 450, 452, 476
Watts, A. J., 152, 214
Weickmann, H. K., 361, 398
Wenger, R., 217, 218, 252, 279
Weston, K. J., 323, 398, 401, 420
Wexler, H., 452, 478
Wexler, R., 112, 122, 143, 145, 146, 147, 152, 156, 163, 164, 214, 424, 439, 477, 478
White, A. A., 12
Whitney, L. F., 289, 290, 308
Wichman, M., 335, 398
Wilhelmson, R. B., 383–387, 394, 398
Wilkinson, M., 61, 79
Williams, R. T., 315, 390
Winston, J. S., 313, 316, 319, 398
Woeikof, A., 146, 214
Wojtiw, L., 317, 318, 398
Woods, H. D., 10, 21, 305, 309
Wooldridge, G., 37, 76, 79
Wurtele, M. G., 37, 61, 79

Yabuki, K., 105, 107
Yamada, T., 71, 79
Yamamoto, R., 288, 309, 472, 478
Yamashita, R., 174, 214
Yanagisawa, Z., 450, 478
Yoshikado, H., 188, 214
Yoshimura, M., 86, 108
Yoshino, M. M., 83, 86, 87, 90, 93, 108
Yoshizumi, S., 423, 477
Young, J. M., 292, 308
Yuyama, Y., 31, 79

Zambakas, J. D., 133, 214
Zarantonello, E. H., 118, 122
Zimmerman, L. I., 112, 114, 118, 119, 122
Zipser, E. J., 36, 78, 81, 107, 364, 365, 398

Subject Index

Airflow over ridges
    types of, 29
Air Mass Transformation Experiment (AMTEX), 403–410
Amplitude,
    lee wave, 36–39, 52, 54–55
    moving gravity wave, 286–287, 291, 297
Anabatic wind, *see* Up slope wind
Anticyclone, *see* Meso-scale high pressure area, Shallow cellular circulations
Antitriptic wind, 15
Anti-winds, 216–217, 235, 241, 244, 247
Aspect ratio, *see* Shallow cellular circulations
Atmospheric motion
    meso-scale, 3–20
    scale of, 6, 16–20

Bands, *see* Rainbands
Barbados Oceanographic and Meteorological Experiment (BOMEX), 401
Bernouilli's equation, 99, 372–374
Bjerknes's circulation theorem, 125–127, 159, 217–218, 252
Bora
    climatology, 87, 89, 90
    comparison with foehn, 92
    nomenclature, 82
    surface characteristics, 86
    theoretical results, 93–104
Boussinesq approximation, 100, 414–415
Brunt–Väisälä frequency, 101, 293–296
Brunt–Väisälä period, 20, 293
Buoyancy waves, *see* Gravity waves

Cell
    closed, *see* Shallow cellular circulations
    convective, *see* Multi-cell severe local storm, Shallow cellular circulations, Single cell severe local storm, Supercell severe local storm
    open, *see* Shallow cellular circulations
    Rayleigh, 412–419
    storm, *see* Multi-cell severe local storm, Single cell severe local storm, Supercell severe local storm
    thunderstorm, *see* Multi-cell severe local storm, Single cell severe local storm, Supercell severe local storm
Cellular circulations, *see* Multi-cell severe local storm, Shallow cellular circulations, Single cell severe local storm, Supercell severe local storm
Chinook, *see also* Foehn, 85
Circulations in cyclones
    extra-tropical circulations,
        climatology, 425–426
        direct observations of meso-scale airflow, 447–450
        effects of instability, 464–466
        numerical models, 466–470
        organization of meso-scale precipitation areas, 426–447
        radar and precipitation evidence, 423–447
    hardware models, 473–474
    observational results, 423–463
        extra tropical circulations, 423–450
        tropical circulations, 450–463
    theoretical results, 464–473
        extra-tropical circulations, 464–470
        tropical circulations, 470–473
        band movement, 459–462
        band structure, 454–459
        band temperatures and winds, 462–463
        effects of gravity waves, 471–473
        radar evidence, 450–462
Circulations in wakes
    observational results, 110–115
    theoretical results, 115–120
        meso-scale lee laws, 115–117
Circulation theorem, *see* Bjerknes's circulation theorem
Cloud streets, 399–403, 412, 418–419

Clusters
  in squall line, 359–362
Cold wind, see Bora
Continuity equation, 14, 174, 176, 181, 187, 272, 302, 376
Convection
  cumulonimbus, 315
  cumulus, 315
  in cyclones, see Circulations in cyclones
  line, 440
  organized, see Severe local storms
  over slopes, 253–254
  severe, see Severe local storms
Convective systems, see Circulations in cyclones, Severe local storms, Shallow cellular circulations
Conveyor belt, 426–428, 433–438, 447
Coriolis force
  effect on sea/land breeze circulation, 146–150, 162–167, 169–175, 178–186
  in equations of motion, 16–19
  in lee wave theory, 46–47
Critical flow, 98–99
Cumulonimbus cloud, 313–389
  effect of wind shear on, 369–387
Cumulonimbus convection, 315
Cumulus convection, 315
Cyclone, see also Meso-scale low pressure area
  circulations in, see Circulations in cyclones
  tornado, 354–356

Dimensional parameters, 18
Direct circulation, see also Bjerknes's circulation theorem, 125–127, 150, 216
Downdraught, see Multi-cell severe local storm, Supercell severe local storm
Downslope wind, see also Thermal downslope wind
  hardware models, 105
  hydrostatic waves, 99–100
  mechanism, see Downslope wind, theoretical results
  observational results, 83–92
  peak gusts, 80–81
  synoptic background, 90–92
  theoretical results, 93–104
  blocking theory, 95–96
  hydraulic jump theory, 97–99, 103–104
  inversion height, 102–103
  large amplitude wave theory, 96–97
  thermodynamic theory, 93
  wave amplification, 102
  types, 81–83
Down valley wind, see also Valley wind circulation
  observational results, 219–233, 238–249
  duration, 224–225
  frequency, 220
  time of onset, 222–223
  theoretical results, 267–274
  analytical results, 267–270
  numerical results, 270–275
Drag theory, 117

Echo, see Radar
Equation
  of state, 14, 272
  shallow water, 97–98
Equations of motion, 13, 159–160, 165, 174, 176, 180, 255, 267, 271–272, 302, 305, 376, 414–415, 467
  dimensional analysis of, 14–20
  orders of magnitude of terms, 16–18
Eulerian wind, 14
Extra-tropical severe local storm
  associated circulations,
    thunderstorm high, 356
    tornado cyclone, 351, 354–356
    wake low, 356
    wake vortices, 356
  observational studies, 316–365
    associated circulations, 354–356
    classification and storm circulation, 323–354
    effect of instability, 318–323
    mass and water budgets, 356–359
    multi-cell storm, see Multi-cell severe local storm
    squall lines, 359–362
    supercell storm, see Supercell severe local storm
    synoptic background, 318–323
  theoretical results, 367–387
    diagnostic studies, 367–371

prognostic studies, *see also* Prognostic studies of severe local storms, 371–387
Fall wind, *see* Downslope wind
Foehn
    climatology, 87–89
    comparison with bora, 92
    definition, 82–83
    duration, 88
    nomenclature, 81–82
    observational results, 84–85, 87, 90–92
    surface characteristics, 84–85
    theoretical results, 93–104
Front
    meso-scale structure in, *see* Circulations in cyclones
Froude number, 18, 20, 70–71, 98, 103, 105, 207, 419
Friction
    effects on sea/land breeze circulation, 159–162, 167–169, 171–175, 178–186

GARP Atlantic Tropical Experiment, 401
Generators, 423–425, 441–447
Geostrophic approximation, 17–18
Geostrophic wind, *see also* Wind, 14
    effect on sea/land breeze circulation, 192–193, 196
    oscillations about, 46
Glacier wind, 234
Gradient wind, *see also* Wind
    effect on sea/land breeze circulation, 151–156
    effect on valley wind circulation, 250–251
Gravity waves, *see also* Lee waves, Wave
    circulations in cyclones, 471–473
    mechanism, 303
    moving, 283–307

Hardware models
    circulations in cyclones, 473–474
    circulations in wakes, 120–121
    downslope winds, 105
    lee waves, 69–73
    sea/land breeze circulation, 206–207
    slope wind circulation, 274–276
    valley wind circulation, 274–276
Heat equation, 14, 174, 183, 187, 193, 272, 376, 467

Hurricane
    meso-scale structure, *see* Tropical severe local storm
Hydraulic jump, 33, 38, 68
Hydrodynamical equations, *see also* Equations of motion, 17

Inertial motion, 15
Instability index, 319
Instability line, *see* Squall line, extratropical
Instability, *see also* Stability
    Kelvin–Helmholtz, 298–301
Inversion, *see* Stability, effect on lee waves, Moving gravity waves, Shallow cellular circulations

Jet
    effect on moving gravity waves, 304–305

Karman vortex street, 117–120, 356
Karman vortex train, *see* Karman vortex street
Karman vortices, 117–121
Katabatic wind, *see* Thermal downslope wind
Kelvin–Helmholtz instability, 298–301
Kelvin–Helmholtz waves, 285, 298–301

Land breeze
    climatology, 135
    depth, 143–146
    duration, 142
    effects on weather, 208–209
    horizontal extent, 146–150
    humidity, 142
    rotation, 146–150
    speed, 137, 141
    temperature, 142
    time of onset, 136
Large meso-scale precipitation areas, 429–441
    types of rain band
        cold frontal, 438–440
        warm frontal, 433
        warm sector, 433–438
        wave, 440
Lee low, *see* Meso-scale lee low pressure area
Lee vortices, *see* Meso-scale lee vortices, *see also* Karman vortices

Lee waves, 25–79
  analytical results,
    two-dimensional models, 50–58
  atmospheric conditions suitable for, 40–43
  clouds, 26–27, 30
  dimensionless parameters, 69
  effects of stability, 46–50
  effects of vertical wind shear, 46–50
  general mechanism, 28
  hardware models, 69–73
  hydraulic jump, 33, 38, 68
  momentum transfer, 73–76
  natural wavelength, 28, 54
  observational results, 30–42
    amplitude, 36–39
    rotors, 29, 31, 33–34, 68
    vertical velocities, 28, 36–37, 39–40
    wavelength, 28, 34–37
    wave trains, 32–34
  rotors, 29, 31, 33–34, 68
    rotor streaming, 29, 31
  Scorer's parameter, 48–66
  Scorer's work, 46–50
  theoretical results, 42–69
    amplitude, 52, 54–55
    boundary conditions, 50, 57
    effect of wind profile, 53–54
    mountain effects, 53–54
    perturbation equations, 46–48
    large amplitude, 67–69
    long wavelengths, 64–67
    numerical studies, 58–67
    small amplitude, 42–67
    three dimensional models, 61–67
    vertical velocity, 64, 66
    wavelength, 58
  wave drag, 73–76
  work of Lyra and Queney, 43–46
Line convection, 440
Line segment, 359
Lin parameter, 118

Meso-cyclone, *see* Meso-scale lee low pressure area, *see also* Tornado cyclone
Meso-high, *see* Meso-scale high pressure area
Meso-low, *see* Meso-scale lee low pressure area, *see also* Meso-scale lee vortices

Meso-scale
  definition, 3–20
  observational results, 3–12
  theoretical results, 13–20
Meso-scale high pressure area, *see also* Thunderstorm high, 111, 115–117
  theoretical results, 115–117
Meso-scale lee low pressure area
  observational results, 110–112
  theoretical results, 115–117
Meso-scale lee vortices
  favourable atmospheric conditions, 112–115
  observational results, 112–115
  theoretical results, 117–120
  vortex street, 112–113
Meso-scale low pressure area, *see* Meso-scale lee low pressure area
Meso-scale precipitation areas
  large, 429–441
    types of rain band, 429–441
  organization of, 426–447
  small, 441–447
Meso-scale pressure systems, *see* Meso-scale high pressure area, Meso-scale lee low pressure area
Moisture equation, 14, 193, 376, 467
Mountain-plain circulation, 249–250
Mountain wave, 25
Mountain wind, *see* Down-valley wind
Moving gravity waves
  characteristics, 288–289
  effect of stability, 292–297
  effect of wind shear, 292–297, 304
  modelling, 305–307
  observational results, 285–297
  theoretical results, 298–307
Multicell severe local storm, *see also* Severe local storm, 325, 327–340
  downdraught, 334–336
  effect of stability, 330–331
  effect of wind shear, 330–331
  gust front, 336–337
  movement, 330, 337–340
  outflow, 336–337
  propagation, *see* movement
  radar evidence, 327–340
  rotation, 340
  splitting, 340
  structure and evolution, 328–329

updraught, 334–335
Multi-cell storm, *see* Multi-cell severe local storm

Non-dimensional parameters, 120
Numerical models
  of circulations in cyclones, 466–470, 472–473
  of lee waves, 58–66
  of mechanically induced downslope winds, 99–104
  of sea/land breeze circulation, 178–206
  of severe local storms, 374–387
  of slope and valley wind circulation, 262–266, 270–274
  of thermally induced slope winds, 262–266
  of thermally induced valley winds, 270–274
  of thunderstorms, *see* Numerical models, of severe local storms

Order of magnitude of terms in equations of motion, 16–18
Orographic influences, *see* Circulations in wakes, Downslope winds, Lee waves, Slope wind circulation, Valley wind circulation

Prandtl number, 418
Precipitation areas, 426–447
  large meso-scale, 429–441
  small meso-scale, 441–447
Pressure systems, *see* Meso-scale high pressure area, Meso-scale lee low pressure area
Prognostic studies of severe local storms
  three dimensional models, 383–387
  two dimensional models, 371–383
    analytical results, 371–374
    numerical results, 374–383

Radar, 5, 38, 285–287, 314–315, 324–326, 328–329, 342–343, 348–351, 357, 417, 423–447, 450–462
Rainbands, 429–441, 450–462
  spiral, 450–462
Rainfall areas, *see* Precipitation areas
Rayleigh cells, 412–419
Rayleigh number, 18, 20, 411–414, 419

Return current, *see also* Sea/land breeze circulation, Slope and valley wind circulation
  depth, 144
  height, 144
  maximum velocities, 145
Reversible circulation, *see also* Bjerknes's circulation theorem, 125
Reynolds number, 18, 20, 118–120
Richardson number, 18, 20, 101, 121, 291–292, 295–298, 374
Rossby number, 18, 20
Rotation
  of land breeze, 146–150
  of sea breeze, 146–150
Rotor
  lee wave, 29, 31, 33–34, 68
  streaming, 29, 31

Scale
  definitions and different processes, 19
  meso, 3–20
  observational studies of, 3–12
  of atmospheric motions, 16–20
  synoptic, 10
  theoretical studies of, 13–20
Scorer's parameter
  effect on lee waves, 52–66
  exponential profile, 58–59
Sea breeze
  convergence, 156
  depth, 143–146
  duration, 133–134
  effect on pollution, 207–208
  humidity, 136–137, 140–141, 154
  maximum velocities, 145
  horizontal extent, 146–150
  potential temperature, 154
  rotation, 146–150
  speed, 137–138
  temperature, 134, 138–140, 154
  time of onset, 131–132
Sea breeze convergence zone, *see* Sea breeze convergence, Sea breeze front
Sea breeze front, 127, 152–156, 197–203
Sea/land breeze circulation, *see also* Land breeze, Sea breeze
  hardware models, 206–207
  horizontal extent,
    coriolis effects, 146–150

Sea/land breeze circulation (*cont.*)
  mechanism, 125–127
  observational results, *see also* Land breeze, Sea breeze, 127–157
    climatology, 127–134
    diurnal variation, 150
    effects of gradient wind, 151–156
    effects of stability, 156
    effects of topography, 156–157
    effects on weather, 156–157, 207–209
    empirical model, 149
    horizontal extent, 146–150
    life cycle, 138–143, 150
    return current, 144–146, 172
    surface characteristics, 128–129, 134–138
    temperature gradients, 151
    vertical velocities, 151, 152, 155
  role, 207–209
  theoretical results, 158–206
    analytical results, 158–178
    asymmetry of circulation, 203–204
    diurnal variation, 197–199
    effects of coriolis force, 162–167, 169–175, 178–186
    effect of eddy viscosity, 171–178, 186–197, 203
    effect of friction, 159–162, 167–169, 171–175, 178–186
    effects of geostrophic wind, 192–193, 196
    effect of land/sea distribution, 197–201
    effect of land-sea temperature, 171–186
    effect of realistic topography, 204–206
    effect of static stability, 175, 178, 203
    effect of vertical heat transfer, 171–175, 186–196, 201–205
    effect on weather, 206
    linear theory, 158–178
    non-linear theory, 178–206
    numerical results, 178–206
    sea-breeze front, 197–203
    three-dimensional models, 176–201
    two-dimensional models, 178–195
Severe local storm, *see also* Multi-cell severe local storm, Single cell severe local storm, Supercell severe local storm, 313–398
  associated circulations, 354–356
  classification and storm circulation, 323–354
  definition, 313–314, 316
  effect of wind shear, 327, 369–387
  kinematic characteristics, *see* Severe local storm, classification and storm circulation
  numerical models of, 374–387
  observational results, 316–365
    climatology, 316–318
    extra-tropical storms, 318–362
    tropical storms, 362–365
  prognostic studies,
    three dimensional models, 383–387
    two dimensional models, 371–383
  propagation, 338–340
  single cell, 325–327
  steady state, 367–368, 372
  theoretical results, 365–389
    diagnostic studies, 366, 367–371
    prognostic studies, 366, 371–387
Shallow cellular circulations, 399–419
  aspect ratio, 404–405, 410, 416
  closed cells, 404–410
  effect of eddy anisotropy, 416–418
  effect of vorticity, 412–413
  effect of wind, 405–408
  observational results, 401–411
    three dimensional cells, 403–411
    two-dimensional cells, 401–403
  open cells, 404–410
  temperature effects, 405–406, 407–408
  theoretical results, 411–419
    flowing medium, 418–419
    medium at rest, 416–418
  three dimensional cells,
    climatology, 407
    description, 404–407
    heat fluxes, 410–411
    synoptic background, 407–410
  two dimensional cells,
    climatology, 402
    description, 401
    synoptic background, 402–403
    vertical motion, 409–410
Shallow water equations, 97–98
Shear, *see* Wind shear

SUBJECT INDEX

Single cell severe local storm, 325–327
Slope wind, *see also* Slope wind circulation, Thermal downslope wind, Upslope wind
  climatology, 219–229
  diurnal variation, 227
  mechanism, 217–219, 253–254
  observational results, 219–251
  role, 276–277
  surface characteristics, 229–233
Slope wind circulation, *see also* Slope wind, Thermal downslope wind, Upslope wind
  hardware models, 274–276
  nomenclature, 215–217
  observational results, 233–238
  theoretical results, 251–274
Small meso-scale precipitation areas, 441–447
Spectra, *see* Velocity spectra
Spectral analysis, 6–12
Spectral gap, 7–12
Squall line
  extra-tropical, 359–362
  tropical, 362–365
Stability
  effect on circulations in cyclones, 464–466
  effect on extra-tropical severe local storm, 318–323
  effect on lee waves, 46–50
  effect on moving gravity waves, 292–297
  effect on multi-cell severe local storm, 330–331
  effect on sea/land breeze circulation, 175–178, 203
  effect on shallow cellular circulations, 407–410
  effect on supercell severe local storm, 342
  effect on valley wind circulation, 250–251
Steady state storm, 367–368, 372
Storm cell, *see* Multi-cell severe local storm, Single cell severe local storm, Supercell severe local storm
Strouhal number, 18, 20, 118–120
Supercell severe local storm, *see also* Severe local storm, 325, 341–354
  downdraught, 345–350, 381–383
  effect of instability, 342
  effect of wind shear, 341–342
  internal airflow, 342–354
  kinematical structure, 342–354
  movement, 351–354
  radar evidence, 342–354
  updraught, 345–350
  steady state, 367–368, 372
Supercell storm, *see* Supercell severe local storm

Thermal downslope wind, *see also* Slope wind, Slope wind circulation, Upslope wind, 219–238, 255–259
  duration, 224–225
  theoretical results, 255–259
  analytical results, 255–259
  time of onset, 222–223
Thermally direct circulation, *see also* Bjerknes's circulation theorem, 125–127, 150, 216
Thunderstorm, *see also* Severe local storm, Multi-cell severe local storm, Supercell severe local storm
  life cycle, 314
Thunderstorm high, 382
Tornado cyclone, 354–356
Tropical circulations, *see* Rainbands, Sea/land breeze circulation, Shallow cellular circulations, Tropical severe local storm, Tropical squall line
Tropical cyclone, *see* Tropical severe local storm
Tropical severe local storm
  movement, 365
  storm circulation, 364–365
  theoretical results, 387–389
Tropical squall line, 362–365
Tropical systems, *see* Rainbands, Sea/land breeze circulation, Shallow cellular circulations, Tropical severe local storm, Tropical squall line
Types of wind, 14–16

Updraught, *see* Multi-cell severe local storm, Supercell severe local storm

Upslope wind, *see also* Slope wind, Slope wind circulation, Thermal downslope wind, 217–219, 221, 226–230, 233–238; 253–254, 259–266
  duration, 226
  theoretical results, 259–266
    analytical results, 259–262
    numerical results, 262–266
Up-valley wind, *see also* Valley wind circulation
  observational results, 221, 226, 238–249
    depth, 241–243
    duration, 226
    velocity variations, 243–246
  theoretical results, 267–274
    analytical results, 267–270
    numerical results, 270–275

Valley wind, *see also* Down-valley wind, Up-valley wind, Valley wind circulation
  mechanism, 217–219
  observational results, 219–251
    climatology, 219–229
    surface characteristics, 229–233
    valley wind circulation, *see* Valley wind circulation
  role, 276–277
  theoretical results, 267–274
    analytical results, 267–270
    numerical results, 270–274
Valley wind circulation, *see also* Down-valley wind, Up-valley wind, Valley wind, 238–249
  along-valley section, 238–239
  cross-valley section, 240–241
  depth, 241–243
  down-valley acceleration, 246–247
  effect of gradient wind, 250–251
  effect of stability, 250–251
  evolution, 247–249
  hardware models, 274–276
  nomenclature, 215–217
  theoretical results, 251–274
  velocity variations, 243–246
  vertical velocities, 247
Vault, *see also* Weak echo region, 325
Velocity spectra, 7–12
Vertical-transverse waves, 283–285
Vertical velocity, *see also* Multi-cell severe local storm, Supercell severe local storm
  in lee waves, 28, 36–37, 39–40
  in sea/land breeze circulation, 151, 152, 155
  in shallow cellular circulation, 409–410
  in valley wind circulation, 247
Viscosity
  eddy
    effect on sea/land breeze circulation, 171–178, 186–197, 203
Vortex
  Karman, 117–120
    period of formation, 118
    propagation speed, 118
    rate of shedding, 118
Vortex pairs, 117
Vortex street
  Karman, 117–120, 356
  meso scale, 112–113
Vortex train, *see* Vortex street
Vorticity
  conservation of, 115
  effect on shallow cellular circulations, 412–413
  in wakes, 117–120
Vorticity equation, 262, 272, 377

Wake
  circulations in, *see* Circulations in wakes
  definition, 109
Wake low, 356
Wake vortices, 356
Wake vorticity, 117
Warm wind, *see* Foehn
Wave, *see also* Lee waves
  gravity, *see* Gravity waves
  Kelvin–Helmholtz, 285, 298–301
  lee, *see* Lee waves
  lee wavelength, 28, 34–37, 58
  lee wave train, 32–34
  mountain, 25
  standing, *see* Lee waves
  stationary, *see* Lee waves
  vertical transverse, 283–285
Wave drag, *see* Lee waves
Wave equation, 43
Wavelength, *see* Lee waves, Moving gravity waves, Wave

Weak echo region, 325–329, 342–344
  bounded, 342–344
Wind, *see also* Wind shear
  anabatic, *see* Upslope wind
  antitriptic, 15
  cold, *see* Bora
  Eulerian, 14
  geostrophic, 14
    effect of on sea/land breeze circulation, 192–193, 196
  glacier, 234
  gradient,
    effect on sea/land breeze circulation, 151–156
    effect on valley wind circulation, 250–251
  katabatic, *see* Thermal downslope wind
  mountain, *see* Down-valley wind
  slope, 215–266
  types of, 14–16
  valley, 215–277
  warm, *see* Foehn
Wind shear
  effect on cumulonimbus cloud, 369–387
  effect on lee waves, 46–50
  effect on moving gravity waves, 292–297, 304
  effect on multi-cell severe local storm, 330–331
  effect on severe local storm, 327, 369–387
  effect on shallow cellular circulations, 405–408
  effect on supercell severe local storm, 327, 341–342